Techn
Slot Canyon
Colorado Plateau
2nd Edition

Michael R. Kelsey

Kelsey Publishing
456 E. 100 N.
Provo, Utah, USA, 84606-3208
Tele & Fax 801-373-3327
Email Addresses--one of these should work:
kelsey@canyoneering (best)
mrkelsey@iveracity.com

For updates on the canyons in this book, or to make updates yourself to any of the hikes featured here, please go to:
kelseyguidebooks.com
or go to
canyoneering.com
for Weather Forecasts - Desert Ecology
Open Forums - Canyoneering Techniques
Gear & Book Reviews

Other websites which may have updated information on canyons or hikes in this book are:
americansouthwest.net
canyoneeringUSA.com (Tom Jones)
climb-utah.com (Shane Burrows)
toddshikingguide.com (Todd & Stephanie Martin)
ajroadtrips.com (Ryan Cornia)
jonjasper.com (Jon Jasper)
adventure-geek.com (AJ Pastula)
http://groups.yahoo.com/group/canyons/ (Ram--Steve Ramrus)

American Canyoneering Association: canyoneering.net
The American Canyoneering Association provides a free online canyon database where canyoneers can post updates to canyon beta and seek out updates posted by others. The database is intended to serve as a clearinghouse of up-to-date information to be used as a resource by canyoneers, rescue teams and guidebook publishers. The database is located at: **http://www.canyoneering.net/forums** (Rich Carlson).
Within this ACA website is a place to document changes in canyons due to floods, public lands policy, etc., etc. If you have the time, please alert people to situations (and dangers) that have changed from the descriptions found in this book.

Also, try Ram's website: **http://groups.yahoo.com/group/canyons/**. (no www). The group has threads where folks talk gear, equipment, technique, safety, trip reports, requests for conditions, trolling for partners, land management issues, environmental debate and anything else canyon related (Steve Ramrus).

1st Edition--July, 2003
2nd Edition--September, 2008
Copyright © 2008 Michael R. Kelsey All Rights Reserve
Library of Congress Control Number: 2007920138
ISBN 978-0944510-23-0

Distributors for Kelsey Publishing

Primary Distributor All of Michael R. Kelsey's books are sold by this distributor. A list of Kelsey's titles is in the back of this book.
Brigham Distribution, 110 South, 800 West, Brigham City, Utah, 84302, Tele. 435-723-6611, Fax 435-723-6644, Email brigdist@sisna.com.

Most of Kelsey's books are sold by these distributors.
Alpenbooks, 4206 Chennault Beach Road, Suite B1, Mukilteo, Washington, USA, 98275, Website alpenbooks.com, Email cserve@alpenbooks.com, Tele. 425-493-6380, or 800-290-9898.
Books West, 11111 East, 53rd Avenue, Suite A Colorado, USA, 80239-2133, Tele. 303-449-5995, or 800-378-4188, Fax 303-449-5951, Website bookswest.net.
Liberty Mountain, 4375 W. 1980 S., Suite 100, Salt Lake City, Utah, 84104, Tele. 800-578-2705 or 801-954-0741, Fax 801-954-0766, Website libertymountain.com, Email sales@libertymountain.com.
Treasure Chest Books, 451 N. Bonita Avenue, Tucson, Arizona, USA, 85745, Tele. 520-623-9558, or 800-969-9558, Website treasurechestbooks.com, Email info@rionuevo.com.

Some of Kelsey's books are sold by the following distributors.
Canyonlands Publications, 4860 North, Ken Morey Drive, Bellemont, Arizona, USA, 86015, Tele. 928-779-3888, or 800-283-1983, Fax 928-779-3778, Email info@clpbooks.net.
High Peak Books, Box 703, Wilson, Wyoming, USA, 83014, Tele. 307-739-0147.
Rincon Publishing, 1913 North Skyline Drive, Orem, Utah, 84097, Tele. 801-377-7657, Fax 801-356-2733, RinconPub@UtahTrails.com.
Recreational Equipment, Inc. (R.E.I.), 1700 45th Street East, Sumner, Washington, USA, 98390, Website rei.com, Mail Orders Tele. 800-426-4840 (or check at any of their local stores).
Online--Internet: amazon.com; adventuroustravelers.com; btol.com (Baker-Taylor); Ingrams.com; Bdaltons.com; borders.com (teamed with amazon.com).

For the **UK and Europe**, and the rest of the world contact: **Cordee,** 3a De Montfort Street, Leicester, England, UK, LE1 7HD, Website cordee.co.uk, Tele. Inter+44-116-254-3579, Fax Inter+44-116-247-1176.

Printed by Press-Media, *press-media.com*. 1601 West, 820 North, Provo, Utah.
All fotos by the author, unless otherwise stated.
All maps and geology cross-sections were drawn/created by the author.

Front Cover

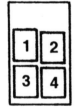

Front Cover Fotos

1. Pothole Fork/East Pasture Canyon, Roost Country, Utah, Page 76 (Jon Jasper foto)
2. The Black Hole, Middle White Canyon Area, Utah, Page 172
3. Clearwater Canyon, South R. Roost and North Lake Powell Country, Utah, Page 126
4. Cowboy/Gravel Canyon, Middle White Canyon Area, Utah, Page 176 (Dan Logan foto)

Back Cover

Back Cover Fotos

5. Gravel Canyon, Middle White Canyon Area, Utah, Page 176 (Dan Logan foto)
6. Middle Fork, King Mesa Slots, Escalante River Drainage, Utah, Page 230
7. West Canyon, South Lake Powell Country, Arizona, Page 192
8. Golden Cathedral, Neon Canyon, Escalante River Drainage, Utah, Page 208
9. Middle Fork/West Butler, Cedar Point Area, Utah, Page 144
10. Birch Canyon, Near Zion National Park, Utah, Page 284

Table of Contents

Acknowledgments

There have been dozens of people contribute information leading to the writing of this book but the following people were most helpful.

In the mid-1990's, German hiker Eberhard Schmilinsky invited the author to explorer some technical canyons near Page, Arizona. This was the first time since the 1960's this writer had done any rappelling. Going through canyons with ropes & rappelling was a step up in challenge and the beginning of this guide. Since then, and since the mid-1990's, Keith Krumbach and Warren Egbert (and later Julia Roetzer) joined us in a yearly reunion of canyoneering mostly on the long 4th of July weekend.

Steve Tyler, long time friend going back to grade school, has been on many hikes with the author, especially in the San Rafael Swell, Robbers Roost & Cedar Point/North Wash Country. Also, the author met Nat Smale going to Anasazi Canyon and later they teamed up for many of the best canyons in this book. Along with rock climber Bill Bees, and others, we did Heaps and Imlay Canyons.

The author's brother Gail (Bill) Kelsey joined him in doing Segers Hole and Eardley Canyons; David Harrison and Rick Ahlander joined in going through the North Fork of Iron Wash for the first time; Mike Kempt & Bob Janzen were along on the 1st hike through Cable Canyon. Scott Patterson accompanied the author in Upper Iron, Larry, Hideout, Triple (San Rafael Swell) and several hikes in the Dinosaur NM area.

Kolob, Imlay & Heaps were done with Leroy Anderson & sons Craig & Matt, plus Newel White, Bryan Anderson, Blake Emett & assisted by Helen White, Dianne & Michael Anderson; Cottonwood & Fivemile were done with Andrew Fitzgerald, and Burro Wash with James Heller, both NPS employees at Capitol Reef NP at the time; John Summerson was along in Maidenwater, Egypt Slots and several other North Wash Slots (and many others over a 5-6 year period); Otis Manson and Tim Martin went with the author into lower Water Holes Canyon; and Tom Jones joined in doing the direct route through Quandary and Little Segers Hole Canyons--on the 2 hottest days of 2002!

The author met Byron Lemay and Herb Taylor in Nasja Canyon, north of Navajo Mountain, and they shared information on many isolated canyons. With Byron we did Kolob, Pine Creek, Icebox, Engelstead Hollow, Great West (Upper Left Fork--The Subway), Cheesebox & Hideout Canyons; and several canyons in the Roost Country along with Tiffany Winget. Steve Brezovec came along in Upper Iron Wash and in canyons of the Roost Country (including Clearwater Canyon)--along with Marty Glaubitz.

For this 2nd Edition, other hiking partners in addition to those already mentioned above were: Jeff Webb, Jon Jasper (jonjasper.com), Tim Barnhart, Stefan Folias, Justin Keener, Ryan Cornia (ajroad-trips.com), Cristina Amat, Tom Tolboys, Bryan Olliver, AJ Pastula (adventure-geek.com), Richard Patterson, Dan Logan, Steven Ho, Stuart Paul and Jonas Fast.

The rangers (several over the years) working at the back country desk at the visitor center in Zion N.P. were most helpful; many if not most are not at all like most NPS rangers you meet. Most recently, Cindy Purcell was most helpful getting the 1st Edition out. For route descriptions, the rangers at the back country desk will allow you to look at their *black book* with route descriptions of technical canyons. Jonathan Zambella who owns & operates Zion Adventure in Springdale, Utah, was always willing to part with ideas & information about his specialty--the canyons of Zion, and nearby.

As usual, and for the 1st edition, my mother Venetta B. Kelsey, was instrumental in watching after the business while I was out having fun.

The Author

The author, who was born in 1943, experienced his earliest years of life in eastern Utah's Uinta Basin, first on a farm east of Myton, then in or near Roosevelt. In 1954, the family moved to Provo and he attended Provo High School and later Brigham Young University, where he earned a B.S. degree in Sociology. Shortly thereafter, he discovered that was the wrong subject, so he attended the University of Utah, where he received his Master of Science degree in Geography (minoring in Geology), finishing classes in June, 1970.

It was then real life began, for on June 9, 1970, he put a pack on his back and started traveling for the first time. Since then he has seen 223 countries, republics, islands, or island groups. All this wandering has resulted in self-publishing 16 books. Here are his books as of 2008, listed in the order they were first published: *Climber's and Hiker's Guide to the World's Mountains & Volcanos (4th Edition); Utah Mountaineering Guide (3rd Edition); China on Your Own, and the Hiking Guide to China's Nine Sacred Mountains (3rd Ed.)* **Out of Print***; Non-Technical Canyon Hiking Guide to the Colorado Plateau (5th Edition--all color); Hiking and Exploring Utah's San Rafael Swell (3rd Edition); Hiking and Exploring Utah's Henry Mountains and Robbers Roost (Revised Edition),* **(Out of Print--will be Reprinted in 2009);** *Hiking and Exploring the Paria River (4th Edition); Hiking and Climbing in the Great Basin National Park (Wheeler Peak, Nevada)* **Out of Print***; Boater's Guide to Lake Powell (5th Edition-full color); Climbing and Exploring Utah's Mt. Timpanogos; River Guide to Canyonlands National Park & Vicinity* **(Out of Print--will be Reprinted in a few years);** *Hiking, Biking and Exploring Canyonlands National Park & Vicinity* **(Out of Print--will be Reprinted in a few years);** *The Story of Black Rock, Utah; Hiking, Climbing and Exploring Western Utah's Jack Watson's Ibex Country; and Technical Slot Canyon Guide to the Colorado Plateau (2nd Edition).*

He has also helped his mother Venetta Bond Kelsey write and publish a book about the one-horse town she was born & raised in, *Life on the Black Rock Desert--A History of Clear Lake, Utah* **(Out of Print, but at some time in the future will be re-printed).**

Map Symbols

Town or Community	⬜⬜	Peak & Prominent Ridge	■✖■
Building, Cabin or Home	▫	Stream or Creek, Running Water	～～
Backcountry Campsite	▲	Backcountry Campsite (NPS)	△
Campsite with Vehicle Access	⬟	Large River	〰〰
Campground	⋏	Dry Creek Bed or Channel	—···—
Cemetery or Grave Site	✚ †	Canyon Narrows	⬙⬙⬙
Ranger Station, Visitor Center	◀⌂	Lake, Pond or Stock Pond	⬭ ◁
Hotel, Motel or Lodge	⬛	Large Pothole	⬯
Airport or Landing Strip	✈	Waterfall or Dryfall	—+—
U.S. Highway 20 21 89		Spring, Seep or Well	o
Utah State Highway 24		Canyon Rim, Escarpment	⊥⊥⊥⊥
Road--Maintained	≡≡≡	Natural Bridge or Arch, Corral	∩ ⌒
Road--4 Wheel Drive (4WD)	-≡≡≡≡	Mine, Quarry, Adit or Prospect	↖↗
Track--Old Road, Unusable	-———	Geology Cross Section	⌐⌐
Trail, Foot or Horse	-----	Pass or Divide	⋈
Route, No Trail	••••••	Rock Art--Pictograph	(PIC)
Cowboyglyph or Campground	(CG)	Rock Art--Petroglyphs	(PET)
Elevation in Meters	-1490	Mile Posts (mp) Markers 30 31	
600 Meters	-600m	Car-Park or Trailhead	(P)

Abbreviations

Campgrounds	CG	Four Wheel Drive (Vehicle or Road)	4WD
Campsites, Chokestone	CS	Two Wheel Drive (Vehicle or Road)	2WD
Kilometer (s)	km, kms	High Clearance Vehicle	HCV
Piñon/Juniper Trees or Forest	P/J	Off Road Vehicle	ORV
2 Bolts & Hangers	2 B&H	All Terrain Vehicle	ATV
Rappel 2 or Rap 2	R2	Low Clearance Vehicle	LCV
Carabiner	biner	Formation, Sandstone	F, SS
Downclimb/Upclimb (s) (ing)	DC/UC ('s) ('ing)	Pothole (s)	PH's
Webbing & Rapide/Quik Link	W&R/QL	Keeper Pothole (s)	KPH ('s)
Knee & Elbow Pads	K&EP	Geologist's Pick	G-pick
50 meters, 1128 meters	50m, 1128m	Descent Rings	D-rings
July 4, 2007	7/4/2007	Mile Posts	mp
Knee, Elbow & Ass Pads	KE&AP	South, Southeast	SSE
Rappelling Gear (Harness & Rap Device)	RG	North, Northwest	NNW
Handline	HL	Pack Rope	PR
Bolt Kit (Bolts, Hangers, Wrench & Bit)	BK	High Water Mark (Lake Powell)	HWM
Runner/Quick Draw/Daisy Chain	R/QD/DC		
Glen Canyon National Recreation Area			GCNRA
United States Geological Survey			USGS
National Park Service			NPS
Bureau of Land Management			BLM
Bureau of Indian Affairs			BIA
Civilian Conservation Corps		Triple C's, 3 C's,	CCC's
Trails Illustrated/National Geographic			TI/NG

Metric Conversion Table

1 Centimeter = 0.39 Inch	1 Mile = 1.609 Kilometers	1 Ounce = 28.35 Grams
1 Inch = 2.54 Centimeters	100 Miles = 161 Kilometers	1 Pound = 453 Grams
1 Meter = 39.37 Inches	100 Kilometers = 62.1 miles	1 Quart (US) = 0.946 Liter
1 Foot = 0.3048 Meter/30.5 Cms	1 Liter = 1.056 Quarts (US)	1 Gallon (US) = 3.785 Liters
1 Kilometer = 0.621 Mile	1 Kilogram = 2.205 Pounds	1 Acre = 0.405 Hectare
1 Nautical Mile = 1.852 Kms	1 Metric Ton = 1000 Kgs	1 Hectare = 2.471 Acres
1 Kilometer = 3281 Feet	1 Mile = 1609 Meters	0.1 Mile = 161 Meters
1 Cubic/Liter = 61 Cubic/Inches	50 C/L = 3050 C/I	100 C/L = 6100 C/I

Meters to Feet (Meters x 3.2808 = Feet)

100 m = 328 ft.	2500 m = 8202 ft.	5000 m = 16404 ft.	7500 m = 24606 ft.
500 m = 1640 ft.	3000 m = 9842 ft.	5500 m = 18044 ft.	8000 m = 26246 ft.
1000 m = 3281 ft.	3500 m = 11483 ft.	6000 m = 19686 ft.	8500 m = 27887 ft.
1500 m = 4921 ft.	4000 m = 13124 ft.	6500 m = 21325 ft.	9000 m = 29525 ft.
2000 m = 6562 ft.	4500 m = 14764 ft.	7000 m = 22966 ft.	8848 m = 20029 ft.

Feet to Meters (Feet ÷ 3.2808 = Meters)

1000 ft. = 305 m	9000 ft. = 2743 m	16000 ft. = 4877 m	23000 ft. = 7010 m
2000 ft. = 610 m	10000 ft. = 3048 m	17000 ft. = 5182 m	24000 ft. = 7315 m
3000 ft. = 914 m	11000 ft. = 3353 m	18000 ft. = 5486 m	25000 ft. = 7620 m
4000 ft. = 1219 m	12000 ft. = 3658 m	19000 ft. = 5791 m	26000 ft. = 7925 m
5000 ft. = 1524 m	13000 ft. = 3962 m	20000 ft. = 6096 m	27000 ft. = 8230 m
6000 ft. = 1829 m	14000 ft. = 4268 m	21000 ft. = 6401 m	28000 ft. = 8535 m
7000 ft. = 2134 m	15000 ft. = 4572 m	22000 ft. = 6706 m	29000 ft. = 8839 m
8000 ft. = 2438 m			30000 ft. = 9144 m

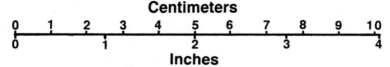

Centimeters / Inches

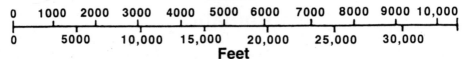

Meters / Feet

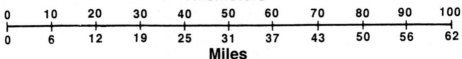

Kilometers / Miles

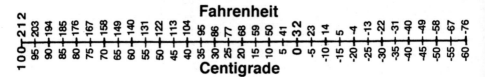

Fahrenheit / Centigrade

Introduction to Hiking Technical Slot Canyons

The Colorado Plateau & Technical Slot Canyons

The Colorado Plateau is a large physiographic region covering roughly the southeastern half of Utah, the northern half of Arizona, the western fifth of Colorado and a small area in the northwestern corner of New Mexico. It basically includes the middle third of the Colorado River drainage. Almost all the canyons in this book are in Utah but with a few--some of the best, in northern Arizona near the town of Page and on Navajo Nation lands.

This is a technical slot canyon guide to the Colorado Plateau. As defined here, **Technical Slot** means a very narrow canyon often a meter wide or less, usually requiring ropes & rappelling, and/or high-stemming, and difficult up/downclimbing to get all the way through.

Why is the Colorado Plateau so Unique?

What makes the Colorado Plateau so unique is the geology--and/or the rocks. Most of the rock formations on the Plateau are horizontal or flat-lying, and in many places looks like a layered wedding cake. Each rock formation has a different color and composition, thus a different rate of erosion. That, plus low rainfall and not much vegetation, is what makes this region so interesting. For example, the Grand Canyon consists of cliffs, benches and terraces making a wide open, but deep and scenic gorge. In other parts of the Plateau you find deep, dark, narrow slots, some of which may be 50 or more meters deep and only a meter or less wide. These narrow slot canyons are among the strangest, most-interesting and fotogenic features on earth. Almost all the hikes in this book are in this type of canyon. There is simply no other place on earth quite like the Colorado Plateau.

Equipment Needed for Technical Slot Canyons

ATTENTION: This book is not meant to teach people how to rappel or tie knots, etc. etc; so it's up to each individual to learn these skills before doing any difficult canyon. Your best choice is to **go with an experienced person** until you learn the ropes. Other options are to **take a class** from someone who might also offer guide services, or contact the American Canyoneering Association at **canyoneering.net**. In recent years, the University of Utah has set up a class geared to canyoneering. Also, on your computer, do a search under **Canyoneering or Canyoning**. At last count, there were 421 websites having to do with this subject.

Perhaps the most important thing to remember is, you must have all or most of the equipment mentioned below on each hike, whether you think you'll need it or not. You must go prepared for any emergency and have the skills to handle any situation. **Canyons change with every flood, and in some absurd cases, hardcore canyoneers remove bolts (but only if there's another option such as a tree, log, chokestone, or other natural anchor in which to place webbing for the rappel); so go prepared. The responsibility is yours.**

To do the canyons in this book, here's a list of things you'll need to get through safely. Please keep in mind this book is for advanced, or at least slightly more adventurous hikers, and for those who seek more challenges than just walking up or down a canyon or on a national park trail.

Rappelling Gear (RG) In this book, the term **rappelling gear** means these things: You'll need a **rappelling harness**, one with loops around the sides and/or back. These are for carrying/storing extra carabiners, hooks, ascenders, baby angle pitons or rapide/quik links (R/QL). Next, you'll need some kind of **rappelling device** attached to your harness to slow you down on the rope using friction. The most popular gadget is a **tube-type device** called an **ATC** (Air Traffic Controller. Try the ATC-XP by Black Diamond). The **figure 8** is another popular device. You'll also need a strong **locking carabiner (biner)** to attach the rappelling device to the harness. This writer now uses a wide-mouthed **steel locking biner** because it out-wears the aluminum kind about 20 times! In addition, the serious canyoneer has with him at all times 6-8 other regular **carabiners, or locking biners**.

For those serious about staying alive, add to these some kind of **ascenders**, either **Jumars** or **Tiblocs**, or even **prusik ropes** or **cords**. You never know when you'll have to ascend a rope that's stuck for whatever reason. If not each person, each group should have at least one set, maybe 2. Read more below.

Figure 8 or ATC These are the 2 most popular devices used in rappelling along with a harness. The **figure 8** works fine, but one disadvantage is, it tends to twist the rope as you go down. Another potential problem is; while rappelling, if your rope is wrapped entirely around the '8, and if the '8 is allowed to make contact with the edge of a cliff when you start the descent, there's a chance the rope will get snagged and forced up on the back side thus creating a **girth hitch**. If this happens you're stuck, and you'll need help from a partner! Or pull yourself up by hand to relieve the load and realign the rope. To avoid this dilemma, always push yourself away from the edge with one hand and/or feet as you're starting to rappel so the '8 doesn't come into contact with the rock or log. Or better still, use a steel biner, then wrap the rope through the '8 & biner, similar to the way ATC's are used. By doing this, there's no way you'll ever be tied up by a girth hitch.

The advantage of the **ATC** is it doesn't allow your rope to twist. ATC's come in various sizes and shapes, which like figure 8's, are used to create friction and control the speed of descent. For the last big rappel in Heaps Canyon, or any long free-rap, most people now use a **variable speed control ATC** which has a tilted face instead of being straight across. This slows you even more when doing a single rope rappel. Using a second locking biner attached to your brake-hand-side leg loop will also cause more friction and slow you down as well. But this also twists the rope and you'll spin like a top coming down.

Back to figure 8's. In canyons such as Imlay, where you rappel into potholes full of water and will be disconnecting your rope & rappelling device while swimming, the author prefers a **steel biner** and a weird-looking '8 called a **Pirana.** The Pirana has a smaller hole than most '8's where it's attached to the biner. With this, rig your rope so it passes through the Pirana and the biner the same way as with an ATC. By doing this, the Pirana always remains attached to the steel locking biner; thus there's no chance of loosing the '8 in water. If using an ATC, you almost have to remove it from the biner when disconnecting, thus a chance to drop it in the pothole.

The Pirana also has a couple of fingers sticking out on one side for wrapping a single rope around. This creates more friction when doing single rope rappels, and is especially good when using a soft or limp rope. Only problem is, it's made of aluminum, and wears out quick, especially when used with a wet rope caked with sand.

Carabiners One or more good quality **locking carabiners** (these have a screw gate or locking mechanism to prevent the gate from opening at the wrong time) is used along with a figure 8 or ATC for rappelling. Regular non-locking or wiregated biners are also used for various other tasks or situations. Serious canyoneers going through potentially risky canyons like Heaps or Imlay, will carry 6-8 carabiners, including several lockers, on their harness loops.

Ascenders (Jumars) Today there are a number of devices on the market which allow you to ascend a fixed rope. The most popular is the one shaped like a handle (sometimes called by the trade name of *Jumars*), but they're also the largest and heaviest of the group. Other smaller devices used for ascending are **prusiks** (2 short ropes at least 2mm smaller than the rope you're ascending) or small metal gadgets such as **tiblocs** which are used with an ordinary carabiner and short piece of rope.
See the ACA website **canyoneering.net** for information on other devices, or ask about these and other gadgets/technics at a climber's supply store. Also checkout Dave Blacks book, *Canyoneering*, for gadgets/technics as well. **Ascenders are standard equipment for canyoneers in technical slot canyons** because all too often someone has to go back up to retrieve or loosen a rope that's stuck. Or, as in the case of this writer who often goes alone into canyons he knows nothing about, several short ropes are taken and one left at each dropoff or rappel. When he reaches the end of a canyon, or decides for whatever reason to turn back, he then returns upcanyon using ascenders.

Ropes There are 2 standard kinds of ropes. One is **dynamic**. These stretch a lot and are used by rock climbers to absorb the energy of a fall--instead of their bodies. The other is **static**. These stretch very little and are the standard type rope for canyoning and/or rappelling. Standard rope lengths are 50 and 60 meters. If you really get into canyoning, the author recommends you buy a pair of 60's; or one 60 and an equally long pull cord which is much lighter than carrying 2 regular ropes. If you choose a use a cord, most buy one that's 6mm thick. With one of these, and in an emergency, you can rappel with a 6mm cord. The author ended up buying a 4mm cord (35m & 65m long); this costs less and weighs almost nothing, but it stretches a lot, and can't be used for rappelling. If it's a hard pull, it's hard to hang onto as well.
Also, most people prefer using a smaller diameter rope like 9mm--as opposed to 10.5mm or 11mm. Most canyoneers now prefer 8mm rope because they weigh a lot less. The only real drawback to using 8mm is, if it gets nicked, there's less good rope left to hold you, thus greater risk. Some manufactures are now making ropes specifically for canyoneering with a kevlar lining. Ask about these.
To avoid getting a nick, cut or abrasion on your rope, be sure to rappel straight down from the anchor. Don't move from side to side once you go over the edge of a cliff, because to do so with your full weight, may do serious rope damage.
When setting up a rappel from any anchor; tree, chokestone or deadman, make sure there's no sharp edges down along the line of descent. If you do rap over a sharp lip or edge, it may put your rope out of business--maybe you too! If this happens you'll have to cut it and use the 2 pieces for short raps or sling material. The author has lost 3 ropes this way, and has seen 2 others destroyed.
Pack Rope Another rope that comes in handy in many situations is what this writer calls a pack rope (**PR**). This is a short rope 8-10m long & 6-7mm in diameter. It's tied to the loop at the top of your pack and inserted in a pack pocket. If you have to upclimb (UP) over something, tie the PR to your wrist, make the climb, then pull the pack up. You can also lower the pack first, then downclimb (DC). Or in a bind, you can use it for a sling/webbing, or to handline (HL) down or make a short rappel. Or if you nick one of your rappelling ropes, use the PR plus pieces of webbing as a pull cord; while rapping down a single rope (using a biner block at the top).
Biner Block Might as well mention this here. A biner block is what you use if you take one regular rope for rappelling, plus a smaller pull cord. Run one end of the rope through the R/QL (descent ring) and tie it to the pull cord (or in an emergency a PR and assorted pieces of webbing). Attach a **locking biner** with a **clove hitch** to near the end of the rope so it can't pull through the R/QL when you rap. Rappel down the rope, then retrieve everything by pulling the pull side or cord. If doing it this way, make sure the pull side is separated from the rope before the last person comes down; otherwise, it may get tangled or crossed. If that happens, you may have to jumar up to straighten things out.
More Tips If your rope gets wet, and it almost always will--especially if you're in Zion, then don't expose it to heat while drying. For example, don't place it in the trunk of your car on a hot summer day. Instead put it inside on the floor where it's cooler. Once wet, all ropes will shrink some, but exposing it to heat makes them shrink more. Letting them dry at room temperature is best. To get sand out of your rope, hand wash it with a hose, in a bath tub, or even a front loading washer.
Recently the author measured all his ropes by laying them on the sidewalk. They were all shorter than when they were bought. His one 50m (which was an even 50m when it was cut) was down to 45m--but it was measured while limp on the sidewalk. Under weight, it would be another meter or 2 longer for sure. Remember too, when you buy a new standard rope advertised as 50m or 60m, it will actually be closer to 55m or 65m to compensate for what will happen once it gets used & wet. It's not a bad idea to measure your ropes occasionally. The best way is to run it through a device which measures ropes at a climber's supply store.

Backpacks For most hikes in this book you'll need at least a **medium sized daypack**, and in some cases, a pack that's big enough to carry 2-3 days of supplies. The 2 daypacks this writer uses are about 35 cubic/Liters, and **40 c/l** (maybe the best all around daypack size). Tom Jones of Imlay Canyon Gear also makes a 30 & 50 c/l packs. You can always buy something cheap at Kmart/Walmart, but they don't last long; best to buy something that's well-built, then if you're dragging it through slot canyons, you can put patches over patches for a long time.
If you're going to buy a new pack just for technical slot canyons, buy a tall thin one without pockets on the sides. If your present pack has external pockets, when it comes to the narrow slots and pulling it in and out of potholes, don't put anything in the pockets--otherwise they'll be shredded to pieces. The author used a full-sized internal frame pack with homemade side-pockets for one trip through Imlay, and by the end of the hike the pockets were unusable.
Another thing to remember, if there's something hard inside your pack, such as a water bottle, waterproof canister, box, or keg, keep it in the middle and pack softer things around it; otherwise you'll

have holes in the pack where the hard object is exposed to abrasion. If dragging a pack through lots of tight slots, expect to replace it in a year or two; or if it's well-built, patch it by using some kind of tough cloth along with **contact cement**. Duct tape works well too. Forget about sewing patches on.

If looking for a new backpack for slot canyons, ask someone at a climbing supply store about one made of new material that resists abrasion. Also, look into what's available in the way of a backpack that's waterproof and built like a drybag, but with shoulder straps. One of these, plus the use of **dry-bags or kegs** (read more below) inside, should guarantee everything stays dry.

Bolt Kit (BK) A bolt kit consists of a **hand drill holder & bit** (standard size is 3/8", or about 1 cm in diameter); a small crescent or adjustable **wrench** to tighten the bit in the drill holder body and for tightening the bolt into the wall; a short piece of **fuel line tubing** for blowing dust out of holes; a sharp **pocket knife**; and **hammer**. The author recommends a **geologist pick (G-pick)** which is much more versatile than the standard climber's hammer.

Be aware that **hardcore canyoneers** despise bolts in every way. In recent years this group has been leading the charge to eliminate bolts altogether. Their argument is, they're unnatural to the canyon (but they don't seem to mind leaving W&R/QL everywhere), and that everyone should learn how to rig natural anchors. This is definitely a good idea, but there are situations where there is absolutely nothing in which to attach a rope. Read about **Natural Anchors** and **Ibis hooks** below.

Cordless Drill One option to the the time-honored bolt kit is the cordless drill. These make a clean smooth hole in sandstone in less than a minute, but they're not officially allowed in wilderness areas or national parks. Disadvantages to using these are, you must keep them safe in a drybag if you're in a swimming situation, and you must carry an extra battery (or have lots of trust in one!). Smart canyoneers often carry a standard bolt kit as a backup (especially when they're with a larger group). Hardcore canyoneers despise cordless drills, as well as bolts--unless there's absolutely nothing else to attach a rope to. Best to learn how to hand drill and place a bolt or baby angle piton, and hope you don't have to use one.

Bolts & Hangers (B&H) Most **expansion bolts** in the USA are 3/8" (about 1 cm) diameter and 7 1/2 cms (3") long (some are as long as about 13 cms or 5"). Bolts preferred by climbers are those with the capabilities to bend rather than snap off if the steel is too brittle. Those bought in hardware stores are likely OK, but you might inquire about their brittleness. **Rawl (blue banded) bolts** are generally considered the best. Some of those from hardware stores are designed differently and won't grab as well in an enlarged hole. When using a hand drill, the hole is always bigger than when using a cordless drill, but Rawl bolts seem to expand & grab better than others when the hole is a little large.

A **hanger** is a flat & bent piece of metal with a hole for the bolt to hold it to the wall, and another hole for a chain, webbing or your rappelling rope. If it's a short rappel, you won't need webbing attached to a single B&H. The latest rage seems to be **eye bolts** that are glued in with some kind of epoxy. These are supposed to stay in longer (?), but maybe these are a lot more trouble than simple anchor building. Read more below.

Baby Angle Pitons If you must drill a hole for a bolt using the standard bolt kit, then in some cases the hole may be too large for a conventional bolt to grab and be secure (Rawls work OK). You can eliminate this by using a cordless drill; or by using a baby angle piton. The preferred size is a 12mm (1/2") piton 8-10cm long. As you pound these in, you can tell by the ring if it's in tight.

Natural Anchors The hottest topic of debate in canyoneering circles today is what some people think is the overuse of **bolts** as opposed to using **natural anchors** such as **trees, boulders, small natural arches/bridges, chokestones or piles of rocks** called a **deadman**. Keep this in mind as you arrive at a dropoff. If there's something strong & safe close by to attach a rope or a sling, then use that. At the same time, if the closest chokestone is 25m away, consider the **trash factor**--that is, leaving a ton of webbing littering the canyon as opposed to one B&H next to the dropoff; or even rappelling off an Ibis hook if it's a short drop.

To make a rappel, the standard procedure is to place a short piece of **webbing**, or part of an **old rope**, around a small chokestone or log next to a dropoff. Attach to the webbing a **Rapide Link or Quik Link (R/QL or W&R/QL)** or Descent Ring (**D-ring**). Read below. Also, be sure to double check any webbing/sling you find. If it's been there a long time and is stiff & faded, it must be replaced.

Anchor--Rock Pile/Deadman Here's the **latest rage in natural anchors**. Build an **anchor** out of a **pile of rocks** called a **deadman**. Look around for a fairly large stone, one that's oblong. Wrap W&R/QL around it, then place it in crack or pothole, or dig a hole in the sand or gravel (a G-pick works great). Bury it as best you can, then place a number of smaller rocks or sand on top so it can't move. Have the end of the W&R/QL reach to, or just beyond, the top of your rappel. Leave everything there; in most cases, it will eventually be washed away--including the W&R/QL which will eventually be **trash** somewhere downcanyon.

Another idea for an anchor is to tie a **knot** in a short piece of rope or webbing, then push the knot in a nearby crack. If it's a **tight crack**, the knot can't move, and should be safe.

Rappelling from a Hook Here's an option to using a B&H or W&R/QL around chokestones, at least on short rappels. Locate a crack or natural hole or depression in the rock immediately next to a dropoff, then place an **Ibis hook** or **Spoonbill** (tested at 450 kgs) already equipped with a **short piece of webbing & carabiner**, in the hole or crack. Tie a loop in one end of a short rope, and attach that to the biner & hook, then either handline (HL) or rappel. Once down, give it some slack, flip the rope, and the hook should pop off, leaving nothing behind. In some cases, and for **safety**, you may want to use a **G-pick** to make a very small hole to keep the hook from migrating or slipping. The best reason for using a hook is, you leave nothing behind--**no webbing, no trash**. Even if you have to use a G-pick to make a small hole, after the next rains or flood, no one will ever see it or know that a hole was made. Some hardcore canyoneers hate the idea of using a G-pick, but at the same time, they think nothing of leaving a ton of webbing around a chokestone, log or building a pile of rocks with W&R/QL as an anchor. Read more about hooks & holes below under **Hooking Kit** and **Getting Out of Keeper Potholes**.

Retrievable Sling Here's an idea among a few experienced canyoneers who are trying to leave as little as possible in a canyon. Use one of several methods of rigging a **retrievable sling**, or learn how to tie a **macramé**, which allows you to rappel down one side of a doubled rope, then pull the other side from the bottom and take everything with you. From the author's limited experience with these techniques, not many situations will allow you to retrieve your sling making it a clean site. **Trees or small arches** work well, but most other places don't. But regardless, it will benefit everyone to learn about these and other methods of how to get through a canyon alive if things go wrong--and to leave as little behind as possible. To see drawings or information on these techniques, visit the website of the **American Canyoneering Association**, canyoneering.net for drawings and techniques.

SAFETY FIRST To this writer, there are 2 main things to keep in mind regarding rappelling and the use of natural anchors. Number 1, and above all else, is **safety**. But even with this in mind, some hardcore canyoneers are removing bolts (which are some of the safest anchors) and building piles of stones in an effort to teach everyone else how to do it right. Or do it their way! Sooner or later someone will rap from a deadman and have the rope rip out, **and die**. And sooner or later, and in some canyons, all the loose material with which to build an anchor, might be gone. This is especially worrisome at the last big 44m rap in Little Middle Fork of Robbers Roost Canyon. In 2007, 2 good B&H were put in there, making it totally safe, but they may have been removed?

The second thing or issue is leaving **trash** (W&R/QL) in the canyons. Everyone needs to learn how to create anchors, if for no other reason than to survive. But it seems in many cases, and for short raps, it's best to use an **Ibis hook**, which means leaving nothing. Even if you have to use an **G-pick** and make a small hole, it will soon blend with all the other holes and cracks and won't be noticed.

Rapide/Quik Link or Descent Rings & Webbing (W&R/QL) Whether you're using a natural anchor such as a boulder, chokestone, log or tree, or a B&H, you normally use a short piece of webbing or rope to be used as a **sling**, and a metal descent ring of some kind. Sometimes these are standard one piece aluminum **D-rings**, or the more preferred **Rapide Links**, sometimes called **Quik Links. R/QL** look like a link of a large chain, but one side can be screwed shut by hand. One way to get a very strong and cheap D-ring is to go to a hardware store selling heavy chains, buy a meter or so, then have them cut every other link. You keep the ones that aren't cut. Running your rope through the D-ring or R/QL makes it easier to pull from below without burning the sling. Make sure the sling/webbing & R/QL (W&R/QL) are close to, or right on, the edge of the cliff for an easier pull from below.

Pack Tethers--Runners, Quick-Draws or Daisy Chains (R/QD/DC) Runners are short pieces of webbing up to about 1 meter in length and used along with 2 carabiners to secure yourself to an anchor such as what you'll find at the last big rappel in Heaps Canyon, or the midway points in Icebox and Englestead Canyons. Or to tether your pack to the front your harness and let it hang down between your legs when making a big free rappel. These can also be used with ascenders to go up a fixed rope, or to hang your pack from your harness when high-stemming in X SLOTS. You can make your own runners with any size webbing, and with any size loops, but daisy chains are factory sewn. The advantage of a daisy chain, which is a meter-long piece of webbing with many loops, you can use it at any length you like to fit the situation. The disadvantage is, in the eyes of this writer, they're made from skimpy material. Some prefer a homemade runner made out of full-sized webbing!

In recent years, starting in perhaps the 1970's, there's been a small army of canyoneers out looking for the most-difficult canyons under the sun--and be the first to descend them! In many cases these are too narrow to get through at the bottom so you have to chimney up as high as 15m. This is called **high-stemming**. To do this type of canyon you often must remove your pack and let it hang down out of the way. Let's call this a a **Pack Tether**. Most use a **daisy chain** (adjustable) for this, others make their own runner, but the author likes a **quick draw** (2 biners with a short piece of webbing between) to attach his pack to a side-loop of his harness. This keeps the pack close by, but others like to let it hung down a meter or 2 and for the most part, out of the way.

Getting Out of Keeper Potholes (KPH) A keeper pothole is one you can easily get into, but difficult to get out of. Here's another big debate among those hiking Zion canyons where most KPH are located; how to get out of a KPH and leave it in a natural condition. Heretofore, the method used most in some big KPH's was to **hook** your way out as described below. But this method leaves man-made holes which some hardcore canyoneers despise. As of 2008, it's still the most-used method in Imlay.

Besides hooking, another popular way out is the **pack toss**. Tie a rope to your pack, a second smaller pack, a sleeping bag stuff sack (or a similar sack specially designed for the job called a **toss bag--by Tom Jones, canyoneeringusa.com**, makes & sells what he calls **Pot Shots**), or even a small log; then before you actually enter the pothole, toss the pack/bag over the pool and down the other side. Then you get in the pool and pull yourself out using the rope. You will be heavier than the pack or toss bag, but the friction created by the rope on the wall and/or lip of the pothole, will counter the weight difference. In some cases, you may have to throw 2 or more packs; or 2 or 3 smaller toss bags filled with sand, gravel or small stones, in order to get enough weight, plus friction, so one person can climb out. With one person out, ropes and/or etriers (aiders) can be used to get others across.

Another possible way is to use a **large #3 Ibis hook**. Duct tape the hook to the end of a sectioned **aluminum tent pole** or avalanche probe (sometimes called a **cheater stick**), reach up to a ledge, crack, hole or in the case of 3 KPH's in lower Imlay Canyon, a B&H & W&R/QL on the lip of the next pool; then using one or 2 etriers attached to the hook, or short rope with knots & loops, climb out.

Tom Jones also makes what he calls a **Happy Hooker** (just described). In a few situations this works fine, but seems limited to just a few select KPH's such as in Heaps or Imlay Canyons, or the 1st KPH in Quandary. Rigging a tent pole seems best.

Another method is to tie a **large knot** at the end of a short rope, throw it across a KPH and wedge it in between a chokestone and canyon wall. Or perhaps tie an **Ibis hook** on the end of a rope and throw that hoping that it gets stuck somewhere, then pull yourself across and out the other side. This works well in Piñon Canyon.

Another time-honored method is the simple **shoulder stand**. With 2 or more people inside the pothole, have one person stands on someone else's shoulders. This works well if one person can hold his breath a long time under water, or if there isn't too much water in the pothole.

Other techniques are as simple as building a **pile of rocks** to stand on, or lean a **log** up against the wall to climb up. Also, in a few cases, have a good look around and it's sometimes possible to **simply climb up and around a KPH**. This may sound silly, but on one occasion, this writer struggled to get through a big keeper, only to find on his second trip, there was a walking route around it.

Just as this book goes to press, Dave Black has published a book called *Canyoneering*. It's a how-to book with suggestions on potholing, anchor making, etc, etc, from Falcon Guides.

Hooking Kit   Hooks such as grappling hooks, cliff hangers, talons or bat hooks are metal devices shaped similar to an inverted "L" or fish hook, some larger than others, and used to place into small holes or cracks or over a tiny ledge.   In canyoneering, these are often used to escape **KPH's**.

Here's how to rig up or create what this writer is calling a **Hooking Kit**.   Take **2 runners** (or the variable-length **daisy chain**) about a meter long (arms length) and attach each to your rappelling harness with a **locking carabiner**.   At the other end of each runner attach another **locking biner**.   To each locking biner, attach a **grappling hook, cliff hanger, talon** or **bat hook** already rigged with a short piece of thin webbing.   Also on each locking biner, attach an **etrier** or **aider** (a rope ladder made of webbing).   At some climbers equipment store, ask for a **pocket etrier** which comes in a small sack and is, for the most part, out of the way until you need it.   Pocket etriers seem designed just for this task.

Also needed will be something to make holes to place the hooks in.   In the past, the time-honored method was to use a bolt kit & hammer, but it sure seems a lot easier to use a **geologist pick (G-pick)**.   Using the sharp end, you can tap out a hole large enough for a **bat hook/grappling hook** in a few seconds.   The best part of using a G-pick is, you can do it with one hand.   Also, attach an arms-length runner to the your G-pick, and attach that to another biner on your rappelling harness.   When you're not using the G-pick, place it down inside one of your belt loops for storage.   Or just let it hang.

The reason **grappling hooks** are recommended over the smaller bat hooks is that the sandstone inside potholes, and especially near the waterline, is much softer than that on a canyon wall or elsewhere.   For grappling hooks, make the outside of the hole less than 2 1/2 cm (1") in diameter and have it angle down a little at the back end, again much less than 2 1/2 cms deep.   This is a larger hole than you need for bat hooks, but it's safer because of the crumbly sandstone.   On the other hand, and depending on the slope you're climbing, it doesn't matter if you fall back into water, so maybe the smaller bat hook holes will be fine.

Hook Placement.   If you're right-handed, and escaping a new undiscovered KPH, place your first hole on the left side of an imaginary track running up the wall about 25 cms (10") wide.   Make each hole at about eye level; that way you can blow out any excess sand.   Place a hook in the hole, step into the etrier, lift up a little, then tap out the 2nd hole 25 cms to the right and roughly half meter above.   Once the 2nd hook is in, you're on your way, but make as few & small holes as possible.

Keep in mind, there are some hardcore canyoneers who think making even the tiniest of holes in a canyon wall, or modifying the rock in any way with a G-pick, is a bigger crime than placing a bolt in the wall!   You'll have to be the judge on that.   A good idea is, make only holes that are absolutely needed, while using cracks or other natural features as much as possible.   Also, reread the headings above, **Getting out of Keeper Potholes & Natural Anchors, so that hooking or using a G-pick will be your last resort.   Most of the KPH's you'll encounter will be in Zion and holes are already there.**

Helmets   Some people, notably those coming from rock climbing circles, insist that everyone going through any canyon should wear a helmet.   Until recently, this writer never used one, except at the last rappel in Heaps Canyon and through the waterfalls in Kolob.   He has always been more concerned about sun protection and skin cancer.   Whether you use a helmet or not in ordinary canyons is up to you.   However, a small group of hardcore canyoneers are raising the bar when it comes to difficult canyons such as the King Mesa Slots in the Escalante River Country.   In these cases where you're chimneying or high-stemming up to 15m, then having a helmet could save lots of pain if you slipped into the crack below.   The author now wears a helmet in risky slots after a minor fall in Scorpion West.

Wetsuits/Drysuits   Since many canyons in this book have pools or potholes to swim through, you'll need a wetsuit or drysuit, at least in the months of April, May, September and October (In recent years, a small group of canyoneers hike throughout the year, and on New Year's Day crawl through the Black Hole!   That one's officially called Freeze Fest).   If you're hiking in summer weather, you can usually get by without a wet/drysuit in most canyons discussed here, but some exceptions are those in Zion.

Wetsuits are made of a tight-fitting porous nylon material.   This allows some water to penetrate, which is then warmed by body heat and held next to the skin.   If buying a wetsuit, be sure to get one that covers from ankle to wrist--shorties are as useless as tits on a boar!   One with reinforced or padded knees, elbows, chest and rear end will last longer.

The thickness of most wetsuits seems to be 2mm in the arms and legs, 3mm around the body.   For the most part, these would be warm enough to handle most hikes from sometime in May through mid-September.   Other wetsuits can be 5mm or more thick throughout.   These would likely be good for dips in late March and April, and in October.   If water is expected to be really cold, some people use what's called a **Farmer John**, a sleave-less wetsuit which covers only the body torso and legs.   On top of that can be added another **jacket-like** upper part which covers from the waist up and sometimes includes a **hood**.   These worn together create extra insulation where it's needed most.

Most people wear nothing on top of or under wetsuits, but the author prefers to wear a T shirt & shorts on the outside.   This protects the more expensive wetsuit when crawling in or out of potholes or downclimbing (DC'ing) through tight slots.   Another nice thing about wetsuits is, they really protect all parts of your body when DC'ing, plus they offer good friction.   Also, even if you wear a hole in a place or two, the suit still insulates--unlike when you have a hole in a drysuit.   Best to wear K&EP's with wetsuits to protect the vital parts.

Drysuits on the other hand are made of waterproof coated nylon, or in some cases, Gortex.   At the neck, wrists and ankles are very tight-fitting rubberized cuffs or gaskets which hopefully keeps water out.   These are especially good when water is super cold.   Under these suits are worn sweat pants or sweat shirt.   The colder it is, the more insulation you use.   These allow hiking the Zion Narrows and other canyons even in winter.   If you're planning to get a drysuit, buy extra Aquaseal (or just apply contact cement) to reinforce parts that get the most wear; knees, elbows, shoulders and rear ends.

The author has used simple **contact cement** from a can to patch holes (in packs as well) and reinforce his drysuit.   And so far it's working moderately well.   Use a small paint brush to spread the cement over the area that's scuffed and let dry.   Do this 3 or 4 times, then let dry for a week before using (or rub sand or dirt on the patched area to make it less sticky).   Clean the brush with paint solvent.   If a small hole develops in a rubber cuff, first try patching it with a small piece of fresh **duct tape**.   One of these patches has been on the author's drysuit for years and it's still doing the job.

Of these 2 types of swimming gear, the vast majority of people prefer a **WETSUIT** to a drysuit.   Wetsuits really protect your body when climbing/crawling through difficult slots because they're great body armor, and you can pee when swimming, and for the most part, it washes away!   They're also maintenance free.   With **drysuits**, you have to half-remove them to relieve yourself, you never stay completely dry, they require lots of maintenance and they're 3-4 times as expensive as wetsuits.

Drybags, Waterproof Plastic Boxes or Kegs Drybags are made of waterproof material, usually some kind of nylon. They come in 3 or 4 different sizes--some just right for a camera, others large enough for a tent and sleeping bag. They open at the top, then roll down and fastened with buckles. These are not meant to be submerged and **not totally waterproof!** Yes the fabric is waterproof when new, but water can and does seep in via the top, especially if it sits in water a long time. For this reason, always double-bag important items like a sleeping bag. This can be done with one or more ziplock bags, or heavy duty trash compactor bags. You can buy drybags in most sporting goods stores, especially those stocking equipment for river running or kayaking. All serious canyoneers have a couple of these because so many canyons have deep wading or swimming pools--mostly in Zion.

Before going through any slot canyon where swimming is required, test your drybags by filling them with water. If you find a leak, one way to patch them is to dab a couple of layers of **contact cement** over the hole. This makes a flexible patch. This same method works well on Therm-a-Rest sleeping pads or air mattresses instead of applying a rubber patch. Contact cement comes in tubes, small bottles or cans and can be found in supermarkets or hardware stores. Take a tube with you on all longer 2-3 day hikes. A small roll of **duct tape** is good to have for rips or gashes in drybags, drysuits or packs (place on the inside if possible, otherwise they can rub off quickly).

Other waterproof gadgets include **plastic boxes** or **cases**. These are built primarily to keep cameras or other electronics dry. Look for these at specialty camera stores. Look for the name **Otter** or **Pelican**. They make boxes small enough for the smallest cameras, all the way up to suitcase size. These are guaranteed to be waterproof for life, but the author once took a brand new camera in a brand new Pelican case down Neon Canyon, and it wasn't water tight. Water totally destroyed the camera! Fortunately, he had bought an extended warranty, so he got a brand new camera out of the ordeal. Pelican refunded his money, plus sent him another box they personally tested to be waterproof. However, it wasn't any more water tight than the first! Submerging the larger cases for just a moment or two is generally OK, but don't leave them under water for long. The smaller boxes with narrower lids, are waterproof.

One nice thing about some cases, you can hang them outside your pack by the handle attached to a carabiner. This allows easy access. A camera buried inside your pack does not take pictures. By hanging the case on the outside, and keeping it on top of your pack and in front of you while swimming, you can reach up, take the camera out, and get fotos even while treading water.

But here's a better idea for keeping cameras dry & dust free. This is what the author has been doing for a couple of years--and this beats all other methods! He has a Canon Power Shot 560 and it fits perfectly in the **Pelican 1010 plastic camera case** with no bouncing around. This case has belt loops on each end and can be worn on a 2nd belt around the waist for easy & quick access. When faced with a pothole or pool, he jumps in, swims through, and the camera stays dry! It also keeps the dust out, as water & dust are a cameras biggest enemy. If you're looking for a camera--and canyoneers always want small cameras--**buy the 1010 case first, then find a camera to fit it.**

Another type of box to keep things dry while swimming is a round plastic container made by **Curtex** of Der Nederlands (Holland). These have screw-on lids and are shaped like an oversized wide-mouthed Mason jar. The one Rich Carlson of *canyoneering.net* traded to the author was definitely waterproof going through Imlay, Neon and South Fork of Choprock Canyons. These are a great place for food too. Tom Jones (**canyoneeringUSA.com**) calls these **dry kegs** and they come in 2 sizes--3 1/2 & 5 1/2 liters; he also sells them from his website. They're worth the money.

Check List for Experienced Canyoneers For the beginner, here's a list of items the **author** or any experienced canyoneer takes attached to his/her **rappelling harness** on all/most hikes: Locking steel carabiner, ATC (or figure 8), 6-8 biners (some locking), a couple of baby angle pitons and several R/QL. Also very imported for tight slot canyon are knee & elbow pads (K&EP).

Also taken in the pack will be a 8-10m-long reach rope (PR) for lowering the pack (DC'ing is much easier without a pack), a hooking kit (only for canyons with KPH's), G-pick & Ibis hook, lots of webbing or short ropes, ascenders (Jumars), short ropes/runners for use with ascenders, waterproof box or drybag for wallet, cameras, LED headlamp, pens, chapstick, small notebook, water purification tablets, emergency first aid kit, cigarette lighters, super glue & duct tape. Also, a small pocket knife & compass worn around the neck & handy, a 2nd knife inside the pack, altimeter watch with barometer, more than one map and extra ziplock bags.

For longer hikes the list includes a stitching awl & waxed thread, contact cement, needle & thread, another pocket knife (best to have an all-in-one tool which includes can opener, pliers, several knife blades, etc). If 2 or more people are going along, then a bolt kit is often included, along with a couple of B&H's. These in addition to the normal camping or hiking supplies.

American Canyoneering Association (ACA) Canyoneering Rating System

The American Canyoneering Association was founded in 1999 in response to canyoneerings' growing popularity and the obvious need to present accurate and timely information regarding safety, ethics and access issues. Membership is open to recreational canyoneers and professional canyoneering guides. See the ACA website at **canyoneering.net**. The Commission Europeene de Canyon (CEC) based in Europe, was instrumental in helping Rich Carlson presently of Cedar City, Utah, and others, in creating the ACA.

In just the last few years, technical canyoning or canyoneering, has become popular almost overnight. This has prompted some in the business to come up with a consistent way of rating or classifying canyons on their difficulty similar to the way mountaineers or rock climbers rate climbs or routes.

The **ACA's rating system** is divided into 4 parts. The **Technical Class** ranks the technicality of the terrain and the type of rope work required. The **Water Rating** indicates possible complications due to flowing, quiet or pothole water. The **Risk Rating** indicates the presence of additional risk factors such as double rappels down a big wall, or the presence of keeper potholes (KPH). The **Grade** indicates the estimated time required to complete the hike.

Here are some examples. At or near the top of the difficulty ladder would be Imlay or Heaps Canyon in Zion National Park which rank **4B V (or VI) R**. Popular Pine Creek Slot, also in Zion, would likely

be **3B II**. North Fork of Robbers Roost Canyon a **3A III**. Every canyon in this book is rated a 3 or 4 on the Technical scale (Some exceptions are slots with difficult UC/DC'ing or high-stemming, but no real rappels. In the case of some **slots**, there's a new classification for them--read more below).
   As this book goes to press, Rich Carlson and the ACA are working to modify the rating system with new classifications, especially at the upper end of difficulty or risk. They're considering ratings from 1-7, instead of 1-4 as it is presently. To get the latest information on this new system, go to the ACA website, **canyoneering.net**, for updates.

TECHNICAL CLASSIFICATION
1. Canyon Hiking Nontechnical, no rope required. Hiking mostly on established routes. Involves some scrambling with the occasional use of hands.
2. Basic Canyoneering Scrambling, easy vertical or near vertical climbing and/or downclimbing (DC'ing). Rope recommended for handlining (HL'ing), belays for novices, lowering packs and possible emergency use. Exit and/or retreat possible without ascending fixed ropes. Scrambling requires the use of hands and arms for pulling oneself up.
3. Intermediate Canyoneering Technical canyoneering and climbing. Route may involve any combination of the following: Problem solving. A basic knowledge of technical climbing and chimneying. Rope and climbing hardware for single pitch rappels and belays. Basic pothole escape techniques. The use of obvious natural anchors for rappels. Retreating upcanyon will require ascending or jumarring up fixed ropes.
4. Advanced Canyoneering In addition to intermediate canyoneering skills, this will require one or more of the following skills: Advanced free climbing. Difficult and/or exposed up/downclimbing. Climbing using direct aid. Multipitch rappels. Complex rope work (example--guided rappels, deviations, rebelays). Obscure or indistinct natural anchors. Advanced problem solving and anchor building. Advanced pothole escape techniques (example--pack toss or hooking out). At least one canyoneer has noted that introducing bolts into what originally would have been a 4 rated canyon (many in Zion NP), permanently reduces it to a 3.

WATER--VOLUME & CURRENT
A--Normally dry or very little water, usually in potholes. Dryfalls.
B--Normally has water with no current or light current. Still pools. Falls normally dry or running at a trickle.
C--Normally has water with strong current. Waterfalls

RISK
No Rating (blank)--Normal risk factors are present in this canyon.
R Risky--One or more extraordinary risk factors exist that could complicate the descent. Solid technical skills and sound judgment are critical.
X Extreme--Multiple risk factors exist that will complicate the descent. Errors in technique or judgment will likely result in serious injury or death. Descent should only be attempted by expert canyoneers.

New Classifications for SLOT Canyons: For tight slots only--with UC/DC & High-Stemming.
PG SLOT Plenty Good, but not life threatening. Will involve simple downclimbing or low-stemming, but no extreme moves. Beginners may sweat a little, but it's all pure fun for the experienced.
R SLOT (Some Risks), X SLOT (Extreme Risks) or XX SLOT (Double Extreme Risks)--Extra tight slots with strenuous up/downclimbing and high-stemming, in some cases up to 15 meters or more above the bottom. Likely will have wide silos to cross, straight/vertical walls without foot/hand holds, and sometimes covered with green, slippery moss. **For experts only.** XX slots would definitely be **life threatening** for one or more moves. These ratings are based on the most difficult move in each slot. Each tight slot in this book has been rated by the author, with input from others. Your rating may vary a little, depending on whether the canyon was dry or if done just days after a flood.
   To the above, some canyoneers are adding **+ or -** to each. If it's a little better/more difficult than a regular **PG**, then push it up part of a notch to **+PG SLOT**. Or if it's not quite a full X SLOT, make it an **-X SLOT**, or perhaps an **+R SLOT**. This should make it easier to hit the mark. Again, all subjective.

GRADE
I--Short. A couple of hours.
II--Requires about half a day.
III--Will take most of a day.
IV--Expect a long day. Get up early, bring a headlamp & mylar space blanket. Possible bivouac.
V--More than one day. Normally done in two days.
VI--Two full days or more.

Hiking in Zion National Park
   Whether you're going into technical slot canyons, backpacking or day-hiking in The Narrows of Zion Canyon, here is information on Zion National Park you should be aware of. Much of this information comes directly from 2 free handouts given to everyone upon arrival in Zion, and from their website. See below.

   ENTRANCE FEES & PASSES FOR 2008: Entrance Fee--$25 per vehicle, good for 7 days. **Individual Fee**--$12 per person, not exceeding $25 per family (this applies to pedestrians--walking in from Springdale, bicycle, motorcycle, and organized groups), good for 7 days. **National Parks & Federal Recreational Lands Pass, Senior Pass**--$10 lifetime pass for U.S. citizens age 62 or older. **National Parks & Federal Recreational Lands Pass, Annual Pass**--$80, good for entrance for one year in all federal fee areas. **National Parks & Federal Recreational Lands Permanently Disabled Pass**--Free lifetime pass for U.S. citizens who are permanently disabled; documentation required.
   Pay fees at any entrance station. These are the **East Entrance** on Highway 9, **South Entrance** located just north of Springdale, and the **Kolob Canyons Entrance** located just east of Interstate 15 at mile post & Exit 40 south of Cedar City.

IMPORTANT TELE NUMBERS OR WEBSITES: Superintendents Office, Zion National Park, Springdale, Utah, 84767, Tele. 435-772-3256; Visitor Center, Tele. 435-772-7616; Backcountry Desk, Tele. 435-772-0170, or Zion's website for currant or late breaking information, **nps.gov/zion.**.

HIKING PERMITS: Permits are required for overnight backpacking trips, climbing bivouacs, all through day-hikes of the Virgin River (Zion Narrows) and tributaries, the Left Fork of North Creek (The Subway--see below), and all canyons requiring the use of descending gear and/or ropes. Permits for technical slot canyons are issued at both visitor centers (Zion Canyon and Kolob Canyons) beginning at 8am (7am in about the 3 months of summer) same day, or the day before your hike. Backpacking permits are issued up to 3 days in advance of your trip. Backcountry fees are based on group size: $10 for 1-2 people; $15 for 3-7 people; $20 for 8-12 people. A vehicle license plate number and emergency contact name & Tele. number are required for all backcountry permits. Be prepared to submit this information.

Here's an important point about late arrivals and/or possible rescues. Even though the NPS will issue a permit and be aware that you're in a particular canyon, they will not come looking for you unless they get a call from a family member or friend. So it's up to you to let someone know where you're going and when you expect to return. If you fail to arrive at a given hour and the NPS gets a call, then someone will go looking for you. **Go to the website nps.gov/zion for the latest updates.**

The emergency Tele. number to have someone call if you should spend an extra night in a canyon is 435-772-3322, or locally call 911.

LATEST INFORMATION FOR CANYONEERS! All the information here will be outdated by the time it hits the press, so go to the website **nps.gov/zion** then click either **Backcountry Planner, Reservation & Permits, Backpacking, Canyoneering or Left Fork (Subway).** Within these pages should be all the information regarding permits, fees, reservations, etc.

Primitive Zone: Limit will be 80 people per day. Areas include Left Fork of North Creek [Great West Canyon] (The Subway), Pine Creek Canyon, Keyhole Canyon and Orderville Canyon.
Pristine Zone: Limit will be 12 persons per day. Areas include Mystery Canyon, Behunin Canyon, and many of Zion's other slot canyons.
Research Natural Areas: As of 2008 recreational use is not permitted. Areas include Goose Creek and lower Parunaweap Canyon.

THE SUBWAY Here's the latest regulations as of 2008 regarding getting a permit for **The Subway** (the following was taken directly from the website **The following information will change so see the website for updates!).** Go to this website and submit the following information:
Name, Number in Party (may not exceed 12), 1st choice of trip date, 2nd choice of trip date, 3rd choice of trip date, E-mail or mailing address where you would like to receive a reply.
Contact the Backcountry Office by either: Email: zion_subway@nps.gov or Fax 435-772-3426.
To request a reservation, visitors may contact the park by fax or e-mail. Up to 3 prioritized dates may be chosen. Trip dates will be awarded by lottery. Applicants will only receive one trip date.
For Trip Dates In: . . . . .Requests Must Be Received During the Month of:
April . . . . .January
May . . . . .February
June . . . . .March
July . . . . .April
August . . . . .May
September . . . . .June
October . . . . .July
A limited number of permits (25%) are available on a walk-in basis the day before an intended trip. Only one application per name or e-mail address will be accepted for each month. The person who is named in the reservation must pick up the permit and must participate in the hike. All applicants must be at least 18 years of age. Everyone will be notified of the status of their reservation no later than one month prior to their trip. Lottery winners can pick up their permit the day before or the day of their trip. Permits will be available at either the Zion Canyon or Kolob Canyons Visitor Centers. To increase your odds of success, select days other than Fridays, Saturdays [Sundays & Holidays]. Large groups increase impacts on the backcountry. Group size is limited to a maximum of 12 people sharing the same affiliation (school, club, scout troop, family, friends) in the same drainage, route, or backcountry trail on the same day. This is strictly enforced; violators will be cited.

The above information is from previous years, but as of 2008, there was too much on the Zion website to print here. So go to **nps.gov/zion** and see if you really want to go canyoneering is Zion. After viewing all the rules & regulations, and the number of hoops to jump through, and you feel like peuking, maybe you'll head of the **Escalante, North Wash, San Rafael Swell,** or better still, the **Robbers Roost Country**. No permits were need in any of these places as of 2008.

ZION CANYON SHUTTLE BUS: Travel into Zion Canyon north of Springdale and Highway 9 from about the 1st weekend in April through the last weekend in October (roughly the period of daylight savings time) is by park shuttle bus only. The reason for this new system, which began in 2000, is the lack of parking space in the canyon. Since you've already paid a rather substantial fee to enter the park, this shuttle bus is free and you can get on or off when ever you like. Each bus has space for backpacks, climbing gear and 2 bicycles, *but no pets.*

Shuttle Buses begin at the Zion Canyon Visitor Center (where the Backcountry Office is located) and stops at Zion Museum (where the old visitor center used to be), Canyon Junction (Highway 9 & Canyon Road), Court of the Patriarchs, Zion Lodge, The Grotto, Weeping Rock, Big Bend, and Temple of Sinawava, the end of the road in Zion Canyon.

Bus Schedule--Spring & Fall: 1st bus leaves visitor center, 6:45am; 15 minute interval service, 7am to 8:30am; 7-10 minute service, 8:30am to 8pm; 15 minute service, 8pm to 9pm; last departure from Temple of Sinawava, 10pm.

Bus Schedule--Summer (June through about mid-September): 1st bus leaves visitor center, 5:45am (runs to Zion Lodge only); 30 minute service, 6:30am to 7:30am; 15 minute service, 7:30am to 9am; 6-10 minute service, 9am to 8pm; 15 minute service, 8pm to 9pm; 30 minute service, 9pm to 9:30pm; last bus leaving visitor center, 10:30pm; last bus leaving Temple of Sinawava, 11pm. These schedules may change without notice, so check the website **nps.gov/zion** for updates.

FREE & PRIMITIVE CAMPSITES NEAR ZION With all the traffic & hubbub around Zion in the warmer half of the year, it's not easy for we cheapskates to pull off the road and camp where we like--and not pay. Here are some places you can crash for a night for free near the park: Just east of the park, drive south on a dirt road from between **mile posts 47 & 48 on Highway 9**. After driving about 1 km look for parking/camping places. Best to continue a ways south past a new home built in about 2006, and camp away from it.

Just west of the park, pull off **Highway 9** at **mile post 25** and drive north & west 200m and park/camp. This is at the mouth of **Coalpits Canyon**. Or continue west, and halfway between **mile posts 23 & 24**, pull off the highway south next to the Virgin River. This is called **Mosquito Cove** which is a large area with cottonwood trees and many campsites. But there's a problem--no toilets, so expect this situation to change if it gets overused--which it will at some point in time!

If you're planning to hike somewhere on the western side of the park, you could also pull off the **Kolob Reservoir Road** between Km 1.5 & 3/Mile .8 &1.8; or just before the park boundary at Km 10.5/Mile 6.5; turn left at Km 12.4/Mile 7.7 and drive 1.6 kms (1 mile) toward **Smith Mesa**; turn left at park boundary at Km 17.2/Mile 10.7; or at about any place north of Kolob Reservoir. Or further west, and between **mile posts 14 &15 on Highway 9**, is the turnoff to the **La Verkin Overlook**. Anywhere in that area is generally OK. There are no garbage cans or collection at any of these sites, so please pickup your own trash and that of others.

Hiking in **Capitol Reef National Park** isn't nearly as popular as in Zion, but there are many good trails & walks and several good technical slot canyons. As of 2008, they didn't have the permit system that Zion has. Yes, if you're doing any backcountry or overnight camping you'll have to have a camping permit, but no permit or permission is required for day-hikes into any of their 3 main technical slots; **Burro, Cottonwood or Fivemile Washes**. Regardless, you'll definitely want to stop at the **visitor center** near **mile post 80** on **Highway 24** to buy maps or get updates on weather, hiking or road conditions. As of the fall of 2007, the **Bullfrog-Notom Road** was **paved** south from Highway 24 to the mouth of Fivemile Wash, a distance of 17.9 kms (11.1 miles).

Hiking on the Navajo Nation

To hike anywhere on the Navajo Nation you're supposed to get a **hiking permit**. Contact the Navajo Tribal Parks & Recreation, P.O. Box 2520, Tele. 928-871-6647, in Window Rock, Arizona, 86515, and see what the latest requirements are. On the website **navajonationparks.org**, it states; *Camping fees: $5.00 per person, per night for anyone over the age of 6. Fees subject to change. You will need to obtain a camping permit from one of the following locations listed below. Backcountry permit fees: $5.00 per person, per day.*

In the northeast part of Navajoland, you can also pick up a permit at the **Monument Valley Navajo Tribal Park visitor center** on Highway 163 in the middle of Monument Valley, Tele 435-727-5870. Or stop at the **visitor center** in **Cameron** (Tele. 928-679-2303), located just east of the Grand Canyon at the junction of Highways 89 & 64 about halfway between Flagstaff & Page. Or stop at the **Leche-e Chapter** (Tele. 928-698-2800, or Tribal Parks Office in a small white building just behind the chapter house at 928-698-2808 or 2809) 4 kms south of Page. Or, the best place to pickup a permit for any part of the Navajo Nation seems to be at the small **Antelope Canyon Tribal Park fee booth** immediately south of **Highway 98** at **Antelope Canyon** just east of Page between **mile posts 299 & 300**. Call them at their office in **Lechi-e** at 928-698-2808. This little booth-type office is open 7 days a week from about 8am to 5pm during the warmer 8 months or so of the year.

Or if you're heading for canyons north of Navajo Mountain, you could at least get information at **Inscription House Chapter** (Tele. 928-672-2337), located at mile post 5 near the beginning of the Navajo Mountain Road. Or continue north and stop at the **Navajo Mtn. Chapter**, which is where the old Navajo Mtn. Trading Post used to be. In the past, some hikers have gotten a permit there, but call first to verify--Tele. 928-672-2957. It's 1.6 kms (1 mile) north of the Utah-Arizona state line and they're open from 8am to 5pm, M-F. At last report, the Navajo Mtn. Road was paved nearly to the stateline.

Now the bad news. The **Kaibito Chapter** (A Navajo chapter is a political area something like a small county. A chapter house is the building where the local government is located. It's also the local social gathering place) which includes about 33 families in the area of **Starting Water, Kaibito, Peach, Piñon and Butterfly Canyons** voted in the late 1990's to close these canyons to hiking. The reason was, the Sheriff of Coconino County out of Page, was called out to help an injured hiker out of upper Kaibito Canyon; and had to locate 2 hikers who under estimated the time it would take to do Starting Water Wash on Christmas Day, sometime in the late 1990's. The Navajos didn't like search parties crawling all over their land, nor do they have the knowledge or equipment to do rescues, so they closed the area to hiking.

Even though the above mentioned canyons are officially closed, it's hoped that one day, people in the Kaibito Chapter will reconsider. It's for this reason these canyon are in this book--hoping that one day we'll all get the chance to visit some of the best slots in the world. If you've seen how many tourists there are at Antelope Canyon just east of Page, and they come by the bus load, one has to think the local Navajos could surely benefit financially if these 5 great canyons were officially open. To see if they are open, call the Kaibito Chapter at 928-673-3407/5850, or the Leche-e Chapter south of Page at 928-698-2807/2808 to see what their plans are. It's strongly recommended you not go to any of these canyons until they are officially opened again. But when that day comes, the information needed is in this book.

Odds and Ends

Emergency Kit for Your Car

Because some of the canyons in this book are in isolated places, notably the Robbers Roost Country, around Navajo Mountain, parts of the Escalante & Paria River drainages and the San Rafael Swell, it's recommended you take a good reliable vehicle--one you can depend on. Always have a full tank of fuel--and in some cases maybe extra fuel. Also take extra water, food, tools, first aid kit, battery jumper cable, a tow rope or chain, a shovel, tire pump & gauge, extra oil and anything else you think might come in handy in an emergency. In recent years and with improved facilities, there is now more traffic on the back roads of the Colorado Plateau, especially on weekends in the spring and fall, but you'll always want to go as well-prepared as possible.

Best Times to Travel and Hike--Weather & Climate
Most people would prefer hiking in the **spring or fall** in most canyons of the Colorado Plateau because it's cooler and most are dry hikes. However, **some canyons in this book** involve wading and swimming through potholes. This type of canyon is best done in summer, the warmest 3, 4 or 5 months of the year.

But there are ways to extend the hiking season for the wetter technical slot canyons. This involves the use of wetsuits or drysuits (not as good). With a wetsuit of varying thicknesses, you can extend the season for the full year. In recent years, several groups have gone through the Black Hole on New Years Day! They call it *Freeze Fest*, for good reason. On 1/1/2008, they found ice for the first time. But the season for most canyons in this book is from **early April to mid or late October**.

As a general rule, the Robbers Roost Country east of Hanksville, and the Cedar Point--North Wash Area south of Hanksville, have dry canyons with few potholes, so you can hike there year-round--if you can stand the cold, and long nights. Slots in the other regions may have some, or lots of pothole swimming. The region with the most swimming and need for wet/drysuits is Zion National Park, and a few places in the Escalante River country. If hiking Heaps and Imlay Canyons, better have a thick wetsuit or drysuit even in summer as there are dozens of potholes at moderately high altitudes.

Warning--Flash Floods: One important thing to remember here is the monsoon season. Generally speaking, it begins in about **mid-July** and continues through **mid-September**. The driest month is June. Each year is different. However, since this is an arid or semi-arid desert area, the wetter season isn't the same as the rainy season in the humid tropics. There are still many dry sunny days even in August. But when heavy rains come, they can turn a dry wash into a raging torrent. When you get close to the area where you'll be hiking, always tune in to local radio stations for the latest weather forecast. Or stop at the nearest **BLM/NPS office** for a look at the weather forecast on the internet.

If you're heading for any of the slot canyons, take extra precautions. In one flash flood in Zion Narrows in the 1970's, 5 hikers were swept away and drowned. On August 12, 1997, 11 tourists perished in a flash flood in Antelope Canyon near Page, Arizona. That was the worst incident of its kind on the Colorado Plateau.

Length of Days. One nice thing about hiking in late spring and early summer, are longer days and shorter nights. Remember, October 21 has the same amount of daylight hours as does February 21; and September 21, the same amount of daylight as March 21. June 21 has about 16 hours of daylight, 8 hours of darkness; December 21, only about 10 hours of daylight--take lots of good reading material during this time of year!

The Insect Season
When it comes to insects in slot canyons, it's mostly good news. The deeper and darker the canyon, the fewer the insects. In fact, there are almost none. In the darker slots, you may find daddy long leg spiders, but they're not a problem. In places with potholes and water, you may find a few **mosquitos**, but they don't bother you while hiking. At night it's a different story as that's when mosquitos in this part of the world come out. You will need a lightweight tent with mosquito netting if camping in most of these canyons, especially in Zion.

Probably the most bothersome insects, but which are in scattered places only, are the large gray **horse flies**. You may run into these along the Escalante River, and on some of the approaches in the San Rafael Swell. These flies are around in early to mid-summer, then they seem to disappear. The common **housefly** and **gnats** are almost unheard of in the canyons in this book.

Drinking Water
Most water in the canyons discussed here will be in potholes. If it's a one day hike, carry your own water from a culinary source. But if you're on an overnight trip, you'll either have to treat the water, or if it's really smelly, which it often is, better carry a **filter**. This should remove most of the bad smell and/or taste. Or, if you're there soon after rains, you can usually drink pothole water as is. At any rate, a small bottle of **iodine tablets** will be an essential part of your emergency hiking kit.

What to Take on Day-Hikes
Here's a list of clothes and other items the author normally takes on day-hikes into most technical slot canyons. These things are in addition to what is suggested earlier in this big chapter under, **Check List for Experienced Canyoneers**, and in addition to thing like K&EP and W&R/QL: A medium sized pack, one 2 liter bottle of water (up to 4 liters in hot summer weather!), 2 digital cameras (one in a waterpoof plastic box on your belt loop, one in a waterproof box big enough for 2 cameras in your pack), a tripod, toilet paper, pen & small notebook, map, chapstick, compass, pocket knife (or Leatherman-type tool), a baseball-type cap with a *sun shield* or *cancer curtain* sewn on around the back, a pair of long pants (for deer flies), mylar space blanket/bivvey sack, LED headlamp and a lunch.

In warmer weather, most people prefer wearing shorts & T shirt; in cooler weather, long pants and a long-sleeved shirt, plus perhaps a jacket and gloves. In cooler weather and with more things to carry like a wet/drysuit, a larger day-pack or full-size pack is required.

What to Take on Overnight Trips
For those with less experience, here's a list of things the author normally takes on overnight hikes which may include Imlay Canyon. The first thing to remember is, take as little as possible, because dragging a full-sized pack in & out of potholes, or through a tight slot isn't easy!

Start with a medium sized internal frame pack, lightweight sleeping bag, short Therm-a-Rest mattress or foam pad, tent (you won't need a rainsheet because you'll only do these canyons in good weather), maybe one small stove & fuel for the group, several lighters (no more matches!), one large water jug, one 2 liter water bottle, a stitching awl & waxed thread, waterproof canister or plastic box with odds & ends (bandaids, needle & thread, patching kit or tube of contact cement for sleeping pad or drybag, super glue, duct tape, wire, pens, etc.), maps, small notebook, reading book, chapstick, compass, toilet paper, pocket knife (or a Leatherman-type tool with pliers), watch with alarm, LED headlamp for light & reading (+backup batteries), tooth brush & paste, face lotion, sunscreen, cap with cancer curtain, a dry pair of long pants & long sleeved shirt, vest, and perhaps a lightweight jacket & gloves.

16

Food might include hot or cold cereal (example--place enough Wheaties, powdered milk, sugar, raisins, peanuts, etc. for one breakfast in a plastic bag and add water later), coffee, Postum or chocolate plus powdered milk & sugar pre-measured in a small bag, cookies, crackers, candy, oranges or apples, Ramen instant noodles, soups, canned beans, tuna or sardines, peanuts, instant puddings, bread, butter, peanut butter, salt, plastic eating bowl & cup, spoon, and small cooking pot.

Boots or Shoes

Most of the hikes discussed in this book are in narrow slots canyons with sand, gravel and/or boulders with occasional potholes and/or wading or swimming pools. Most people prefer to wear some kind of simple athletic running shoe. These are comfortable and seldom if ever cause blisters. Most of us have a pair of these in the closet, so there's no need to go out and buy a new pair. But if you have to buy shoes for hiking slot canyons, then the best are the light weight low cut hiking boots or trail running shoes.

If you're serious about spending time in Zion, which means lots of watery-type hikes, look for something designed especially for water. There are now several new wading-type shoes available. One is seen in several stores, including Zion Adventure in Springdale, just south of the park boundary. These carry the brand name **5-10's** and are made of synthetics with sticky rubber soles and built with wading the Zion Narrows in mind. These are a little expensive, and the buckles eventually give out, but may well be worth the money if you're headed for canyons with lots of wading or swimming.

If you're planning to do lots of **high-stemming** in really **tight slots**, consider spending more money for a better than average pair of shoes with sticky rubber soles--flexible trail runners with lugs seem to gripe best.

Here's something few people think about; are you tired of dumping **sand** out of your shoes? To prevent most sand from getting inside in the first place, buy slightly heavier shoes without the light weight, almost see-through mesh surrounding the toe. Also, look for shoes with tongues sewn up along the side, which also keeps sand out. One last thing, almost every canyon has water someplace, even if it's in just one pothole, so buy shoes that **aren't made of leather**. Leather eventually shrinks and gets hard, but synthetics don't, and are less effected when wading. Years ago, it was easy to give up going through a canyon because of pools, but today almost all shoes are made of synthetics.

Take a Good Map!

The author has done his best to make the maps in this book as accurate as possible, but no matter how carefully they're drawn, sketches are no substitute for a USGS topographic map.

To have a successful hike, always have **a state highway map** in your car. This will get you to the San Rafael Swell or the Escalante River Country. Next, and for zeroing in to your canyon, either have one of the **1:100,000 scale metric maps;** or in the case of Zion or Capitol Reef National Parks, or the Escalante or San Rafael Swell regions, buy one of the **plastic maps** put out by **Trails Illustrated** (now owned by **National Geographic**) or **Earthwalk Press**. These maps can be used for both driving to a canyon trailhead and for hiking.

And finally, if you prefer more detail, buy and take with you, the newly created **USGS maps at 1:24,000 scale.** These are commonly called **7 1/2 minute (7 1/2') quads.** These are available on paper or CD. The author prefers to make a fotocopy of the area to be hiked, leaving the original paper map at home. CD's from TOPO! National Geographic were shot at 72 dpi, and have crapy quality.

The maps you'll need are listed under each hike. If you get lost using just the maps in this book, the author assumes no responsibility for your predicament! If you can't read a map or know how to use a compass, then you've chosen the wrong hobby. Stay home and watch TV instead!

Off Road Vehicles (ORV's ATV's, Motorcycles and 4WD's)

In recent years with increased traffic of all kinds on the Plateau, there has also been a dramatic increase in the number of off road vehicles (ORV's). Since these are called "off road vehicles", naturally the owners want to test drive them "off the road". This indiscriminate use & abuse, and the destruction of public lands has caused a backlash from environmentalists. This is the primary reason why there has been a move in recent years to protect the more scenic regions by making them into wilderness areas. **Just one thought for ORV owners; it's you who have been using public lands as test tracks for your noisy toys, and which has triggered a war over public land issues including the battle over wilderness areas.** Slowly but surely the Forest Service and BLM are closing off areas formerly destroyed by ORV traffic. So please, everyone drive on existing roads!

Mountain Bikes

Rather than spend $20,000 to $50,000 for a 4WD or SUV, why not use a simple mountain bike in the few places in this book where you might need a 4WD.

In the second half of the 1980's, there was an increase in the use of mountain bikes. Fortunately, these can seldom if ever be ridden off a road or trail, so they aren't as destructive as ORV's. You simply can't ride them in sandy areas or in most dry rocky creek beds. However, if a lot of people start riding them in areas where they're forbidden, then they too will be on the same s--- list as ORV owners. The Colorado Plateau is crisscrossed with old mining or survey roads which offer plenty of mtn. biking opportunities. Please keep bikes on established roads or tracks.

For hikers with cars only, a mtn. bike is a cheap way to extend the range of the family car. There are many places in this book where a mtn. bike will save a lot of walking (and gas), because you can park your car a few kms from the end of a 4WD road, and use the bike to reach the trailhead. In some situations you can lock up a bike to a tree or fence at one end of a canyon, drive to the other end, do the hike and bike back to your car. This will eliminate backtracking, a lot of road-walking, or the need for a second car. Store the mtn. bike inside your vehicle if possible, rather than on the roof. One bike on the roof will increase fuel consumption up to 10% at highway speeds.

Driving Tips for Back Roads

As you leave the paved highways and head out on graveled or dirt roads, here are some things to keep in mind. First, **lower the air pressure in your tires.** The author normally runs his vehicle on highways at about 38 lbs. When he leaves pavement for an extended drive, he lowers it to between 26-28 lbs. This does 3 things. First, it gives you a smoother ride. Second, it helps prevent sharp stones from puncturing or bruising your tires. And third, it gives you better traction, whether it be in sand, snow or mud. This means you'll have to carry a tire pump & gauge to reinflate your tires when you return to pavement. This can be a **hand or electric pump**, which is run off the car battery. The electric ones are easier to use which means you're more likely to lower the pressure in the first place.

If you're in sandy areas, lower the pressure even more, and you can drive to places you couldn't ordinarily, whether it be a 2WD car or 4WD SUV. This is what dune buggy drivers do, and they never get stuck. Simply lower your tire pressure to somewhere between 5 &10 lbs in your drive tires. This puts more rubber to the road, thus increasing traction. Each person will have to experiment a little with this technique in order to gain confidence; in the meantime, always carry a shovel and a tow rope! To prevent most flats on back roads, buy **LT or 6 ply tires**; this is what most BLM pickups have.

Also keep in mind, when you're driving **sandy roads**, a little rain or moisture helps. It makes the sand more firm and you don't sink in as far. But, if you're in an area with roads made of **clay**, and a storm is coming, better high-tail-it back to pavement quick! Roads made of clay become very slick when wet, making it difficult, if not impossible, even for 4WD's. If you get in such a situation and can't move, simply wait an hour or two, and you can normally drive away OK. If heavy rains come, then you may have to wait overnight for the road to firm-up. This is one of the best reasons for always carrying more water and food than you think you'll need.

Hiking Times

In the past, the author has received numerous complaints regarding the time it takes people to get through a canyon. In all his books, there is a **Time Needed** heading where he estimates the time needed for the average person to complete a hike. In the **Author's Experience**, it always states exactly where he went and how much time it took. In the past, he almost always hiked alone, therefore his times were normally faster than the average hiker.

However, in this book and when it comes to doing technical slot canyons, he normally looks for partners so he can get better pictures. This is done because if you don't know for sure what's ahead, and not exactly sure what equipment you'll need to get through successfully, you'd better load up heavy on ropes & gear. This can make a very large pack for one person. On the other hand, if a big load of ropes & gear can be divided into say 3 packs, it's a lot smaller load for everyone. The point here is, the time it takes the author and his group to do a hike through a technical slot canyon is very close to what the average group will need to do the same hike.

Also, when you're in a canyon with many rappels, the actually distance in kilometers/miles is meaningless. Instead, the time it actually takes is more important.

Respect the Land

Some people are becoming alarmed at the slow destruction of parts of the Colorado Plateau and want to lock it up into wilderness areas. One reason for this movement is the overuse & abuse by ORV's and 4WD's. Another reason is the amount of trash left behind by thoughtless individuals. Around some of the more heavily-used campsites and along some roads one can see the sign of the times; the aluminum soda pop and/or beer can. There are very few if any sites at or near the canyons covered by this book with garbage collection service, so it's up to all of us to pick up our own refuse, and in some cases, the trash of our less-concerned neighbors, and dispose of it properly. This writer always arrives home with a sack of aluminum cans and other trash. Hopefully, others will do the same.

Here are a few things to keep in mind:

1. Protect soils and vegetation by keeping motor vehicles or mtn. bikes on existing roads and trails.
2. If you camp somewhere other than at a designated campground with toilets, bury all solid human body waste 15 cms (6") deep at least 30m from water sources.
3. Carry out what you carry in--such as wrapping materials, tin cans or anything unnatural. While packing up, police the area and pick up trash left by others.
4. Use existing fire pits and clean them after you're through. If you make a fire, let it burn to ashes rather than burying charred stubs. Better to take & use a small stove.
5. Leave prehistoric and historic artifacts as you find them for others to enjoy. Besides, it's the law.

Metrics Spoken Here

As you can see from reading thus far, the metric system of measurement is used almost exclusively in this book. The only place miles is used is the distance you'll have to drive your vehicle. This is because all car odometers in this country are still in miles. Thus driving distance is given in both kilometers and (miles).

In 1975, the US Congress passed a resolution to begin the process of changing over to metrics. They did this because at the time the USA, Burma, and Brunei were the only countries on earth still using the antiquated British System. This progressive move ended with the Reagan Administration in 1981. Today it seems the USA is the only place on earth where inches, feet, yards and miles are still used. Go north to Canada, or south to Mexico, or anywhere else in the world, and you'll be required to use kms & liters, not miles & gallons! If the rest of the world has adopted metrics, and we haven't, who are the stupid ones?

When it comes to rope lengths & diameters, just go into a climbing equipment store. The ropes are almost always marked 50 or 60 meters, and 8mm, 9mm, 10.5mm or 11mm in diameter--seldom in feet and inches. All pack sizes are now measured in cubic liters, not inches. Also, we all know how much 2 liters of soda pop is; all automobile engine displacement throughout the world is now measured in cubic liters--not inches; in high school and college track & field events, they now run the 100m, 400m and 1500m--not 100 yards, 440 yards or the mile! Since we're about 10% there, why not continue learning metrics?

Use the **Metric Conversion Table** on page 6 for help in the conversion process. It's easy to learn and use once you get started. Just keep a few things in mind; one mile is just over 1 1/2 kms, 2 miles is about 3 kms, and 6 miles is about 10 kms. Also, 2000 meters is about 6600 feet, and 100 meters is about the same as 100 yards. A liter and a quart are roughly the same, and a US gallon jug is 3.78 liters. One pound is 453 grams, and one kilogram is about 2.2 pounds. The author stands 6 feet tall, or 183 cms, and weighs about 150 lbs, or less than 70 kgs.

Fotography in Slot Canyons (Updated June, 2008)

Here are some tips on cameras, carrying cases, and how to come home with good pictures on your first trip down a deep, dark, slot-type canyon.

Film cameras are all but history now, so in this edition, it's **digital fotography only**. First thing to consider when buying a camera is the size. In today's world you can get really good pictures with a small camera--there's no need to buy something that weighs a ton or is dreadfully bulky. If you're a hiker, you'll want a small camera--period! Start with one that fits in your shirt pocket.

Next thing to look for in a camera is one that fits into a small, plastic carrying case. This writer now has **2 Canons, Power Shot A550 & A560**, both of which shoot at **7.1 mega pixels (mp)**; most pictures in this book were taken with these (or an older 2mp Canon A-60). Both these newer cameras fit perfectly into a **Pelican #1010 plastic case** that can be carried on a second belt around the waist for easy and quick access; remember, a camera in the pack does not take pictures! This is the smallest case Pelican makes, and it, and both the A550 & A560 cameras are a perfect fit. Buy this case first, then find a camera to fit it. **Otter** also makes small plastic cases for cameras & cell fones.

The reason for a plastic case is to protect a very sensitive piece of electronic equipment. **Dust** and **water** are the 2 biggest enemies of all cameras, and the #1010 box is the best thing this writer has ever seen. He can swim through potholes worry-free with it attached to his belt, and it doesn't scratch the LCD screen. Plus, damage from dust is almost a thing of the past. The only negative thing about this way of carrying a camera, if you're day-dreaming, you can drop it while taking it out of the case.

Something to look for in a camera is its ability to take good pictures in **low light situations**. For example, in darker slots, all cameras automatically go down to lower shutter speeds, but you also want one that raises the ISO (the old ASA) settings automatically to adjust to the low light.

Here's an example; when set on **automatic**, the Canon A550, goes down to 1/8 of a second in low light and with an ISO setting of perhaps 100 or 200. With a little less light, it automatically pushes the ISO up to as much as 400 or 800--but the shutter speed stays at 1/8. If you're careful, you can hold your hand & camera against a wall, and get good sharp pictures at 1/8 of a second--but not less than that. If the shutter speed goes down to 1/4 second or less, you'll need a tripod, or set it on a rock.

If the A560 camera is in low light conditions and set on **auto**, it wants to push the ISO up to 800, 1200 or even 1600. Along with the higher ISO setting, the shutter speed then goes up as well to maybe 1/15 or 1/30 of a second. This is good for hand-held shooting, but although it might look good on the LCD screen, the quality isn't as good as pictures shot at low ISO's such as 100 or 200--same as with the old fashioned film cameras. The higher the ISO, the more **"noise"** there is in digital images, which translates to the *equivalency* of *graininess in film*, and a lower quality picture. **The lower the ISO, the better the foto quality.**

In very low light conditions, if your camera wants to automatically jack up the ISO to say 1600 (like the Canon 560), then switch to **Manual** mode. Then you can set you the ISO at between 80 & 1600--best to set it at 100 or 200--or anything less than 1600! This also means your shutter speed will drop, perhaps making it necessary to use a tripod, or place the camera on a rock or pack.

Here's something few people think about when choosing a camera; the **battery**. The smallest & thinnest cameras come with a one-of-a-kind size battery that fits only one camera and can only be bought in specialty fotographic stores. They also cost a small fortune. You must always carry one or more backup batteries, so it's best to have a camera that uses **AA's**. These can be purchased in any store, gas station, kiosk or supermarket in the world!

When it comes to AA's batteries, best to spend a little extra in the beginning and get the **rechargables** which often come with a **recharging kit** that works at home or in your car. That way, you'll always have fresh batteries. Or buy new alkaline batteries anywhere as backups.

If you enjoy taking pictures with people in most scenes, then prior to your trip, inform everyone to wear **colorful clothing**. The author prefers red, something like the University of Utah's color. Others prefer bright yellow, orange or blue. This not only gives your fotos additional color, but it helps to separate the person from the background. If you're end product will be B+W, sometimes you can't tell a person from a rock.

Squeeze the Trigger! When ever you're taking pictures, regardless of how bright or dark the scene is, press the **shutter release button slowly**. Otherwise your picture will be blurred. It's just like shooting a gun; if you jerk the trigger, you'll miss the target. Remember, your finger does not determine how fast the camera's lens opens & closes! This is the biggest difference between good & bad pictures, or between professionals & amateur fotographers. **Hold your breathe, squeeze the trigger slowly, and in between heart beats!**

When in any **slot canyon**, avoid taking a foto when there's direct sunlight in your subject area. If you do, part of the picture will be washed out with too much light; the other part will be dark or totally black. Instead, take a picture where the sunlight is being bounced off an upper wall or from around a corner, and diffused down or back into the dark corners. One exception to this rule would be if you're after special effects. In the case of **Antelope Canyon**, there is one place **(The Crack)** where a shaft of sunlight reaches the bottom at around high noon. Some fotographers set their camera on a tripod, then throw sand in the air which creates dust in the shaft of light. This technique exaggerates the small sunny part adding a pleasing effect to the picture.

Another way to get a good foto with bright sun covering half the scene is to wait for a **cloud** to cover the sky, then the light is diffused, eliminating the harsh difference or contrasts between sun & shadow. The best time to take fotos in slot canyons is generally from **late morning to early afternoon**. At that time you'll have to find a place where the sun isn't shining directly into the slot, but instead is coming from around a corner, and the light from the strong mid-day sun is bounced into the darker recesses. The **lighter the subject area** (but not direct sunlight), the **brighter & richer the colors will be. Darker scenes in dark slots subdues colors** and you won't be as happy.

If you're in a slot at or near high noon, you can get really good light & colors, and no sun, if you back up or walk forward a little, or around a corner. If at all possible, move around just enough to keep that one streak of sunlight out of your picture. Or, if you do get a small bright sunny steak in a corner of the frame, you can sometimes clone-stamp-it-out at home on your computer using **Photoshop**. One more idea when shooting in slots and making vertical frames, aim the camera down a little so you don't include the sky in the image, otherwise that part will be totally washed out. Doing this also reduces the harsh contrast between the lighter sky above, and the darker slot below.

If you should slip while wading, or somehow drop your **camera in water**, here are the steps to take. Immediately take out the batteries. Open the camera (if possible) and shake and blow out any water. Allow it to sit in the hot sun to dry, turning it occasionally to help evaporate any water inside. If you're near your car, start the engine, turn on the heater, and hang the camera in front of a vent. The warmer the camera gets, the quicker the water evaporates. The quicker the water evaporates, less corrosion there will be on the electrical system and less rust on metal parts.

If your camera is under water for just a nano second or so, there may not be any water deep inside.

In this case, by following the above steps, you may be back in business again in half an hour, especially if the water is clear, no sand has gotten inside, and if the sun is warm. The author had several of these little accidents with each of his older mechanical Pentax K-1000's film cameras. The last several times, no repair work was needed because he did the right things to get the camera dry fast. But he has lost about 4 digitals since 2003. Fortunately, each camera had an no-questions-asked **extended warranty** (had he been using the **1010 Pelican case**, most of those cameras would still be alive!).

Regarding the author drowning several of the older Canon A60 digitals, the good thing about those accidents was, he never lost any images because **memory cards** are well sealed. With his extended warranty, the cameras were replaced without charge, but the pictures were saved.

However, if you drop a digital into water, it likely can't be repaired, so they throw it away. When buying a new camera, pay a little more and get an **extended warranty**; that way if you drown a camera in the warranty period, they'll replace it free (read the small print!). Ritz or affiliated camera stores offer such extended warranties, but the cost of the warranty for digital cameras is almost double that of film cameras. At this stage in fotographic history, it seems fewer people know how to fix digitals.

New Mile Post Markers on Utah Highways

Sometime in late 2003 (?), the Utah Department of Transportation (UDOT) made the decision to update the mile post markers throughout the state. The main reason given was that parts of the Interstate highway system were built in segments over the years, and that some sections were off as much as 2 miles (3 kms). Another reason given was that some law enforcement agencies were using GPS systems, and they wanted to join the 21st Century--even though some mile posts had been there since that system began, some 40-50 years (?) ago. So, without telling anyone, most notably guide book writers, they began replacing old mile posts with new ones. By December of 2005, all State and Federal highways in Utah had new markers. The new ones are round metal posts set in cement. If you have any of this writer's other books, every mile marker will be off a little. In some cases, there is little change, but along I-70 running through the San Rafael Swell, they're now about 2 miles (3 kms) different from before. What used to be Exit 129, is now Exit 131; and so on. This required the author to re-drive every pertinent highway in the state to make this book and others as up-to-date as possible. Please make note of this change and tell your friends.

Important BLM Offices--Colorado Plateau

Before starting any hike, consider stopping at one of these BLM offices and get the latest information on road, trail, water, flood or weather conditions. In some cases, the information in this book may be outdated the minute it goes to press! This is especially true in slot canyons where conditions can change dramatically with every flash flood. People at the BLM likely can't help you much in the field of technical canyoneering, but they normally have good information on road conditions, and can go online to get an updated weather forecast. Here's a list of offices on the Colorado Plateau.

Utah
Cedar City, 176 East, D. L. Sargent Drive, 84720, Tele. 435-586-2401.
St. George, 345 East, Riverside Drive, 84790, Visitor Center/Arizona Strip Interpretive Association, Tele. 435-688-3200 or 3246.
Escalante, Visitor Center, Tele. 435-826-5499; **BLM & Grand Staircase-Escalante N.M. Office,** Tele. 435-826-5600; **Glen Canyon National Recreation Area Office,** Tele. 435-826-5651. All these offices are in the same building complex located on Highway 12, west side of Escalante, 84726.
Kanab (Field Office), 320 North, 100 East, PO Box 459, 84741, Tele. 435-644-2672. BLM & Grand Staircase-Escalante N.M. **Headquarters Office,** 190 E. Center Street, Tele. 435-644-4300. Or **GSENM Visitor Center,** east side of Kanab on Highway 89, Tele. 435-644-4680.
Richfield, 150 East, 900 North, Tele. 435-896-8221.
Hanksville, Southwest part of town, P.O. Box 99, 84734, Tele. 435-542-3461.
Moab, Multiagency Visitor Center, Center & Main Streets, Middle of Moab, Tele. 435-259-2468.
Price, 125 South, 600 West, 84501, Tele. 435-636-3600.
Monticello, 435 North, Main Street, 84535, Tele. 435-587-1500.
Vernal (Field Office), 170 South, 500 East, 84078, Tele. 435-781-4400. Or **Dinosaurland Welcome Center,** 55 E. Main (Center of Vernal), Tele. 435-789-6932.

Colorado
Grand Junction, 2815 H Road, 81506, Tele. 970-244-3000.
Montrose, 2505 South, Townsend Avenue, 81401, Tele. 970-240-5300.
Durango, BLM & Forest Service Visitor Center, 15 Burnett Court, 81301, Tele. 970-247-4874.
Dolores, BLM & Forest Service Visitor Center, Southwest of Dolores at 29211 Highway 184, 81323, Tele. 970-882-7296.

New Mexico
Grants, Northwest New Mexico Visitor Center, 1900 E. Santa Fe Avenue, 87020, Tele. 505-876-2780.

Arizona
St. George, Utah (for the **Arizona Strip District** north of the Grand Canyon), 345 East, Riverside Drive, 84790, Tele. 435-688-3200. Visitor Center, Tele. 435-688-3246.
Kingman, 2475 Beverly Avenue, 86401, Tele. 928-692-4400.
Phoenix Field Office, 21605 N. 7th Avenue, 85027, Tele. 623-580-5500.

National Park/Monument Offices & Visitor Centers
Arches National Park, Visitor Center, North of Moab, Tele. 435-719-2299, **nps.gov/arch.**
Bryce Canyon National Park, Visitor Center, Tele. 435 834-5322, **nps.gov/brca.**
Canyon De Chelly National Monument, Visitor Center, Tele. 928-674-5500, **nps.gov/cach.**
Canyonlands National Park, Headquarters at 2282 S.W., Resource Blvd., Moab, Tele. 435-719-2100; Visitor Information, Middle of Moab, Tele. 435-719-2313; Backcountry Reservations, Tele. 435-259-4351; Island in the Sky Visitor Center, Tele. 435-259-4712; Needles Visitor Center, Tele. 435-259-4711;

Maze (Hans Flat) Visitor Center/Ranger Station, Tele. 435-259-2652; **nps.gov/cany.**
Capitol Reef National Park, Visitor Center, Fruita, Tele. 435-425-3791, **nps.gov/care.**
Dinosaur National Monument, 4545 E. Highway 40, just east of Dinosaur, Colorado, Visitor Center, Tele. 970-374-3000, **nps.gov/dino.**
Grand Canyon National Park, South Rim Visitor Center, Tele. 928-638-7888, **nps.gov/grca.**
Grand Staircase-Escalante National Monument (run by the BLM), Escalante Interagency Visitor Center, West Side of Escalante, Utah, Tele. 435-826-5499; Kanab Visitor Center, 754 E, Highway 89, Kanab, Utah, Tele. 435-644-4680; GSENM Monument Headquarters, 190 East Center Street, Kanab, Tele. 435-644-4300; Cannonville Visitor Center, Cannonville, Utah, Tele. 435-679-8981; Big Water Visitor Center, 100 Upper Revolution Way (immediately south of Highway 89), Big Water, Utah, Tele. 435-675-5868, **ut.blm.gov/monument.**
Natural Bridges National Monument, Mailing address HC-60, Box 1, Lake Powell, Utah, 84533, Visitor Center, Tele. 435-692-1234, Headquarters, Tele. 435-719-2100, **nps.gov/nabr.**
Vermilion Cliffs National Monument, Interagency Information Center, 345 Riverside Drive, St. George, Utah, Tele. 435-688-3200, **az.blm.gov.**
Navajo National Monument, Visitor Center, Tele. 928-672-2700, **nps.gov/nava.**
Zion National Park, Tele. 435-772-7616/3256; Backcountry Desk 435-772-0170, **nps.gov/zion.**

Other Important Websites--US Government
Bureau of Land Management (BLM) . . . . .www.blm.gov
United State Forest Service (USFS) . . . . .www.fs.fed.us
National Park Service (NPS) . . . . .www.nps.gov

Canyoneering Information and Possible Updates

One way to get updated information on various slot canyons is to contact one of the following websites. If you go through any canyon in this book, but a flood has rearranged boulders, or taken out an anchor or B&H, then please post a bulletin on one or more of these website. This way everyone has a chance to read it and go better prepared. Or you can use your computer's search engine and type in the name of a canyon. Also Google Earth might show you some roads or routes.

American Canyoning Association . . . . **canyoneering.net**
Canyoneering Information & Updates . . . . **canyoneering.com**
Michael R. Kelsey's Guidebooks & Updates . . . . **kelseyguidebooks.com**
Shane Burrows--Mountaineering & Canyoneering . . . . **climb-utah.com**
Todd & Stephanie Martin--Desert Hiking Guide--Mostly Arizona . . . . **toddshikingguide.com**
Tom Jones--Utah Canyoneering Guide--Mostly Zion N.P. . . . . **canyoneeringUSA.com**
Jon Jasper's Canyoneering Foto Album . . . . **jonjasper.com**
Ryan Cornia's Canyon Experiences & Fotos . . . . **ajroadtrips.com**
AJ Pastula's Trip Reports . . . . **adventure-geek.com**
Steve (Ram) Ramrus' Trip Reports & Locate a Partner . . . . **groups.yahoo.com/group/canyons/**

Warning: Don't Blame Me!

People should keep a few things in mind when it comes to hiking in challenging & potentially dangerous slot canyons. This writer has done his best to collect information and present it to readers as accurately as possible. He has drawn maps as carefully as possible, and encourages everyone to buy the USGS topo maps suggested for each hike in this book. He has tried to inform hikers that canyons change with every flash flood, and that many of the hikes in this book are in isolated wilderness regions. Also, that **most of the canyons in this book are for EXPERIENCED CANYONEERS ONLY.**
Also, every hike in this book is through a technically difficult canyon most of which involves the use of ropes and rappelling. Some rappel anchors are chokestones or trees, others are bolts in the canyon walls, still others are deadmen that you build, or you rappel off a hook of some kind. If you choose to take on these challenges, it's your responsibility to check each anchor carefully, especially knots, slings or webbing and bolts. Floods can alter, damage or wash away these anchors, especially chokestones or logs, and webbing, so proper tools, equipment and supplies must be taken to insure you get through alive! No matter what, in some canyons always take a bolt kit, bolts & hangers (B&H)(unless you have recent information stating you won't need them), extra ropes & webbing, several Rapides or Quik Links (W&R/QL), ascenders (Jumars), an Ibis hooks, G-pick (sometimes a hooking kit), etc., to meet any kind of new challenge or emergency. **WARNING** New to this edition are some canyons so tight, that you must chimney up as high as 15m above the bottom to get through. These are classified as **easy (PG SLOT), difficult (R SLOT), very difficult (X SLOT), or extremely difficult (XX SLOT).** If you get into one of these that's over your head, and slip & fall or are stranded and **have to be rescued**; it's your tough luck and your fault, not mine! For those who will somehow get lost, stranded, or have to spend an extra night in a canyon and/or be rescued, all I can say is, I've done my best. The rest is up to you, so don't blame me or this book for your mistakes, lack of experience, poor planning, or just a little bad luck.
Also keep in mind, all hikes in this book are a step up in challenges from hikes in other books written by this author, and since many of these canyons are in isolated regions any rescue would be difficult, time consuming and/or very expensive. All or most canyons discussed here should only be undertaken by people who are better physically fit than average. People who never exercise or are overweight should never enter many of the canyons discussed here.
Before doing any hike, always stop at the nearest national park or BLM visitor center, and get the latest information on road, trail, water, flood or weather conditions & forecast. If bad weather means wasting a weekend, better to waste it than get caught in a flash flood. And in some cases, such as in Zion National Park, ask about rappel anchors, when the last group went through, and if there's been a flash flood since--before venturing out. Floods take out anchors, so go accordingly, and be prepared.
Another very important thing to remember, **tell someone where you're going and when you expect to return.** That way, friends or family can call for help if you don't return on time. Even Zion NP rangers won't go looking for you unless someone calls them stating that you're overdue.

In some cases, the information in this book is different than in previous editions and can be outdated the minute it goes to press! For that reason, you're asked to post canyon updates on **kelseyguidebooks.com**. Or post changes on other websites or with canyoneering discussion groups.

USGS Maps for Slot Canyons

One thing to remember while using this book--or any other guidebook--you must have one or more maps in addition to those shown here to locate a particular canyon and have a successful hike. What some people don't realize is, if you use just this book you'll get lost for sure.

The first map everyone needs is an ordinary **state highway map** of either Utah or Arizona, and maybe Colorado. This should get you close to what ever canyon you want. Next, it might also be wise to have a small scale map, one that shows less of Utah or Arizona than a state highway map. One series of charts you might look for is put out by the **Utah Travel Council**. This includes 5 maps covering the state at a scale of 1:357,000. Each map covers about 1/5 of the state and shows all the major highways, plus most of the main graded county roads.

Arizona doesn't seem to have anything to match the Utah Travel Council maps, but you can buy something that might be even better. At some supermarkets, office supplies stores or highway truck stops, look for the **Arizona or Utah Atlas & Gazetteer** put out by **DeLorme Mapping** (or perhaps some other company?). In the Utah Atlas are 63 maps, all at 1:250,000 scale. These are all topo maps; that is, they have contour lines showing equal elevations. These maps show any road that would be considered a road and will get you close to a canyon better than anything else.

Another set of maps that can be used for both hiking and/or driving are from the USGS at **1:100,000 scale**. These would be classed as intermediate scale and are commonly called **metric maps**, because elevations and contour lines are shown in meters, not feet. These are fairly detailed maps and quite new, all originating since the 1980's. The small squares on these maps are called **sections**, each of which is **one square mile**. This makes it easy to calculate approximate distances.

These metric maps come in 2 versions; one is printed by the **USGS**, the other specially designed for the **BLM**. The USGS version shows forests as light green while everything else is white. With the BLM maps, yellow means public domain or Federal lands, white is private property, green is Forest Service land, the blue one mile squares are state sections (Sections 2, 16, 32 and 36 are almost always blue in color and are called **State Sections**--unless the state and BLM have done some horse trading, swapping state sections for federal lands in other more advantageous locations), and national parks or monuments are a purplish color. The USGS maps may be slightly easier to read, but with the BLM version, you'll know beforehand if you're walking or driving on someone's private property.

Another scale of maps you might run into are those at **1:62,500**, sometimes known as **15' quads**. These were put out by the USGS, but sometime in the 1980's, they stopped production. Some are still available though, and they make good maps for hiking if the right one can be found. One problem with these, some are rather old with most going back to the 1950's. This means some are out dated, especially those in the San Rafael Swell which don't show Interstate Highway 70, or even the paved Highway 24 running between I-70 in the north, and Hanksville & Torrey in the south & west.

The latest or newest maps put out by the USGS are at **1:24,000 scale**, more commonly called **7 1/2' quads**. These show a very small piece of land up close. If it's detail you want, these are the ones. For doing slot canyons, or if you're out looking for slots, these are the best. One main disadvantage to these, many times you'll need 2 or 3 maps to do just one hike; whereas with the metrics, you can usually do many hikes with just one map. What some people do is fotocopy all parts of an anticipated hike, perhaps from 2 or 3 different maps, then tape the pieces together. Take the fotocopy on the hike and leave the originals safe at home to be copied and used another day.

These same maps are now available on CD or the internet, but quality is somewhat reduced. Many people like and use those put out by **Trails Illustrated** (now owned by **National Geographic-- TI/NG**). The logo they use is **TOPO!** These are organized, produced and sold by individual states; the ones you'll need are for Utah & Arizona. These maps look nice on the computer screen, but they were shot at 72 dpi, so when you run them out on paper for hiking, they're about 1/10 the quality as the original paper maps.

If you're hiking in Zion, perhaps the best all around map is the plastic **TI/NG** version simple called **Zion National Park**. This map is double-sided at 1:37,700 scale and is highly recommended because it has good detail, and being plastic, it doesn't matter if you get it wet--something to consider with all the swimming you'll be doing in Zion's pothole canyons.

If going to the Escalante River Country, you'll want to invest in another **TI/NG** map called,

Cowboy Steps/Trail out of **Pasture Canyon** immediately east of where the East & West Forks meet. These steps were hacked out of the Navajo Sandstone with some kind of pick, likely by one of the local cowboys. There are several places like this along the steeper parts. The **Point** on the left (see Map 13) is near the trailhead for **Pothole Fork** of **East Pasture Canyon**. (Jon Jasper foto)

Map1A, Index Map of Hikes

Canyons of the Escalante. It's also 2-sided, made of plastic, covers the entire Escalante River drainage, and is at 1:70,500 scale. Use this one for driving and hiking, but some may want to take along a corresponding 7 1/2' quad for greater detail while hiking.

If you're planning to spend a little time in Capitol Reef National Park, best to buy the **Earthwalk Press** publication called, **Hiking Map & Guide--Capitol Reef National Park**. It's made of plastic at 1:62,500 scale. It's also double-sided and covers the entire length of this very elongated park. For hiking the slot canyons there, you'll want to buy and use the 7 1/2' quads as well.

Since the year 2000, **TI/NG** has printed a map called **San Rafael Swell**. It's made of plastic, double-sided and at 1:90,000 scale. They apparently used old maps to create this one because there are lots more roads on the ground than are shown on this map (in this respect, the Huntington & San Rafael Desert metric maps are better, but not quite as easy to read). But the TI/NG map isn't bad over-all and is easy to use. It's made especially for recreationists and shows popular hiking, mtn. biking, rock art sites and other points of interest.

As you read this book, under each hike will be 11 headings, one of which will be **Maps.** Listed there will be the best maps to use for that particular hike. That will include the metric or 1:100,000 scale map, maps at 1:24,000 scale or 7 1/2' quads (the old 1:62,500 maps are not mentioned in this edition), and the TI/NG map if applicable.

Scoggins & Outlaw Canyons, Dinosaur National M., Colorado

Location & Access Shown here are 2 canyons in the middle of **Dinosaur National Monument** and in the extreme northeastern corner of Colorado. Both drain into the lower **Yampa River Gorge** from the north roughly halfway between the **Mantle Ranch & Cave**, and **Echo Park** (where the Green & Yampa Rivers meet). Let's call them **Outlaw Canyon**--because **Outlaw Park** along the Yampa is nearby; and **Scoggins Draw**, after Charles Scoggins who did research in the area before World War II. This is a name the locals call the canyon.

To get there, use a Utah or Colorado state highway map while driving along Highway 40 just east of the stateline and the small ramshackle Colorado town of **Dinosaur** (mile post 3). Between mile posts 4 & 5, turn north, then immediately east into the **DNM Visitor Center & HQ's** parking lot. After getting last minute information on roads, Yampa River water flows, buying maps or books, etc., drive north on the paved **Harpers Corner Road (HCR)** toward Harpers Corner and/or Echo Park. After 41.3 kms (25.65 miles), turn east at the information board and head downhill on the graded **Echo Park Road** (EPR). In dry conditions, this road is good for cars, but it may have a rough place or two.

After driving down through **upper Sand Canyon** you'll come to a junction (Km 12.2/Mile 7.6 from the HCR), turn right. Continue eastward on the graded **Yampa Bench Road** to **Km 18.7 (Mile 11.6)** and park where a side-road comes down from the abandoned **Red Rock Ranch** and meets upper Red Rock Canyon with willows & tamaracks. This is **Red Rock Canyon TH**. Or, continue east to **Km 22.9 (Mile 14.2)** and park at the beginning of a little side-road that used to run north to a point overlooking the Yampa River. You may want to park there at the **Mantle Cave TH**.

Rating Most people will want to backpack in, then from a campsite it's **3A III** for each canyon.

Equipment RG, ropes (**Scoggins Draw**--one 60m; **Outlaw**--one 30m), W&R/QL, & water filter.

Route From the **Mantle Cave TH**, walk north along an old road which is west & above a short canyon holding **Mantle Cave**. At the end of that point, follow tracks east over a couple of ledges (it's a little tricky with a large park) and up to the south end of the canyon. After seeing some ruins, walk north out of this little side-canyon and along cattle trails running west & north along the Yampa River.

Now from the **Red Rock Canyon TH**. Walk north along a fence and into the canyon heading east. There's a cattle trail on the north side; or walk in the dry creek bed. At the mouth of the canyon turn left or north, go through a gate (close it behind you), and head north on one of several cattle trails. Be aware, in this area there are several **rock art panels**, plus at least one **Fremont granary**. At the first big bend north of Red Rock Canyon, is a cattle trail running near the river, then you bench/beach-walk to the sand bar shown. Make a base camp there, or anywhere nearby; lots of sites to choose from.

To do **Outlaw**, you could head up from the bottom end and after 400m, climb a big horn sheep trail east to the rim, then route-find north looking for a way down into one of the upper forks. Or do as we did, cross the river, head left or downstream to about the middle of **Outlaw Park** and climb a minor canyon east, then north between 2 minor drainages to a point immediately east of **Outlaw Peak** (author's name) marked **6486** (1978m) on the *7 1/2' Canyon of Ladore South quad*. At one point you'll be on a meter-wide ledge that looks impossible to get around, so follow this map carefully--**this is a route-find!** Once at a **high pass**, head down to the east on some steep slickrock and into the 2 forks of Outlaw. From there you can go upcanyon as far as possible, or head downcanyon and do an **11m rap** near the lower end. Or you could stay high from the **high pass** heading northward and into the **East**

From the top of **Outlaw Peak** looking southwest at the **Yampa River**. In the middle is a pyramid-shaped peak; behind it is a *steep route* you can walk down to reach Outlaw Park & Scoggins Draw.

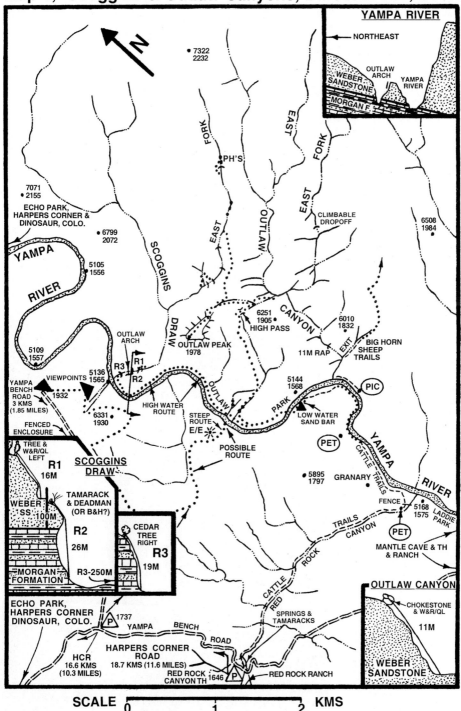

Fork of Scoggins Draw. We went upcanyon a ways in the east fork of the 2 main canyons. We didn't run into any dryfalls on our short trip, but there likely will be some in the upper ends.

Scoggins Draw has 3 rappels in its lower end, so you must climb the route running past Outlaw Peak as suggested above. From the **high pass**, go east & down a little, then double back west to get on a P/J-covered flat between East Outlaw & Scoggins (climb Outlaw Peak for fine views along the way). As shown on the map, you can walk up the **East Fork** to its headwall with lots of big open potholes, then head downcanyon bypassing about 4 dryfalls enroute. When you're within 800m of the Yampa, you'll walk through a deep narrow gorge, then you'll soon come to a dropoff (backup from that first drop and bench-walk west to see the top of what may be the 9th biggest arch in the world. Let's call it **Outlaw Arch**). If you have good knees, you might jump this, or reenter 75m further along on the right or north side. **Rappel 1** is on the left from a tree & W&R/QL and down **16m**. About 150m below that is the **big rap of 26m** into a fotogenic alcove. We built a **deadman anchor** next to a small tamarack bush and got down OK, but it's highly recommended someone install 2 B&H's to make it completely safe. Remember, safety first! Finally the **19m R3** from a cedar tree on the right.

Once at the mouth of the canyon, walk out into the Yampa and upstream--if it's not too deep. We walked right in the river for about half of the 3 kms back to camp. It was full of sand and only ankle to knee deep; at other times we walked on sand bars on one side of the other. On that day the river was flowing at **130 cfs**. If the water is a lot higher, you can walk back upcanyon to the east 250m from the river, climb up on a minor bench, then walk west, south & east along one of 2 benches above the river all the way to Outlaw Park, then the river crossing to camp. This could be done with an inner tube.

Elevations Red Rock Canyon TH, 1646m; Mantle Cave TH,1660m; Yampa campsite, 1568m; end of Scoggins Draw, 1565m; Outlaw Peak, 1978m.

Time Needed About 2 hours backpacking from either trailhead, then a full day to explore and descend either drainage. There are still lots of canyons in this area waiting to be explored.

Water Start with your own, then from the **Yampa River--but filter or purify first!** The Yampa flows through farming & ranching country! After rains, there's lots of pothole water in the slickrock canyons.

Maps USGS or BLM maps Canyon of Ladore & Rangely (1:100,000); Hells Canyon & Canyon of Ladore South (1:24,000--7 1/2' quads) for hiking; and the plastic Trails Illustrated/National Geographic map, Dinosaur National Monument, Utah/Colorado (1:78,125) for driving, orientation & maybe hiking.

Flash Flood Danger Low to moderate in the lower ends of both canyon; low elsewhere.

Best Time to Hike About mid-August to mid-September, when Yampa River flows are roughly 200 cfs or less. To hike either of these canyons, you must wade the Yampa, so you never want to go in spring or early summer with high water; or in cold weather. Check the website **www.cbrfc.noaa.gov/river/ station/flowplot/flowplot.cgi?YDLC2** for currant daily river flows. With higher water, you could also cross by using an inner tube, air mattress or some other float device.

Author's Experience After a couple of scouting trips, he along with Scott Patterson, Stephen Ho & AJ Pastula backpacked down Red Rock Canyon to the campsite suggested in 1 2/3 hours, then quickly set off exploring the route between the canyons just south & east of Outlaw Peak.   From there it was down into **Outlaw** and going up each of the major forks a ways, then downcanyon and back to camp in 6 hours. **Day 2:** climbed Outlaw Peak for views & fotos, and into Scoggins; then upcanyon to near the headwall and finally down the 3 raps to the Yampa.   From there we walked beside or in the river back to camp--a 10 hour day. **Day 3:** headed up from the bottom end of Outlaw and exited east to a high point, then back to camp; 3 1/2 hours. In the afternoon we backpacked out in 1 hour 35 minutes. This was Labor Day Weekend, 2006, and we didn't see another soul.

Looking east from one of 2 viewpoints along the rim of the **Yampa River Gorge**. Above center is the top of **Outlaw Peak**; in the lower half of this picture, and in the middle with shadows is the end of **Scoggins Draw**, along with all 3 rappels & Outlaw Arch.

Above After going through **Scoggins Draw**, the best way to get back to camp, or a way to exit the gorge, is to walk right up the middle the Yampa River. The lower end of **Outlaw Park** is just ahead to the left.

Left This is Rappel 2 in **Scoggins Draw**. As of 9/2006, the only thing to rappel from was a small bush and a deadman anchor we built around it. Expect to redo that anchor or install a couple of B&H's.

Buckwater Draw, Dinosaur Land, Colorado

Location & Access This canyon is located in the northwest corner of Colorado, just east of the Utah stateline, and south of Dinosaur National Monument (DNM). To get there, use any Utah or Colorado state highway map while driving along Highway 40 just east of the stateline and the small ramshackle Colorado town of **Dinosaur** (mile post 3). Between mile posts 4 & 5, turn north, passing the DNM visitor center & HQ on the right. If you like, stop there for information on roads, buy maps or books, etc., then drive north on the paved **Harpers Corner Road** (HCR) which leads to Harpers Corner and/or Echo Park in DNM. After 7.1 kms (4.4 miles) turn left into the **Plug Hat Picnic Area (PHPA)** parking which is equipped with a toilet. This will give you access to **Lower Buckwater Draw (LBD)**. Or, continue up the HCR to Km 11.3/Mile 7 which is immediately west of the **Blue Mountain Spring No. 1** & stock pond. Just across the road to the west is a gate and a seldom-used heading west. Park next to the gate without blocking the road. This **Buckwater TH** gives access to **Buckwater Draw**.

Rating Both drainages rate **3A II** or **III**. Here are 2 short slot canyons with just a little bushwhacking.
Equipment RG, **Buckwater Draw**--60m rope; **LBD**--30m rope, W&RQ/L and K&EP.
Route Here's one of several route options. For **Buckwater Draw**, walk west downhill through sagebrush along an old 4WD track (this was in a WSA, but late word is you can now drive it--but it's rough). After about 5 kms, and about 300m before you come to an obvious Entrada Sandstone buttress in front of you, veer left or southwest and route-find down into the drainage. Once at the bottom, and after another 350m or so, will be your first rappel of about 6m from W&R/QL around a chokestone. This is the beginning of a slot lasting for about 350m. About 200m below **Rappel 1** will be **R2**. If the W&R/QL is gone, install a new sling around a chokestone about 4m above the actual dropoff. This will be a 2-stage rap of 27m into the deepest & darkest part of the canyon. About 200m below R2, the canyon begins to open up. About 600m below R2 will be several possible escape routes up to the left (just below that will be several routes out to the right or north and back to the old track you used earlier). Climb this slope, then route-find east toward another obvious **Entrada buttress** where the **old cattle trail** is shown. From there, rim-walk to the **PHPA** to a 2nd car; or road-walk the old road back to Buckwater TH.

To enter **LBD** from near the old cattle trail, and do both drainages on the same hike, make your way down in just north of the old cattle trail, then head west downcanyon. You will have several DC's, then right in the middle of *Section 20* on the *Plug Hat Rock 7 1/2' quad*, will be **R1**, a drop of about 3m. With a small group, you may help each other down without a rappel. Below this, the canyon slots-up pretty good, then just before you reach Buckwater, will be **R2** (or **HL**) of 6m. Or, everyone could be belayed, then the group leader might **DC** this one--if he/she is competent.

To get back to Buckwater TH, walk downcanyon about 1 1/2 kms below the confluence and route-find out of the drainage **north** to the **old track** you came in on, and road-walk back east to the HCR. Or use part of the old cattle trail and rim-walk to **PHPA**; if you've left a 2nd car there. Or maybe the **best option** is, park at PHPA and do LBD first, then swing around and do the main fork, and finally high-tail-it back to you car at PHPA. This way no 4WD needed.

Elevations PHPA, 2134m; Buckwater TH, 2327m; canyon bottom, 1920m.
Time Needed Both slots can be done together in 7-9 hours--less for either one if done separately. Or maybe 6-8 hours starting at PHPA.
Water No water around, so take your own. The nearest water is at DNM HQ or Dinosaur.
Maps USGS or BLM map Rangely (1:100,000) for driving & orientation; and Plug Hat Rock (1:24,000--7 1/2' quad) for hiking.
Flash Flood Danger Potentially high right in the short slots, non-existent elsewhere.
Best Time to Hike Late spring or fall, but given the higher altitudes, summer isn't so hot.
Author's Experience With Scott Patterson, who originally found this place, and Sam Cox, we parked one car at PHPA, then drove to, and started from, the Buckwater TH. We did both canyons as suggested above and ended our hike at Plug Hat Rock. Round-trip took 7 hours.

From the halfway point of **Rappel 2**, looking down at Sam Cox at the bottom. **Buckwater Draw** has a pretty good slot, but it's short.

Map 2, Buckwater Draw, Dinosaur Land, Colorado

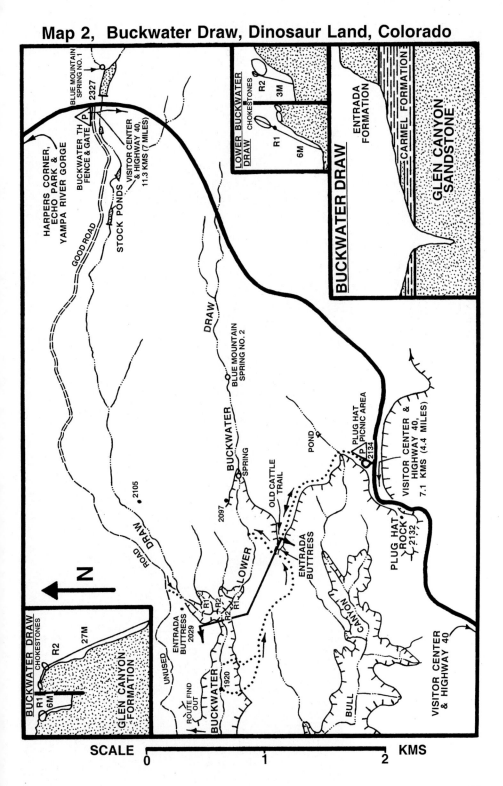

Above Left AJ Pastula starts the rappel from the log in the **North Fork of Iron Wash**. Jonas Fast watches from below.

Above Right The nice slot in Upper Chute Canyon. (Steve Tyler foto)

Right This is what greets you at the end of the technical section of **Little Iron Wash**. You first slide down a steep, narrow chute about 10-12m and right into this swimming pool. From there, it's about a 10 minute walk to your car.

Map 3A, Area Map--San Rafael Swell, Utah

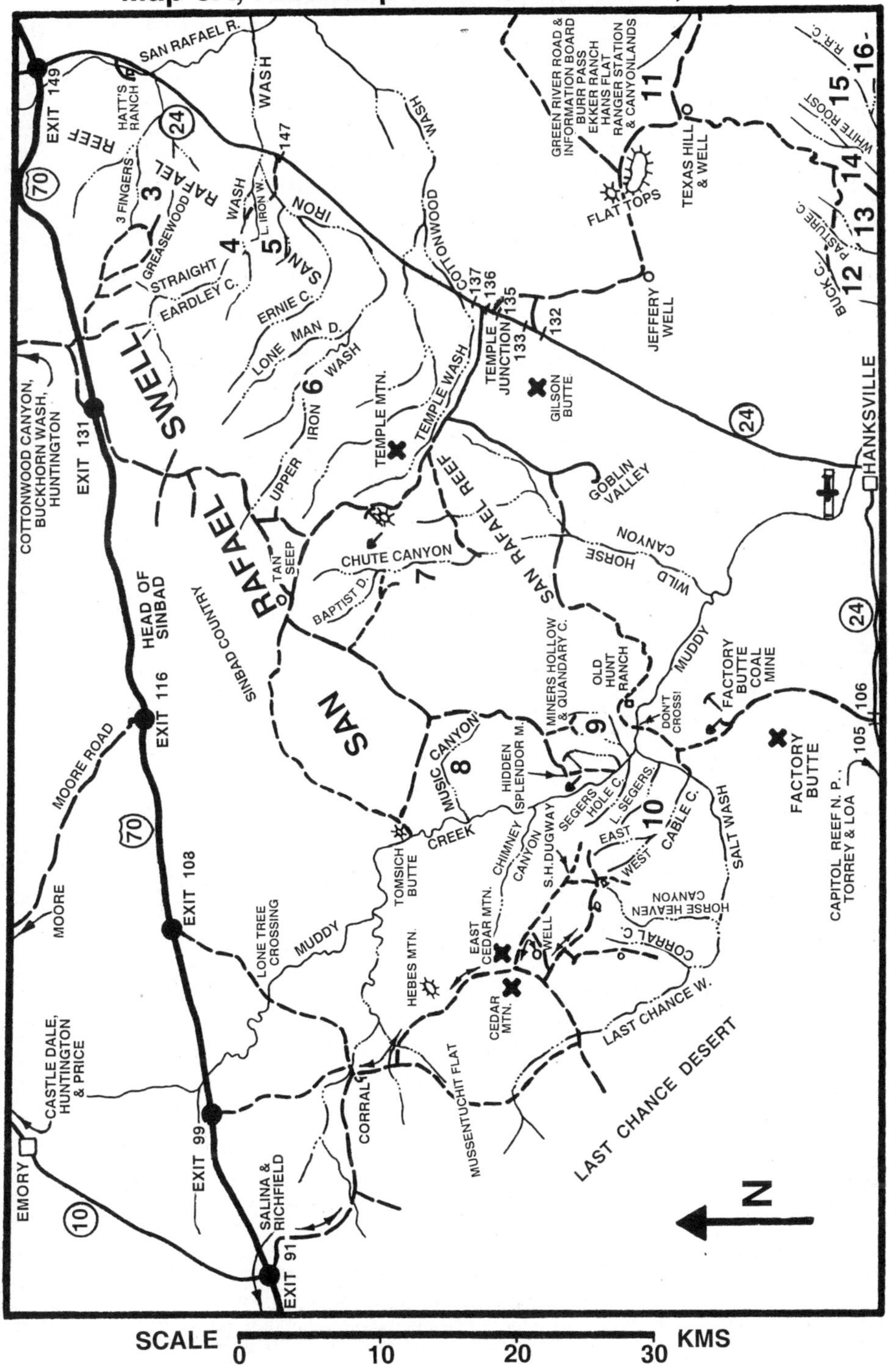

Upper Three Fingers Canyon and Greasewood Draw, San Rafael Swell, Utah

Location & Access Featured here are the upper ends of **Greasewood Draw & Three Fingers Canyon**, those parts that are above or west of the San Rafael Reef, and east of Cliff Dweller Flat. To get there, drive I-70 across the San Rafael Swell (this is west of Green River). Leave the freeway at **mile post & Exit 131 (formerly Exit 129)**. From there, and setting your odometer at Km & Mile 0, drive east on the north side of I-70 to Km 5.9/Mile 3.65; turn southeast, drive under the freeway and turn left or east at Km 6.5/Mile 4; at the junction marked **6684** (2037m) at Km 9.5/Mile 5.9, veer left; at junction **6650** (2027m) & Km 10.8/Mile 6.7, you can go either way on a road that circles what this writer is calling **East Cliff Dweller Flat**. But for this description, let's use the best route, the one that should be OK for HC cars. Continue east on this graded road until you reach Km 16.6/Mile 10.3, then look carefully for a very faint track heading southeast running just north of the **butte marked 6558** (1999m). See **Hiking Map**. Follow this track southeast for 1.8 kms (1.1 mile) and stop at the road closure sign. This place is about 500m before the **butte marked 6403** (1952m); and 18.4 kms (11.4 miles) from the exit on the freeway.

If you use the other route to **Upper Hyde Draw**, it can be sandy in places (a 4WD/AWD may be required when conditions are dry!). Just beyond **Hyde Draw Reservoir** and at Km 14.2/Mile 8.8, veer left or east; at the junction marked **6439** (1963m), and at Km 15.6/Mile 9.7, is the faint vehicle track leading to the trailhead. Using this route, it's 17.4 kms (10.8 miles) from the freeway exit to the trailhead, but because of the possible deep sand, this is not the recommended route. Use the *Drowned Hole Draw & Arsons Garden 7 1/2' quads* or the *metric San Rafael Desert map* to reach the trailhead.

Rating Both drainages **3B II** or **III.** Both are "V" shaped canyons with a number of potholes. Both are in the Cedar Mesa SS and similar to Eardley Canyon. **Three Fingers** could have one or 2 KPH's.

Equipment RG, one 30m rope, a couple of shorter ropes, Ibis hook & G-pick, W&RQ/L, K&EP, wet/drysuits in spring or fall, and drybags/plastic box for camera, etc. In **Three Fingers,** be prepared for a possible pack/bag toss to get out of one or 2 possible KPH's.

Route While you can enter either canyon at many points, here's the quickest way to the best parts of each. For **Greasewood Draw**, walk east along an old faint vehicle track just south of the **butte marked 6403** on the *Arsons Garden quad*. About 300m southeast of the elevation marked **6211AT** on the *Greasewood Draw quad*, route-find to the bottom and head downcanyon. In the first km or so, you'll pass some pools then comes some dropoffs. The technical section lasts for about 750m. **Rappel 1** can be from a ledge & Ibis hook on the right and down 3m into a pool; but some might jump if water is deep enough (?). Test the depth, or help each other. **R2** can be done from an Ibis hook on the top of a huge chokestone and down 5m into a pool. **Caution:** When rapping off an Ibis hook into deep water, don't place it in a crack because if may get stuck and be hard to flip off while swimming.

Rappel 3 is down a crack, but good climbers can **DC** this (?). We rapped 6m from W&R/QL around a chokestone into a pool. **R4** is from a short rope (sling) tied to a small tree on the right and down **10m.** From there, walk 8m and **rap/HL** about **6m** into an Olympic-sized pool using a shorter rope and an Ibis hook from natural ledges (or DC/slide this part). Just beyond the **big pool**, is another 5-6m drop, but we skirted it on the right or south side. Not far below that was a crack that most can likely DC, but we rapped off a chokestone and down about 5m. After a swim, the last rap (**R5**) is from W&R/QL & small boulder/chokestone on the far right side. This is a **10m** drop down a crack into a swimming pool. Immediately below that is a good slot & DC into a pool, then a 70m drop to the San Rafael Reef part of this drainage. Immediately before that big drop, scramble up to the left or northwest, and out of the canyon. Follow wild burro trails back toward Butte 6403 and your vehicle.

Three Fingers Canyon From the trailhead, walk ENE & north of Butte **6403** aiming for the elevation marker **6035T** (1839m). From there, head due north and scramble down into the main drainage along a number of different routes. Going right down a sizable side-drainage is OK. Once into the main canyon, you'll almost immediately come to pools & DC'ing. The technical section here is also about 750m long. As the canyon turns north, northwest, you'll come to a DC into what could be a KPH (?). The author was alone, so he put his Ibis hook in a crack and HL'ed down into this one, then was barely able to crawl out of the pothole using the rope. About 400m below that is the **Rappel 1**. You should see part of an old rope around a long flake boulder on the left to rap **8m** under a large boulder into a big swimmer. Just below that is **R2**; at least it was for the author. If the pool is deep enough, you can slide, then jump 2-3m into a deep possible **KPH** (always belay someone down to test water depth first). Or, use an Ibis hook from an iron-rich lip on the right and rappel about **6m**. However, if the water level is just right, getting out of this pool could be a big problem! A shoulder stand or pack/bag toss should work? A hooking kit is likely not needed, because if all else fails, you can climb out on the left or north avoiding this obstacle altogether.

About 100m below this rappel is an Olympic-sized swimming pool. Best way past this is to slide right down the chute about 5m into deep water. If you're squimish, use a hook set in a tiny crack and HL down, but just before you actually get into the water, flip the hook off. Then swim 20m. Just below that is a 2m drop--help each other over this. The author used a hook and HL'ed down into the cobblestone-filled pothole (bad knees!). From there, it's 500m down to at least 3 easy exits on the right or south side. Immediately below that are 2 more PH's before the big drop off the face of the cliff--but one of those is a long KPH! Best to forget this part--exit and make your way southwest back to your vehicle.

Elevations Trailhead, 1902m; approximate altitude of both exit points, 1660m.

Time Needed Each canyon should take from 5-8 hours.

Water Take your own drinking water, but there should always be some in potholes.

Maps USGS or BLM map San Rafael Desert (1:100,000) for driving & orientation; Drowned Hole Draw, Arsons Garden & Greasewood Draw (1:24,000) for hiking; and/or the plastic Trails Illustrated/National Geographic (TI/NG) map, San Rafael Swell (1:90,000) for driving.

Flash Flood Danger Moderate in the narrower parts, low elsewhere. Lots of possible escape routes.

Best Time to Hike Given the higher altitudes and some pools probably year-round, summer may be best. Otherwise in spring or fall, but take and use wet/drysuits then. Right after a good storm means more water in the pools, and possibly more jump-ins, fewer rappels (?).

Author's Experience He and Steve Tyler did **Greasewood** in about 6 3/4 hours, but we installed W&R/QL in 4 places (2 of which may not be needed). After 2 scout trips, the author did **Three Fingers** solo in 6 3/4 hours.

32

Map 3, Upper Three Fingers Canyon & Greasewood Draw, San Rafael Swell, Utah

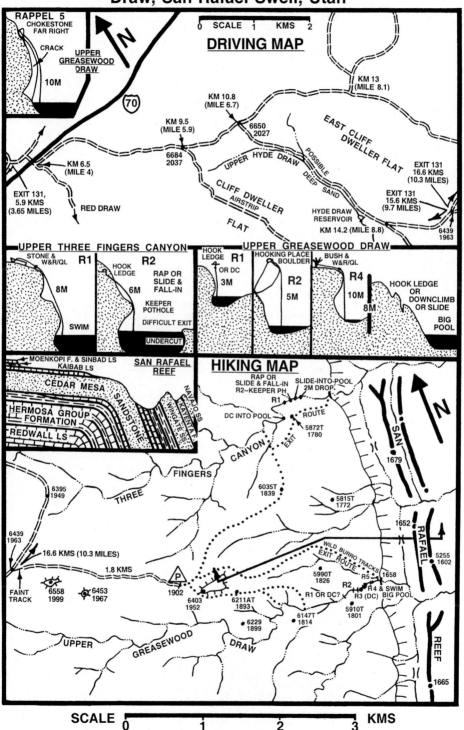

RAPPEL 5
CHOKESTONE FAR RIGHT
CRACK
UPPER GREASEWOOD DRAW
10M

SCALE 0 1 KMS 2

DRIVING MAP

N

70

KM 13 (MILE 8.1)

KM 10.8 (MILE 6.7)

KM 9.5 (MILE 5.9)

6650 2027

EAST CLIFF DWELLER FLAT

6684 2037

UPPER HYDE DRAW

POSSIBLE DEEP SAND

EXIT 131 16.6 KMS (10.3 MILES)

KM 6.5 (MILE 4)

CLIFF DWELLER AIRSTRIP

EXIT 131 15.6 KMS (9.7 MILES)

EXIT 131, 5.9 KMS (3.65 MILES)

RED DRAW

FLAT

HYDE DRAW RESERVOIR

KM 14.2 (MILE 8.8)

6439 1963

UPPER THREE FINGERS CANYON

STONE & W&R/QL R1
8M
SWIM

R2
HOOK LEDGE
6M
RAP OR SLIDE & FALL-IN
KEEPER POTHOLE
DIFFICULT EXIT
UNDERCUT

UPPER GREASEWOOD DRAW

HOOK LEDGE R1
3M
OR DC

HOOKING PLACE & BOULDER
R2
5M

BUSH & W&R/QL
R4
10M
8M

HOOK LEDGE OR DOWNCLIMB OR SLIDE
BIG POOL

SAN RAFAEL REEF

MOENKOPI F. & SINBAD LS
KAIBAB LS
CEDAR MESA SANDSTONE
HERMOSA GROUP FORMATION
REDWALL LS
NAVAJO SS
KAYENTA F.
WINGATE SS.

HIKING MAP

RAP OR SLIDE & FALL-IN
R2--KEEPER PH
SLIDE-INTO-POOL 2M DROP
R1
DC INTO POOL
ROUTE
5872T 1780

CANYON EXIT

FINGERS
THREE

6395 1949

6035T 1839

5815T 1772

6439 1963

16.6 KMS (10.3 MILES)

1.8 KMS

FAINT TRACK
6558 1999
6453 1967

P
1902

6403 1952

6211AT 1893

WILD BURRO TRACKS
EXIT ROUTE
5990T 1826

R1 OR DC?

6147T 1814

6229 1899

5910T 1801

R2
R3 (DC)

R5
R4 & SWIM BIG POOL
5255 1602
1658

1679

1652

1665

SAN RAFAEL REEF

UPPER GREASEWOOD DRAW

N

SCALE 0 1 2 3 KMS

Above Swimming through the big pool just below **Rappel 4** and the DC in **Upper Greasewood Draw**.

Right This is the pool immediately below **Rappel 2** in **Upper Three Fingers Canyon**. This must be considered a **KPH** because if the water level is just right, one person will never get out alone! The author rapped off an Ibis hook set in the rough surface of the wall in the upper right side (the left side of this picture).

Above Getting ready to rappel into **Pool 1**, in the Technical Section of **Lower Eardley Canyon**. It's been a while since the author was there, but some people might be able to DC this dropoff.

Left Tim Barnhart getting ready to DC the **final chute** and into a swimming pool in the lower end of **Little Iron Wash**. It may look difficult to some, but the width is just right for the body to fit in and squeeze down without fear of slipping.

Lower Eardley Canyon & Little Iron Wash, San Rafael S., Utah

Location & Access **Eardley and Little Iron Wash** (AKA-Zero Gravity) **Canyons** are located in the east central part of the San Rafael Swell between Interstate 70 & Highway 24, and about 30 kms (18 miles) west of Green River. You can get to the upper parts of Eardley by leaving I-70 at mile post & Exit 131, but in this book only the lower technical part of Eardley is featured. For information on the interesting upper end, plus history of the area, see this writer's other book, ***Hiking and Exploring Utah's San Rafael Swell***.

To reach the lower end of Eardley & Little Iron Wash, leave I-70 at **Exit 149** and drive south about 19 kms (12 miles) on **Highway 24**. Right at new **mile post 147**, turn west onto a good dirt road (you can also get there from the south and the Hanksville/Lake Powell/Capitol Reef National Park area). After about 2.6 kms (1.6 miles) you'll come to a grove of tamaracks & Iron Wash (just before that is one rough spot--car drivers should take a shovel). In dry conditions any vehicle can make it across Iron Wash, then there's a short steep place before you cross **Little Iron Wash** which is 4 kms (2.5 miles) from the highway. Park on the west side road where an old track heads west (possibly with a sign stating *No Vehicles*).

For **Eardley**, continue northwest past Little Iron Wash. After a couple of rough spots you'll come to the mouth of **Straight Wash**, which is 5.9 kms (3.65 miles) from Highway 24. The road ends as Straight Wash emerges from the San Rafael Reef. Park and/or camp at one of several sites near the end of the road. Or you might drive down into the dry wash and camp under some cottonwood trees. Or you could camp in the tamaracks at Iron Wash crossing.

Rating **Lower Eardley** rates 3B III. The lower 150m of Eardley Canyon has up to 6 rappels (or HL's or jump-ins), all into big potholes. It's recommended you go with no less than 2 people, because if the water level is low, it may be difficult to get out of Pool 2 alone. For experienced slot canyoneers, **Little Iron Wash** rates a 2B II (DC'ing only, no rappels); less experienced hikers, maybe **3B II** (?).

Equipment **Eardley:** RG (or possibly HL or do Dulfersitz rappels), one rope at least 30m, W&R/QL, dry-bags, K&EP, and wet/drysuit for spring or fall. A BK & a couple of B&H or baby angle pitons might be optional because the canyon is bolted-up. **Little Iron Wash:** Experienced people need only K&EP; others may need a rope of 20m or so, and wet/drysuits in cooler weather.

Route **Little Iron Wash** Walk almost due west along the south rim of the canyon. After about 2 kms, drop down into an area where 4 drainages meet. Once there, head east downcanyon. In the first 500m or so, you'll have a couple of nice short slots, then open walking. About 600m into the canyon, it slots-up for another 200m. In this last section will be a **near-KPH** (most can climb out alone, or use a partner assist), then a 5m DC which **some have rappelled** (from W&R/QL on a chokestone). Just below that is a **very tight part** lasting about 2m which goes under a small chokestone with webbing hanging down. Heavy weight hikers will have to somehow chimney over this chokestone. Below that is a 10m-DC'able-chute into a huge swimming pool. From there it's an open canyon back to your car.

Eardley From the car-park, and with an early start, head west up **Straight Wash** through the **San Rafael Reef**. Once through the big cliffs of the Reef, continue to follow the same dry creek bed as it makes a big round gooseneck turn and heads southwest (there is a shortcut route or trail shown on the map). Watch carefully for the mouth of Eardley which will be about 1 km from the big round bend of Straight Wash. Once there, walk northwest up along the sloping northeast side of the **canyon rim** to the pass marked 1725m. From there, head down the steep gully (be careful not to send rocks down on your partners!) to the bottom of Eardley (marked 1475m) and turn left downcanyon. After about 2 kms, and near the bottom of this steep "V" shaped canyon, you'll come to at least 6 big swimming pools and 6 dropoffs. If water levels are high, rappel or HL down into 4 of these pools--the other 2 are jump-ins. If water is low in the potholes, you'll rappel/HL into all 6.

Rappel 1 has 2 B&H and one piton up to the left. Rappel down a steep ramp. Above Pool 2 are 2 B&H to the right just above your head--**R2**. If water levels are very low, you might build a pile of rocks to get out, but if water is chest deep, one person may have a difficult time escaping. Two people can help each other out with no problem. When Pool 3 is full, you can jump down half a meter into the water; or rappel **(R3)** from one B&H on the right.

To reach Pool 4, rappel 7m from 2 B&H on the right if water is low **(R4)**. To reach Pool 5, you can use the one B&H to your right **(R5)**. Otherwise with a full pool of water, it's a jump-in of about 1 1/2m into 2m of water. At the last rappel, **R6**, are some chokestones with W&R/QL in the lower end of Pool 5. If a big storm should rearrange those rocks, you could be in trouble, so have several short ropes or webbing so you can install a new sling around another rock. From there you have a 5m vertical drop, then a steep ramp down to the last pool. From Pool 6 to the mouth of Eardley is a simple walk of about 250m.

Elevations Both trailheads or car-parks, 1350m; high point for Eardley, 1725m.

Time Needed Doing only the technical part of lower **Eardley** via Straight Wash will take from 5-8 hours round-trip. **Little Iron Wash** will take from 2 1/2-4 hours.

Water Always carry water in your car and pack. Use potehole water for emergencies only.

Maps USGS or BLM map San Rafael Desert (1:100,000) for driving & orientation; and Arsons Garden & Greasewood Draw (1:24,000--7 1/2' quad) for hiking.

Flash Flood Danger Low overall, but higher in the 150m technical section of **Eardley**. Beware too, this has a fairly long drainage and flood waters from a storm way upcanyon could roar down and surprise you. **Little Iron Wash** drainage is short, but you could get trapped in the last slot.

Best Time to Hike Do **both canyons** in summer--from late May or early June through mid-September. A wet/drysuit will extend the season for a month or two into the spring or fall. If doing **Eardley** in the spring or fall, it's recommended you camp nearby, get an early start, and arrive at the swimming part at mid-morning so you'll have the sun for warmth.

Author's Experience **Eardley**--With Steve Tyler and one day after a flood, we went through 11 pools--6 of which were swimmers (some were 5m deep!)--before reaching the technical section! We went through rather quickly with all potholes brim full and returned to our cars, all in 6 hours. His last hike through the technical section was with his brother Gail on 6/25/2000. Water levels were low and there were no jump-ins--only 6 rappels. We installed 2 more bolts at the pools you can jump-in to only when water level is high. Total round-trip time was 5 3/4 hours. **Little Iron Wash**--With Jon Jasper & Tim Barnhart we did this hike with medium water levels. We had 3-4 swimmers, and did the hike in 2 1/2 hours round-trip.

Map 4, Lower Eardley Canyon & Little Iron Wash, San Rafael Swell, Utah

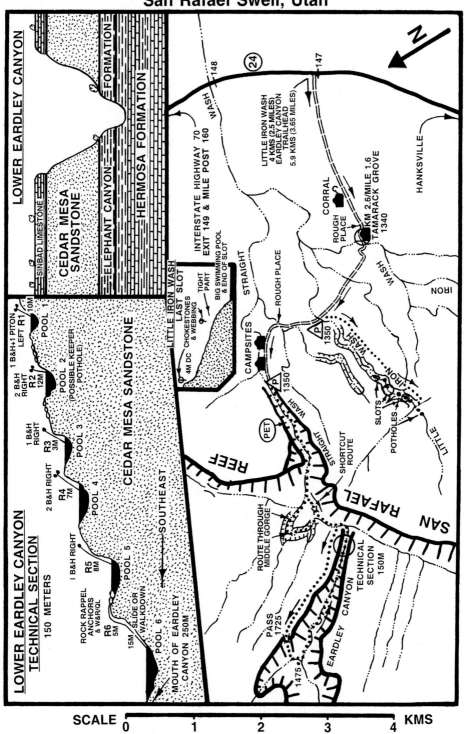

SCALE   0   1   2   3   4   KMS

North Fork of Iron Wash, San Rafael Swell, Utah

Location & Access The **North Fork of Iron Wash** is located along the eastern San Rafael Reef about halfway between Interstate Highway 70 and Temple Mtn. To get there, drive along **Highway 24**, which runs between Hanksville and I-70. Until recently, here is/was the **normal way** there; right at **mile post 147**, turn west onto a good dirt road. This runs due west past a corral, then drops down into the bottom of **Iron Wash** (2.6 kms/1.6 miles from Highway 24). There's one rough place on that road just before the tamaracks, but cars driven carefully, should make it OK (?). Park and/or camp along the creek bed in a **thicket of tamaracks** (with mosquitos in summer!), or nearby. This is the same access route you'd use to reach the mouth of Straight Wash and **lower Eardley Canyon**.

If you have a HC/4WD, here's another way with a longer rougher drive, but a much shorter walk. Between **mile posts 141 & 142** on **Highway 24**, turn west, go through a gate (close it behind you), and drive down into **Iron Wash** at **Lost Spring**. Continue up the other side and turn right or north at the junction (Km 1.6/Mile 1). See the map, *Shortcut Route to North Fork Iron Wash*, below. Continue north through one **very sandy dry wash** (Km 2.7/Mile 1.7) to where the road turns left or west and drops into lower Ernie Wash (Km 4.7/Mile 2.9). From there, drive straight ahead to the north another 300m (Km 5/Mile 3.1)) and park on top of a hill to the right. Stop there at the **4843T (1476m) TH.** Beyond it can be very deep sandy gravel and fit **for really tough 4WD's only!**

Rating It rates **3A (or B) III**. Best part is a 1 km-long narrows & short-but-sweet 75m-long technical slot. This one could have a rating of **3A II -PG SLOT**? because you climb up a slanted slot before rappelling.

Equipment RG, 30m rope, W&R/QL & K&EP.

Route Old Route Start at the **tamaracks** and walk southwest up along **Iron Wash**. You'll pass some running water, cottonwood trees, and after about 1 km, will see the **North Fork** come in from the right or west. Enter it, but after about 100m, climb upon the bench to the right or north and rim-walk above the shallow canyon west and southwest. The reason for rim-walking is to avoid a big dryfall 400m upcanyon.

Now follow the route symbols on this map. If you want to see just the **Lower North Fork**, then follow the shallow drainage into the lower canyon. By doing this, you'll come to the lower end of the slot and the **5m-deep Big Keeper Pothole**, but you'll have to stop there.

If you want to go around the slot and come down into it from above, then again follow the route symbols on the map staying on a route just east, and outside, of the lower canyon. Actually, a better way might be to follow Iron Wash SSW; about 2 kms above the tamaracks, and in a big wide valley, turn west into the lower end of **Ernie Wash**. Watch closely, and you'll see a nice panel of **rock art** on the right 2-6m above the ground. About 1 km above that is an easy exit. Leave the canyon, and head west toward the elevation **4902AT** (1494m) and join the other possible route to that same point. You'll eventually turn northwest above the lower end of the slot. From there, rim-walk about 1 km northwest on slickrock immediately next to the slot. You may see a few cairns. When you reach the vicinity of a **12m dryfall**, reenter the canyon at the upper end of the narrows.

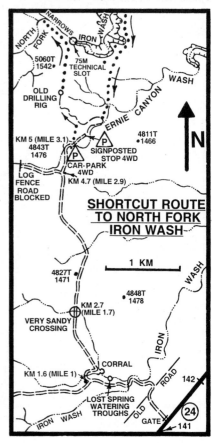

The first 500m has lots of sculptured potholes, some of which may have water, but no swimming. Finally, you'll come to a 5m drop and the beginning of a 75m-long technical slot (see the 2 cross-sections). At that point is 1 B&H on the left. With a short rope, you could **HL** down to a shallow pool; however, anyone worth their salt can chimney down a crack on the right side. After 2 possible shallow pools, you'll be in a dark slot, for which you may need a **headlamp**. Crawl down to the same **5m-deep Big KPH** mentioned above. If you're there right after a flood, all you have to do is swim across and you're out of the tight slot. If it's dry, a good climber with the aid of an Ibis hook & pole & etreir and/or 3 toss bags or packs thrown up and over the lip to the other side, can climb out (always keep one person & rope out of the KPH). Few will want to try this.

The normal way through this slot is to retreat 15m from the **Big KPH** and climb up onto some tiny ledges and chimney your way across the Big KPH. Beginners may be challenged. As you climb across, you'll always have the opposite slanted wall about 1m away, so spanning or chimneying is easy. Directly above the Big KPH is a small platform, and 2m above that is an even larger 2x3m flat place (upper ledge). Chimney up to that. From there, it's an easy walk, with your hands on the opposite wall, of about 10m to a log & sling along the upper ledge. Once on the log, rappel **11m** to a point 10m beyond the Big KPH.

Another more challenging way across the Big KPH, is to chimney down from the upper ledges to a point just beyond the pothole. It's fairly easy until you're about 2m above the bottom and just beyond the Big KPH, then you'll have to jump about 2m with a hard landing. Tall people will have a shorter jump.

Beyond the rappel and/or the Big KPH, the slot gradually opens up. Continue east downcanyon, turn north and walk through Lower North Fork. Follow the drainage back to your car at the tamaracks.

If you've parked at the **Shortcut Route Trailhead, 4843T** (1476) TH, look north and you may see an **old drilling rig** about 800m away. Route-find north from the TH down across **Ernie Dry Wash** and be looking for an old road heading northwest to the drilling rig. Once there, continue north and just east of the dome marked **5060T** (1542m), then route-find into the North Fork as described above.

Once you're out of the slot, and for those who came in

Map 5, North Fork of Iron Wash, San Rafael Swell, Utah

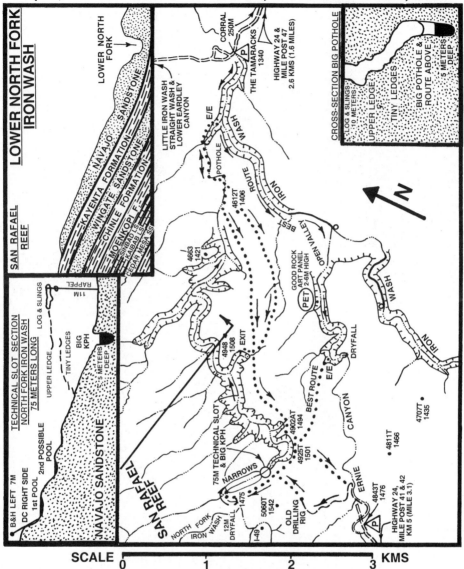

SCALE 0 1 2 3 KMS

via the drilling rig, watch your map closely and use the exit as shown, then walk south cross-country using a compass back to your vehicle.

Elevation Car-park, 1340m; dropin point in the upper slot, 1475m; 4WD parking for new route, 1476m.

Time Needed From the car-park, walking outside the Lower North Fork, entering the narrows from above, then down the slot and back to your car should take from 6-9 hours. The **Shortcut Route**, 2 1/2-4 hours.

Water Always carry water in your car and pack. On a hot summer day, you'll need up to 4 liters of water!

Maps USGS or BLM map San Rafael Desert (1:100,000) for driving; and Arsons Garden, Old Woman Wash, Crows Nest Spring & Greasewood Draw (1:24,000--7 1/2' quad) for hiking.

Flash Flood Danger No danger except right in the 1 km slot, which has high risk.

Best Time to Hike Late spring or early fall, but if there has been rains & floods just before your hike, take a wetsuit. Summers are hot, but it's cool inside the slot.

Author's Experience One trip was with David Harrison & Rick Ahlander. We went straight to the narrows, took the easy route above the Big KPH, and returned, all in 8 hours; however 2 hours were lost because one member became ill and had to rest. In 5/2008, along with AJ Pastula and Jonas Fast, we came in via the **Shortcut & drilling rig route** and were at the beginning of the narrows in just 35 minutes. But AJ was hell-bound & determined to conquer that Big KPH--which took 2 hours--so that trip took an even 4 hours overall.

Above Left AJ Pastula climbing up the slanted slot above the **Big Keeper Pothole** in the **North Fork of Iron Wash**. The log & W&R/QL used for the rappel is about 20m away.

Above Right AJ gets a boost from Jonas Fast as he tries to climb out of the **Big KPH** in the lower end of **North Fork of Iron Wash**.

Right AJ spanning the narrow slot while going above the **Big KPH** in **North Fork of Iron Wash**. Just beyond is the log with W&R/QL which you can rappel from.

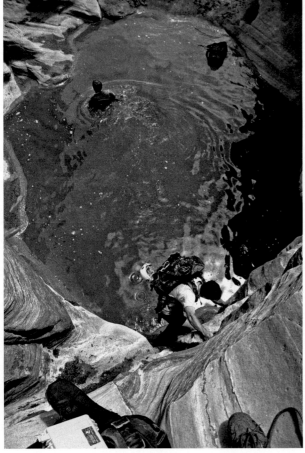

Above Steve Bresovec watches as the author raps off an Ibis hook at the last rappel in **Upper Iron Wash**.

Left Bresovec & Scott Patterson swimming through a pothole in the lower technical section of **Upper Iron Wash**. We found only one pothole that could be a KPH, but you can walk around that.

Upper Iron Wash, San Rafael Swell, Utah

Location & Access The upper end of **Iron Wash** is located near the center of the San Rafael Swell. The best part of this canyon is located about 14 kms (9 miles) due south of Exit 131 on Interstate 70; and about 8 kms (5 miles) northeast of Temple Mountain.

To get there, have the plastic *Trails Illustrated/National Geographic (TI/NG) map*, or the *USGS/BLM San Rafael Desert metric map* in hand. One way there is to drive I-70 and leave the freeway at **new mile post & Exit 131**. From there, first head west on the south-side frontage road, then south toward **Taylor Flat**. Stay on the most-used road for 16.6 kms (10.3 miles), then turn left at a **major junction**. Continue south for another 1.1 kms (.7 mile) for a total of 17.7 kms (11 miles) from the freeway, then turn left or east onto a less-used road signposted *Motorcycle Trailhead* (the sign is about 100m east of the main road).

Or if you're coming from the south and Hanksville, drive to **new mile post 136 on Highway 24** and turn left or west at **Temple Junction** (marked by a parking area, one old lamp post & a tree). From there, head west on a paved road toward Temple Mtn., but from just south of this obvious peak, continue west up a steep grade toward **Flat Top** and beyond to the beginning of **Taylor Flat**. At the first major junction (26.7 kms/16.6 miles from Highway 24), turn right or north and proceed to the junction & *Motorcycle Trailhead* mentioned above which is 29.8 kms (18.5 miles) from Temple Junction and Highway 24.

From the **Motorcycle Trailhead** turnoff, head east for a total distance of 5.5 kms (3.4 miles). Cars with higher clearance can make it. At that point you'll be at the trailhead called **Iron Divide**, which is out in the middle of a big treeless flat (if you want to park in the shade, regress about .6 km (.4 mile) and stop next to some large piñon pine trees). In the past, some 4WD's used to continue east past Iron Divide, but now everyone stops there and either uses a motorcycle or mtn. bike. At the official BLM sanctioned motorcycle/mtn. bike trail & trailhead is a register.

Rating This canyon has a number of potholes which retain water even in the driest of droughts! Its rating is **3B III or IV**. Upper Iron Wash cuts deep into the Cedar Mesa Sandstone (formerly known as the Coconino Sandstone) and has the look & feel of lower Eardley Canyon.

Equipment RG, one 30m rope, W&R/QL, K&EP, one Ibis hook, and in spring or fall, a wet/drysuit. In addition, a mtn. bike for each person will shorten this round-trip hike by 2-3 hours.

Route From the trailhead, ride a mtn. bike (or walk) east on what the TI/NG map calls the *Palisades Trail*, an old 4WD track. Using the 7 1/2' quads listed below, ride roughly 6 kms to a trail junction located about 500m south of the elevation marked **6131T** (1869m). From there head north, then east, on a good trail to the butte marked **6274T** (1912m). Just northwest of that butte is a **2nd trail junction**; if using a mtn. bike leave it there to be picked up later, then walk left or north and follow the trail first west, then north and into the bottom of **Iron Wash**. This 2nd motorcycle trail begins/ends at Twin Knowles, which could be another starting point, but the road going out there is very rough and not recommended.

Once in **Iron Wash**, head east downcanyon. After 1 km, the canyon narrows and you'll come to potholes which can be bypassed. Just below that, a side-canyon comes in from the left or north. About 500m below that is a deep pool you'll likely have to swim, then 2 more pools you can avoid by climbing up to the right. Soon after that will be the **1st rappel**. The author rapped 6m (with very low water) from 2 small hook holes on the left. It will almost certainly be a swimmer, so the group leader could belay everyone down, then jump if the water is deep. About 500m below that will be a 10m drop into a pool, but you can climb up, then down, on the left side with several people helping each other over ledges.

After another 500m will be an exit on the left, then 2 big swimmers. Just below that will be a climb-out on the right or west. Soon you'll climb high on the left to avoid a big pool. Immediately after that will be a 2-tiered KPH that will be 8-9m deep if full, but you can completely avoid this by walking down a ramp on the right or west side. On the downcanyon-side of the KPH will be a **2m drop**, but you can help each other down and/or jump. Just below this will be **R2** of about **12m**. The author rapped from a natural hole on the right using his Ibis hook (give your rope lots of slack when flipping it off)--his partner Steve built an anchor of stones & a log and used a retrievable sling. The Ibis hook is better and faster. See picture.

The bottom 1/4 of the canyon is at its deepest & narrowest with many potholes, none of which are KPH's. Many of these potholes can be bypassed on one side of the other. Also, there are many places where you'll help each other down; or sometimes it's faster to jump--if you have good knees! Near the end, the canyon becomes shallow and there are many escapes, but you're not finished until you come to one **big 30m-long swimming pool**. Immediately after that, turn right and climb up to the next bench. Set your compass on northwest and head for the small butte marked **6274T** (1912m). On the eastern side of that butte is another motorcycle/mtn. bike trail you can follow around on the north side and back to your mtn. bike, then the 6 km ride back uphill to your vehicle.

Elevations Trailhead, about 2000m; mtn. bike parking, 1850m; last pool--end of canyon, 1600m.

Time Needed From the trailhead, be prepared to spend from 9-12 hours round-trip doing this canyon; without a mtn. bike it will take 2-3 hours longer (?).

Water Take plenty in your car and in your pack. Only stagnant pothole water is in the canyon.

Maps USGS or BLM map San Rafael Desert (1:100,000) for driving; Twin Knolls, Temple Mountain & Old Woman Wash (1:24,000--7 1/2' quads) for hiking; or possibly the Trails Illustrated/National Geographic map San Rafael Swell (1:90,000) also for driving.

Flash Flood Danger Generally low. The head of this drainage is about 15 kms northwest of where you first enter the canyon, and in most situations, you'll have some place to climb to in the event of a flood.

Best Time to Hike Late spring might be best, but you'll need a wet/drysuit then. During the heat of summer, wet/drysuits are not needed, but you will need 4 liters of drinking water.

Author's Experience The author scouted this canyon twice, once getting to the big KPH. On his 3rd trip, he was joined by Scott Patterson and Steve Brezovec and all used mtn. bikes. We entered just above the big KPH because the lower canyon was the only part unknown to us. Round-trip time from the trailhead was about 7 3/4 hours.

Map 6, Upper Iron Wash, San Rafael Swell, Utah

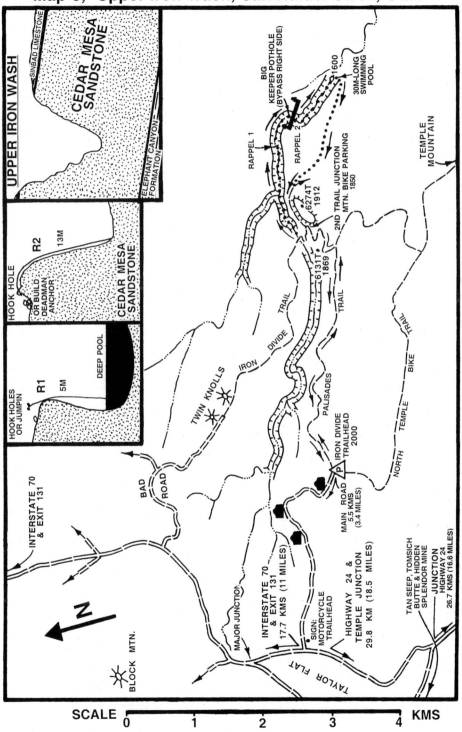

SCALE

0   1   2   3   4   KMS

Upper Chute Canyon / Lower Baptist Draw, San Rafael Swell, Utah

Location & Access These canyons are located right in the middle of the San Rafael Swell in an area called Sinbad Country. This is just west of **Temple Mountain**.

To get to the **upper trailhead**, have the *TI/NG* (see below) or *San Rafael Desert* metric map in hand, then leave **I-70** at new **mile post & Exit 131** and head southwest toward **Tan Seep**, a distance of about 22.7 kms (14.1 miles). See map. Or if you're coming from the Hanksville or Goblin Valley side, leave **Highway 24** at **Temple Junction** & mile post 136, and drive west toward Temple Mountain on a paved road. From there, continue northwest past **Flat Top** to **Tan Seep** (31.1 kms/19.3 miles from Highway 24), which has a corral & metal shed just to the north. From Tan Seep head west, then south, toward **McKay Flat**. About 7.7 kms (4.8 miles) from Tan Seep is a fence & **cattle guard**; from there continue south another 250m and turn southeast on a less-used & ungraded road. Drive for another 4.5 kms (2.8 miles) and turn left or northeast. Continue for 1 more km and park in some cedars at **2051m altitude**. Any car can make it there if driven carefully.

To reach the lower end of the canyon, which is just inside the big Wingate Sandstone walls of the San Rafael Reef, drive to **Temple Mtn.** (11.6 kms/7.4 miles from Highway 24 & Temple Junction), then turn southwest on the graded **Chute Canyon Road**. After another 9.6 kms (5.95 miles), and 1300m before you reach the end of the road where Chute Canyon cuts through the Reef, park on the right at a little campsite (1583m).

Rating Depending on which hike you do, either **3A II or III** (after floods this will be a **B**). This is one of the best all-around fun hikes in the San Rafael Swell.

Equipment RG, one 50m rope, a shorter 15m rope, W&R/QL, and K&EP.

Route To hike the entire length of this canyon, you'll need 2 vehicles. However, the best slots are in the upper end, and you can make a nice loop-hike there; then on a another day, do the lower end of the canyon from the Chute Canyon Road, eliminating the need for a car shuttle.

Instead of following the route symbols marking the old route to Lower Baptist Draw, here's the latest trail info from Scott Patterson. From the car-park at 2051m, walk north but a little west and over the hill and down and around the Teepee-like rock as shown on the little **insert map**. This is now apparently a good trail. Follow it until you can get down into **Baptist Draw**. Head downcanyon. After about 750m, you'll come to a tight walk-sideways slot where you'll have to crawl under several chokestones. Soon you'll come to the first of 2 rappels. **Rappel 1** is about **6m**. Rappel from W&R/QL that's between a big boulder and the wall on the left (advanced hikers/climbers can **DC** this). About 100m below R1, the slot opens and you look straight down into **Chute Canyon**. At that point you'll be standing on top of a chokestone-filled slot. There you should find several W&R/QL around a chokestone which makes a safe anchor for rappelling. **R2** is **22m** into a crack. Be aware that the big floods of 10/2006, may have changed things at this point.

From R2 in lower Baptist Draw, walk north **up Chute Canyon**. This drainage is deeper than Baptist, and just as dark in several places. There will be logs & chokestones to grunt & climb over, but nothing too difficult; but this changes with every flood. When the canyon opens and the slot is behind you, climb up the slope to the left or west, then walk north a little and get up on top of a limestone bench or point. From there you can see where you've just come from. Head west about 1 km, then using a little side-drainage, walk down into **lower Baptist Draw** until you can exit at the same place you entered earlier. Backtrack to your car.

If you've made arrangements to walk down the entire canyon, head down **Chute Canyon** after R2 in lower Baptist. Immediately, you'll climb around a large boulder, then after about 150m you'll come to the 2nd of 4 big boulders in this part of Chute. You can climb over or under these, except for the 2nd one; but of course this can change with every flash flood. If necessary, rig some W&R/QL to a log or chokestone and **rappel/HL** over the 2nd boulder, a drop of about **8m**. Beyond the 4-5 boulders, it's basically a walk through a deep, narrow gorge. All of this middle section is like the best parts of upper Eardley Canyon. It's up to 100m deep in Cedar Mesa Sandstone (formerly known as Coconino Sandstone).

When you reach the **faultline canyons** (see map cross-sections), the main drainage opens some and there are several possible entries or exits. About 1 km below that, the canyon again tightens-up and there are 6 places where you'll have to **rappel, HL or jump/slide** down (see the Technical Section--Chute Canyon cross-section below). The reason this 2nd narrow part exists is that south of the faultline canyons, the Cedar Mesa has dropped down, re-exposing the upper layers, which are more conducive to slot making.

The **1st Rap/DC** in this **lower section** is about **3m**, perhaps from a log. **R2** can be from a short log, or better still, wrap your rope over one of the big chokestones. The last person should rappel down one rope with the other end held by a partner below. **R3** has a B&H on the left; rappel about **7m**. This part may have changed because of big floods (?).

Next obstacle is going down a steep **5m slippery-slide groove**. You'll probably want to use a rope here--if you have bad knees! The last person down may want to rig up a small chokestone or log as an anchor. Or, you might climb up and around it on the left. The **last problem** is a **2m slide**, then a drop of nearly 2m into a pothole which will likely have some water or mud. Or, you can climb up and around on the right (south) side. After this pool, it's a 2 hour walk to the lower car-park.

Another option would be to hike upcanyon from the lower car-park on the Chute Canyon Road, exit the canyon below the technical section, reenter at the faultline canyon, then head back downcanyon. Or, you could go down Baptist & Chute, then at the faultline canyon head west for 3 kms or so, then climb out to the north and walk across the flats back to your car. You'll need a good map & compass for this, but it's easy.

Elevations Upper car-park, 2051m (or 2042m); lower car-park, 1583m.

Time Needed Hiking from the upper car-park to the lower one can be done in one full day for most people, but it could be a very long 10-12 hour day for some. To do the short loop-hike using **lower Baptist & upper Chute** is only about 5 kms round-trip and can be done in from 3 1/2-6 hours. To walk up from the lower trailhead, do the **Technical Section/Chute Canyon**, and return takes 6-9 hours.

Water None in the canyon so carry plenty in your car and pack. Take 3-4 liters each on a hot summer day!

Maps USGS or BLM map San Rafael Desert (1:100,000) for driving; Horse Valley (1:24,000--7 1/2' quad) for hiking; and/or the plastic Trails Illustrated/National Geographic (TI/NG) map, San Rafael Swell (1:90,000).

Flash Flood Danger Always high in the tight slot, but low to moderate otherwise.

Best Time to Hike Spring, summer or fall, or whenever the roads are dry. Some people do this in winter.

Author's Experience After 2 days of scouting, and with brother Gail & Steve Tyler, we did the loop-hike in the upper end in a leisurely 5 hours. Next day it was up from the bottom to do the technical section. That took 6 hours round-trip. Later, the author and Keith Krumbach & Warren Egbert did the upper canyon loop in 3 1/2 hours. Tyler later took a group of 20 Boy Scouts through the upper loop in about 9 hours. In 2004, yours truly and Tyler parked on the main road near elevation 2059m, hiked down Little Ocean Draw, through upper Chute and exited at the faultline canyon going east to a second car just north of Flat Top; 7 3/4 hours.

Map 7, Upper Chute Canyon & Lower Baptist Draw, San Rafael Swell, Utah

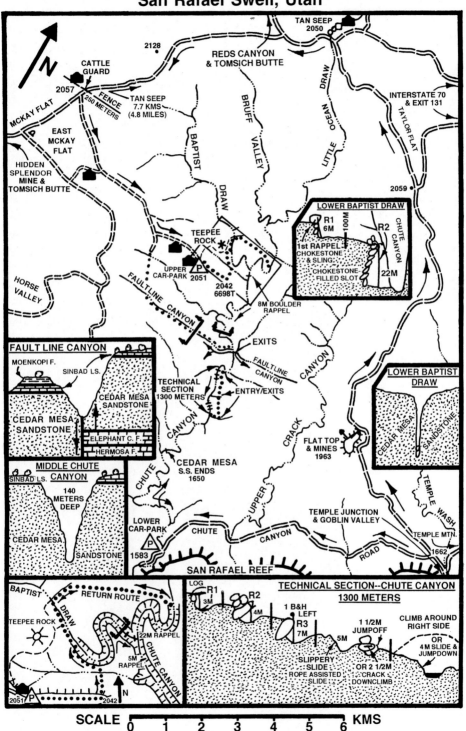

Above and Right
Both of these fotos were taken in the area just below where **Baptist Draw** enters **Upper Chute Canyon**.

Above Left The tight upper part of **Upper Chute Canyon**. It's normally dry, but wet & muddy after rains.

Above Right HL'ing off an Ibis hook in the lower end of **Music Canyon**. With a large group, you can sometimes do this canyon without any rappels by helping each other down these short dropoffs (?).

Left This is the 8m boulder rappel in **Upper Chute Canyon**. This seems to have changed since the 10/2006 floods (?). (All fotos on this page by Steve Tyler)

Music Canyon, San Rafael Swell, Utah

Location & Access On USGS maps this is an unnamed drainage, but hikers call it **Music Canyon**. It's located in the southern San Rafael Swell and drains into the middle part of **The Chute of Muddy Creek**. To get there, drive along I-70 inside the Swell. Leave the freeway at **Exit 131** and head southwest along a good graded road. At **Km 16.4/Mile 10.2 (from I-70)** is a junction at **Taylor Flat**; veer right. At **Km 22.5/Mile 14** is another junction with **Tan Seep** immediately to the west--turn right or west and proceed southwest toward **Tomsich Butte/Hidden Splendor**. Drive across McKay Flat and at **Km 38.2/Mile 23.7** is **Tomsich Butte Junction**; to the right is Tomsich Butte & Reds Canyon, but turn left instead toward **Hidden Splendor**. Drive another 6.3 kms (3.9 miles) and park on the right. This is where you begin hiking. This place is also about 400m (.25 mile) north of an **old junk car** on the west side of the road.

You can also get there from **Highway 24**. At **Temple Junction**, located about halfway between Hanksville & I-70, and at **new mile post 136**, turn west onto the paved Temple Mtn. Road and head for Temple Mtn., Flat Top and Tan Seep. From there, head south as described above. See the **Area Map, San Rafael Swell** on **page 31**, and carry one of the maps listed below.

Some people hike down Music and return the same way (for winter hikes), but the preferred way to finish the trip is to walk down **The Chute of Muddy Creek**. To do that, drive a second car (or perhaps mtn. bike) down to Hidden Splendor & the Delta Mine and park at the top of the hill about 300m south of the **airstrip**. The road down that dugway is now blocked off. This parking place is 16.6 kms (10.3 miles) from the Tomsich Butte Turnoff.

Rating 3A III This is quite an easy hike with 3-4 possible rappels/HL, but some groups, under the right conditions, and with the right shoes, have ascended with help from friends.

Equipment RG, one 30m rope (if going back upcanyon, take 4-5 short ropes of 10-15m to be left in place, plus ascenders), an Ibis hook, and a small supply of W&R/QL just in case.

Route The first 9/10's of **Music Canyon** is uninteresting, then about 1 km from Muddy Creek it narrows and turns into a pretty good slot. The technical part begins with about 3 pretty good DC's; help each other over these. After another 200m or so, you may see on the right, a piece of W&R/QL with a knot on the end stuffed into a crack. This is **Rappel 1**, about **7m**. Immediately after that is a **6m drop**. With one long rope, or several shorter ones, you could do both raps from the one anchor. Or use an Ibis Hook placing it in a small hole in the wall behind. Or with help from tall friends, you could belay everyone, then help the last person down (?). Just before reaching Muddy Creek, we **HL'ed** from an Ibis hook in a crack on the left and down **4m**. However, we had lots of pools and wet feet; in dry conditions, some do this with a fast slide & hard landing! Once at Muddy Creek, look for a walking stick, then head downstream. In the first 3 kms you'll see the best part of **The Chute of Muddy Creek** as it cut deep into the Cedar Mesa Sandstone (formerly known as the Coconino). If you're really fit and have a full day, you can apparently make a loop-hike by walking up **Mud Canyon** to get back to your car.

Elevations Trailhead, 1902m; Music & Muddy Creek, 1494m; end of hike, Hidden Splendor, 1463m.

Time Needed Going down Music to the creek and back the same way, may take 6-8 hours or more (?); but continuing down Muddy Creek to Hidden Splendor may take from 8-10 hours.

Water Take your own. Muddy Creek water has to be purified or filtered.

Maps USGS or BLM map San Rafael Desert (1:100,000) for driving; Hunt Draw & Tomsich Butte (1:24,000--7 1/2' quads) for hiking; and the plastic Trails Illustrated/National Geographic map, San Rafael Swell (1:90,000) for driving.

Flash Flood Danger High in the last km of Music; moderate to low in Muddy Creek.

Best Time to Hike If returning back up Music, spring or fall. If going down Muddy Creek, avoid the higher spring runoff (sometime in April or May?) by going in early summer or fall (or whenever water levels are low).

Author's Experience With Steve Tyler, we left a car at Hidden Splendor, then walked down Music taking 5 hours to reach Muddy Creek; we took lots of fotos which slowed the trip. Arrived at Hidden Splendor after 9 1/2 hours. This was on 5/3/2004 with high water chest deep in Muddy Creek.

If you go all the way through **Music Canyon**, then continue making the loop down **Muddy Creek**, this is what you'll see in the narrowest part of **The Chute**.

Map 8, Music Canyon, San Rafael Swell, Utah

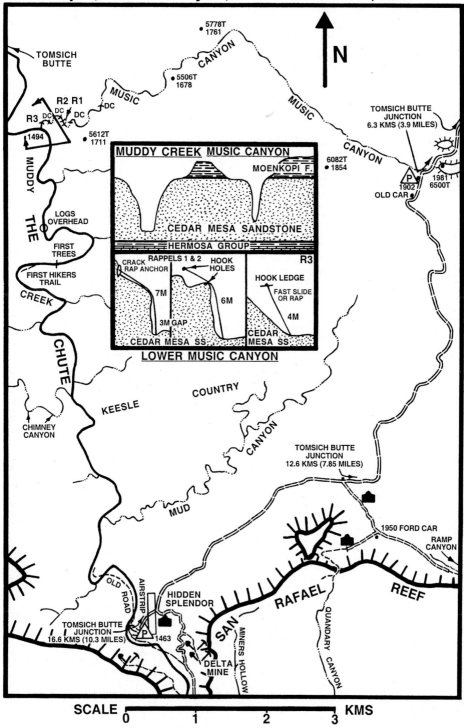

SCALE

0   1   2   3   KMS

Quandary Canyon & Miners Hollow, San Rafael Swell, Utah

Location & Access These 2 canyons are located immediately east of **Hidden Splendor** and the **Delta Mine**, and where **Muddy Creek** cuts through the southern part of the San Rafael Reef. Both are very short drainages which begin right on top of the Reef and flow little more than 3 kms. Both are unnamed on USGS maps; **Quandary** is what hikers call one; **Miners Hollow** is this writer's name as opposed to **AKA Knotted Rope Canyon**--which has nothing to do with the canyon itself, its geography or the mining history at the head of the drainage. Read more below.

To get there, drive along **Highway 24** north of Hanksville to **Temple Junction** at **mile post 136**, and turn west onto the paved **Temple Mtn. Road**. Head northwest past Temple Mountain and **Flat Top** to the center of the Swell, then turn south at **Tan Seep** (Km 31.1/Mile 19.3), and head for the Hidden Splendor airstrip 63 kms (39.1 miles) from Highway 24.

If you're coming from the north and **I-70**, leave the freeway at **mile post & Exit 131**, and drive southwest past **Tan Seep** (Km 22.7/Mile 14.1) to the **Hidden Splendor**, a total of 54.6 kms (33.9 miles) from I-70. See these routes on the **Area Map, San Rafael Swell, page 31**. All roads to the mine are graded and excellent for any car, except right after heavy rains.

If you plan to use the **Muddy Creek Gorge** as a return route, drive as far as you can along the mining road leading into the gorge. Higher clearance cars can usually make it to the **miner's bunkhouse** (use a shovel or throw rocks off the road occasionally), a distance of 2.4 km (1.5 miles) from the airstrip as shown on this map. Leave one car or mtn. bike there, then drive back up the main road from the airstrip about 3.5 kms (2.2 miles), and turn southeast onto a fairly good track with one or 2 rough spots (LCV drivers take a shovel). About 1 km from the main road, you'll come to a **1950 Ford Coupe**; turn right or southwest, and go as far as you can and park and/or camp. This will be the starting point for both hikes.

If you plan to return via **Ramp Canyon**, it's possible to drive a second vehicle about 1 km southeast from the 1950 Ford Coupe to a parking place marked 1630m right at the head of that canyon.

Rating For the normal routes through both canyons, **3B II or III** (sometimes an **A**). Or **4B R III** for the usually-bypassed **Technical Keeper Pothole Section (TKPS)** in Quandary. These canyons are short & sweet and gives hikers a good look at canyons with potholes.

Equipment To do the normal route in **Quandary Canyon**: RG, one 50m rope, W&R/QL, K&EP, compass (for every hike!) & drybags. If doing the **TKPS** in **Quandary**, add a *happy hooker* (Tom Jones--canyoneering USA.com), which is a large Ibis hook bolted/taped to the end of an avalanche or tent pole, plus etrier (or lots of extra webbing), and/or a hooking kit which includes a G-pick, among other items. Also, be prepared to **pack/bag toss** or do **shoulder stands** to get out of at least **3 KPH**. For **Miners Hollow**: RG (?), one 15m rope, W&R/QL, drybags and K&EP. In spring or fall, take a wet/drysuit for both canyons.

Route Quandary Canyon From where you park near the 1950 Ford Coupe, walk southwest up the old washed-out mining track as it winds its way up to a notch & **pass** in the big Wingate Sandstone wall marked **1767m**. Beyond the notch, and in about 1956, a road was blasted down through the upper end of Quandary Canyon to the drill hole site shown. Since then, floods have reamed out the canyon and there's now two, 4m dropoffs. At the first, hang a rope around a rock outcrop and **rap/HL** down. At the second, you should find a heavy-duty mining timber about 3m long. Wedge it into the rocks to form a safe anchor, then **rappel 4m**.

Beyond, you'll pass through an open area (with an escape route to the northwest), a drill hole pipe, 2 natural bridges, then the narrows. When you come to a pothole & dropoff, you can retreat 15-20m and climb up the right side (looking downcanyon) of the drainage over steep slickrock to a bench. Shortly after, reenter the canyon bottom, then comes a 75m-long section with a number of potholes with perhaps some wading. If this isn't to your liking, scramble up on a bench on the right and walk about 100m. Then descend a short steep pitch, to a second level where you'll see a small arch in the wall on the right, and W&R/QL. **Rappel** from there about **18m** to the canyon floor.

Or, if you choose to do the direct route down the canyon, instead of the bypass just suggested, you'll find on the left above the big pothole, a tiny arch & W&R/QL for **R3**, a **14m rap/HL**--or for some perhaps a steep DC. Next will be some DC'ing through a nice 75m slot with likely some wading. After that, you'll be at the bottom of the 18m rappel mentioned above.

Shortly after the 18m rappel (or direct route), you'll come to 2 shallow potholes then a drop into the big **1st KPH**. To continue down the main canyon through the **TKPS**, you'll need the equipment mentioned above and someone with the experience to use it (see this books Introduction & Equipment), and do pack or bag tossing. Don't attempt this unless you're prepared to get out of some very deep keeper potholes. **This part is for experienced & well-equipped canyoneers only!**

In the **TKPS** (see cross-section, page 52), **R4** will likely be from a stone with W&R/QL buried in sand and into a **3 1/2m-deep KPH** with vertical walls. As of 2002, everyone either tossed a loaded pack or two; or 2 or 3 toss bags full of sand--over the opposite lip before getting into the KPH, then used the rope attached to the packs/bags to get one person out. Or, they used a **happy hooker** to hook into a groove on the lower lip of the KPH, then with a couple of attached etriers, climbed out (if it's full of water just swim across).

Further downcanyon, use an Ibis hook to do **R5** from a small natural hole on the right under a tiny alcove. **R6** is from one piton on the right and into the **2nd KPH**. Before rappelling, throw one pack/toss bag (attached to a rope, right!) over the other side of the pothole so one person can pull him/herself out. Then everybody **raps 14m** down the other side of the **2nd KPH**. DC to reach **R7**, which is W&R/QL around an arch inside the **3rd KPH**. Before going in, do another pack toss to the other side, then use that rope to pull one person out, while everyone uses that W&R/QL to rap down the other side. **R8** is from one piton on the right and down **13m**. After that, climb right, then left and join others doing the **normal route** down through the boulders.

If you want to do the **normal route** around the TKPS, regress about 40m from the **1st 3 1/2m-deep KPH** and climb up a slickrock gully heading east. Stay high on the left or northeast side of the canyon for about 400m, then descend steep slickrock, and finally over & under boulders (joining the TKPS route) down to the west and the canyon bottom. From there, walk downcanyon past several potholes, then use a trail at the Navajo-Carmel contact until you come to a seep, trees and the end of the canyon.

From there, route-find northeast about 2 kms and hike up **Ramp Canyon** back to your car; or head south to **Muddy Creek** and walk up through the gorge and back to your car at the **miner's bunkhouse**.

Miners Hollow The first part of this hike is the same as going to Quandary Canyon, but just after you pass through the notch at 1767m, and enter a small valley, look straight ahead (as the canyon veers left), to see a steep gully called **Knotted Rope Crack** with a big chokestone at the top. Climb up an emerging trail, past one chokestone, and under a second. It's steep, but there are lots of handholds. Beyond this crack, route-find southwest across the top of the San Rafael Reef near the rim, as shown on this map.

At the head of this drainage, are the remains of tent cabins, drill holes, pipe for pumping Muddy Creek water up, a winch site, and telefon wire left behind by the Atlas Corp. when they abandoned uranium mining & core

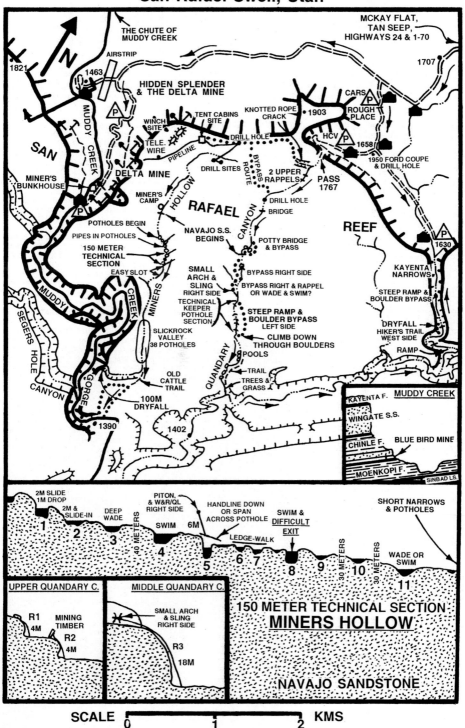

SCALE 0   1   2 KMS

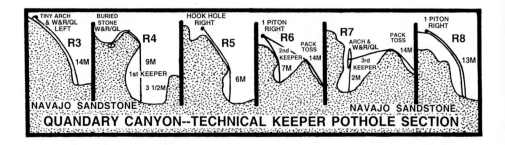

TINY ARCH & W&R/QL LEFT — BURIED STONE W&R/QL — HOOK HOLE RIGHT — 1 PITON RIGHT — 1 PITON RIGHT

R3 14M   R4 9M 1st KEEPER 3 1/2M   R5 6M   R6 2nd KEEPER 7M PACK TOSS 14M   R7 ARCH & W&R/QL 3rd KEEPER 2M PACK TOSS 14M   R8 13M

NAVAJO SANDSTONE   NAVAJO SANDSTONE

QUANDARY CANYON--TECHNICAL KEEPER POTHOLE SECTION

drilling in about 1956--thus the name of this canyon. All the clutter you see on top was ferried up by helicopter. Exploring these sites is interesting and an easy half-day hike with great views in all directions.

From the tent cabins site, head south downcanyon--using your *compass*. In 20 minutes, you'll come to slickrock and potholes. This will be the beginning of a challenging 300m section. Bypass the first group of PH's on some narrow ledges to the left, then you'll come to a 150m section with several deep PH's. The first 2 are 3m-deep KPH's; pass these on the right with tiny ledges, perhaps with a belay (?). Be careful though, in 2007, a woman was being belayed around the 1st pothole, slipped and broke an arm. She later drowned due to hypothermia. If you have dry shoes, climbing around this pothole isn't too bad. When that KPH is low on water, you can get someone out with a shoulder stand according to Scott Patterson. Best not to do this canyon alone! After that, DC over a small PH and into a second. Next is a tight slot with PH's. Chimney up 3-4m and over a very narrow place, more wading, then comes the more challenging 150m technical (for some, the word **"technical" can be removed**) section shown on the cross-section.

PH 1 is a deep slide-in; a taller person can easily get out the other side. **PH 4** may be a swim; then someone has placed a piton & sling on the right with W&R/QL to help walk along a ledge and rappel over **PH 5**-- but anyone worth their salt can do this without the anchor or rope. **PH 8** could be a problem for those not prepared. If full, swim across. If the water is low, do a **partner assist/shoulder stand** to get out.

After **PH 8**, it's easy. Route-find back & fourth down the **Slickrock Valley,** then when the main drainage turns right, walk straight ahead up an **old cattle trail** (few people have ropes long enough to make the near 100m rappel over a dryfall into a pool at the bottom of Miners Hollow) and down the slope to Muddy Creek and back to your car near the old miner's bunkhouse.

Elevations Hidden Splendor (Delta Mine) airstrip, 1463m; lower end of gorge, about 1390m; pass at head of Quandary Canyon, 1767m.

Time Needed Hiking down **Quandary** and returning via Ramp Canyon can be completed in from 5-10 hours, depending on group size, which route, fitness and/or abilities. To go through **Miners Hollow** and return via Muddy Creek to a second car left near the miner's bunkhouse can take 6-10 hours.

Water There are lots in potholes, but carry water from a culinary source in your car and your pack.

Maps USGS or BLM map San Rafael Desert (1:100,000) for driving; and Hunt Draw (1:24,000--7 1/2' quad) for hiking; or the plastic Trails Illustrated/National Geographic map, San Rafael Swell (1:90,000) for driving.

Flash Flood Danger Both are very short drainages, so danger exists only if it rains hard right on you or immediately upcanyon.

Best Time to Hike Late spring through early fall. For the **Technical Keeper Pothole Section** in Quandary, the ideal time would be soon after rains when potholes are full. Take wet/drysuits if you go in spring or fall.

Author's Experience The author and Steve Tyler did the normal route down Quandary and up Ramp in 5 3/4 hours. Later, and on the second try, they went through Miners Hollow, returning to a 2nd car at the miner's bunkhouse in 8 1/4 hours. On Sunday, 7/14/2002, with 44°C temps in Hanksville, the author and Tom Jones did the TKPS of Quandary in 5 1/4 hours from the trailhead to a 2nd car at the miner's bunkhouse. The canyon was bone dry at the time.

Opposite Page Richard Patterson climbs around the first KPH in **Miner's Hollow**. In 2007, a woman fell into this pothole and broke an arm. Her husband couldn't get her out and she eventually died of hypothermia. (Scott Patterson foto)

Above Steve Tyler spans across a pothole in upper **Miner's Hollow**.

Left Tom Jones climbing out of the 1st big KPH in the **Technical Keeper Pothole Section** of **Quandary Canyon**. He's using a Happy Hooker--an Ibis hook attached to a sectioned avalanche/tent pole, plus an etrier. The hook is set in a groove on the lip, then you climb up the etrier.

Segers Hole, Little Segers Hole & Cable Canyons, San Rafael Swell, Utah

Location & Access These 3 canyons are located at the southern end of the San Rafael Swell and in the area known as the **Moroni Slopes**. Here you'll find big deep potholes, some of which may be keepers.

One of the best ways to approach **Segers Hole (SHC)** (AKA The Squeeze) **& Little Segers Hole Canyons (LSHC)** (in the San Rafael book the author calls this the South Fork of Segers Hole Canyon and Tom Jones calls it Segers Window Canyon) is to drive to **Hidden Splendor** and the **Delta Mine**. To get there, drive along **Highway 24** north of Hanksville to **Temple Junction** at **new mile post 136**, and turn west onto the **paved Temple Mtn. Road**. Head northwest past Temple Mountain and Flat Top to the center of the Swell, then turn south near **Tan Seep** at Km 31.1 (Mile 19.3). Head south to the Hidden Splendor airstrip, another 31.9 kms (19.8 miles). This is 63 kms (39.1 miles) from Temple Junction & Highway 24.

Or, if coming from the north and I-70, leave the freeway at **Exit 131**, and drive southwest to **Tan Seep** at Km 22.7/Mile 14.1, then continue south to the Hidden Splendor airstrip, a total of 54.6 kms (33.9 miles). See all these routes on the **Area Map, San Rafael Swell, page 31**, and **Map 8 showing Quandary Canyon & Miners Hollow**. All roads to the mine are graded and excellent for cars--except right after heavy rains.

From the Hidden Splendor airstrip, drive southeast as far as you can along the road leading into the **Muddy Creek Gorge**. Higher clearance cars can usually make it to the **miner's bunkhouse**, if drivers throw rocks off the road occasionally. From the airstrip this is a distance of 2.4 km (1.5 miles) as shown on **Map 8**. This may be the best starting point for the **SHC & LSHC hikes**.

One long way to upper **Cable Canyon (CC) (and SHC & LSHC)** is to drive along Interstate 70 to the area due south of **Emery**, which is roughly halfway between Green River and Salina. See the **Area Map** again. Leave the freeway at **new mile posts & Exits 91**, 99 or 108, and head south as shown. **Exits 91** (best) & 99 have the best roads to the corral not far north of **Mussentuchit Flat**. Drive past **Hebes Mtn.**, between the **2 Cedar Mtns.**, past the **Last Chance Well** and toward the **Moroni Slopes Catchment** and **Horse Heaven Reservoir (HHR)**. If you have a 2WD, park at the 2012m car-park (6602' on *The Frying Pan quad*). This is 60.7 kms (37.7 miles) from I-70 via **Exit 91** and taking the better road running past the **Last Chance Well**. We'll call this the **northern route**. Or it's 55.6 kms (34.55 miles) from I-70 if using the same exit, but the route down **Carlyle Wash**, which is slightly rougher, at least part of the time. From this one trailhead, you can enter the **North Fork of CC** (or an optional much longer route into **SHC** or **LSHC**).

Or, for a shorter hike down **Cable**, turn southeast about 75m beyond or north of HHR (which is 58.8 kms/36.5 miles from I-70 & Exit 91 using the normal route), and park 2WD's at the bottom of a hill about 300m from the HHR; or with a 4WD, continue about 2 more kms (1.2 miles) along a flat bench and park at about 1951m, or nearby. Parking there will allow you to find a way into the upper end of the **South Fork of Cable**, which will make a shorter descent and quicker return to your car.

In **dry conditions**, the **BEST ACCESS ROUTE** to the technical sections of **all 3 canyons** would be to drive east from Hanksville on Highway 24. Using the *Hanksville & San Rafael Desert metric maps*, turn north from between **new mile posts 105 & 106** and drive past **Factory Butte** and the turnoff to **Factory Butte Coal Mine**. At Km 18.2 (Mile 11.3) is a **Y junction**; turn left or west and drive down a graded road over gray clay beds to **Salt Wash** 21.3 kms (13.2 miles) from Highway 24. At the bottom, park & camp on the east side of Salt Wash and across from the mouth of **Cable Canyon**. In dry conditions, a car can be driven to Salt Wash. Or, from the **Y junction**, turn right or northeast and continue to the canyon bottom and **Muddy Creek** (Km 22.6/Mile 14 from Highway 24). This seems to be the best place to start if doing **Segers Hole & Little Segers Hole Canyons**. Any car can get here too, but only in dry conditions. **When wet, 4WD's can't make it to either trailhead!** But it's dry most of the time.

Rating For **SHC: 4B IV, CC: 3B IV** in a worst case scenario, and for those less-prepared, both might jump to **4B R IV** because of possible KPH's. **LSHC** ranks about **3A or B III**.

Equipment For all canyons, RG (ropes--for **SHC** one 60m, or two 50's, or a 30m & pull cord; for **LSHC** two 60's, or a 60m plus several shorter ropes tied together to equal 60m for your pull rope; in **CC** one 50m), one or more 15m ropes for short raps, BK & several B&H, W&R/QL, hooking kit & G-pick (likely not needed, but....), Ibis hook, drybags, K&EP and if doing these in cooler weather, a wet/drysuit.

Route Segers Hole Canyon You could begin at the **2012m trailhead**. Walk eastward over the rim on a cairned route, then go around the obvious small butte in the notch on its south side. On the east side of the butte, route-find northward down steep slickrock into Segers Hole and to the old washed-out road shown. Road-walk southeast to the road's end at 1768m. From there, walk around the corner to the left to an inclined bench and head into the canyon above the drainage bottom. After about 1 km, you'll drop into the dry creek bed to the left and into a nice slot. This part lasts about 1 km with a few easy DC's over chokestones.

After a ways, you should come to a big log. Crawl under it, then after 100m you'll see a bench on the right or south. This is an escape route for those not going all the way downcanyon through the technical part. For a description of this exit route, see the author's other book, *Hiking and Exploring Utah's San Rafael Swell*.

Let's stop here for a moment. Perhaps the **BEST PLACE TO START HIKING** into **SHC** (and **LSHC**) is at **Hidden Splendor** and the **Delta Mine**, or rather down near the **miner's bunkhouse** at the beginning of the **Muddy Creek Gorge** (the route from the Factory Butte side is discussed below). From somewhere near the old miner's bunkhouse, simple walk down the creek and through this spectacular canyon. Further along, and as you emerge from the gorge, make a hard right turn and locate an ATV/4WD track heading southeast. About 100m from the creek, and as you pass **2 half-dead cottonwood trees**, look to the right again to see an old **cow trail** zig zagging up the steep slope. This is now becoming a pretty good hiker's trail. Follow it up the **Moroni Slopes**. After about 1 km, this emerging hiker's trail seems to veer to the right or north. Regardless of the trail, head for the elevation marker **5729T** (1746m). Somewhere on that rim, route-find down into a large half-bowl-shaped feature and to the canyon bottom near where the log is, as mentioned above. In this part, you should be able to follow a big horn sheep/hiker's trail-of-sorts, or perhaps stone cairns.

Now back in the canyon again. You'll crawl under another chokestone, then DC into a pothole. After another 300m or so, is what used to be the 1st rappel. Someone has removed a B&H, so just slide in.

This canyon can have up to 16 rappels (7 or 8, or less, if potholes are full of water) and as many potholes to swim. Some of these are very deep and if water levels are just right (or wrong!), some could be difficult to get out of. This is where you may need a **hooking kit & G-pick**, or use a **pack/toss bag** or the time honored **partner assist/shoulder stand method** to get out. The technical section is only about 1 1/2 kms long, but it's one pothole after another and the going is slow.

One of the first groups through (first time was in the mid-1980's?) this canyon bolting it up, placing B&H's high on one side or the other, and out of the way of most floods. But some could be washed out or damaged, or become loose--so a **BK** could be handy; or in an emergency, use a **G-pick & Ibis hook**. Also, there are 5

Map 10, Segers Hole, Little Segers Hole & Cable Canyons, San Rafael Swell, Utah

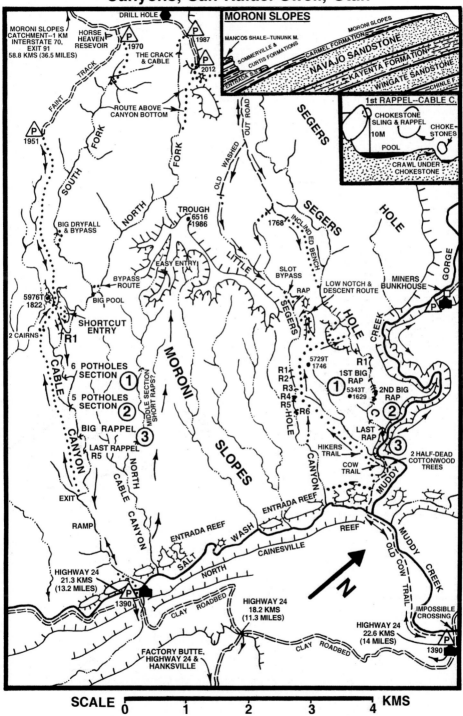

or 6 jump-ins/slide-ins of 1-2m into pools, which could be risky if water levels are low, so in some situations you could add to the 16 raps by rigging up Ibis hook raps from a single rope. **Go prepared for anything!**

Along the way, you'll find **2 deep KPH-type potholes,** each having a B&H on either side. On the author's trip with brother Gail, one pothole was equipped with a rope between the 2 B&H's; the other was not. That one took a lot of energy to get out of, but using a partner assist, it's easy. Getting out of the 2nd KPH, is more difficult; use a pack toss, partner assist, hooking kit, or maybe Tom's happy hooker (a simple tent pole with Ibis hook taped to one end to reach the B&H or ledge, plus an etrier).

Now a brief route description. The first rappels are easy, then some jump-ins/slide-ins. Then the first of 2 bridges; and shortly the 2 deep KPH (on our 2nd trip with all potholes full, we didn't even see the 2nd one!). **R10** is the first of 2 raps where you'll need two 50m ropes (or one 60m). Here you'll drop into one pool, walk/swim across and make a 2nd rap from the same anchor. These 2 together are about 30m. Further along will be **Rappel 12.** This has 2 B&H on the right and you rappel into a very large & deep pool, bigger than some city water tanks. This is where you may need a partner assist/shoulder stand to get out.

Immediately below that is another rap into another large swimming pool, then you'll have some regular walking for a change. In the last km or so, you'll have 3 short raps or slides-ins. The last of these is from 2 B&H high on the right wall. Use it to descend over one pool and land next to the very last rap, perhaps **R16.** This one has at least 3 pieces of W&R/QL wrapped around a big chokestone for a **15m** rappel to a ledge next to a pool & spring.

Fill up with water at this pool, but filter or purify it first because of all the cow poop around. From the pool, walk **200m to Muddy Creek,** turn left, and walk upcanyon back toward the **Hidden Splendor or miner's bunkhouse** and your car. Or walk south, then eastward along the south side of **Muddy Creek** and back to your car at the end of the road coming down from **Factory Butte & Hanksville.**

Or if you're coming from the top end of the canyon, turn right and walk completely out of the gorge, then turn right or southwest and follow the **cow or hiker's trail** back up the **Moroni Slopes** as described earlier. Higher up the Slopes, you'll come to where the sloping mesa pinches down to a ridge between SHC & LSHC. At that point you'll see an emerging hiker's & big horn sheep trail heading down to a low notch or divide. Once there, veer left or south and following cairns, make your way **down into LSHC.** At the bottom, walk upcanyon a ways until you come to a slot and maybe a pool. From there backup a ways and look for a **slot bypass** to the south and up to a bench. There should be some cairns & footprints showing the way up around the short slot. Once you've bypassed the slot and return to the dry creek bottom, continue upcanyon and back to your car at the trailhead marked 2012m.

Little Segers Hole Canyon If beginning at the **miner's bunkhouse** below the **Hidden Splendor & Delta Mine,** walk downcanyon through the gorge as described above, then make a right turn and follow the newly emerging hiker's trail up the Moroni Slopes (what might be an even **better starting point** is the one on **Muddy Creek**--the one you get to from Hanksville & Highway 24). About halfway up, instead of veering right to enter SHC, continue upslope just about due west. Two kms from the bottom, you'll come to the rim of LSHC, and will have a peek at what you're in for. Continue northwest along the rim, then when you come to where the Slopes pinch down to a narrow ridge, follow a hiker's trail & cairns west down to a notch, then turn south, again following cairns, and into the bottom of the upper part of LSHC.

Proceed downcanyon. After about 1 km, you'll come to **R1.** You can bypass this high to the left, but if you want the whole canyon, might as well use the W&R/QL from a small arch. About 50m below this is **R2** which begins on a bench to the right. There you'll find W&R/QL around a large bush to do a **28m rap** down a low angle slope that could be a steep & risky DC for some (?).

After a ways, will be a dropoff and another sling wrapped around an arch/chokestone **(R3)** on the right. Not far below that will be a big pothole--possibly in the **keeper category?.** If this one is bone dry, you can slide in, then climb out on the left side. Or you may have to do a pack toss; or possibly a shoulder stand/partner assist, to get out. On the far side of this pothole and to the left, is W&R/QL around several chokestones and a **12m** rap to the bottom of **R4.** Not far below that is **R5,** a **10m** rap from a chokestone on the right.

R6 is a big 2-part rap from a chokestone in the bottom of the dry creek bed. A **long W&R/QL** extends up through an **arch, or window** as Tom Jones calls it (you may have to reinstall the W&R/QL). From inside the arch, you rap down about **40m** into a huge **near-keeper pothole**--but don't pull your rope yet! Help each other out of this pothole, then continue the rappel another **12m** making the total rap about **52m.** From there walk downcanyon to near Salt Creek, then north over a low divide to Muddy Creek and back to your vehicle, either at the **miner's bunkhouse** or east along **Muddy Creek** and the road coming from **Factory Butte.**

You can also begin this hike from the car-park at 2012m, but getting there involves a longer drive on dirt roads, plus the hike itself is longer. If you do the canyon from the top of the Moroni Slopes, then about 500m after the last rappel, look left or north for a route out and up to the top of the Navajo & Carmel, and the Moroni Slopes. From there, simply walk northwest upslope to where you can descend back into the upper part of LSHC and return to your car the same way you entered.

Cable Canyon The traditional entry method has been right down The Crack at the upper end of the **North Fork,** which begins about 150m from the car-park marked 1987m. This crack was created by a black volcanic dike and has 2 short dropoffs you must rappel over. One has a **cable** attached, thus the name of the canyon. However, an easier and faster way in--if you want to do the entire canyon, is to begin at the upper parking place marked 2012m (**6602'** on *The Frying Pan* quad). There's an emerging trail running southeast over the rim and down some steep slickrock. You'll have to stay high above the dry creek bed until the slope eases up, as shown. This is an easy way, but requires route-finding.

The North Fork is an easy walk with a couple of bypasses around dropoffs. When you finally reach the **confluence of the North & South Forks** with a 7m dropoff into a **big pool,** regress about 125m, and walk southwest up through a crack. This puts you into the lower end of the South Fork where you turn downcanyon to reach the same big pool. For hikers without ropes and rappelling experience, this is where you stop. You can return the way you came, or head up South Fork.

Now the **South Fork of Cable** from the top. If you intend to do the entire canyon, it's best to use this drainage. You could park and begin at or near **Horse Heaven Reservoir,** but you can shorten the hike considerably by driving along a very faint & rough track which heads almost due south from the reservoir. Cars driven with care might make it to the end of the track at 1951m altitude, but a couple of hiking/starting points begin 600-800m before the end of the track. There are several places you can enter the upper end of South Fork, but you'll have to route-find in. Consider checking this out the afternoon before your hike.

There's not much to see in the upper South Fork, but 2/3's the way down, you'll come to a big dryfall & bypass on the left over steep slickrock and reentering the bottom about 250m below the dryfall. Below that, you'll walk through a pretty good slot for about 1 km before reaching the big pool mentioned above. Much of the time you can wade this pool, but it can be a swimmer.

For the first 3 kms below the confluence pool, there are some pretty good narrows and several chokestone

Cross-Sections for Cable and Segers Hole Canyons

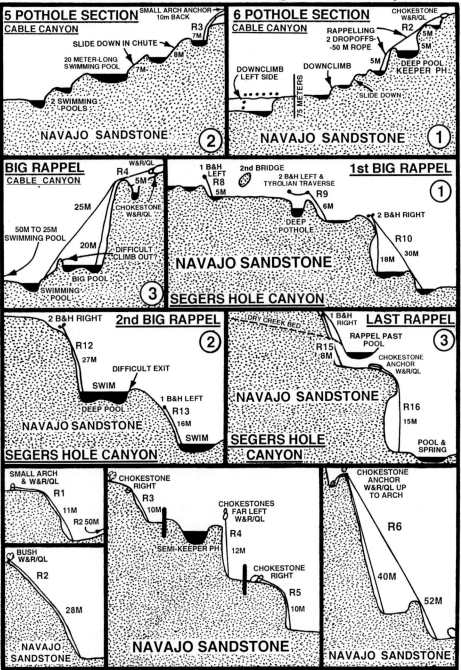

LITTLE SEGERS HOLE CANYON--RAPPELS 1-6

& DC's where you may need a short rope or a partner assist to get over safely. Let's stop here for now and describe what is for most people the **best and/or shortest way** into the **technical part of Cable Canyon**. To do this **Shortcut Route**, start along the road running down from **Factory Butte to Salt Wash**. From the car-park at 1390m, walk due west into the mouth of lower Cable Canyon. About 2 kms from the Salt Wash Road, look for one of several routes out of the canyon to the west & south, then walk along the south rim up to the point marked **5976T** (1822m) on the *7 1/2' Hunt Draw quad*. At that point, which is at the head of a little side-canyon, walk along an emerging hiker's trail out along the top of the ridge north of the side-canyon. Do a zig zag into that short drainage until you reach the main Cable Canyon.

Now back to the description of Cable. About 100m below the little side-canyon & **Shortcut Route** in, you'll come to the **1st rappel**. This is from a chokestone & W&R/QL on the left--with an awkward start. You'll drop about **8m**, perhaps landing in water. From this point on, you're in the canyon with no going back.

About 3/4's the way through Cable, you'll come to the first of **3 technical sections**. The first part has 6 potholes & dropoffs as shown on the cross-section. The reason for taking the hooking kit is, if the water level is just right (or wrong!), it might be difficult to get out of one **KPH**--the first **Deep Pool**. Upon the author's first trip in 5/1999, everyone needed to be pushed or pulled up out of that pothole. Another option after a **partner assist**, would be to do a **pack toss** from the upper rim over the other side; or possibly use a **hooking kit** to get out (last option). Always go prepared to handle the worst case situation! Rappel down the drop on the far side of the KPH before pulling your rope. DC or slide down the last 2 drops in this section.

The **2nd technical section** has **5 potholes**. Just above the first pothole & rappel is a long piece of W&R/QL tied to a mini arch about 10m back upcanyon. After that, you can easily slide down into 4 more pools, the last 3 of which may be swimmers.

The last technical part has the **Big Rappel**. At the beginning are 2 chokestones, one of which has W&R/QL. From there, DC 5m to the dropoff point to begin a **20m rappel** into a large sand-filled pool. If you've placed a long sling at the top, then one full-length 50m rope will get you down a 3m drop on the other side of the first large pothole and into another pool, as shown on the cross-section. Then a short swim, and 50m below that, a possible 25m-long swimming pool. After another 400m you'll come to the last rappel, which is down **3m** from a chokestone. This ends the technical part. If you started at the top of the canyon, about 500m below the last rappel, turn right or southwest, climb up to the canyon rim, then walk up along the Moroni Slopes parallel to the canyon back to your vehicle. Or if you started on **Salt Wash**, continue downcanyon back to your car.

There are several canyons in between LSHC & Cable, but they're not as deep as the 3 discussed above. However, the one just north of Cable has what appears to be a moderate technical middle section that might be interesting to explore. Let's call it **North Cable**. The author walked up from the bottom about 1/3 of the way, then had to exit to the south and bench-walked up along the Navajo rim past the middle 3rd of the canyon. He could see several short drops, and a number of slide-ins or short jumps, but apparently no big potholes or raps (?). The upper 3rd of the canyon was seen from down in, then he exited north and walked back to the bottom while looking into the next canyon north--not really a great canyon either. Take a 50m rope, RG, lots of W&R/QL, and G-pick & Ibis hook as life insurance.

Elevations Moroni Slopes car-parks, 1951m to 2012m; about 1390m at Muddy Creek & Salt Wash; and 1413m at the miner's bunkhouse below the Hidden Splendor.

Time Needed To do **SHC** from the 2012m car-park, 12-16 hours, or maybe 9-11 hours from the Hidden Splendor miner's bunkhouse, or from Muddy Creek. Do **LSHC** from the 2012m car-park in 10-12 hours, or maybe 7-10 hours from Salt Wash and the miner's bunkhouse. Do **CC** from the 2012m car-park in 12-14 hours; or going down **CC's South Fork** and return to the car-park at or near the 1951m mark, 9-12 hours; or if doing **CC** from Salt Wash, maybe 7-10 hours--**this is the best option**.

Water Take plenty in your car and in your pack. Salt Wash is salty, Muddy Creek usually muddy; or in some cases, dry below the mine, but it always has water in the lower gorge. If doing SHC, you can load up on water at the spring & pool at the bottom of the last rappel, but take a filter or pills.

Maps USGS or BLM maps Hanksville (showing southern approach), Salina (western approach) and San Rafael Desert (1:100,000) for driving; The Frying Pan, Factory Butte & Hunt Draw (1:24,000--7 1/2' quad) for hiking; or the plastic Trails Illustrated/National Geographic map San Rafael Swell (1:90,000) also for driving.

Flash Flood Danger The danger is high in the narrowest parts, but these are short drainages, so you'd have to have a heavy downpour immediately above you to be in trouble.

Best Time to Hike May, June and early July, when the days are long; you'll need all the time you can get in all 3 canyons, especially if doing SHC and North Fork of CC from the top. August is also good, probably with more water in the KPH. September days are getting short; October, shorter! If you go in any canyon during April, May or late September & October, you'll need a wet/drysuit, especially in SHC & CC.

Author's Experience After exploring, and with Scott Patterson, Mike Kempt and Bob Janzen, we did all of CC from the 2012m car-park in 12 2/3 hours. The author and brother Gail, went down SHC, also from the 2012m car-park. It took us 7 1/2 hours from the 1st to the last rappel (technical section), and 14 hours round-trip. Much later, he and Tom Jones did LSHC from the miner's bunkhouse in 7 2/3 hours round-trip. Best to do SHC & LSHC from the Hidden Splendor or lower Muddy Creek side, and CC from Salt Creek. In 8/2006, along with Jon Jasper & Tim Barnhart, we did SHC & CC one & 2 days after a big flood, and swam through everything with no sign of any KPH! We did them from lower Muddy Creek & Salt Wash in 9 1/2 & 7 1/2 hours (CC).

Opposite Page This is the same pothole in **Segers Hole Canyon** as seen above, with Jon Jasper crawling out. When it's full, it's easy to get out of. With a lot less water, more difficult: but it should never be a KPH.

Above Tim Barnhart & Jon Jasper crawling across a deep pothole (also shown on the opposite page) in **Segers Hole Canyon**. If water is low, you can either climb out, or get out with a partner assist.

Left The 2nd bridge in **Segers Hole Canyon**, but it may be mislocated on the cross-section (?).

Above Jon Jasper starts rappel about #10 in **Segers Hole Canyon**. Tim Barnhart is at the bottom. All the fotos from Cable and Segers Hole Canyons were taken a day or two after a flood and all potholes were brim full. We jumped or slid into most pools, and only made 6-8 rappels. In times of low water, you may have up to 16 rappels in Segers Hole Canyon.

Right Tim makes Rappel 2 in **Cable Canyon**. He is rapping into a KPH, but it was full on this day. If water levels are low in that pool, you'll need a partner assist & shoulder stand to get out.

Jon Jasper starts **Rappel 3** in **Cable Canyon**. This hike was done 2 days after a big flood and all pools & potholes were full. We did lots of sliding & jumping into pools, and swimming.

Looking down into the last big pool from the top of **Rappel 4** in lower **Cable Canyon**. We did this hike in mid-August, but because of all the swimming, wetsuits were welcome!

Above Left Steve Tyler raps down the 10m Rappel 5 in **Upper Greasewood Draw**.

Above Right HL'ing down the 2nd half of Rappel 4 in **Upper Greasewood Draw**. Some people can probably DC this part. (Steve Tyler foto)

Right An old drilling rig on the newer **Shortcut Route** to the **North Fork of Iron Wash**. (AJ Pastula foto)

Above Left The last rappel in in **Segers Hole Canyon**. You rap from a big chokestone and land on a ledge next to a pool & spring. This could be a water supply, but cows get in there in the cooler months of the year. (Jon Jasper foto).

Above Right Jon Jasper & Tim Barnhart at the beginning of Rappel 3 (?) in **Cable Canyon**. Notice the rope is tied to W&R/QL which is attached to a small natural arch to the upper left in this foto. The arch is about 10m away from the dropoff (see upper foto, page 61).

Left MRK makes the last Big Rappel in **Cable Canyon**. This is a bit of trick fotography by Jon Jasper. Two fotos were stitched together using special software.

Above Left Part of the super neat slot just below the confluence of the **Middle & East Forks of Bull Canyon**.

Above Right Steve Tyler examines the W&R/QL at the next-to-last rappel in the **Main East Fork of White Roost Canyon**. This webbing is tied to a very small arch in the bottom of this pothole. This same pothole has water at least half the time.

Right The 7m rappel in the lower end of the **East Fork of Blue John Canyon**. In 2006, this rope was left attached to 1 B&H seen above. If it's still there when you arrive, please leave it; it should last a long time.

Map 11A, Area Map--Robbers Roost & Spur County, Utah

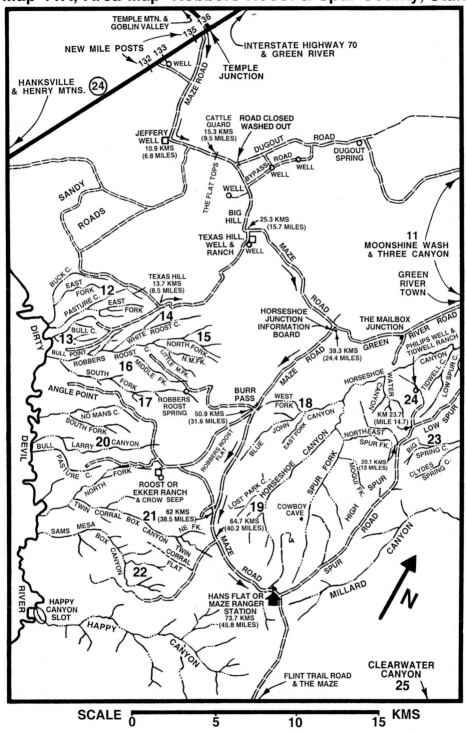

TEMPLE MTN. &
GOBLIN VALLEY
136
135
NEW MILE POSTS
132 133
WELL

INTERSTATE HIGHWAY 70
& GREEN RIVER

TEMPLE
JUNCTION

HANKSVILLE
& HENRY MTNS. (24)

MAZE ROAD

CATTLE
GUARD
15.3 KMS
(9.5 MILES)

ROAD CLOSED
WASHED OUT

JEFFERY
WELL
10.9 KMS
(6.8 MILES)

DUGOUT ROAD
BYPASS ROAD

DUGOUT
SPRING

WELL
WELL

THE FLAT TOPS

WELL

BIG
HILL
25.3 KMS
(15.7 MILES)

TEXAS HILL,
WELL &
RANCH
WELL

MAZE

ROAD

11
MOONSHINE WASH
& THREE CANYON

GREEN
RIVER
TOWN

SANDY

ROADS

BUCK C.
EAST
FORK
PASTURE C.
EAST
FORK

TEXAS HILL
13.7 KMS
(8.5 MILES)

12

BULL C.

13

14

WHITE ROOST C.

15

NORTH FORK
N.M.FK.

LITTLE M.FK.

HORSESHOE
JUNCTION
INFORMATION
BOARD

39.3 KMS
(24.4 MILES)

THE MAILBOX
JUNCTION

GREEN

RIVER ROAD

PHILIPS WELL &
TIDWELL RANCH

DIRTY

BULL POINT
ROBBERS ROOST C.

16

SOUTH
FORK

MIDDLE FK.

ANGLE POINT

17

ROBBERS
ROOST
SPRING
50.9 KMS
(31.6 MILES)

BURR
PASS

WEST
FORK

18

BLUE

JOHN

CANYON

EAST FORK

HORSESHOE

CANYON

WATER

CANYON

TIDWELL

CANYON

LOW SPUR C.

24

KM 23.7
(MILE 14.7)

NORTHEAST
SPUR FK.

BIG

LOW SPUR

SPRING C.

23

DEVIL

NO MANS C.

SOUTH FORK

BULL

LARRY

20

CANYON

PASTURE C.

FORK

NORTH

ROOST OR
EKKER RANCH
& CROW SEEP

ROBBERS ROOST FLAT

LOST PARK C.

HORSESHOE CANYON

SPUR FORK

MIQUI FK.

SPUR

HIGH

20.1 KMS
(13 MILES)

CLYDES
SPRING C.

TWIN

CORRAL

BOX CANYON

21

NE FK.

62 KMS
(38.5 MILES)

19

64.7 KMS
(40.2 MILES)

COWBOY
CAVE

ROAD

SPRING C.

CANYON

SAMS MESA

BOX CANYON

22

TWIN
CORRAL
FLAT

MAZE

ROAD

SPUR

MILLARD

ROAD

N

RIVER

HAPPY
CANYON
SLOT

HAPPY

CANYON

HANS FLAT OR
MAZE RANGER
STATION
73.7 KMS
(45.8 MILES)

CLEARWATER
CANYON

25

FLINT TRAIL ROAD
& THE MAZE

SCALE
0
5
10
15
KMS

Moonshine Wash & Three Canyon, North Roost Country, Utah

Location & Access These 2 canyons sit side by side about 40-50 kms (25-30 miles) due south of the town of Green River. They're also immediately south of the lower end of the San Rafael River, and west of the Green River & Junes Bottom. The best way to get there is to begin immediately west of **Ben's Cafe** in the western part of Green River and drive SSE, then south under the freeway at Km 1.4 (mile .9), and finally southwest on the **Airport Road**. About 4.7 kms (2.9 miles) from Ben's, turn south onto what is called the **Green River Road** (GRR), which is graded often by the county.

To enter the upper end of **Moonshine Wash**, drive south on the GRR to Km 48.6/Mile 30.2 (from Green River town & Ben's) to the **3-way Junction** immediately east of **Saucer Basin**. This is marked **4627T** (1410) on the *Moonshine Wash 7 1/2' quad*. From there drive west about 1.1 km (.7 mile) and park beside the road on the west side of Saucer Basin. Parking there allows you access into the best part of **Moonshine**, including an **old historic sheep bridge**. There are no raps in Moonshine, but it has a nice slot with a couple of DC's. (Just as this book goes to press, it was learned that Shane Burrows has found what is likely a shorter route to the upper slot. Approximately 45 kms (28 miles) from Green River, turn west onto a 4WD track labeled **Shortcut Route** on this map. Drive as far as your vehicle can and walk from there.)

To reach the **Northwest** and **North Forks** of **Three Canyon**, drive south on the GRR a total of 42.7 kms (26.5 miles) from Ben's. There you'll see a rough track heading eastward ending at an old drill hole site 2.7 kms (1.7 miles) from the GRR. To enter the **Northwest Fork**, drive only 150m along this track, and park. To enter the **North Fork**, drive this side-road 1.6 kms (1 mile) and park in the dry wash as shown. In dry weather, this side-road is generally good for cars. Or, to enter the **upper (south) end** of **Three Canyon**, park at the **3-way Junction** mentioned above.

You can also leave **Highway 24** north of Hanksville between **new mile posts 135 & 136**, and drive southeast, then at **Horseshoe Junction** with an information board 39.4 kms (24.5 kms) from the highway, veer left or east, then north to reach this same area. The main roads mentioned here are on any Utah State Highway map.

Rating All 3 rap entries to **Three Canyon** are rated **3A II** or **III**. **Moonshine Wash: 2A III**.

Equipment For all 3 rap entries to **Three Canyon**, you'll need RG, a 60m rope (maybe two 50's for the North Fork--read more below), W&R/QL and K&EP. For **Moonshine Wash**, K&EP only.

Route Moonshine Wash From **Saucer Basin**, set your sights/compass on NNW & walk for nearly 2 kms. Once in the shallow drainage of **Moonshine**, continue north downcanyon. After about another 2 kms, you'll come to the beginning of the slot (if you've come in via the **shortcut route**, you'll enter from the east at about this point?). In the slot, you'll find several chokestones & DC's, but anyone should be able to navigate these minor obstacles. Near the end of this km-long slot, look up to see an **old 3m-long sheep bridge** about 30m above. Soon after that, the canyon begins to open. After yet another km is a steep scramble to an exit on the left, or west. You can leave the canyon at that point and rim-walk back to your car, seeing the sheep bridge along the way. The lower or northern 3/4's of the canyon is confined, but not nearly so deep or interesting as the slot just described.

A short side trip, would be to climb up the bottom end of **Moonshine's West Fork** about 600m to find **Moonshine Spring** and **3 cement tanks** once used by Wylie Mecham of Green River for making moonshine whiskey during the prohibition days of the 1920's. There are no cattle trails to it, so it's good drinking water, if you get there.

Three Canyon To enter Three from the upper (south) end, walk northeast from the **3-Way Junction**. After about 500m route-find into the shallow drainage. After another 500m, you'll come to a **29m, 2-stage rappel**. At the top is 1 B&H & piton (baby angle) with W&R/QL. Rappel in and pull your rope (an alternate would be to leave your rope so you can jumar back up after seeing the upper end of the narrows). Walk Downcanyon. After 1 km, water begins to flow, and for the next km or so, you'll find scenery similar to the lower end of Buckskin Gulch. Altogether, water flows off & on about halfway through this canyon. As you near the lower (northern) end of Three, you'll be walking on trails made mostly by boaters or rafters coming up from the Green River.

There are **3 ways to exit** Three Canyon. One route is out of the **Northwest Fork**, but that will be discussed later. The **traditional way** is to walk to the Green River, turn left or north and walk along a narrow band of slickrock just above the waterline. In places, it's now an emerging trail. After about 1 km, and as the river begins to veer east, the bench widens. Get up on the next level and route-find northeast passing 2 shallow drainages in the slickrock. When you reach the 3rd little drainage, turn northward then veer back to the west. There you'll see the remains of an **old horse trail** blasted out of slickrock. Once on top of the slickrock bench, head west & southwest until you come to the old drill hole site. A mtn. bike or 2nd car left there would get you back to the **3-way Junction**.

A 3rd way out would be to walk east from the mouth of Three Canyon along an **old cattle trail** just above the river. This goes to **Junes Bottom**. From there, you can use an old wagon road to get back out as shown on the map. The route is all east of Three Canyon For a complete history of June Marsing's ranch on the river, read this author's other books, **Hiking, Biking and Exploring Canyonlands National Park and Vicinity,** or his **River Guide to Canyonlands** (both are out of print as of 2008, but may be reprinted at some time in the future).

Northwest Fork From where you park, walk south 200m and enter the shallow drainage. Head east downcanyon. Soon you'll come to some DC'ing, then head south to **Rappel 1**, which some very confident canyoneers might be able to DC (?). If you're not that confident, you can tie W&R/QL to a chokestone and rappel from that. Or, the leader could belay everyone down, then DC. The author left part of a rope & R/QL there in 2004.

Just below R1 may be a pool, then after another 125m will be a big dropoff and **R2**. Actually a really good climber could conceivably DC this, but with big time risks! However, about 12m back from the top of the dryfall is a large bush with W&R/QL. Using this with a 60m rope gets you down **29m**, and perhaps into a shallow pool (?).

About 300m below R2, and on the left or north, is an **exit or escape route**. Route-find up the slickrock to find 1 old bolt sticking up; just above that is a 2m-high ledge with moki steps; then after another 20m or so is a 3rd bold sticking up. These old bolts may have been put there by June Marsing who homestead Junes Bottom. If you're planning on using this exit, you could go to the upper bolt beforehand and attach a rope of 25-30m to make it easier getting back up. Or, from this exit walk downcanyon, bypass 2 dropoffs on the right or west, and into the main drainage. Then exit as discussed above.

North Fork From the parking place, walk south down the shallow drainage. After 400m, will be a dropoff of 8m or so. You can downclimb this, but to get back up, you'd need help from a friend, or leave

Map 11, Moonshine Wash & Three Canyon, North Robbers Roost Country, Utah

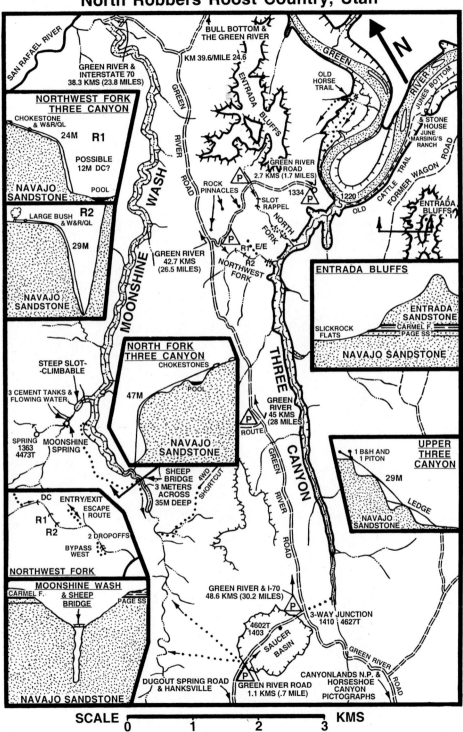

SAN RAFAEL RIVER

BULL BOTTOM & THE GREEN RIVER

KM 39.6/MILE 24.6

GREEN RIVER & INTERSTATE 70
38.3 KMS (23.8 MILES)

GREEN RIVER

OLD HORSE TRAIL

GREEN RIVER

ENTRADA BLUFFS

JUNES BOTTOM

& STONE HOUSE JUNE MARSING'S RANCH

NORTHWEST FORK THREE CANYON

CHOKESTONE & W&R/QL

24M **R1**

POSSIBLE 12M DC?

POOL

NAVAJO SANDSTONE

LARGE BUSH **R2**
& W&R/QL

29M

NAVAJO SANDSTONE

MOONSHINE WASH

GREEN RIVER ROAD

ROCK PINNACLES

P

GREEN RIVER ROAD
2.7 KMS (1.7 MILES)

1334

P

1220

OLD CATTLE TRAIL

FORMER WAGON ROAD

ENTRADA BLUFFS

SLOT RAPPEL

P R1 E/E

NORTH FORK

R2

NORTHWEST FORK

GREEN RIVER
42.7 KMS
(26.5 MILES)

ENTRADA BLUFFS

SLICKROCK FLATS

ENTRADA SANDSTONE

CARMEL F.
PAGE SS

NAVAJO SANDSTONE

STEEP SLOT- CLIMBABLE

3 CEMENT TANKS & FLOWING WATER

SPRING
1363
4473T

MOONSHINE SPRING

NORTH FORK THREE CANYON

CHOKESTONES

47M

POOL

NAVAJO SANDSTONE

THREE

GREEN RIVER
45 KMS
(28 MILES)

P ROUTE

GREEN RIVER ROAD

CANYON

UPPER THREE CANYON

1 B&H AND 1 PITON

29M

LEDGE

NAVAJO SANDSTONE

SHEEP BRIDGE
3 METERS ACROSS
35M DEEP

4WD SHORTCUT

DC ENTRY/EXIT

ESCAPE ROUTE

R1

R2

2 DROPOFFS

BYPASS WEST

NORTHWEST FORK

MOONSHINE WASH & SHEEP BRIDGE

CARMEL F.

PAGE SS

NAVAJO SANDSTONE

GREEN RIVER & I-70
48.6 KMS (30.2 MILES)

P

3-WAY JUNCTION
1410 4627T

4602T
1403

SAUCER BASIN

GREEN RIVER ROAD

DUGOUT SPRING ROAD & HANKSVILLE

P

GREEN RIVER ROAD
1.1 KMS (.7 MILE)

CANYONLANDS N.P. & HORSESHOE CANYON PICTOGRAPHS

GREEN RIVER ROAD

N

SCALE
0   1   2   3   KMS

a rope in case you want to return the same way. Just below this, the slot begins and you'll find several pretty good DC's. You may find up to 4 pools, one of which could be a swimmer (?). Below the last chokestone, it's **47m** to the bottom. This means leaving a 50 or 60m rope in place, to be picked up later; or use two 50's (or a 50m & pull cord) attached to W&R/QL on the last chokestone, and pull the rope from below. If doing a single rope rappel, use a variable speed ATC with a leather glove on your brake hand because part of this is a free rap. Further downcanyon are 2 dryfalls you can skit on one side or the other with a little scrambling.

Elevations Saucer Basin, 1403m; 3-way Junction, 1410m; drill hole, 1334m; Green River, 1220m.

Time Needed To enter at the head of **Three Canyon** and walking out to the drill hole site may take from 8-10 hours. If leaving via the **escape route** in the **Northwest Fork**, this can shorten the hike by a couple of hours. Going down the **North or Northwest Fork** and using the **escape route**, could take as little as 2-3 hours (?). Or, do both in one day. Doing just the upper end of **Moonshine Wash** may take 3-5 hours; add another hour or 2, if you explore the lower end of **Moonshine's West Fork** and the moonshiner's tanks.

Water Take your own water into Moonshine, but if you make it to Moonshine Spring, that's good water--and no cows! Three Canyon has good water (best in the upper end) for half of its length.

Maps USGS or BLM map San Rafael Desert (1:100,000) for driving & orientation; Moonshine Wash & Tenmile Point (1:24,000--7 1/2' quads) for hiking; and a recent Utah State Highway map.

Main Attractions Good slots & water, historic sheep bridge, and 2 deep & narrow fotogenic canyons.

Best Time to Hike Spring through fall. It's cool in these narrow canyons even in summer.

Author's Experience The author was in this area as early as 1990 doing footwork for both of his guides on Canyonlands N.P. and vicinity, then returned in the mid-1990's to finally re-explore all of Moonshine and use a rope to rappel into upper **Three Canyon**. In one day he explored upper Three, then drove to a parking place on the Green River Road and walked to the drill hole site and to the lower end of the canyon. That took about 3 1/2 hours round-trip. Later that same afternoon, he hurried down **Moonshine** to the exit, but then walked back up the slot to his car in Saucer Basin. Round-trip was 2 1/2 hours. In 2004, he rehiked Moonshine from Saucer Basin and the lower West Fork to both springs (Moonshine Spring shown on USGS maps is not the one near the cement tanks), then back in 5 1/3 hours. He also rappelled into the **Northwest Fork**, explored all of Three Canyon to it's upper end (including the bottom part of North Fork) and left via the drill hole site, all in about 7 hours. He rapped to the bottom of **North Fork** and returned. He jumarred back the same way in 2 hours.

The top of Rappel 1 in the **Northwest Fork of Three Canyon**.

Above The sheep bridge over **Moonshine Wash**. The bottom of the slot is about 30m below.

Left The bottom of the only rappel in the **North Fork of Three Canyon**. There are many choke-stones in the slot above to form an anchor.

Buck & Pasture Canyons, Robbers Roost Country, Utah

Location & Access **Buck & Pasture Canyons** are located northeast of Hanksville, north of the Dirty Devil River, and northwest of the main Robbers Roost Canyon complex. To get there, drive along Highway 24 to a point about halfway between Hanksville & I-70. Between **new mile posts 135 & 136**, turn eastward at the sign & road running into the Maze District of Canyonlands National Park. See the *Area Map, Robbers Roost Country* on *page 65*. From Highway 24 drive south on the **Maze Road** 10.9 km (6.8 miles) to the Jeffrey Well; turn east and go between the Flat Tops at Km 15.3/Mile 9.5; continue east & south to the top of **Texas Hill** at Km 25.3/Mile 15.7. At that point look south 550m to see the old **Moore Ranch** (now Vachers) which has one tree, several buildings, corrals & a well.

From the Maze Road proceed south past the old Moore Ranch on a good graded road. After about 13.7 kms (8.5 miles) turn right or west instead of going south into the White Roost Trailheads. Continue west to Km 15.9/Mile 9.9 which is where the road makes a big curve from southwest to northwest. Park there to access the **East Fork of Pasture**. Or continue west to Km 16.6/Mile 10.3 and park somewhere just west of the crossing to the upper main drainage of the main or **West Fork of Pasture Canyon**.

For **Buck Canyon**, continue west to about Km 20/Mile 12.4 (from the Maze Road). One exit option would be to park one car somewhere just before dropping into the upper West Fork; that is if you have 2 vehicles, and if you decide to go down the **West Fork** and exit via the **East Fork**. Read about another option below.

And finally, continue west & southwest to about Km 23/Mile 14.3. There, the road is starting to veer right or west, then north. To the right you may see a cairn just off the road to the north. At that point, turn left or south, and drive 200m on a newly-developing road to the canyon rim. This should be about 200m east of where a fence ends at the canyon rim. It's also 125m west of the entry point for the main or **West Fork of Buck Canyon**. Park/camp there. It's the beginning for all hikes down the West Fork, and also the end of your hike if using the **Alternate Return Route** if you go down the **West Fork** and up the **East Fork** (the author learned of this route possibility 4 days before this book went to press; see the map). The road to both canyons is periodically graded and a little sandy in places, but is normally good for any 2WD.

Rating **Buck Canyon** has 2 forks; the **East Fork** is an interesting walk-up with 2 arches; the main or **West Fork: 3A II or III**. This one has only 1 rappel and 1 steep DC. **Pasture Canyon's West Fork** (it's main drainage); this hike is down to where the technical part (maybe a -PG SLOT?) ends after 4 dropoffs and the canyon opens up. Going down and returning the same way is rated **3A II** (if the upper canyon is choked with tumble weeds, it could be a **3A III**). The **East Fork of Pasture: 3A II or III PG SLOT**, that is, if you come back the same way instead of going all the way through to the exits shown. This fork has one of the better short slots around, and is similar to the main fork of White Roost Canyon.

Equipment RG, K&EP and W&R/QL for all canyons. For **Buck**, one 50m rope and maybe a shorter one to help beginners over one dropoff. For **Pasture's West and East Forks**, one 30m rope, long pants for the West Fork (it could be choked with tumbleweeds!?). In spring, fall or winter, you can likely get by without a wet/dry-suit because if there are pools, they will be small and not very deep.

Route **West Fork of Buck Canyon**. About 125m east of where you park on the rim, is a walk-down route to the bottom. Immediately, you'll be in the Navajo (maybe Page SS?) Sandstone with an easy DC, then walking. After about 450m, and right below where the fence ends at the canyon rim, will be your only **rappel**. This is a drop of about **8m**. If you want to take your rope with you, install a long piece of W&R/QL to 2 boulders just above the dropoff. Or, just tie any rope 30m or longer to the boulders, make a single rope rappel, then at the end of your hike, have one person run down and retrieve the rope. This way you'll carry almost nothing through the lower slot.

From the rappel, you'll have several easy DC's, then a 3m drop. With dry shoes, most people reading this book can DC this, but some may want to be belayed. The last person might HL down from an Ibis hook set in a small natural hole. Less than 100m from there, the bottom drops out. At first it looks like you may want to rappel from a chokestone, but don't; it's an easy but steep DC of about 15m. At the bottom is an undercut, then the canyon opens quickly and soon you'll see cottonwood trees and high walls.

About 3 1/2 kms below the slot, will be your first chance to escape the canyon. When you see a sizable side-canyon coming in from the right or west, turn right into this narrow corridor and walk maybe 50m. From there, climb up on the first **gravel bench** right at the northwest corner of the confluence, then be looking upwards to where the Navajo slickrock comes down to the gravel bench. Climb up right at the corner beginning on the lower right, then left a little, and finally straight up to the northwest. It's a little steep, but nothing serious. This is the shortest & best escape route for those having only one car.

However, if you have a 2nd car or mtn. bike to be used as a shuttle (park one vehicle as suggested above at Km 20/Mile 12.4), you can continue downcanyon about 1 km and turn left or northeast into the lower end of the **East Fork**. At first you'll wade through tall joint grass & willows, then after 125m, you'll see an **arch** on the left. About 600m further along will be the 10m-high **Buck Canyon Arch** on your right. After another km, you'll come to a slot on your left and a 20m dryfall which is the main drainage. However, to the right will be a sand & talus slope and a walk-up **bypass trail** past a fence. Just after that, you reenter the canyon.

Continue upcanyon through a short shallow slot, then the canyon opens up into a wide valley. Right at the head of this valley is a steep walk-up slickrock route on the left. Once out, walk up the broad drainage about 500m, but instead of veering right or east with the dry wash, continue straight north over a low divide and into another very shallow valley. Once in it, head updrainage northeast 2 kms, then veer left or north a little and back to the road marked 1481m and your 2nd vehicle. Have a map & compass.

A **better option** to hiking back to the road and car-park marked 1481m, is to turn left into the little side-drainage in upper **East Fork** and route-find out and NNW back to the starting point or trailhead at 1433m. The author hasn't done this--just learned of it--but it's shorter, and you won't need a car shuttle.

West Fork of Pasture Canyon Going down the middle of the 3 short upper forks of Pasture is easy with some DC'ing into shallow slots. In this upper part, the author swam through a sea of **tumbleweeds**, evidently for lack of a good flood in a couple of years, so take long pants just in case (the floods of 2003 or '04, and 2006 cleaned them out, but they could return)! After about 3 kms, the canyon tightens-up & gets deeper; this will be the beginning of a **150m technical section**.

At the first dropoff, tie one rope to a small chokestone and **rappel** (if you intend to return the same way instead of going all the way through the canyon--otherwise rappel off W&R/QL) down about **6m** (some can DC this, but the top is too wide for most). Walk under some large chokestones, then not far away is **R2/DC** of about **7m**, and more DC'ing. **R3/DC** is only about **3m**. Experienced slot rats can DC all of these, except for maybe the first one. A little further along you may want to leave a 4th rope going down a steep tight slot (only if you plan to regress the same way). Then the slot opens into the typical Navajo Sandstone canyon with high vertical walls. Retreat the same way; or walk to the the west-side exits below the confluence of the 2 forks, and walk back along the rim, or along a very old, unused & sandy track as shown. That track ends 18.75 kms

70

Map 12, Buck & Pasture Canyons, Roost Country, Utah

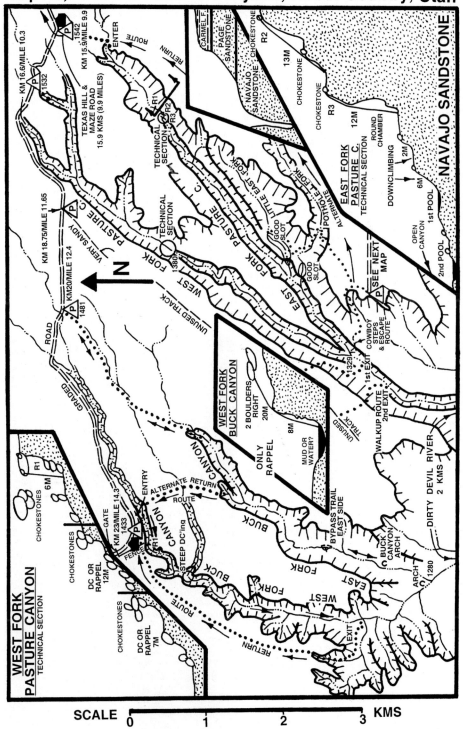

SCALE 0 1 2 3 KMS

(11.65 miles) from The Maze Road.

East Fork of Pasture From the curve in the road & car-park at Km 15.9/Mile 9.9 marked 1542m, head due south about 600m. Follow this map carefully. Enter the main shallow northeast drainage and head downcanyon. You'll soon come to 2 dropoffs about 2 1/2m high. Some have helped each other over these obstacles, but there's an obscure walk-down route on the south side. From the 1st dropoff, climb up to the rim on the south, then 40m downcanyon from the 2nd dropoff, climb down a very minor ridge made of soft clay. When you reach the 1st little rock band, turn west under this ledge and walk along a narrow meter-wide bench. Further along, head down to the bottom of what appears to be the Page Sandstone (until recently, this was referred to as the top of the Navajo Sandstone). From there, head downcanyon.

About 500m into the drainage, the canyon begins to slot-up. In the next 200m or so, you'll have about 4 pretty good DC's, then walking for another 500m with one good DC. Then **Rappel 1** of **8m** from a chokestone and W&R/QL.

About 350m below R1 is **R2**, again from a chokestone and W&R/QL. At the bottom of this **9m rap**, is another short drop of 4m (making it a **13m rap**). Five meters below that is another chokestone and **R3** and W&R/QL. From there you'll drop **11m** into a **round chamber**, then immediately another easy DC. This puts you into a very deep dark slot. From there, you'll come to a narrow place and a tight DC of about 6m. At the bottom of that will likely be a wading pool. Just a few meters beyond that are a couple of DC's, then you can chimney over the last long pool to where the slot opens into the lower canyon.

To get out of the **East Fork**, walk down to a point about 300m below the confluence with the West Fork and climb a wall to the west rated about 5.6; or continue another 500m downcanyon and walk out to the west and north. Observe the map carefully. Once on top, return to the main road via an old unused sandy track. Or if you're good at chimneying, and fit & tough, head up the West Fork and do 4 or 5 pretty good UC's--the last of which will be the most difficult because it's widest at the top. If doing this, attach an Ibis hook to a short rope and throw it up over a chokestone, catching on something, to be used as a security line for the last step.

Still another option is, right at the confluence of the East & West Forks of Pasture, look left or east for a cairn on a low bench. Head that way and route-find east up a series of **cowboy steps** cut with picks. This takes you up to the rim near the end of one road in the **Bull Canyon complex**; see next map. This is the escape route for those doing the **Little East** or **Pothole Forks** as shown.

Elevations Buck: main West Fork Trailhead, 1433m; parking for East Fork entry/exit, 1481m; lower canyon bottom, 1280m. **Pasture:** main West Fork Trailhead,1532m; bottom of technical section, 1380m; East Fork Trailhead, 1542m, confluence of East & West Forks, 1329m.

Time Needed For **Buck Canyon**, 3-4 hours round-trip if returning via the west-side escape route; or 5-6 hours if returning via the East Fork. For **Pasture's West Fork**, to go through the 150m-long technical part and return the same way will take about 4-5 hours--if there are no tumbleweeds to slow you down. To go down the **East Fork** leaving rope and jumar back up from the bottom of the slot, should take about 4 hours. To go down either fork and return via one of the west-side exits below the confluence may take 5-7 hours. About the same amount of time using the **cowboy steps route** out the east side of East Pasture and back to the trailhead at 1542m.

Water None around so take plenty in your car and in your pack. Water from Texas Well is not so good.

Maps USGS or BLM maps Hanksville & San Rafael Desert (1:100,000) for driving; Point of Rocks West, Point of Rocks East, Angel Point & Angel Cove (1:24,000--7 1/2' quads) for hiking.

Flash Flood Danger High in the tight slots but otherwise moderate to low. The East Fork of Pasture is a very short drainage, so your only danger is if a big storm dumps right in this slickrock canyon (there is basically no drainage above the canyon itself).

Best Time to Hike Spring or fall, but it's cool in the slots in summer. Also possible in winter dry spells.

Author's Experience After scouting both forks of **Buck Canyon**, he and Byron Lemay & Tiffany Winget went down the West Fork on 11/2/2002 and found mud knee deep. We returned via the west-side escape route in a total of 3 1/4 hours. On his scouting trip, he went down the East Fork to the confluence and back to his car in 3 2/3 hours.

He went down the middle of the 3 upper forks of **Pasture Canyon's West Fork** alone, and immediately found the shallow slot filled with tumbleweeds. That slowed things down. In the 150m-long technical part, he left 3 ropes at the first 3 dropoffs to insure a retreat. He continued down another 2 kms, before turning around. He ended up climbing all 4 without using the ropes, but for safety, it's best to use them, especially at the 1st one going downcanyon. Round-trip time was 9 hours.

After an afternoon scouting trip of **Pasture's East Fork**, he went alone down to the end of the slot, which included wading 2 pools waist deep. He returned to his car in a total of 4 hours. In 6/2004, with Steve Tyler, we did the East Fork and tried exiting up the West Fork, but floods had changed things a little, so we went up the lowest west-side exit and back to the cars. Round-trip was 8 1/2 hours, but you should go faster.

Opposite Page Steve Tyler DC's in the upper end of the slot in **East Pasture Canyon.**

Above Left Doing **Rappel 1** in the upper end of **East Pasture Canyon.**

Above Right At the very bottom end of of the technical section of **East Pasture Canyon** is this long pool you'll either wade through or chimney/span across.

Left One of many DC's in the upper end of **East Pasture Canyon**.
(Steve Tyler foto)

Above If you don't want to go all the way through the **West Fork of Buck Canyon**, you can leave a rope at the rappel and jumar back up.

Right Spanning a pool in the upper end of **West Fork of Buck Canyon**. This part is between the rappel and the steep DC'ing section. (Steve Tyler fotos)

Above Left Nat Smale chimneys through parts of **Little East Fork of East Pasture**.

Above Right Jeff Webb & NS in the lower end of **West Fork of Bull Canyon**.

Left Ryan Cornia DC'ing in **Pothole Fork of East Pasture Canyon**.

Below Ryan Cornia watches MRK make a fast slide in **Pothole Fork**. (Jon Jasper foto)

Bull & Little Bull Canyons, and Pothole & Little East Forks of East Pasture Canyon, Robbers Roost Country, Utah

Location & Access The **Bull Canyon complex** is located in the northwestern part of the **Robbers Roost Country** and in between Pasture & Robbers Roost Canyons. It got the name because at one time local Hanksville ranchers fenced off the mouth of the canyon and kept their bulls there--until breeding season. Bull Canyon has 5 forks which flow south & southwest into the Dirty Devil River. Also included here is a short unnamed slot called here **Little Bull Canyon (LBC)**, and **2 short side-drainages** to the **East Fork of Pasture Canyon**.

To get there, drive along Highway 24 to a point about halfway between Hanksville and I-70. Between **mile posts 135 & 136**, turn southeastward at the sign & road running into the **Maze District of Canyonlands National Park**. See the *Area Map, Robbers Roost Country on page 65*. From Highway 24, drive south on the **Maze Road** 10.9 km (6.8 miles) to the Jeffrey Well; turn east and go between the Flat Tops & cattle guard at Km 15.3/Mile 9.5; continue east & south to the top of **Texas Hill at Km 25.3/Mile 15.7**. Look south 550m to see the old **Moore Ranch** (now Vachers) with one tree, several buildings, corrals & a well.

From the **Maze Road (Km & Mile 0)** proceed south past the ranch buildings on a good graded road. After about 13.7 kms (8.5 miles), instead of turning right or west toward Buck & Pasture Canyons, continue straight ahead to the southwest on a good but ungraded track. At Km 15/Mile 9.3 will be a flat parking place on the left or east immediately before you start down a steeper section of road. This is about 400m west of elevation 5190T (1582m) on the *Point of Rocks East 7 1/2' quad*. This place, marked 1585m on the White House map, is easily accessible with a 2WD car. Beyond that, the road is sandy in places, so a 4WD/AWD is recommended (required when conditions are dry and after being used a lot!). Continue southwest toward **Bull Point**. At Km 17.7/Mile 11 is the **2nd major junction** below the 1585m car park. To reach the 3 forks of Bull to the west, and the 2 little drainages of **East Pasture Canyon**, veer right and follow this map carefully as there are more roads on the ground than on USGS quads (this map is correct!). At about Km 20/Mile 12.4 (elevation 1487m) stop and park. This is a good place to begin hiking the **Middle Fork**.

To reach the **North** and **West Forks** of **Bull Canyon**, and the **Little East** and **Pothole Forks** of **East Pasture Canyon**, and again following this map carefully, proceed west on the road just north of the Middle Fork's parking place. Soon you'll come to the head of one drainage, then veer right and go down one slanted, sandy slope which may be difficult to get back out of (Km 20.4/Mile 12.7, & 1490m). That short section is definitely for 4WD's. Drive west to the end of that road at **Km 21.6/Mile 13.4 & 1490m**.

Now back to the **2nd major junction**. Continue south down the other track (as you progress southwest down along Bull Point, you'll have 2 tracks criss-crossing each other several times--taking either one should get you there). At about 20.4 kms (12.7 miles) from the Maze Road is one possible place to park if entering the **East Fork**. Or, continue southwest to another **fork in the road (Km 21.6/Mile 13.4)**. Park there if going to **LBC** (to your left or east at that point is a small **low hill**), which is completely east of Bull Canyon. Or veer right (veering left is a **deadend**) and drive to the end of **Bull Point** which is **24.2 kms (15 miles)** from the **Maze Road & Texas Hill**. The road ends at elevation marker **4866T** (1483).

Rating The **Little East (PG SLOT) and Pothole Forks** of **East Pasture** both 3A or B II. The **West Fork** of Bull Canyon: **2A II PG SLOT** (no raps, just fun DC'ing); while the **North Fork** is a walk-through bypassing one big drop, and can be an escape route. Below the confluence of **Middle & East Forks** (AKA Big Bad Ben) down a short distance is rated **3B III PG SLOT**. The **South Fork** of Bull (AKA Chambers): **3A II PG SLOT** (some may rate it a little higher?), or **3B II (?)**. About half the time, there will be shallow water in pools in each fork (except North). **Little Bull Canyon, 3A II or III**.

Equipment For **all slots K&EP**. The **West Fork**, one 15m rope, and possibly an Ibis hook for one short HL (or the leader belay everyone, then DC one 6m pitch)--no packs needed. For **Middle & East Forks** and **Little East & Pothole Forks**, RG, one 30m rope, W&R/QL, drybag or box for cameras, wet/drysuits for fall, winter or spring hikes, smallest of packs, and map & compass as usual. For **South Fork**, only K&EP & camera--it's easier to do this slot without a pack (a 15m rope just in case, and in summer take water)! For **LBC**, RG, one 60m rope & W&R/QL.

Route **West Fork of Bull Canyon**. From the parking place, walk north (or west) and down one of 3 routes into the upper part of the West Fork as shown. Head south downcanyon about 2 kms to find the beginning of a tight 150m-long slot. You can get into the beginning of this slot by doing a full-length body span over the 1st pothole. Below that, it tightens fast so chimney up to where your feet are 1-2m up. After a short distance, get back to the bottom to find a steep 6m DC & slide into a possible shallow pool. To get down this, the leader can belay everyone, then he can DC, & slide the last 2m. Below that, walk under a bridge, then the confluence with an eastern fork (which requires at least one short rap) and another 75m or so of tight slot where you must chimney up 3-4m. To get out of this canyon below the slot, turn right and climb west, then route-find north back to the beginning of the slot. From there, retrace the route back to your car--or head downcanyon and exit via the North Fork.

Little East & Pothole Forks of East Pasture Canyon These 2 short side-drainages are included here because access is easiest from the Bull Canyon side. From the end of the road at the head of West Fork, head northeast about 850m, then walk down a sandy slope a ways (perhaps on a new trail?) and turn east (this will be in the lower part of Pothole Fork). After 125m, route-find down steep slickrock to the bottom. This will be the easiest & fastest way to the best parts of both forks which is their lower ends. Once down in, walk west in the sandy dry wash. At the first dropoff, skirt right and route-find back in from the right or north side and continue west downcanyon. About 200m before the end of **Pothole Fork**, you'll come to the first of several DC's--one is a **fast slide of about 6m** (see foto). Help each other if needed, but there's no need to rappel. Less than 50m before the end, will be the first of **3 potholes**; DC into knee-to-waist-deep water. Right after that is **Pothole 2**; DC this as well, again into shallow water. On the far side of this pothole is a chokestone with W&R/QL and the beginning of a **10m rappel** into the **3rd pothole**. We landed on sand, then waded waist-deep water to reach the far side (see foto). This last pothole seems to be brim full all the time, perhaps seep-fed from the upper potholes.

From this last (3rd) pothole, walk along a bench and into the **East Fork of Pasture Canyon**; then head left or southwest downcanyon. After about 1200m, and right where the **West Fork of Pasture** comes in on the right or west, walk left upon a slickrock bench (look for a cairn on top). Once on top of this bench, zig zag up to the east. In several steeper places will be steps cut with a pick, likely the work of a cowboy, thus the name **Cowboy Steps Trail**. Aim for the protruding **point** sticking out 500m due east of where you started up this trail. Near the top, walk around to the south side of the point to find one of 2 ways up through the last layer of Carmel to the top, then circle around to the east & south back to your car. Since this hike is short & sweet, consider heading south from the point and do the West Fork

Map 13, Bull & Little Bull, and Pothole & Little East Forks of East Pasture Canyon, Roost Country, Utah

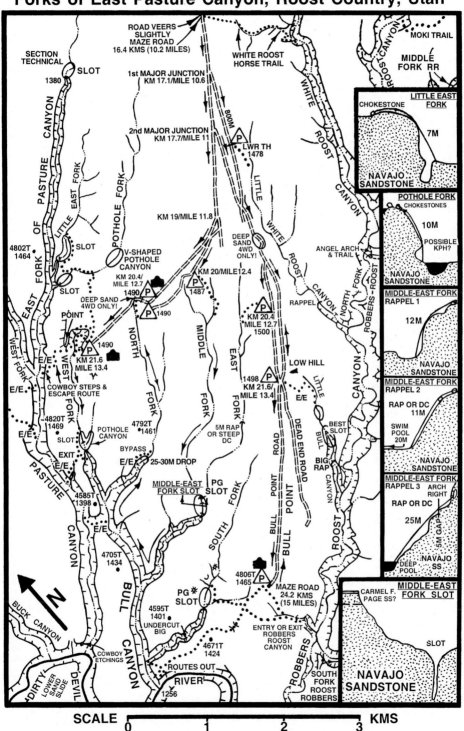

SCALE
0   1   2   3   KMS

of Bull before returning to your vehicle. Besides the Cowboy Steps Trail, there are 2 more exits further down the main Pasture Canyon. Observe map.

To get to the best part of the **Little East Fork**, enter via the lower end of the **Pothole Fork** as described above. Just before Pothole Fork slots-up, climb north up on the bench and circle around to the east on another bench just above the slot of the Little East Fork. From there, head north and finally drop into the last 600m of the slot. At first this slot isn't too deep, but a little tight, then in places you have to walk sideways. About 200m before the end, you'll do some DC'ing into a couple of potholes, then a tight section which gets darker with every turn. Soon you'll come to a round chamber, then it begins to open with more light. After another 40m, is the only rappel in the canyon. We left W&R/QL on a small chokestone in a crack, but that would have disappeared in the next flood. Be prepared to reinstall W&R/QL around a small chokestone; or maybe use an Ibis hook to rappel from. Below that is more low stemming & DC'ing in a tight slot, then the open East Fork of Pasture Canyon. Walk downcanyon and exit the same as described under Pothole Fork.

North Fork This is the easiest (and non-technical) of all the **5 forks of Bull Canyon**. From the parking place at 1490m, walk west and route-find down into the upper part of the drainage. In the upper end are 2 sections with deep potholes--skirt around both of these and continue downcanyon. Not far south, southwest of elevation marker **4792T** (1461m) is a 25-30m dropoff and no slot whatsoever. You can skirt around this on the north side, and re-enter the canyon at one of at least 3 places. Once down in, you can return the same way, walk up and use the West Fork, head over a low divide into lower Pasture, or head down to the Dirty Devil River. Or, use North Fork as an escape from other routes/forks.

Middle Fork From the parking place at 1487m, walk southeast about 400m and into the most-easterly of the 2 upper short forks. Once in the canyon, you'll find a broad open drainage with 4-5 short, shallow slots that don't amount to much. The good part comes **just above** where the **Middle & East Forks** meet. There you'll find some small chokestones where you can place W&R/QL for a 6m drop right at the **confluence**. **Rappel 1** puts you into a dark slot of about 225m. You'll have 2 possible shallow pools to begin, then comes a tight 11m sloping drop. If you have beginners in your group, the leader can belay them, then DC with no problem. Just beyond that is a short DC into a pool that will likely be there all the time. It was a 5m-long swim for the author (or chimney up & over it). Then more slot and an open place and **R2 or steep DC** (?). On the far right side of this open place, is a small arch in the wall where you could place W&R/QL, then rappel. Or, the leader could belay everyone down 8-9m into the last pool; then if the water is deep enough, he could DC and drop a meter or so into water. Or perhaps rap/HL from an Ibis hook (?). Anything is better than leaving a 10m-long sling. After the pool the canyon opens and it's a walkout. You can **exit** up the **North Fork**, or walk out to the end of **Bull Point** using the **South Fork** exit. Or leave several ropes and UC/jumar back out.

East Fork From the East Fork parking at 1500m, walk northwest 300m and route-find into the drainage. For the first 1 1/2 kms, it's boring, then comes a dropoff of about **5m**. You can place W&R/QL around one of several chokestones; or if you're with a small group, it's possible to belay everyone down, then the leader might DC a ways, then be helped by others for the last 2m. From there it's an easy walk with a couple of short, shallow slots down to the confluence with the Middle Fork. To go through the nice slot, walk north and enter the lower half of **Middle Fork**.

South Fork This is easily one of the **neatest, tightest & funniest slots** on the Colorado Plateau. Begin at the end of the **Bull Point Road** marked **4806T** (1465m). Walk northwest down steep slickrock beside a small drainage to about the middle part of the 400m-long slot. Best to do this with no pack, but in hot weather, better take a very small pack with water (and maybe a 15m rope for beginners).

From the middle & open part of these narrows, walk east for about 200m to where you can get into the beginning of the slot. Don't chimney up above; get right down in, and for the most part, stay on the bottom. It's tight and you will have to chimney up 2-3m in several places, but if you're skinny, you can stay on the bottom most of the way. **Bigger guys** will have to **chimney up** more, thus a possible **-R SLOT** rating for some. There may be a shallow pool in the upper dark slot, then after 200m, it opens into a **subway-type narrows or chamber** with regular walking for about 75m. Just past some large boulders, it's DC'ing for a ways, then 6m down a vertical chimney into a pool (it was knee deep for us a week after a big flood--but figure out a way to protect your camera in case it's deeper!). After this pool, it pinches down again with more really tight places. Two of us went up about 4m in one section, but a bigger Jon Jasper (a caver), stayed on the bottom and got through! Just after that, the slot ends abruptly coming out of a crack in the sheer vertical wall with a 2m DC/jump into soft sand.

Downcanyon 400m from the end of the slot, is the **biggest & deepest undercut** in the Roost. About 600m below that, climb left or south up a sandy slope right at the corner of South Fork & the main Bull Canyon and onto a bench, then head east as shown along the escape route.

By placing one vehicle at the end of **Bull Point Road**, it'll save some walking if you're going down the full lengths of **Middle or East Forks**. Another escape or exit possibility is to head up the **North or West Forks** and back to a vehicle somewhere in that area.

Little Bull Canyon From the fork in the road at 1498m, walk southeast 250m to the rim. Look for a route down to the little bowl which is the head of LBC. An **upper slot** begins immediately, it's a walkthrough of 150m. It opens momentarily, then the **Best Slot**. About 50m into this, the bottom drops out and you'll see a chokestone just beyond the drop. Chimney out to it and tie on your 60m rope, then rap 13m, walk 10m, then drop another 10-12m into a very dark chamber--at least one person should have a **headlamp**. The whole drop is about 30m. From there it's a tight fotogenic walk-out of about 400m.

Once out of the slot, return to the head of the Best Slot on the west side and retrieve your rope. From there you can return to your car; or head south to where it drops into the **lower canyon complex**. If you decide to rappel into the lower canyon, going right down the stream channel means a drop of about 57m. However, it should be a little less if you build a deadman anchor to the right side and rap from there (this writer rapped about 7m down the dry slot from an anchor to get a measurement, but didn't do the rappel to the lower canyon). Once in the lower canyon, walk another 500m to the main **Robbers Roost Canyon**. From there, continue southwest downcanyon about 2 kms to where the **South Fork of RR** comes in on the left; from that point right and head north into another little side-canyon. Walk up an old cattle trail to a bench, then northwest and curling around to the east, ending at the trailhead at the end of the road on **Bull Point** and a 2nd vehicle, or road-walk back to the fork in the road and your car.

Elevations TH's: 1490m, 1487m, 1500m, 1465m & 1498m; end of Bull Canyon & Dirty Devil, 1256m.

Time Needed **West Fork** of Bull Canyon may take 2 1/2-3 hours round-trip; or, add to it the Little East or Pothole Fork, and it may take 5-7 hours, or more. Doing the **Middle & East Fork Slot** and exit via Bull Point, could take from 5-8 hours; less with 2 vehicles. Count on 4-5 hours for the **South Fork**. In

one 8-10 hour day, you can likely do both the short forks to **East Pasture Canyon** making it in 2 separate loops. Doing just the slots in **LBC** should take maybe 3 hours, but to build an anchor, do the last rap into the lower canyon, and return via Robbers Roost Canyon, could take 5-6 hours.
Water Have plenty in your car & pack; there may be water in the lower main canyon, and in potholes.
Maps USGS or BLM maps San Rafael Desert & Hanksville (1:100.000) for driving & orientation; and Point of Rocks West, Point of Rocks East, Angel Cove & Angel Point (1:24,000--7 1/2' quads) for hiking.
Flash Flood Danger High in the short, tight slots, but nonexistent elsewhere.
Best Time to Hike Spring or fall, but you may need a wet/drysuit then depending on air temperatures. Mid-summer is pretty hot, but it's cool in the slots, and at that time any pothole swimming would be welcome for the most part.

Author's Experience After several scout trips (including a solo trip through the Middle-East Fork Slot), and with Nat Smale & Jeff Webb, we did the **West Fork of Bull** in 2 1/2 hours; the **Little East Fork of East Pasture** in 3 1/3 hours. With Jon Jasper & Ryan Cornia, we did the **South Fork** in 6 1/4 hours, but that included lots of scouting & wandering! Next day we did **Pothole Fork** in 3 1/3 hours using the **Cowboy Steps Trail** exit coming out. Later, and with Brian Olliver, we did **Little Bull Canyon** in just over 8 hours; but we did a lot of wandering, along with making several probes into the various slots before finally getting through the Best Slot, and measuring the last drop.

The last & only rappel in the lower end of **Pothole Fork of East Pasture Canyon**.

Trick fotography (stitching 2 digital images) by Jon Jasper

Above This is one way to keep your feet dry. **Middle Fork of Bull Canyon** above the confluence of the Middle & East Forks. (Ryan Cornia foto)

Right DC'ing out of the tight upper slot of the **South Fork of Bull Canyon** and into what some are calling the *Chamber*. (Jon Jasper foto)

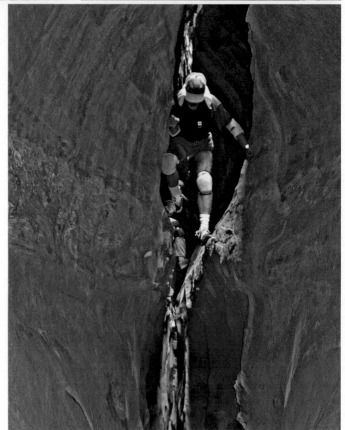

Above Left Jon Jasper DC'ing into the pool in the lower part of **South Fork of Bull Canyon**. This can be done without a rope.

Above Right Ryan Cornia doing some moderately high-stemming in the lower part of **South Fork of Bull Canyon**.

Left The end of the very nice slot of **South Fork of Bull Canyon** comes right out of a wall. It's a 2m jump into soft sand.

White Roost & Little White Roost Canyons, Robbers Roost Country, Utah

Location & Access White Roost and it's smaller counterpart occupy the northwestern corner of the greater Robbers Roost Canyon complex. To get there, drive along **Highway 24** about halfway between I-70 and Hanksville. Between **mile posts 135 & 136**, turn east, then south on the **Maze Road** leading into The Maze District of Canyonlands N.P. Drive 25.3 kms (15.7 miles) to the top of Texas Hill and turn south toward the old Andy & Chad Moore ranch (now Vachers) buildings, corrals and one tree about 550m to the south.

From the **Maze Road (Km & Mile 0)** drive south past the ranch buildings. After 13.7 kms (8.5 miles) you'll come to a junction. If you turn west, you'll end up at Pasture & Buck Canyons; but instead, continue straight ahead southwest on a good but ungraded road. At Km 15/Mile 9.3 will be a flat parking place on the left immediately before you start down a steeper section of road. This is about 400m west of elevation **5190T** (1582m) on the *Point of Rocks East* 7 1/2' quad. This parking place marked **1585m** on the map, is easily accessible with a 2WD car and is the trailhead for **White Roost Canyon.**

If you plan to come out, or enter, via the old **White Roost Horse Trail**, continue southwest down the steeper & rougher road (**Warning:** This can be done with a 2WD car under normal conditions, but it's sandy at the bottom and if used a lot, can become soft & deep!) to **Km 16.4/Mile 10.2** and park right where the road begins to veer slightly to the right or west (marked 1530m). This is the best landmark to look for close to the horse trail.

If you're going to the little unnamed canyon west of lower White Roost, and what this writer is calling **Little White Roost**, then continue southwest along the sandy road to **Km 17.1/Mile 10.6** where the road splits at the **1st major junction**. Take the left or eastern track for about another 800m (.5 mile) and park at 1487m, as shown on this map. There are no landmarks at all here. This can be a soft, sandy road so it's better to have a 4WD (in winter, or if the sand is a little wet, 2WD's do fine).

Rating East Fork of White Roost: 3A or B III PG SLOT; the **West Fork** and **Little White Roost: 3A II** (one short section in each may rate **PG SLOT**). The upper end of each canyon contains some of the better slots around, and with little or no wading.

Equipment For each canyon, RG, one 50 & 15m ropes, K&EP, W&R/QL, and as usual ascenders, compass & map.

Routes East Fork of White Roost Walk ENE about 1 1/2 kms from car-park at 1585m to where the "**C**" is in **C**anyon on the *Point of Rocks East 7 1/2' quad.* Walk in there. For the first km, walking is easy with several minor dropoffs, then it tightens and you'll have to slide & footjam sideways 1-2m above the slot bottom while dragging your pack behind. K&EP really help.

After that the canyon opens some and it's easy walking until the bottom drops out. At that point you're instantly in a dark slot and at the 1st of 4 rappels. You may have to install your own W&R/QL from a chokestone for a tight **20m rap**, then comes W&R/QL attached to another chokestone for the awkward start of the **15m R2**. After that is more W&R/QL attached to a tiny arch which gets you down **R3**, a drop of **10-14m**, depending on the length of sling. After that it's more DC'ing for 30m or so and finally the last rappel. **R4** is from W&R/QL around a chokestone which lets you down a **10m** slope into a perpetual pool that should never be more than chest deep. Just beyond the pool is the end of this **100m-long dark slot**. Finally, walk downcanyon about 2 kms and look to the right or west for the bottom end of the old **White Roost Horse Trail**. It's the only place in the canyon with any chance of escape. Observe map carefully. By route-finding up this historic trail, you'll arrive near the 1530m 4WD parking place.

There's a 2nd way into the upper East Fork from a side-canyon on the west, but you'll miss one of the nicer tight slots around. You'll also have to DC a 2 1/2m pitch which seems easy for 2 people, but more challenging for one.

West Fork of White Roost Head southeast from the car-park at 1585m. After about 750m, look for a cairn and possibly tracks of other hikers leading northeast down a minor ridge into the bottom. From there, head south to find the **1st dropoff** of about **2m**. Two or 3 people can help each other down; or wrap a rope around a deadman anchor and do a body wrap rappel (perhaps leave a rope there to be picked up later). After another 45m, is the **2nd dropoff** of about **2m**. Jump over this into sand; or help each other down. After another 60m is a **3rd dropoff** which is a steep **12m DC**. Beginners may want to be belayed (or, as of 6/2004, you could **rappel** from W&R/QL wrapped around a boulder on the left).

From there it's about 250m to the only real rappel in the canyon. This is a near-vertical **22m rap** from W&R/QL on a chokestone. Beyond this it's an open canyon for 400m, then a **dark slanted slot (PG SLOT)** which goes on for about 100m. DC'ing this slot is challenging enough to be fun; just keep your back on the east wall and slide down half a dozen pitches.

Alternate: Here's one idea on how to do both forks in one day together. Have your best climber go down the West Fork and install a single rope at the 22m rappel; others may have to help him back up the last 2 dropoffs. Then head into the East Fork, but instead of walking out via the horse trail, climb up through the **slanted slot** in the West Fork, then jumar up the 22m rappel, and finally help each other up the 3 dropoffs. If doing this, be sure to have K&EP for the squeeze-up through the tight slanted slot. **Warning:** If jumarring up the 22m rap, have your toughest & skinniest climber go up first, as it's narrow and difficult climbing up the last meter! Then that person can help others. Don't attempt this upcanyon trick if you have out-of-shape or inexperienced hikers in your group.

Little White Roost Start at the site marked 1487m on the map (2WD owners will be wise to park at Trailhead 1585m and walk from there, especially if the sand is dry!). Walk southeast 300-400m and enter the obvious drainage. Use a compass. Route-find downcanyon. Along the way are several short & shallow slots you can DC; or bypass them and rim-walk the slickrock bench above. Near the lower end, you'll have to DC into a possible pool. Not far below that the canyon turns south, slots-up and is dark for about 100m. In the dark **PG SLOT** are several DC's and 2 pools you can body-span across keeping your feet dry.

At the end of the slot is a dropoff and **rappel** of about **15m**, but you'll need all of one 50m rope to make it down next to a pool. The anchor is 2m of W&R/QL tied to a chokestone. To retrieve your 50, you may have to tie a short rope to one end, walk to the other side of the pool to reduce friction, and pull from there.

Below the rappel, there's always a pool and a little running water in a wider canyon. About 600m beyond the rappel, you'll enter the main Robbers Roost Canyon. From there, you could walk 1 km downcanyon and exit via the **Angel Arch Trail**; but for now **head upcanyon** about 2 kms and enter the lower end of White Roost. Walk up White Roost 2 1/2 kms and exit west via the old horse trail and return to your vehicle.

Elevations Trailheads/car-parks 1585m, 1530m & 1487m; junction of Little White Roost and the main Robbers Roost Canyon, about 1300m.

Time Needed To hike down the **East Fork of White Roost** and exit via the horse trail and back to the same

Map 14, White Roost & Little White Roost Canyons, Robbers Roost Country, Utah

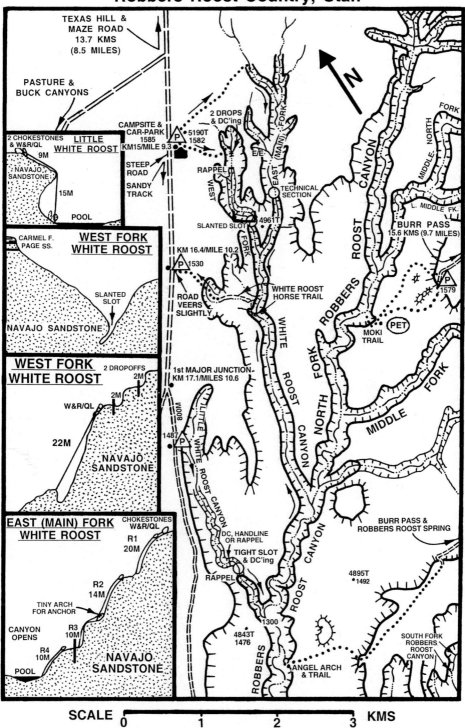

TEXAS HILL & MAZE ROAD
13.7 KMS
(8.5 MILES)

PASTURE & BUCK CANYONS

2 CHOKESTONES & W&R/QL

LITTLE WHITE ROOST

NAVAJO SANDSTONE

9M

15M

POOL

CAMPSITE & CAR-PARK
1585
KM15/MILE 9.3

STEEP ROAD

SANDY TRACK

P 5190T 1582

2 DROPS & DC'ing

E/E

RAPPEL

WEST FORK

EAST (MAIN) FORK

TECHNICAL SECTION

SLANTED SLOT

4961T

CARMEL F. PAGE SS.

WEST FORK WHITE ROOST

SLANTED SLOT

NAVAJO SANDSTONE

KM 16.4/MILE 10.2

P 1530

ROAD VEERS SLIGHTLY

WHITE ROOST HORSE TRAIL

WEST FORK

WEST FORK WHITE ROOST

2 DROPOFFS

2M

2M

W&R/QL

22M

NAVAJO SANDSTONE

1st MAJOR JUNCTION
KM 17.1/MILES 10.6

800M

1487

P

LITTLE WHITE ROOST CANYON

WHITE ROOST FORK

NORTH FORK

ROOST CANYON

MIDDLE FORK

EAST (MAIN) FORK WHITE ROOST

CHOKESTONES W&R/QL

R1 20M

R2 14M

TINY ARCH FOR ANCHOR

CANYON OPENS

R3 10M

R4 10M

POOL

NAVAJO SANDSTONE

DC, HANDLINE OR RAPPEL

TIGHT SLOT & DC'ing

RAPPEL

1300

4843T 1476

ROBBERS

ANGEL ARCH & TRAIL

4895T 1492

ROBBERS ROOST CANYON

BURR PASS & ROBBERS ROOST SPRING

SOUTH FORK ROBBERS ROOST CANYON

BURR PASS
15.6 KMS (9.7 MILES)

MIDDLE-NORTH FORK

L. MIDDLE FK.

NORTH FORK

P 1579

PET

MOKI TRAIL

ROBBERS ROOST CANYON

N

SCALE 0   1   2   3   KMS

trailhead, will take most people 5-7 hours--slightly less with 2 cars. To do the **West Fork of White Roost**, may take only 3-4 hours. **Little White Roost**, about 5-7 hours round-trip.

Water Always take your own, but there should be a seep & running water in the middle parts of White Roost, plus some water in the lower end of Little White Roost.

Maps USGS or BLM maps Hanksville & San Rafael Desert (1:100,000) for driving; Point of Rocks East & Angel Point (1:24,000--7 1/2' quad) for hiking.

Flash Flood Danger High in the slots but little danger otherwise. These drainages are short & sandy (above the slickrock part), so danger would exist only if a heavy storm dumped right on you or immediately above.

Best Time to Hike Spring or fall with cooler temperatures, but it's possible year-round.

Author's Experience He parked at 1585m and headed down the East Fork of White Roost. He had 3 ropes, so he left them in place at the first 3 raps but didn't do the last drop into the pool (he had been upcanyon to that point from the bottom several times). He then explored parts of the West Fork and returned to his car in 5 1/2 hours. After a scouting trip to **Little White Roost**, he returned later and did the hike down to the main

Roost Canyon and back out via the old horse trail. Round-trip time was just over 4 1/2 hours. Later, with Nat Smale, we went down the **East Fork**, then up-climbed the slanted slot in the West Fork to the bottom of the 22m rap, then headed up the horse trail and back to the cars at 1585m. Round-trip took 7 1/4 hours. Next day it was down the **West Fork**, up the horse trail and back to the cars in 2 3/4 hours. Later, in 6/2004, and with Steve Tyler, we rigged up the West Fork with a rope at the 22m rap, then headed down the East Fork. It took longer getting up the slanted slot than expected, then even more time getting out at the top of the jumar on the fixed rope. It turned into a 9 1/2 hour day.

Opposite Page Immediately before the end of the slot in **East Fork of White Roost**, is this seemingly year-round pool.

Above Left Steve Tyler in the darkest part of the Slanted Slot in the lower end of **West Fork of White Roost**. This has to be rated a PG SLOT.

Above Right Part of the Slanted Slot in the **West Fork of White Roost Canyon.**

Left The beginning of the rappel at the lower end of the slot in **Little White Roost Canyon.** Just below and out of sight is a pool, usually some running water, and lots of greenery. This picture was taken in 2004, so hopefully these 2 chokestone anchors will still be there (?). If not you'll have to rig something up to make the rappel.

North Forks of Robbers Roost Canyon, R. Roost Country, Utah

Location & Access Discussed here is the main drainage of the **North Fork of Robbers Roost Canyon**, plus its **East Fork** which is nearly as long. The best way to get there is to drive along **Highway 24** about halfway between I-70 and Hanksville. Between **mile posts 135 & 136**, turn east, then south onto the **Maze Road** leading into The Maze District of Canyonlands N.P. Drive about 25.3 kms (15.7 miles) to the top of Texas Hill and turn south toward the old Moore Ranch (now Vachers) with one tree, several buildings, a well and corrals, a distance of 550m. From the turnoff at Texas Hill (Km & Mile 0), drive south **9 kms (5.6 miles)** to a parking place at the beginning of a side-road and **fence & gate** on the left. This is near the head of White Roost Canyon and the road to that point is good for any car (but 2WD's can get stuck if they pull off the main road!). This parking place is about halfway between elevations 5497T (1676m) & 5438T (1658m) on the *Point of Rocks East quad.*

Rating The **North Fork: 3A II or III PG SLOT**; **East Fork: 3B III PG SLOT. North** is easy and fun, with a good fotogenic slot. **East** is more challenging with interesting DC'ing. Exiting up the **Crack Route** has lots of easy UC'ing.

Equipment North & East Forks, RG, K&EP, one 50m rope, W&R/QL & PR. For **East Fork**, wet/drysuit for the cooler weather, BK and at least 3 B&H; otherwise someone will die using small chokestones in a down-sloping crack as an anchor!

Route North Fork From the car-park, walk east through the gate and down along the **very sandy 4WD track**. This track heads east, south, and east again crossing the upper part of the White Roost drainage. From there the track heads northeast for a km then turns south to the shallow drainage of North Fork. Walk downcanyon less than 2 kms to find a **10-15m dropoff**. To get down in without rappelling, rim-walk south on the left or east side of the canyon about 200m and route-find down to the bottom. After another 100m is a 2m dropoff which 2 people can handle easily without rappelling.

Or if you're alone, walk east from the 10-15m dropoff to the head of a **2nd drainage** about 200m over a ridge, then route-find down in. About 200m below where the 2 upper forks meet, you'll come to **Rappel 1**. This is down about **6m** from W&R/QL on a small chokestone with an awkward start. After about another 100m you'll come to **R2**. As of 6/2004, this was from 1 B&H plus 2 baby angle pitons in one hole on the right and down about **5m** (if someone should remove the B&H or pitons, rappel from your **Ibis hook**). Immediately below that is a low angle slide into a pothole.

Below R2, you'll find a nice fotogenic slot and narrows with some resemblance to Zion's Subway, but there's no water. About 600m below R2 is a walkup escape route to the right or west which heads northwest up to the rim. In the 200m below this exit and just before R3, is still another rather dark walk-through slot. **R3** drops about **11m** from **3 B&H** on the left. Below that is another walk-through crack-like slot, then the canyon slowly opens up. For the moment, let's stop here and talk about the East Fork.

East Fork Use the same trailhead & route as if going to the North Fork, but continue southeast after crossing the North Fork's main drainage in the middle of **Section 33**. Further southeast, you can enter East Fork at **one of 2 upper branches**; look for the elevation markers 5470T or 5368T on the quads listed below. You can walk down either of these upper branches to the technical area above the big dropoff (a **shortcut route** exists passing over the **top of high point 5261T (1604m)**, then via a long ridge right to the beginning of the technical section--**see this map**). At the beginning of the slot is an easy **3m DC**, then a steep **9m DC**, which some prefer to rappel (look for W&R/QL under a chokestone). To do this DC, lower your pack first, then comes a short walk, more DC'ing perhaps into a shallow pool, then a second pool (hop across on the right--see picture), then a **7m rappel** into a **3rd pothole** (this was knee deep for us, but it could be much deeper). We built a deadman anchor in a small pothole above with W&R/QL tied to the bottom rock. However, if this looks risky, or there is insufficient building material handy, install 1 B&H nearby.

After that, you'll have 2 more interesting/challenging DC's and perhaps chimneying over a long pool, then the big **20m R2** into the lower part of the canyon. We found 3 small chokestones in the last groove and tied W&R/QL to 2 of them, but it wasn't the most secure anchor! Installing 2 B&H would insure R2 is a lot safer, and you wouldn't be leaving so much trash--5m of W&R/QL--in the canyon! It's your choice--and your life.

Once in the lower part of the canyon, you can walk almost forever and camp (big packs are not recommended in either upper fork); and exit via the **Moki Trail** (see Middle Forks map), at the **Angel Arch Trail**, or go down to the **Dirty Devil** and leave. Or you can exit up the **Crack Route** and do either upper canyon fork as a short day-hike. To do that, walk about 2 kms downcanyon from R3 in the main **North Fork** and turn right or north into a side-canyon (if coming down the **East Fork**, walk about 1600m beyond R2). About 500m into this side-canyon, you'll come to what looks like a dead end--but it's not. From there, chimney & UC about 5m with some fairly good handholds. It's easier than it first appears, but you'll likely want to pull your pack up with a PR after you climb it. From there to the rim it's about 250-300m up a fairly tight crack. Along the way, you'll find 6-7 places where you chimney up 2-3m over chokestones. It's just difficult enough to be interesting & fun. But in summer you'll sweat like hell!

When you finally get out on top, you'll see several cairns; from there head a little west of due north about 600m (there's now an emerging hiker's trail--follow it) until you come to a **fence** running east-west. Head west and northwest along the fence line as it climbs up a broad **hogsback ridge**, then down into the upper part of White Roost Valley. After that, follow the same fence north and northeast back to your car. Or, if you can see your car on the distance horizon, make a slightly shorter beeline directly toward it.

Elevations Car-park, 1675m; lowest part of this hike in North Fork, about 1400m.

Time Needed From 5-8 hours in the main **North Fork**, maybe less; **East Fork**, 6-9 hours round-trip.

Water Take you own water. Texas Hill Well water is awful tasting, but cows drink it.

Maps USGS or BLM maps Hanksville & San Rafael Desert (1:100,000) for driving; Point of Rocks East & Whitbeck Knoll (1:24,000--7 1/2' quad) for hiking.

Flash Flood Danger Above the first dropoffs, these are very sandy drainages, so the only risky part is between the 1st rappel down about 400m past R3 in the main **North Fork**, and only if a storm dumps right on you or just upcanyon. High risk only in the technical section of **East Fork**, low above that.

Best Time to Hike Spring or fall, but can be done in winter or summer. Wet/drysuits may be needed (?) in the East Fork during cooler weather.

Author's Experience After several scouting trips, he did the **North Fork** in about 8 hours round-trip from the car-park mentioned, but he did lots of exploring and fotography with a tripod. In 6/2004, and with Steve Tyler, we did this same loop-hike in 6 1/2 hours. After a scout trip, and on 9/25/2004, the author, Jeff Webb & Nat Smale, did the **East Fork** in 5 2/3 hours.

Map 15, North Forks of Robbers Roost Canyon, Robbers Roost Country, Utah

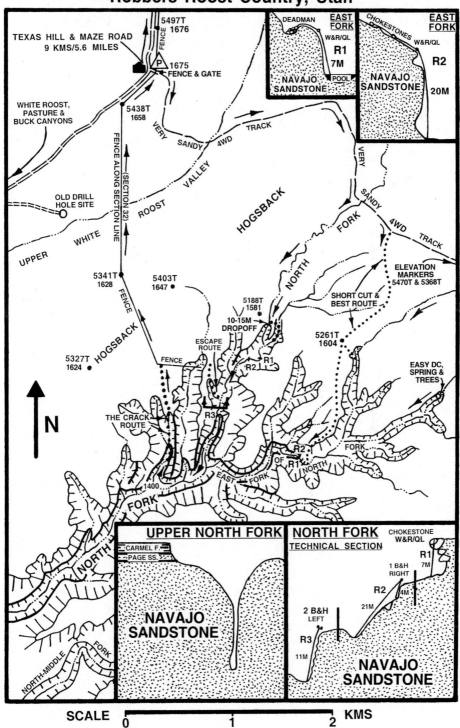

TEXAS HILL & MAZE ROAD
9 KMS/5.6 MILES

5497T 1676

FENCE

P 1675
FENCE & GATE

DEADMAN

EAST FORK
W&R/QL

R1
7M

POOL

NAVAJO SANDSTONE

CHOKESTONES

EAST FORK
W&R/QL

R2

NAVAJO SANDSTONE

20M

WHITE ROOST, PASTURE & BUCK CANYONS

5438T 1658

VERY SANDY 4WD TRACK

FENCE ALONG SECTION LINE

(SECTION 32)

OLD DRILL HOLE SITE

ROOST VALLEY

HOGSBACK

NORTH FORK

VERY SANDY 4WD TRACK

UPPER WHITE

5341T 1628

5403T 1647

ELEVATION MARKERS 5470T & 5368T

FENCE

SHORT CUT & BEST ROUTE

5188T 1581

5261T 1604

10-15M DROPOFF

EASY DC, SPRING & TREES

N

5327T 1624

HOGSBACK

FENCE

ESCAPE ROUTE

R1

R2

THE CRACK ROUTE

R3

R2

FORK

OF

R1

NORTH

1400

EAST FORK

FORK

NORTH

NORTH-MIDDLE FORK

FORK

UPPER NORTH FORK

CARMEL F.

PAGE SS.

NAVAJO SANDSTONE

NORTH FORK
TECHNICAL SECTION

CHOKESTONE W&R/QL

R1
7M

1 B&H RIGHT

R2

4M

21M

2 B&H LEFT

R3

11M

NAVAJO SANDSTONE

SCALE   0   1   2   KMS

87

Above Left Nat Smale & Jeff Webb rig a deadman anchor for Rappel 1 in the **East Fork**. One well-placed bolt would save building an anchor every time someone goes down this canyon. **Above Right** Jeff hops across the 2nd pothole in the **East Fork**. **Below Left** Nat high-stemming between the 2 rappels in the **East Fork**. **Below Right** Steve Tyler does the last rap in the North Fork.

Above Left Below Rappel 3 in the **North Fork** is a nice dark slot some call The Crack.

Above Right The normal way out of both the East and North Forks is via **The Crack Route**.

Left Rappelling off an Ibis hook at Rappel 2 in **Little Middle Fork of Robbers Roost Canyon**.

Middle Forks of Robbers Roost Canyon, Roost Country, Utah

Location & Access Shown here are 3 tributary canyons in the greater Robbers Roost Canyon complex, but only one has a name, that's the **Middle Fork**. Between the Middle and North Forks is one substantial canyon this writer is calling **North-Middle Fork (N-MF)** (AKA Not Mindbender). Between the N-MF and the Middle Fork is a shorter drainage but with a pretty good slot; we'll call this the **Little Middle Fork (LMF)** (AKA Mindbender). Also discussed is a new route out of these canyons called the **Moki Trail**.

To get there, drive along **Highway 24** about halfway between I-70 and Hanksville. Between mile posts 135 & 136, turn east, then south on the **Maze Road** heading into **The Maze District** of **Canyonlands National Park**. Drive past **Jeffrey Well**, through the **Flat Tops**, to the top of **Texas Hill & Well** and continue east to **Horseshoe Junction** with an **information board** (39.3 kms/24.4 miles) from Highway 24. Turn left to go Green River, but instead turn right or south and drive to the highest point around called **Burr Pass**. This is **50.9 kms (31.6 miles)** from Highway 24. At that point is a sign stating distances to various points including the **Ekker Ranch** (also known as the Robbers Roost or Biddlecome Ranch)--7 miles [11 kms].

At Burr Pass veer right, then 75m from the sign, turn west onto the road running to the **Robbers Roost Spring (RRS)**. This is a good road which runs all the way to the RRS and the beginning of Middle Fork and the Moki Trail, but after about 2.4 kms (1.5 miles) & 2.9 kms (1.8 miles) are **2 migrating sand dunes** across the road. If these haven't been graded recently, it will be for 4WD's only. Assuming you get through somehow (a mtn. bike is an alternative to 4WD), drive a total of **9.7 kms (6 miles)** to the **RRS turnoff**. About 300m down that road to the south is the RRS and a possible campsite.

From the **RRS** turnoff, continue west to **Km 10.9 (Mile 6.7)** and an open parking/camping place on the right marked 1640m. This will be the best place to start if going to **N-MF**. Continue west to **Km 12.5 (Mile 7.8)** and a low divide marked **1609m**--park there for the entry to **LMF & Middle Fork**. If doing the **LMF**, it might be best to park/camp at the **drill hole site**, which is another 1200m west. Or, continue to the end of the road at **Km 15.6 (Mile 9.7)** from Burr Pass. This is just above the **Moki Trail** and your exit from several canyons. It's marked 1579m. This is a good place to camp too, but it can be windy in unsettled weather.

Rating The **North-Middle Fork: 3A III or IV PG SLOT** coming out the Moki Trail. **Little Middle Fork: 3A III**. Going down **Middle Fork: 3A II** or **III** if going over the big rappel and returning via the Moki Trail.

N-MF and **LMF** both have a nice slot above the big rappels, then a deep, sometimes well-watered canyon below. **Middle Fork** pretty much lacks a slot (a short one exists a ways above the drop-in point), but does have a deep, open gorge below the big rap. It also has 2 stands of douglas fir trees.

Equipment RG, W&R/QL, K&EP for all. Ropes--**N-MF**, one 50m; **LMF**, two 50's (or one 50 & pull cord), plus a large #3 Ibis hook to make one rappel; **Middle Fork**, two 60m (or tie one 60 off to be picked up later).

Route North-Middle Fork First, park one vehicle/mtn. bike at the end of the road marked **1579m**; then return and walk north from the trailhead marked **1640m**. With map & compass in hand, head for the hard-to-see high point marked **5325T** (1623m) crossing upper LMF along the way. From that high point, walk northwest about 800m and route-find north into a short side-drainage, then into the main canyon. After another 500m or so, you'll drop into a nice **PG SLOT**. After about 300m of pretty interesting DC'ing/down-squeezing, you'll come to a short dropoff; DC the first part, then HL down the 2nd part, a **5m** slide (or just let go for a fast slide!; or put an **Ibis hook** in a crack on the left and **HL** down). Soon after that will be **Rappel 1** from a chokestone and down about **11m**.

Soon after R1, the canyon opens up some and walking is easy for nearly 2 kms. Then you'll have a short slot where you'll have to span (or rap-HL down 3m) one round pothole, then the big **R2** of **22m**. There used to be 2 B&H there, but they got bad and were removed. On 3/15/2007, believe it or not, there was 4 lengths of nearly 20m of W&R/QL anchored to a small chokestone back upcanyon, plus another 6-7m tied to another rock jammed into a side-crack. **Commentary:** This is absolutely ridiculous. These anchors seemed safe, but with all that trash, someone really should take it out and install 2 good B&H's high above most floods with an easy pull from below. Sooner or later all that W&R/QL will be washed away to clutter the canyon bottom.

From the bottom of R2, continue downcanyon and exit at one of 4 places. By far the **best place** will be what this writer is calling the **Moki Trail**. Walk down N-MF to where the **North Fork** comes in on the right. About 800m below that confluence, look to your left or south. There you'll see a break in the cliffs and the only place around with a snowball's chance of climbing out. Walk up the sandy bench and just as you step onto slickrock, notice some rounded **moki steps** where the sand & slickrock meet. This indicates the sand was much lower during Fremont Indian times. Continue up the slickrock to where there are natural ledges and what may be more moki steps. At the wall, climb up the **5m** face angling up from lower left to upper right.

Once above this 5m wall, veer left or east and as you cross a minor drainage, walk south 40m to see a small panel of **rock art**. From there, walk along a meter-wide ledge to more moderately steep slickrock and route-find eastward, and south of a big dome rock, up to the end of the road marked 1579m (2 routes here).

Other route options for leaving the **Robbers Roost Canyon** would be to head downcanyon to the Dirty Devil River, turn south and exit at the **Angel Trail**, either east or west; or climb out via the **Angel Arch Trail** 3 kms below where White Roost Canyon enters. Either of these routes would take at least 2 days. Or walk up White Roost Canyon and leave via the **White Roost Horse Trail**; or head up North Fork and climb up what this writer calls the **Crack Route** (see the previous map for North Forks, Robbers Roost Canyon).

Little Middle Fork You could enter this drainage by parking at 1640m and walking north, but most of the upper part of the canyon is uninteresting. Best to park at the **drill hole site** or the divide marked **1609m**, which are the easiest landmarks to locate. From there, head northeast (or north) and route-find down into the canyon. Soon you'll come to **Rappel 1**, an **11m** drop from one of 2 chokestones. You may have to reinstall W&R/QL. About 500m below that will be **R2**. By attaching an **Ibis hook** under a natural ledge, you can make this **6m** rap easy and safe without leaving trash in the canyon; then with a little slack, flip the single rope to retrieve your hook. In the next 500m, will be 3 good DC's, including a double drop into a widening canyon. There may be some W&R/QL there, because it is an awkward & tight start of a pretty good DC!

Finally it opens up just before the final big drop and **R3**. About 25m above that, is a short slot with a 3m HL or partial-DC & jump into sand. However, the first person down should check the anchor situation before others follow. As of 3/2007, there were **2 B&H's** on the left safely out of flood waters. If some bolt chopper has removed them, you'll have to reinstall the bolts; or build an anchor of stones from upcanyon and **rappel 45m!** Rappelling 45m from a pile of stones seems ridiculous for many, so please leave the 2 B&H's there. Otherwise, sooner or later someone will die on this one!

At the bottom, you'll descend through small trees landing next to a spring & pool. To get out of the canyon, walk 600m downcanyon to N-MF, then about 1 1/2 kms further until North Fork comes in on the right. From there walk 800m downcanyon and exit via the **Moki Trail** as described above.

If doing the **Middle Fork**, park and begin hiking from the place marked **1609m** (you could head downcanyon from the parking place marked 1640m, but there's not much to see in the upper shallow canyon).

Map 16, Middle Forks of Robbers Roost Canyon, Robbers Roost Country, Utah

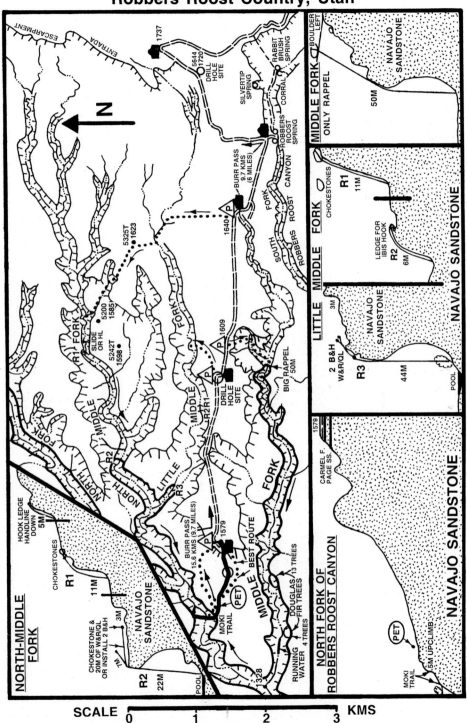

SCALE 0 1 2 3 KMS

From your car, route-find south into a side-drainage of Middle Fork. Once at the bottom, head west to the only rappel in the canyon.

At that point, you can either rappel down a single 60m rope tied to a boulder on the left; or use a 50 tied to another short rope to rappel down the 2-stage **50m drop** to the bottom. This is a partial free-rap, so it's recommended you tether your pack to the front of your harness and have it hang between your legs. Also, better use a variable speed control ATC, or wrap the single rope around a locking biner (a munter hitch works well) attached to your brake-hand-side leg loop to slow you down. At that point you can leave the ropes, take a look downcanyon a km or two, then return and using ascenders, jumar back up and return to your car. A 2nd choice would be to use two 50's and W&R/QL, then pull your ropes and take them with you for the entire hike. Or, **best** leave a single rope, finish the hike downcanyon, and retrieve your rope at the end of the day. It took the author only 35 minutes round-trip to retrieve his rope.

After the big rappel, head downcanyon past another little **3m dropoff**, which you can DC or jump. Beyond that, it's walking in a semi-narrow and impressive gorge. About 4 kms below the big rappel, look to the left or south, and you'll see **13 douglas fir trees** in a north-facing alcove; 600m below that are **4 more doug firs** in another alcove. About a km below that will be a stretch of running water. In the terrible drought year of 2002, it was running for only 100m, so water should be there no matter how dry the drought.

Perhaps 2 1/2 hours below the big rappel will be the confluence with the **North Fork**. From there, walk up North Fork less than 2 kms and look to your right or south for the break in the canyon wall. Climb the **Moki Trail** to the end of the road marked 1579m altitude, then the 2 1/2 km/30 minute road-walk back to your vehicle; or an easy & short car/mtn. bike shuttle.

Elevations Parking for N-MF, 1640m; parking for LMF and Middle Fork, 1640m & 1609m; trailhead for the Moki Trail, 1579m; confluence of Middle and North Forks, 1328m.

Time Needed Going down **North-Middle Fork** and returning via the Moki Trail, 6-8 hours--less with a car-shuttle. To do **Little Middle Fork** returning via the Moki Trail, about 5-7 hours; going down **Middle Fork** returning via the Moki Trail, also about 5-7 hours round-trip.

Water Always carry water in your car and pack. Below the big dropoffs in all these canyon may be minor seeps or pools for much of the year, but at the end of a long dry spell, some seeps can dry up. Go prepared!

Maps USGS or BLM maps Hanksville & San Rafael Desert (1:100,000) for driving; Point of Rocks East, Whitbeck Knoll, Robbers Roost Flats and Angel Point (1:24,000--7 1/2' quad) for hiking.

Flash Flood Danger Moderate to high for short distances in the tight narrows, especially those short slots in N-MF or LMF. No danger in Middle Fork.

Best Time to Hike Spring or fall, but summer isn't bad--just carry lots of water. Winter is possible too.

Author's   Experience
Years ago, the author and Nat Smale drove to and camped at the tight bend marked 1737m. Next morning we went into the northern-most branch of the upper forks of **North-Middle Fork**. After making the last rappel, we walked downcanyon and up the bottom end of LMF to the big dropoff, then returned, jumarred up and out of the canyon, returning via the BM 5325T and straight back to the cars. Round-trip took 9 2/3 hours (much better to start at the 1640m car-park and return via the Moki Trail).

The author & Steve Brezovec parked at the drill hole site and walked north into the **Little Middle Fork**. At R2, a deadman was constructed and we rapped off that. After lunch at the spring & pool at the bottom, we exited via the Moki Trail and were back at the 2nd vehicle in less than 5 hours.

After a scouting trip, the author went down **Middle Fork** from the 1609m car-park, tied a 60m rope to the boulder on the left, rappelled and headed downcanyon. He returned via the Moki Trail in about 4 1/2 hours, then the 30 minute road-walk back to his car---5 hours round-trip. After lunch and drinks, he retrieved his rope in about 35 minutes. In March, 2007, with Steve Tyler, we did LMF in 6 3/4 hours. Next day, plus Brian Olliver, we did N-MF in 7 1/4 hours.

Opposite Page Steve Tyler starting the big 45m rap from 2 B&H in **Little Middle Fork of Robbers Roost Canyon.**

Above Left Lots of DC'ing in the **North Middle Fork** of Robbers Roost Canyon above Rappel 1.

Above Right Steve Tyler & Brian Olliver in the area below Rappel 1 in the **North Middle Fork**.

Left Making the last big 22m rappel in **North Middle Fork**. There's always good drinking water at the bottom.

South Fork of Robbers Roost Canyon, Robbers Roost Country, Utah

Location & Access The **South Fork of Robbers Roost Canyon (SFRR)**, and a short side-canyon called **Lost Spring Canyon (LSC)** (author's name), are located in the middle of the Roost Country. South Fork is the major drainage beginning just southwest of Burr Pass, and running west past Robbers Roost Spring. LSC begins at 2 separate seeps called **Lost Spring**, and runs west joining the upper part of South Fork.

To get there, drive along **Highway 24** about halfway between I-70 and Hanksville. Between **mile posts 135 & 136**, turn east onto the **Maze Road** and drive past the Jeffery Well, Texas Hill, turn right or south at **Horseshoe Junction** (information board at Km 39.3/Mile 24.4) and head for **Burr Pass at Km 50.9/Mile 31.6**. See **Area Map, page 65**. Right at the pass is a road junction; veer right toward the Ekker (old Robbers Roost & Biddlecome) Ranch, then make a hard right turn 75m south of the sign and head due west on the **Robbers Roost Spring (RRS) Road**. After about 2.4 kms (1.5 miles) you may find the first of **2 migrating sand dunes** which can cover the road. Over the years, this means 4WD country about half the time! Stop at each sand dune and walk it off first, then decide if you can make it across. Beyond these **2 possible sand traps** (the county puts a blade on this road about once year, but strong south winds can pile up sand across the road in a hurry) the road is good to as far as **RRS Junction (Km 9.6/Mile 6)** and beyond. Follow map carefully. You'll see some large cottonwood trees just above the spring, but cattle come daily for a drink, so don't scare them or block their way to water by camping there. Also, there are some historic Butch Cassidy/Wild Bunch sites just upcanyon to the east. For more information on these, see another book by this writer, *Hiking and Exploring Utah's Henry Mountains and Robbers Roost* (3rd Ed. will be ready in 2009)

Rating SFRR: 3A III or IV (or V) PG SLOT. You can do this in one day by leaving 5 ropes and ascending back up; or pull your rope at each rappel and wander throughout the greater Robbers Roost Canyon Complex for 2 or more days. **LSC's North Fork: 3A III**; the South Fork is shorter and easier.

Equipment For SFRR, RG, W&R/QL, K&EP, one rope at least 30m long, 1 PR per/person for lowering/lifting. Or if going down the technical section and returning the same way, take **5 ropes** at least **15m long**, plus **ascenders**. For LSC, one 50m, if returning from the 37m rappel, or two 50's if going down into SFRR.

Route South Fork of Robbers Roost Canyon From **RRS** walk west into **SFRR**. After 40m look at the wall on the right to see Pearl Biddlecome (Baker's) signature from 1920 (she was raised at the Robbers Roost-Biddlecome Ranch). From there continue downcanyon about 100m to **Rappel 1**. From where it drops off, look up to your left and behind a little to find 2 B&H and W&R/QL (if someone has removed these bolts, drag 2 or 3 large stones over to the dropoff and place W&R/QL around them). The 1st rappel is only about 5m, but a single 50m rope will get you down it, plus the first of 4 steep DC's immediately below that (halfway down this section is more W&R/QL around a natural arch on the left, which for some could be R2)

After about 20 minutes, and several DC's, you'll come to the **R2**. You should see W&R/QL around some chokestones and a drop of about **13m**. Below that are about 5 more interesting DC's in a tight & slanted slot before the canyon widens a little. This upper slot is roughly 1200m long.

After walking 3 more kms, you'll come to **R3**. This will be from W&R/QL around 2 bushes on the left and down about **8m** (a good climber can DC/UC this). About 300m below this is another dropoff. On either side of this are large bushes you can rappel from, but the best way is to **ledge-walk left** about **125m** to find W&R/QL around a **boulder & large bush**. This is actually just inside the lower end of lower **Lost Springs Canyon**. **R4** is a free rap of about **8m**.

From R4, walk up the side-canyon to find running water, and after 200m a big dryfall. This side-drainage the author calls **Lost Spring Canyon**, is a good one with a rappel of **37m** at the end. Read more below.

Going down SFRR again. Right at the junction of these 2 canyons is a good campsite, then another site after 350m under an overhang. About 1 km below R4 is the **5th rappel**. This one had W&R/QL around a chokestone with a total drop of about **8m** to the edge of a shallow pool.

After this dryfall, it's walking downcanyon all the way to the Dirty Devil River, then downstream a ways and up the eastern or western end of the **Angel Trail** to a 2nd car. Or, from the very end of SFRR, walk north across the drainage bottom and into another short side-canyon. Walk up this to find a **cattle trail** heading up to a bench. From there, route-find up to the end of **Bull Point** at the elevation point **4806T** (1465m) and the end of that road. See the **Bull Canyon map, page 77**.

Here are 2 other possible escapes/exits. When you reach the main Robbers Roost Canyon, walk upstream to the **old cattle trail** which runs under **Angel Arch**, then after several kms, you'll come to the end of an old driller's road which you can use to get back to **RRS**. You could also spot a mtn. bike or 2nd 4WD/AWD vehicle somewhere on the road to Angel Arch to make a quicker return. A better exit would be to head up the main RR Canyon and exit via the **Moki Trail**, as described under the **Middle Forks map, page 91**.

An alternative to making a rather long one day 2-3 day backpacking trip, would be to take 5 ropes at least 15m long, and leave one at each of the 5 rappels. After the last rap, use **ascenders** to return upcanyon to the Roost Spring. You'll grunt & sweat UC'ing the dozen or so 3m-high dropoffs, but any experienced climber or slot rat can do each alone. Climb up, then pull your pack afterwards using a **pack rope (PR)**.

Another way of entering SFRR would be to come in via **Lost Spring Canyon (LSC)**. This short drainage coming in from the east has 3 upper forks, 2 beginning at 2 different Lost Springs. See the *Robbers Roost Flats & Angel Point* quads below. To come down this canyon, read the driving instructions for **No Mans Canyon**, which is the next mapped area. Briefly, make your way to Burr Pass, then proceed 16.1 kms (10 miles) along the **Angel Point Road**. At that location, and as you're turning left or south and going downhill, you will see a minor 4WD track heading northward. Follow this map & watch carefully as there are several roads crossing the 1st shallow drainage. Best to park at or near the Angel Point Road.

About 1300m from the **Angel Point Road**, you'll pass the 1st of 3 drainages. You can walk into and through the **South Fork**; the **Middle Fork** has some good DC'ing, but we still haven't seen 300m of it. The **North Fork** is more difficult with some rappelling. It has 4-5 raps up to 10m high in the upper end in one 300m section, plus some interesting DC'ing.

Below where the 3 forks meet, there's some easy walking, then several good DC's, and finally the **big dropoff of 37m**. At that point is a chokestone, perhaps with W&R/QL. This is mostly a free rappel, so hang your pack from the front of your harness. If going to the last rappel and returning, get back to the 4WD track by walking up between the North & Middle Forks, or up along the South Fork of LSC.

Elevations Robbers Roost Spring, 1646m; bottom of 5th rappel, 1439m.

Time Needed During June, with long days, fast hikers can likely walk from **RRS** to the end of **Bull Point** in one long day (?), but you'd have to setup a car shuttle the day before. But most people will want at least 2 days, or 3 shorter days, depending on where they park, and exit. The one-way hike to the **east end** of the **Angel Trail** is just a little longer than going to Bull Point, but with large packs, just getting through the upper technical part will take most of the 1st day. Or leave a 2nd 4WD/AWD at the **west end** of the **Angel Trail**, but setting up that shuttle would take half a day. If you have only one car, the best plan would be to come out ei-

Map 17, South Fork of Robbers Roost Canyon, Robbers Roost Country, Utah

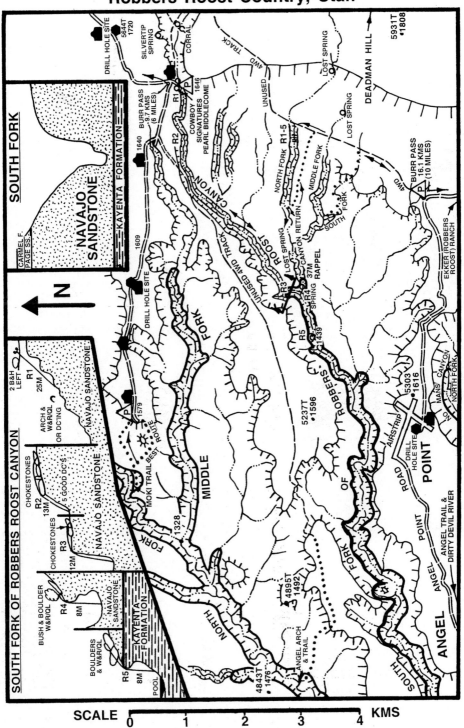

SCALE   0   1   2   3   4   KMS

ther the **Angel Arch Trail** or **Moki Trail**, then walk all the way back to the **RRS**.

If you choose to do only the upper **technical part** of **SFRR leaving 5 ropes** in place & returning the same way, this can be done in one day, perhaps 6-9 hours. Another thing to consider is where you actually park. The 2 places where there may be sand dunes crossing the road is about 5 kms east of the RRS using a short-cut cow trail. If you have a car and the sand hasn't been cleared from the road in a while, you may have to walk this extra distance. If so, plan on 8-12 hours, depending on the size and skills of your group.

Water Cattle have been milling around Robbers Roost Spring for a century and the water could be suspect, so have lots of water in your vehicle. There should be running water in the lower end of LSC above R4, and there is running water off & on throughout the lower part of SFRR and the main Robbers Roost Canyon. Expect to find more running water in winter than in the heat of summer.

Maps USGS or BLM maps Hanksville & San Rafael Desert (1:100,000) for driving & orientation; Angel Point, Robbers Roost Flats, Angel Cove and possibly Point of Rocks East (1:24,000--7 1/2' quad) for hiking.

Flash Flood Danger Moderate to high danger in the slot parts of both canyons, but low risk elsewhere.

Best Time to Hike Spring or fall, but can be done year-round when the roads are dry.

Author's Experience First trip down **SFRR**, he had to park west of Burr Pass at the 1st sand dune and walk the extra distance. From his car to the bottom of the 5th rap and back took 9 hours. From the 1st rap to the 5th and back, took 6 2/3 hours. On 3/28/2003, with Scott Patterson & Ryan Cornia, we went down the South Fork of **LSC** to the end. Yours truly rapped down 37m for fotos and measuring, jumarred back up, then we went out via the North Fork. Second trip down **SFRR**, he went down past the 4th rap & into lower LSC and returned in 5 3/4 hours from RRS.

Above The bottom of R4 in the **South Fork of Robbers Roost Canyon.** If your plan is to return back up South Fork, this is a free-hanging 8m jumar. It's always easier to jumar up when there's tension on the rope, so hang your pack like this before ascending.

Right Starting Rappel 4, which is actually in the lower end of **Lost Spring Canyon** just before it reaches **South Fork of Robbers Roost**.

Above Left Robbers Roost Spring located at the upper end of the **South Fork**. About 40m below the spring are old cowboy signatures, among them the late Pearl Biddlecome (Baker), who was raised at the Robbers Roost Ranch. All these fotos were taken on 10/30/2006, and some deep wading was expected, thus the wetsuit. But all pools were less than knee deep.

Above Right This is what the **South Fork of Robbers Roost Canyon** looks like between Rappels 2 & 3. This part is a nice walk-through narrows.

Left Looking up at the last rappel in **Lost Spring Canyon**. This is in between Rappels 4 & 5 on the South Fork of Robbers Roost Canyon hike.

Blue John Canyon, Robbers Roost Country, Utah

Location & Access **Blue John Canyon,** including several tributaries, is located due east of **Hanksville,** northwest of the **Maze** or **Hans Flat Ranger Station** (Canyonlands National Park), and southwest of the **Horseshoe Canyon** part of Canyonlands. It's in the Roost Country near where Butch Cassidy hid out, and near the old Robbers Roost-Biddlecome-Ekker Ranch.

To get there, drive along **Highway 24** about halfway between Hanksville and I-70. Between **mile posts 135 & 136,** turn southeasterly and drive toward **Burr Pass,** the highest point in that part of the country, a distance of about 50.9 kms (31.6 miles). See *Area Map, Robbers Roost Country.* Stop, park & camp at one of these places: 300m north of (before) Burr Pass near the **metal culvert;** or 4.2 kms (2.6 miles) south of Burr Pass on **The Maze Road,** turn left or east at the sign and park/camp at **Granary Spring & Motel 6;** or drive the sandy road north from Burr Pass and camp at the **trailhead** for the **West Fork of Blue John.**

Rating The Main Fork (PG SLOT), Squeeze Fork (PG SLOT), Little West & East Forks: 3A III; West Fork, 2A II.

Equipment If going down Blue John and coming out via **Horseshoe Canyon** & the rock art sites, you'll need RG (K&EP & W&R/QL for all slots), a 50 & 15m ropes. **Little West Fork,** one 50m rope, or a 30 & a 30m pull cord. **Squeeze Fork,** one rope 20m & headlamp. Nothing special for West Fork; in East Fork, a 30m rope.

Route Main Fork There are **3 ways** to enter the **Main Fork;** from **Burr Pass Camp (BPC)** walk southeast to the beginning of the slot; walk east to the slot from the **Maze Road** about **1.6 kms (1 mile) south of Burr Pass;** or walk north in the drainage from **Granary Spring.** See map. With care, most people can chimney down **6m** in the first part of the slot, but for safety, use a short rope to aid beginners. See **Cross-section 1.** Below the 6m drop, the canyon is tight & steep, with some challenging DC'ing. Experienced **slot rats** can do this fine, but others may need a little help. This steep & tight slot is less than 75m long. At the bottom, is a deep & dark walk-through narrows about 325m long and about 40m deep--one of the nicer slots around. After that, the canyon gradually opens, and in the area where the 4 tributaries of Blue John come together, are many places you can route-find out of the canyon to the west; or exit via the **West Fork.**

West Fork One way out of all the other forks of Blue John would be to walk up the West Fork. To go down it, start at the **metal culvert & BPC.** Or with a 4WD, drive the sandy track northeast until you see slickrock and a **pinnacle or sandpipe.** Park or camp there and head downcanyon in a shallow drainage. After about 2 kms is a dropoff. Walk along the right (east) side on a little trail, then route-find down 2 steep sections. See **Cross-section 2.** Climbing up is easier than going down.

Little West Fork This is a short tributary to the **Main Fork.** From the **BPC,** walk due east over the top of a rise marked **5764T** (1757m) and into the upper drainage. **Rappel 1** is from a cedar tree **6m away (down 13m)**--so don't cross your ropes at the top. Then the **16m R2,** and a nice slot. Exit up Main or West Forks.

East Fork Start at **Granary Spring/Motel 6.** With a compass & map in hand, head due east. You'll cross the heads of 3 drainages, then turn north in the 4th. About 2 kms into the upper East Fork, you'll come to a 6m dropoff. Don't rap, just rim-walk on the south side for 100-200m and walk into the 500m-long **Upper Slot.** Soon it's more open with many escape routes. Later, you'll come to a series of potholes, some perhaps with water. At the **Big Pothole** (see **Cross-section 3**), you'll find a B&H on the left, add rope & **HL** down about **7m.** Below that it's easy for 40m, then a 4m drop. This one doesn't have an anchor, so with about 3 people, 2 can be belayed, then they can reach up to help the last person down. Or **HL** down from an **Ibis hook,** which the author did on his last solo trip. Exit up Main or West Forks--or if you're good, upclimb Squeeze Fk.

Squeeze Fork Get there by walking north in the 3rd drainage east of **Granary Spring;** or cross-country from the **BPC.** Enter at one of several places. There's **one 8m rap,** then 2 nice DC's. About 80m before the confluence with **East Fork** is an open place, then a **5m DC/rap,** a wading pool, and a vertical 3-4m DC into a dark crack about 25 cms wide (this writer has DC'ed & UC'ed this). Do this by spreading out like a frog, using K&EP. This might be the best part of all Blue John's tributaries! Walk sideways 50m to East Fork. Walk through the lower slot in East Fork, and up West or the Main Fork, or any of several other routes out.

If you want still more adventure and have 2 cars, park one at the **West Side Trailhead** to **Horseshoe Canyon** near where the pictograph panels are located. To get there, drive south & east 39.3 kms (24.4 miles) from Highway 24 to **Horseshoe Junction & information board** (see *Area Map, Robbers Roost Country).* From there, continue left or northeast on the **Green River Road** 8.4 kms (5.2 miles), then turn right or east at the sign and continue for another 2.9 kms (1.8 miles) to the West Side Trailhead (this last part has a couple of rough sections but cars driven with care can make it).

To start the hike, head down any of the upper forks, but instead of returning the same way, continue the full length of Blue John. About 2 kms below where all the upper forks of Blue John meet, you'll find another nice slot and the **Big Dryfall.** Getting through this 200m-long slot means passing **Aron Ralston's chokestone** (he was trapped by a round chokestone and lost his right hand in April, 2003) and a couple of pretty good DC's. Beginners may need a little help getting down in a couple of places.

At the **Big Dryfall,** is a platform and **2 B&H** on the left. These are about 1 1/2m from the edge (make sure the W&R/QL extends to the edge), but with one 50m rope it shouldn't be a problem pulling it from the bottom--just don't cross your ropes. From there, walk downcanyon about 3 kms to Horseshoe Canyon; another 8 kms to the normal **West Side Trailhead.** Along the way, you'll pass 5 of the best pictograph panels around.

Elevations Trailheads, 1728m, 1762m, 1777m, 1798m; top of Big Drop, about 1550m; Horseshoe Canyon below the West Side Trailhead, 1425m; West Side Trailhead, 1600m.

Time Needed Down the **Main Fork** or **East Fork** and back up **West Fork** to the road, about 5-8 hours, but that's with 2 cars, or a mtn. bike as a shuttle. Down Main Fork, over the **Big Dryfall** and exit at Horseshoe Canyon's West Side Trailhead, 8-10 hours. Down **East** and up **West,** 5-8 hours; down **Squeeze** & up **Main** (a little difficult for beginners), 5-8 hours; down **Little West** & up **Main,** back to BPC, 4-6 hours.

Water Take all the water you'll need for the trip in your car and in your pack. Granary and Blue John spring water is suspect or undrinkable because of all the cow pies around.

Maps USGS or BLM maps Hanksville & San Rafael Desert (1:100,000) for driving; Robbers Roost Flats, Whitbeck Knoll & Sugarloaf Butte (1:24,000--7 1/2' quad) for hiking; and a Utah state highway map for the approach.

Flash Flood Danger As always, possible high danger in the short tight slots, little or no danger elsewhere.

Best Time to Hike Spring or fall, but can be hiked in summer or winter, or whenever roads are dry.

Author's Experience The author & Steve Tyler left one car at Horseshoe Canyon, then parked on the main road west of Blue John Spring. They hiked down the Main Fork, installed the 2 B&H at the Big Dryfall, then continued to Horseshoe Canyon Trailhead in 8 1/2 hours. In late 4/2003, he went down the West Fork to check out Ralston's chokestone. In 3/2007, with Brian Olliver, went down Little West Fork, up Main, down Squeeze, but chickened-out (no information & no headlamp) and upclimbed the whole canyon, back to BPC in 8 2/3 hours. On 5/23/2007, he went down East Fork, up Squeeze to the pool (which means he's done the entire canyon), back down and up West to the Maze Road in just over 5 hours.

Map 18, Blue John Canyon, Robbers Roost Country, Utah

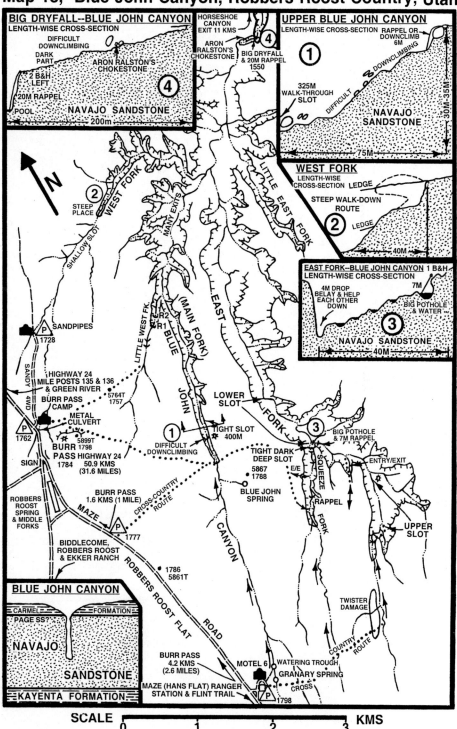

BIG DRYFALL--BLUE JOHN CANYON
LENGTH-WISE CROSS-SECTION
DIFFICULT DOWNCLIMBING
DARK PART
ARON RALSTON'S CHOKESTONE
2 B&H LEFT
20M RAPPEL
POOL
NAVAJO SANDSTONE
200m
④

HORSESHOE CANYON EXIT 11 KMS
ARON RALSTON'S CHOKESTONE
BIG DRYFALL & 20M RAPPEL 1550
④

UPPER BLUE JOHN CANYON
LENGTH-WISE CROSS-SECTION
RAPPEL OR DOWNCLIMB 6M
①
DOWNCLIMBING
325M WALK-THROUGH SLOT
DIFFICULT
NAVAJO SANDSTONE
75M
30M-35M

WEST FORK
LENGTH-WISE CROSS-SECTION LEDGE
STEEP WALK-DOWN ROUTE
②
LEDGE
40M

EAST FORK--BLUE JOHN CANYON 1 B&H
LENGTH-WISE CROSS-SECTION
4M DROP BELAY & HELP EACH OTHER DOWN
7M
BIG POTHOLE & WATER
③
NAVAJO SANDSTONE
40M

N

WEST FORK
②
STEEP PLACE
SHALLOW SLOT
MANY EXITS
LITTLE EAST FORK
EAST
LITTLE WEST FK.
(MAIN FORK) BLUE JOHN
R2
R1
SANDPIPES
P 1728
SANDY 4WD
HIGHWAY 24 MILE POSTS 135 & 136 & GREEN RIVER
5764T 1757
BURR PASS CAMP
METAL CULVERT
5899T
BURR 1798
PASS HIGHWAY 24
1784 50.9 KMS (31.6 MILES)
P 1762
SIGN
ROBBERS ROOST SPRING & MIDDLE FORKS
MAZE
BURR PASS 1.6 KMS (1 MILE)
P 1777
CROSS-COUNTRY ROUTE
BIDDLECOME, ROBBERS ROOST & EKKER RANCH
ROBBERS ROOST FLAT
1786 5861T
ROAD
①
DIFFICULT DOWNCLIMBING
TIGHT SLOT 400M
LOWER SLOT
CANYON
TIGHT DARK DEEP SLOT
5867 1788
BLUE JOHN SPRING
WEST FORK
③
SQUEEZE FORK
E/E
BIG POTHOLE & 7M RAPPEL
ENTRY/EXIT
RAPPEL
UPPER SLOT
TWISTER DAMAGE
CROSS COUNTRY ROUTE
BURR PASS 4.2 KMS (2.6 MILES)
MOTEL 6
WATERING TROUGH
GRANARY SPRING
MAZE (HANS FLAT) RANGER STATION & FLINT TRAIL
CROSS
P 1798

BLUE JOHN CANYON
CARMEL FORMATION
PAGE SS?
NAVAJO
SANDSTONE
KAYENTA FORMATION

SCALE
0   1   2   3   KMS

Above Left Starting Rappel 2 in **Little West Fork of Blue John**. Don't let your ropes get crossed. (Brian Olliver foto)

Above Right This is the nice slot in the main fork of **Blue John Canyon**. This is in the 325m-long walk-through section.

Right Brian Olliver UC'ing the 75m upper section of the main fork of **Blue John Canyon**. You can upclimb or downclimb (UC or DC) this part.

Above Left Middle of **Squeeze Fork**. **Above Right** See the log just ahead of the author? This is where you slide down 3-4m in total darkness. No large people here! (Brian Olliver fotos) **Left** The Lower Slot of **East Fork of Blue John**. **Below** Cristina Amat in the upper part of the Best Slot of **Lost Park Forks**.

Lost Park Forks, Upper Horseshoe C., Roost Country, Utah

Location & Access Shown here are 2 unnamed but major forks, plus several short side-drainages, of **upper Horseshoe Canyon** (it's this same canyon complex that has the big pictograph panels further north). These 2 canyons are east of the Biddlecome, Robbers Roost or Ekker Ranch, west of the Hans Flat Ranger Station and north of Twin Corral Flats. They all run north from **Lost Park**, thus the name.

To get there, drive along **Highway 24** about halfway between I-70 & Hanksville. Between **mile posts 135 & 136**, turn east onto the **Maze Road** and drive past the Jeffery Well, Texas Hill, turn right or south at **Horseshoe Junction** (information board at Km 39.3/Mile 24.4) and head for **Burr Pass at Km 50.9/Mile 31.6**. From the Burr Pass sign, veer left or southeast, and head toward the **Hans Flat or Maze Ranger Station**. About 8.2 kms (5.1 miles) from Burr Pass, you'll come to **Goat Park Junction (Km 58.8/Mile 36.5)**. This junction is hard to see, and the BLM has tried to close it to all motorized vehicles; please don't try to drive this old track--all that country needs is more f'ing ATV's running around everywhere! Parking there is the best place if you want to see only the best slots. Or, continue south to another prominent junction 500m southwest of **Runts Knob**. This is 61.8 kms (38.4 miles) from Highway 24. Park there if you want to see the less-interesting upper ends of these canyons.

Rating For this complex of canyons: **3A III**, plus a couple of **PG SLOTS (one gets close to R)**.

Equipment RG, one rope at least 35m long, W&R/QL, K&EP and R/QD/DC.

Route Start at **Goat Park Junction**. With **map & compass** in hand, head southeast for roughly 2 kms. At a point about 300m north of **6070T** (1850m), turn left or northeast and make a beeline for a Navajo dome marked **5975T** (1821m) on the *Robbers Roost Flats quad*. Route-find to the bottom of the drainage immediately west of that dome. Heading north, you should come to a short little slot, then immediately after that, will be the **2nd Best Slot** (maybe PG SLOT?) in this canyon complex. About 350m north of that, you'll come to the main **West Fork of Lost Park Canyon**. Head north downcanyon. After another 1500m, you'll come to a short slot with a **7m drop** (rig a chokestone in a crack for this rap; or use an Ibis hook), then a DC into a pool where the canyon opens quickly. Or bypass this short section on either side and come in from the bottom end. This place must have water most all the time, because on either side of that pool, are **rock art (petroglyphs) panels**. Let's call this **Moki Hole**. Remember this place because you will leave the canyon at one of 2 nearby routes.

Continue north. After 600m is the **Best Slot**, a nice **PG SLOT**. This one involves 35m of low stemming & interesting DC'ing, then tight walking for about 250m. The slot ends with a **17m rappel** into a **pool**. We rigged for the first time, W&R/QL around one **flaked slab** of rock, backed up by a **knot choke** (a big knot at the end of a piece of webbing) in a crack.

From the pool at 1640m, turn the corner and head southeast in the lower end of **North Faultline Canyon**. About 200m before reaching the main **Horseshoe Canyon**, turn right or south, climb up on a bench, then walk northwest up a minor drainage. Match this map to your topo, and route-find back to the beginning of the slot you just went through (pickup anything you may have cached). From there, head back upcanyon to **Moki Hole**, then exit heading either due west up easy slopes; or northwest up the **South Faultline Canyon**. Once out of the canyon and in the southern part of **Goat Park**, head west past Joe Biddlecome's brush corral to the old vehicle track and back to your car.

Elevations Goat Park Junction Trailhead, 1871m; Runts Knob TH, 1913m; below the rappel, 1640m.

Time Needed Depending on exact route taken, and time wandering, 5-9 hours.

Water Carry it with you, but there may be water (in winter) in some places, as shown on the map.

Maps USGS or BLM maps Hanksville & San Rafael Desert (1:100,000) for driving; Robbers Roost Flats & Head Spur (1:24,000--7 1/2' quad) for hiking; and a Utah state highway map for driving the approach.

Flash Flood Danger Only in the Upper & Lower Slots and for a short distance in each.

Best Time to Hike Spring or fall, but summer & winter have their own appeal.

Author's Experience In 3/2007, he spent 2 full days scouting almost all the upper forks & branches, then on 4/27/2007, with Brian Olliver, Cristina Amat & Ryan Cornia, we went directly to the Best Slot (the only part the author hadn't done) and returned in less than 6 hours.

Brian Olliver & Ryan Cornia (fixing W&R/QL around a flake stone) at the top of Rappel 2 in the **West Fork of Lost Park Fork**.

Map 19, Lost Park Forks, Upper Horseshoe Canyon, Robbers Roost Country, Utah

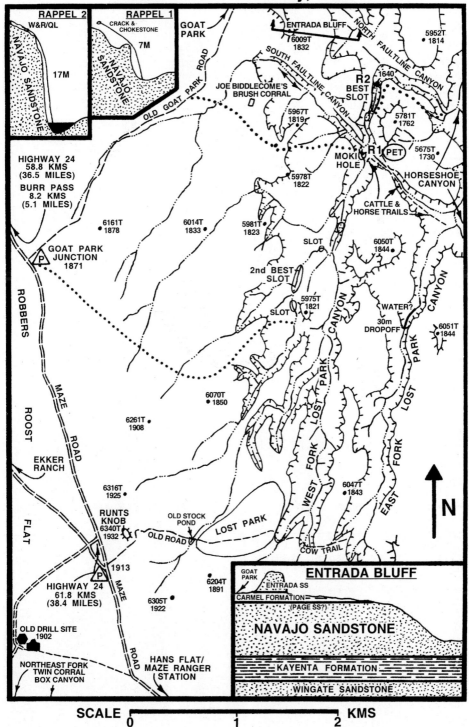

RAPPEL 2
W&R/QL
NAVAJO SANDSTONE
17M

RAPPEL 1
CRACK & CHOKESTONE
7M
NAVAJO SANDSTONE

GOAT PARK

OLD GOAT PARK ROAD

ENTRADA BLUFF
6009T 1832

NORTH FAULTLINE CANYON

5952T 1814

SOUTH FAULTLINE CANYON

JOE BIDDLECOME'S BRUSH CORRAL

R2
1640
BEST SLOT

5967T 1819

5781T 1762

HIGHWAY 24
58.8 KMS
(36.5 MILES)

BURR PASS
8.2 KMS
(5.1 MILES)

R1 PET
MOKI HOLE

5675T 1730

HORSESHOE CANYON

5978T 1822

CATTLE & HORSE TRAILS

GOAT PARK JUNCTION 1871

6161T 1878

6014T 1833

5981T 1823

SLOT

6050T 1844

ROBBERS

2nd BEST SLOT

WATER?

30m DROPOFF

6051T 1844

5975T 1821

SLOT

ROOST

MAZE ROAD

6070T 1850

LOST PARK CANYON

6261T 1908

EKKER RANCH

WEST FORK

EAST FORK

6316T 1925

6047T 1843

FLAT

RUNTS KNOB
6340T 1932

OLD STOCK POND

OLD ROAD

LOST PARK

N

1913

COW TRAIL

HIGHWAY 24
61.8 KMS
(38.4 MILES)

6204T 1891

6305T 1922

MAZE ROAD

OLD DRILL SITE 1902

ENTRADA BLUFF

GOAT PARK
ENTRADA SS
CARMEL FORMATION
(PAGE SS?)

NAVAJO SANDSTONE

KAYENTA FORMATION

NORTHEAST FORK TWIN CORRAL BOX CANYON

HANS FLAT/ MAZE RANGER STATION

WINGATE SANDSTONE

SCALE 0   1   2 **KMS**

103

Above Left Ryan & Brian in the lower part of the Best Slot in the **West Fork of Lost Park Canyon**. **Above Right** Top of Rappel 2 near the end of the **West Fork of Lost Park Canyon**. **Right** Cristina, Ryan & Brian at the bottom of Rappel 2 in the West Fork of **Lost Park Canyon**. **Below** Brian in the upper part of the Best Slot of the **West Fork of Lost Park Canyon**.

All fotos on this page are from **North Fork of No Mans Canyon**. **Above Left** Brian Olliver DC'ing between R1 & R2. **Above Right** Rappel 3. **Left** Starting the last Rappel which is down a crack. **Below** A telefoto lens shot of Brian coming out of the crack at the beginning of the last rap. Before the last guy comes down, separate your ropes for an easier pull from below. Right at Brian's feet is a crack which could jam-up your rope.

Larry, No Mans & Bull Pasture Canyons, Roost Country, Utah

Location & Access The 3 major drainages, and 2 smaller ones on this map, are located immediately west of **Crow Seep** and the historic **Biddlecome-Ekker Ranch,** sometimes referred to as the **Robbers Roost Ranch**. They're also east of Hanksville and the Dirty Devil River.

To get there, drive along **Highway 24** halfway between Hanksville & I-70. Between **mile posts 135 & 136,** turn east and south and drive the **Maze Road** heading for **The Maze District of Canyonlands**. Have in hand the metric maps listed below, and see the *Area Map, Robbers Roost Country,* page 65. You'll drive past Jeffery Well, The Flat Tops, Texas Hill & Ranch and later turn south at **Horseshoe Junction** with the **Information Board**. After about 50.9 kms (31.6 miles) you'll arrive at **Burr Pass**. At that point you'll see a sign; straight ahead left is the Hans Flat (Maze) Ranger Station, but veer to the right and head toward the **Ekker Ranch**, 11 kms (7 miles--closer to 6.8). Drive this good graded road southwest, but about 300m before the ranch buildings & corrals, turn right or west and drive 800m (.5 mile) to a **junction**.

Larry & Bull Pasture Canyons From the **junction**, and if you have 2 cars and will be doing a shuttle, turn right toward **Angel Point**, but after another **1.8 kms (1.1 miles)** you should be at the main shallow drainage of **upper Larry Canyon**. Up to this point you're normally OK with a 2WD car. Park one car to begin hiking.

Now to reach the exits on **Larry Point**. Go back to the **junction**, and turn right or west. Head southwest always staying on the most-used road. Soon you'll be beside a shallow & sandy wash (this could be a problem for 2WD's if a lot of people start driving it or it gets flooded, but the author made it once in his VW Golf TDI in dry conditions!). When you reach slickrock and a big dropoff into the **North Fork of Twin Corral Box Canyon** (this is 5.3 kms/3.3 miles from the ranch turnoff, and 15.9 kms/9.9 miles from Burr Pass), turn right or west to drive upon the flat Carmel bench on a good road. Drive southwest on **Larry Point** to a place **18.6 kms (11.5 miles) from Burr Pass** and park next to a **juniper tree** on the right (maybe a stone cairn too?). This is the closest parking place to **Exit 1**; it's also **1.1 kms (.7 mile)** back from the end of the road on **Larry Point**. If you have but one car, park here and walk cross-country to the head of the canyon to begin your hike. Or drive to a point 19.3 kms (12 miles) from Burr Pass; this is the beginning of an **old road**, now a **trail**, leading down to **Bull Pasture**, and the end of **Exit 2**.

The other shorter minor drainage here has no official name, but it's the main drainage into Bull Pasture, so this writer is calling it **Bull Pasture Canyon**. Park at the beginning of the **old road/trail** going down into Bull Pasture, which begins **320m** (.2 mile) from the end of the Larry Point Road.

No Mans Canyon--North (& South Branch) and South Forks Start at the **junction** mentioned above. From there, drive to a point about **16.8 kms/10.5 miles** from **Burr Pass**; as you're driving west, look for a minor road & perhaps a small cairn to the left or south marked **1664m** on this map. Drive due south about 250m and park/camp right at the beginning of the **North Fork of No Mans Canyon**. The head of the **South Fork** is about 2 kms south of the 1664m trailhead.

Rating Upper Larry Canyon: 3A III or IV PG SLOT, depending on your exit: **3A III** for **Lower Larry**. Larry Canyon has a deep, dark slot in the upper end, and one rappel in the lower part. **Bull Pasture Canyon: 3A II or III** is short & quite easy. It has 2 rappels and some pretty good DC'ing in one tight section.

No Mans Canyon (NMC) Both upper forks have some nice slots, DC'ing, rappels, then a big rappel where each canyon begins to open up. You can go down to the last big rappel in each, and return; or make a big rap, go downcanyon and come out one of **2 old horse** or **cattle trails**. The **North Fork: 3A II or III; South Branch: 2A II,** or **3A III** if doing the last big rap; the **South Fork: 3A II, III or IV**, depending on where you come out. At times each could be **3B** with some pothole wading, but likely never any swimming.

Equipment RG, W&R/QL, K&EP, **ropes:** one 60m for each canyon except for NF/NMC take one 60m & pull cord (or several shorter ropes equal to about 30m for a possible hard pull on a 33m rappel).

Route For **Larry Canyon**, regardless of where you park, walk to the 1st dropoff in the main drainage. The normal way in is to walk left from that dropoff maybe 40m, and look for a **steep DC**. Help one person down to the next level, then he/she can help others--it's a little tricky. From there look for some **moki steps** on the south side just west of the DC; go down facing the wall. You can also rappel in after installing W&R/QL.

Once in, walk down into a shallow slot. Soon the slot gets deeper & deeper, and you'll come to several DC's, mostly from chockestones. About 1 km into the slot you'll come to the first of 4 rappels. DC a ways to find a chokestone with W&R/QL. From there **Rappel 1** is **24m** down a twisting groove. At the bottom is a pothole, but we found no water even though it flooded about a week earlier.

Immediately below R1 is a chockestone lodged above with W&R/QL hanging down; this is above flood waters, and always out of the sun. From there, do **R2**, the first part of which is **16m**. Immediately below that is a 2nd drop, then a 3rd (but you can DC the last 2 parts). Together, these add up to 25m and can be done with a 50m rope. You'll likely find some water at the bottom of R2, and may get your feet wet.

At that point the canyon opens a little, then slots up again. This is the deepest & darkest part and no other canyon is better! The lower part of the next section is **tilted** and you have to walk holding yourself up with one hand, arm or elbow, while carrying your pack in the other hand. Use K&EP in this part.

Finally you'll get out of the tight slanted slot and walk normal for nearly 500m, then comes another short narrow section with another good **DC**--or possibly rig a rappel of **4m** to be safe. Just below that are **R3 & R4**. This is where you leave the Navajo SS and drop off the Kayenta Formation benches into the widening canyon below. You should find a long sling rigged up in the drainage bottom, then the **16m rappel** to a short level place, then with the same rope and anchor, the 2nd drop can be made to complete **R3**. Do this with one 60m rope, or a 50 and any other shorter rope (use a beaner block--this means rapping down the single unknotted side). The **2 drops** add up to about **28m**. Immediately below this is **R4**. There you'll find 2 big chockestones, one with W&R/QL for a **22m rappel**. This ends the upper technical section.

Or, you can skip these last 2 raps by **walking along the bench** on the left or south side. After about 550m, route-find down over a ledge or two, then walk west a little before going down a slope to the bottom.

A ways below R4, or about where you come off the bench, you may see a seep and pools (?). Beyond that, it's a boring walk along the dry creek bed down to where **Exit 1** is shown on this map. Carry a map in hand and observe side-canyons carefully. There's no way to describe where it's at, but it is just a crack in the vertical-looking wall facing northeast. Look for tracks, trail & cairns on the left or southwest side. This is a steep climb, but anyone can do it by foot-jamming up the crack.

Or continue downcanyon to **Exit 2**, which is on the line (and in the middle parts) between *Sections 24 & 25, T29S, R13E,* as shown on the *Hanksville metric map*. This is also the **entry** to **Lower Larry Canyon**. Leave the dry creek bed before you reach this exit, then angle up to the base of the cliff. The site will be obvious when you get there; it's the only place around where you'll have a chance of getting out of the canyon. At the first ledge, you may have to have one or 2 people push another one up, then that person may have to lower a rope for the last person to climb out. From there, climb up and to the right or west. You'll come to a short easy pitch, then curl left and climb up a steep place with your boots in a crack. Once on top, walk south-

Map 20, Larry, No Mans & Bull Pasture Canyons, Robbers Roost Country, Utah

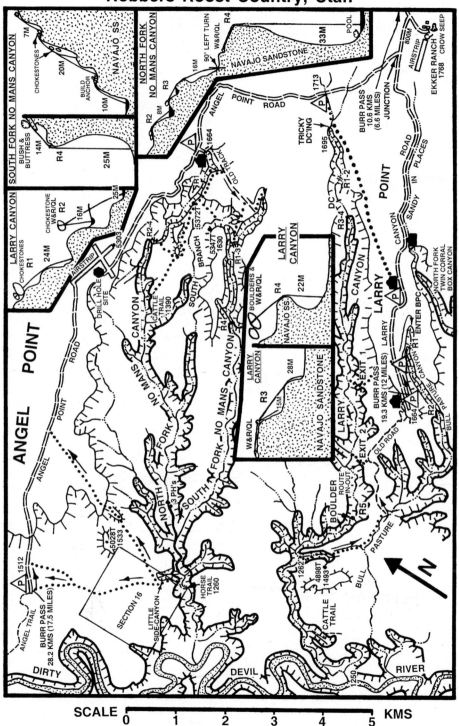

east across some Navajo domes to the **Bull Pasture Trail** (the remains of an old road) and up to your vehicle at the beginning of the road. Your car will be about 1 1/2 kms from the Exit 2.

Lower Larry Canyon. From the beginning of the **Bull Pasture Trail**, walk down the old road about 1 1/2 kms. When the road begins to take a southerly course, route-find northwest over Navajo slickrock to find the **Exit 2** route described above. Help each other down the 2nd tricky DC, then head for the drainage bottom and downcanyon. In the first 2 kms, you'll bypass 2 dryfalls, then come to a 3rd. Once there, bench-walk to the right about 40m to find W&R/QL around a boulder and make a **22m rap** (**R5** on map). From this rappel, continue downcanyon about 100m to find a DC just above a small permanent pool. With one hop across, you should keep your feet dry. Immediately below that is a house-sized **boulder**.

Continue downcanyon about 2 1/2 kms and when you're about 300m beyond a **prominent side-canyon** on the right, turn left or south and climb up a minor drainage (marked 1262m on this map). Climb up between the drainage and terraces on the right (you can also continue downcanyon to within 1 km of the Dirty Devil River, then climb up an **old cow trail** in a short side-canyon to the south and into the western end of **Bull Pasture**). Higher up, you may see some cairns marking the route. Once out of the canyon, walk southeast into Bull Pasture (so named because bulls were once placed there and kept away from cows for 11 months of the year). About 3 kms from the bottom of Larry Canyon, watch carefully on the left or north for the old road. Walk up this track about 3 kms back to your car on Larry Point.

Bull Pasture Canyon It's best to park at the beginning of the **Bull Pasture Trail**, which is a natural landmark, then walk back east to the rim and finally down in at the head of slot. Soon you'll come to **R1**. Rig some W&R/QL around a small chokestone in a groove and make a low-angle 2 stage rap (or maybe HL), or maybe DC (?) about 15m. Below that are some good DC's in a slot, then the canyon gradually opens. A little more than 1 km below R1 is **R2**. This is a **25m free rappel** from W&R/QL around a large bush--this means you tether your pack to the front of your harness. About 600m below R2 may be a seep (?) with pools and good drinking water, then the walk out through the upper part of Bull Pasture and up the **old road** to your car.

North Fork of No Mans Canyon From the trailhead 250m south of the main **Angel Point Road**, circle around to the south side where you can walk down in via a little side-canyon and a sandy slope. Head downcanyon in first a shallow drainage, then it gradually gets deeper and in places turns into a slot with some DC'ing. After about 600m is **R1, 7m** down from a chokestone. After several short jumps/slides, and about 3 kms into the canyon, you'll come to the last 3 rappels one after another. **R2** is down about **8m** from a good chokestone and into a possible pool (never more than chest deep); then **R3**, also from a chokestone and down **16m**. About 5m left or west from the bottom of R3 will be a couple of cracks where people have placed small chokestones as anchors for the last rappel, **R4**.

To begin, DC the first **6m** in a tight, low angle **chute**, then do **R4** which is **33m**. **First person down** should separate the ropes and walk the pull-rope-side up to a bench on the south side of a pool. **Last person down** should place the rappelling rope to one side of a crack for an easier pull from below. From the bottom, head downcanyon about 1200m and exit left or south up an **old cattle trail** (it has piled-up stones & steps cut in the slickrock). From there, route-find southeast around the south side of some Navajo domes, then east along the south side of the ridge just south of the North Fork and back to your car. There's a little route finding here.

Or if you like longer hikes, you could continue downcanyon and exit to the north via an **old horse trail** beginning in the extreme **southeastern corner** of **Section 16** (a Section is one square mile on any USGS map), T29S, R13E. Once out of the canyon, route-find NNW and after 3 1/2 kms, you'll reach the end of the road on **Angel Point** and the parking for the **Angel Trail**. The end of the road is 28.2 kms (17.5 miles) from Burr Pass. Read more below on this trail.

South Branch of North Fork This is the obvious southern fork or branch of the North Fork. This one has 3 or 4 entries/exits, so perhaps the appeal isn't as good as the North Fork itself. To enter this one, walk south, then west from the TH following the exit route/trail for the North Fork. Head for **5372T** (1637m), but once over the first rim, head southwest and enter about 100m east of **5341T** (1630m); this will be the first of 5 short but not-so-difficult slots. You may find one deep pool, then a **Tunnel Slot** is interesting, if you can squeeze down inside it (it can walk around this part); later you pass through one **bridge**, then finally a 600m slot before the big drop. As of 4/2007, it would seem that no one had made this last rap, so if you do, tie a long piece of webbing around a slab of rock with a slight separation from the wall on the right. One 60m rope will likely do, but take a pull cord just in case (or several shorter ropes to be your pull rope). Return via the **old cattle trail** shown, the same way as if doing the North Fork of No Mans. Or backtrack up part of the slot. You can do all or part of this canyon as you return from the North Fork.

South Fork of No Mans Canyon From the same trailhead as discussed above, walk around the head of North Fork, then veer a little west of due south and after walking about 2 kms along an old straight-line **surveyor's track**, you'll come to the head of South Fork. From the canyon rim, turn east, walk 300m and enter right at the head of the drainage. Walk downcanyon. The slot begins almost immediately with some DC'ing and tight sections.

About 1 km into the slot, the canyon opens momentarily and you'll find a walk-out exit to the right or north, then eastward. At that point the canyon steepens and you'll have 3-4 short rappels from chokestones. Now if you decide not to make the last big rappel further downcanyon, and return the same way, you could tie a 60m rope to a chokestone, rappel over all dropoffs and jumar back up later. Or, leave W&R/QL on chokestones for each of 2-3 rappels. If doing this, you have to pile up stones in one pothole creating a deadman anchor. No need for bolts.

At the bottom of these rappels, the canyon turns left and it's a 300m walk until you come to a short tight slot with an interesting DC, passing into a pool. After that are several other DC's, then the canyon opens for good with one section having a flat slickrock bottom, perhaps the top of a limestone lens (?).

The **last rappel** in the canyon is about **25m**. To do this, you'll need to tie W&R/QL to a rock slab and a small tree. You'll need at least 15m, maybe a little more, to tie it, then HL down to the beginning of the last rappel. Be sure the webbing extends to the edge. You can start the actual rappel by stepping down 1m to a small ledge. This way you can pull your rope easily from below. We didn't do this rappel, but did lower a rope from the edge and got a good measurement of the drop.

After the last rappel, walk downcanyon about 4 kms to where the **North Fork** enters from the right or north. About 1 km before you arrive at the North Fork, and as you come to 3 closely spaced potholes & dryfalls, stay on the bench to the left or south (if you stay in the bottom, you'll have several other minor dropoffs, pools, and thick vegetation to deal with. Also, at the very end, you'll have to climb down a tree or make a short rappel into the North Fork). From where the 2 forks meet, bench-walk downcanyon about 300m, then as the canyon is turning left, route-find down over a series of minor ledges to the small stream below. About 60m below where you enter the **main canyon**, is a little year-round **waterfall** & deep pool.

Once down in the main canyon, you have a choice of 2 exits. The 1st would be to hike up the North Fork and exit via the **cattle trail** marked 1390m, then walk & route-find east back to your car. A 2nd way would be to continue downcanyon about 1 km below the **waterfall** and look to the west on the east-facing wall for a 2nd man-made **horse trail** marked 1260m. There you'll see where someone stacked up rocks just as the trail is climbing up through the highest part of the Wingate Sandstone to the bottom of the Kayenta Formation. The trail then runs north along a bench for 150m, ascends another bench, and heads northwest up and out the western wall of a **little side-canyon**.

Once out of this side-canyon, you could, with map & compass in hand, walk a little west of due north and climb out the final part at or near the **end of the road** on **Angel Point**. If you have a 2nd car (or mtn. bike), leave it there before your hike for a ride back to the head of the canyon. Or once out of the main canyon, head northeast and make the final climb out through the Navajo (and Page SS) walls somewhere east of the Navajo dome marked **5028T** (1533m) and back to the road. This would be the best route if you have only one car and will be road-walking back to the head of the canyon.

Elevations Crow Seep--Ekker Ranch, 1768m; head of Larry Canyon, 1695m; end of Larry Point Road and beginning of Bull Pasture Trail, 1664m; bottom of 1st exit to Bull Pasture, 1262m; Dirty Devil River, 1250m; trailhead for both forks of NMC, about 1664m; bottom of cattle trail in North Fork, 1390m; bottom of horse trail out of lower NMC, 1260m.

Time Needed For **upper Larry Canyon** and using 2 cars & Exit 1, will take 4-7 hours; but the same trip using 1 vehicle could take 6-9 hours round-trip. For **lower Larry** or **Bull Pasture Canyon**, it'll take roughly 5-7 hours round-trip each. To do the **North Fork NMC**, exiting the cattle trail at 1390m, will take 4-5 hours. **South Branch NMC**, 3-5 hours, but 5-8 hours if doing the big rappel and returning via the old cattle trail. Going all the way down **South Fork NMC**, then up North Fork and out via the cattle trail, will take 8-10 hours--or a little less time if you exit via the horse trail at 1260m and walk back to a 2nd car on Angel Point. To go down to the big dropoff in each fork, then return the same way, may take from 4-6 hours.

Water Take all you'll need in your car and pack. There may be drinkable water in pools in lower **Larry**, and maybe a seep below the last rappel in **Bull Pasture Canyon**. You may find pools & seeps below the 2 big rappels in the **North & South Forks of NMC**, plus there will always be running water and/or pools in the lower parts of both forks down to the waterfall and nearly to the horse trail.

Maps USGS or BLM maps Hanksville & San Rafael Desert (1:100,000) for driving & orientation; Robbers Roost Flat & Angel Point (1:24,000--7 1/2' quad) for hiking.

Flash Flood Danger High in the slots of all canyons, but there's not a lot of drainage above each so only if a storm dumps right on you would there be any danger.

Best Time to Hike Spring or fall (Easter time can be crowded!), but it can be done in summer or winter. If you have a 2WD, you may have better luck with sandy roads in winter than in summer; that's because there's more moisture in the sand making it firmer, and with less traffic to churn or soften it up.

Author's Experience With Scott Patterson, we went to **Larry Canyon** on 3/17/2000 and did some scouting that afternoon. Next day, we did the upper end of the canyon from our vehicles left near the beginning of the Bull Pasture Trail (and using Exit 2), in 9 1/2 hours round-trip. With John Summerson we did the lower end of Larry in about 7 hours, which included exploring lower Bull Pasture Canyon. Next day, **Bull Pasture Canyon** was completed in 5 2/3 hours. With Ryan Cornia and 2 vehicles, we did **Upper Larry** bypassing of the last 2 raps and using Exit 1 in just over 4 hours.

In 2002, he and Scott Patterson hiked down the **North Fork NMC**, but didn't do the last big rap. They retreated, got out of the canyon and walked back to their cars in just over 4 hours. In 3/2007, with Brian Olliver, we did the entire North Fork in just over 6 hours. In 4/2007, along with Ryan Cornia & Cristina Amat, we did the **South Branch** of North NMC to the big drop and back in 3 1/2 hours.

Another time, the author parked at the head of North Fork, walked into the canyon via the cattle trail at 1390m, then hiked all the way **down North Fork, up South Fork** to the big dryfall, and finally back down and out the horse trail to the north and back to the **Angel Point Road**. From there it was road-walking back to his car, all in a long & tiresome 9 1/2 hours.

Later after some scouting trips, and with Marty Glaubitz & Steve Brezovec, we camped at the 1664m TH and walked to and into the **South Fork**. Three ropes were left at the 3-4 rappels, then stopped at the top of Rappel 4, took a measurement and fotos, and backtracked to the cars. Round-trip took 5 1/2 hours.

The last rap, about 33m, in the **North Fork of No Mans Canyon**. Be sure to separate the pull rope or cord from the rappelling rope before the last person comes down so they don't cross each other.

Twin Corral Box Canyon and the North and Northeast Forks, Robbers Roost Country, Utah

Location & Access Twin Corral Box Canyon (**TCBC**) is the 2nd longest & largest canyon complex in Robbers Roost Country. It drains the area around the historic Robbers Roost-Biddlecome-Ekker Ranch and flows southwest to the Dirty Devil River. It's also in between **Larry & Sams Mesa Box Canyons**.

To get there, read the driving instructions for the Middle Forks of Robbers Roost Canyon. Make your way from **Highway 24** to **Burr Pass** along the **Maze Road**, a distance of 50.9 kms (31.6 miles). From there, to reach the **North Fork of TCBC**, veer right at the sign stating the *Ekker Ranch, 7 miles [11 kms]* and proceed in that direction. After 10.6 kms (6.6 miles) and about 300m before arriving at the ranch, turn right or west and drive 800m (.5 mile) to a **junction**. From there, continue straight ahead instead of turning right onto the Angel Point Road. Drive southwest **toward Larry Canyon**, but 5.3 kms (3.3 miles) from the ranch turnoff, part of which will be in a dry sandy creek bed, you'll come to slickrock and a **big dropoff** into the upper North Fork (15.9 kms/9.9 miles from Burr Pass). Park/camp there on slickrock.

To reach the short & sweet **Northeast Fork TCBC**, veer left at the **sign at Burr Pass** and drive south on the **Maze Road** going toward the Maze or Hans Flat Ranger Station. After driving another 11.1 kms (6.9 miles) and at a major junction (61.8 kms/38.4 miles from Highway 24), turn right for 300m, then left or south onto an old track running an additional 1.4 kms (.85 mile) to an old **drill hole site**. Park/camp there, or with a 4WD continue south uphill for another 250m and park at **6268T** (1911m).

To reach the main or upper **Twin Corral Box Canyon**, veer left again at the **sign at Burr Pass** and drive toward the Maze or Hans Flat Ranger Station. After driving **64.7 kms/40.2 miles** from **Highway 24**, turn right or south and drive another 1.6 kms (1 mile) to the shallow drainage of upper TCBC. This is where you'll begin hiking if doing the entire canyon; or just an upper section.

To shorten the hike by eliminating the upper canyon, which isn't that interesting for slot enthusiasts, turn west from just south of TCBC's dry creek bed onto a faint, winding, and sandy-in-places track. Drive about 1.9 kms (1.2 miles) and park in an open cactus meadow near a shallow drainage at 1860m with piñon/juniper trees nearby. Starting there will get you to the **Big Rappel** quicker along a **big horn sheep trail**.

To get to the **middle part** of TCBC and its short **South Fork**, continue south past TCBC for a total of **4 kms (2.5 miles)** from the Maze Road. In that area look for a very faint track heading due west across **Twin Corral Flat** (this is shown as a trail on *The Pinnacle 7 1/2' quad*). With a 2WD, you can drive about 4 kms to a steep rough downgrade which is about 500m east of the letters *4WD* on *The Pinnacle quad*. If you can get your 4WD down, and back up that rocky hill, then you could drive another 3 kms before running into **deep, deep sand** which begins about 500m south of the elevation marked **6003T** (1830m). If you have two 4WD's, camp there, then backtrack with 1 vehicle to begin the hike.

Rating North Fork: **3B II or III PG SLOT** is short & sweet and tight & dark in places; Northeast Fork: **2A I or II PG SLOT; Main Fork of TCBC: 3A IV or V**, or in the minds of some, maybe **4A R IV or V**, because of one 75m free rappel. The entire Main Fork is a very long all day hike; or it could be a 2 day affair; or a bit shorter if beginning at the big horn sheep trail and at the top of the Big Rappel.

Equipment North Fork--RG, K&EP, W&R/QL, R/QD/DC, one 60m rope for 1st rap, one 30m rope, 1 PR per/person, map & compass, 1 headlamp per/person, and small pack. In cooler weather, a wet/drysuit. Northeast Fork--camera & K&EP, and no pack. **Main Fork of TCBC**--RG, K&EP, one 100m rope (tie it off at the big rappel and pick it up later), leather gloves, W&R/QL, if going down the watercourse through the lower canyon, one 50m rope (or bypass the 2 rappels section and you won't need a rope), lots of water in summer, and map & compass.

Route North Fork is one of the best better slots around. From where you park on slickrock next to the dry creek dropoff, walk southeast along the rim about 125m, climb down to the bench below, turn right, and continue to the head of the slot. Tie a 60m rope to a boulder and do **Rappel 1**. Leave your rope to be picked up later--don't take it down the slot. Another option is to rap off a car bumper into the very head of the slot, but you'll have a couple of good DC's at the beginning. Continue downcanyon. About 75m below R1 will be a small **semi-keeper pothole** (climb or help each other out--no big deal!) with an 8m vertical drop & **R2** on the other side. If no W&R/QL is there, install a new piece on a small chokestone in a crack, and rap.

Further along, it gets dark and a headlamp will be handy, and you'll chimney up 2-3m over tight places. This is where a R/QD/DC comes in handy so you can chimney back-to-feet and suspend your pack from your harness. In the middle of this slot, you'll come to **R3**, a 5m rap from **1 B&H** at eye level on the left. It's possible this can be a **DC**'ed with a short jump at the bottom and likely into water up to chest deep. We were there 3-4 days after a flood, and were in shallow pools wading all the time, including this chest-deep pool.

Not far beyond R3, the canyon gets lighter but is still very deep. Also navigation gets easier. Finally, you'll come to a corner and turn right into **possibly a swimming pool**. After that, the technical part ends.

Once you turn left or southwest, the canyon opens some and you'll notice alluvial benches on both sides up to 8m high. After about 400m of this, you'll come to big boulders and some DC'ing through what remains of a dam created by a **wall collapse** of epic proportions. After the collapse, which created a dam right where you would normally find a dropoff at the Navajo/Kayenta contact, all flood waters were backed up about 400m. Slowly the area behind the dam filled in with sand, then finally one flood overflowed and started downcutting into the sand & debris. Walk fast through the dam rubble and hope the right side doesn't fall on you!

Once through the dam, continue downcanyon for 1500m and turn left or east. Follow the map carefully east over one Kayenta bench/ridge and up to another ridge (**top of the Kayenta**), then turn northerly and skirt to the east side of a **Navajo dome**, then north along a ridge and up to the top of the Navajo/Page/Carmel and the final bench. Rim-walk about 1200m back to your car, picking up the rope you left on the way down.

Northeast Fork From the **drill hole site**, or the **hill top** just to the south, walk SSE down an old very sandy track--don't drive down, you may not get back up! You'll soon come to an **old washed-out stock pond**. Continue south. Enter the main drainage about 1 km above the big dropoff and head WSW down the slot. This is a good one with several tight parts and good DC's. About 9/10's of the way through, are 2 escapes to the right or north, then DC'ing in a tight slot and into a down-sloping 35m-long tunnel with what looks like limestone veins. A cross-section of this looks like a figure "8", which you DC in the middle with one foot on each side. In winter take headlamps. Interesting & fun. At the bottom of this, walk about 100m in narrows, then a 60m+ dropoff. Retreat the same way. Having a 15m rope might help beginners back up to the escapes.

Main Fork of TCBC Since this canyon has one really long rappel, then a very long hike to get out, it might be best to do the upper section in half a day, then do the Big Rap and lower canyon on another long day. First, the upper 2 1/2 kms of TCBC. With a small pack, or maybe no pack, head west down into the upper canyon. You'll immediately get into a slot, but you can DC quite easy, then the canyon opens up and is U-shaped. About halfway through is a 60-70m section which will likely have a pool or two, maybe even some swimming. Fur-

Map 21, Twin Corral Box Canyon and the North & Northeast Forks, Robbers Roost Country, Utah

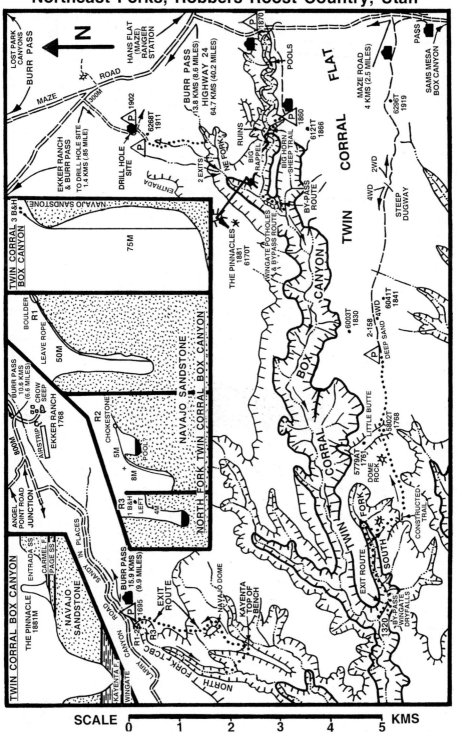

SCALE

0   1   2   3   4   5   KMS

ther along are several easy DC's and probably another shallow pool, then the **Big Rappel**.

From the **Big Rappel**, walk back upcanyon 200m, and route-find up through the benches to the south. The author calls this the **big horn sheep trail** in or out of the canyon. On top you'll be near the end of the track leading to the 1860m car-park or trailhead.

Now for the **Lower TCBC**. Start at the 1860m trailhead, route-find down the **big horn sheep trail** to the **Big Rappel** which this writer measured at **75m**. Right where the drainage drops off are now **2 B&H**, surely put in during 2002 (?), even though there are trees and boulders just above where W&R/QL could be installed. Since this will likely be a **free rap** down a single rope, use a **slanted variable speed control ATC**, and wrap the rope around a **locking biner on your brake-hand-side leg loop**. Some people prefer a **munter hitch** on the leg loop. The reason for the different tactic is, a single rope rappel may be too fast! It all depends on the rope; stiffer ropes make slower descents. Also, tether your pack, regardless of size, to the front of your harness and let it tangle down between your legs; this will keep you sitting upright in the saddle and allow you to use both hands as brakes if needed. It's also a lot less tiring. **Leather gloves** help a lot. You may find some drinking water at the bottom (?). Leave the rope to be picked up later.

From the bottom of the **Big Rappel**, walk downcanyon. After about 1 1/2 kms, you'll make an abrupt turn to the left or south as the lower **Northeast Fork** comes from the right. After about 200m, you'll come to 2 rappels of **25 & 10m** and a pool (you will bypass this pool on the 10m rap) where the Kayenta & Wingate Sandstone (K-W) meet. At that point you could continue right down the chute which means installing W&R/QL to rappel in what appears to be a virgin section (as of 3/2006). A 50m rope & W&R/QL will handle everything here as there are lots of anchors.

Another option would be to walk back upcanyon a ways and route-find up through about 3 ledges, then bench-walk south just east of the 2 raps and around the corner to another side-canyon coming in from the east. Near the mouth of that canyon is a talus slope facing south, which you can scramble down nearly to the bottom, then turn right or west, and again route-find down over several minor ledges to the dry creek bed.

Once back in the bottom of the drainage, head downcanyon. There are no more impediments to deal with. About 11 kms from the pothole section, turn east into what this writer calls the **South Fork**. Read the **Author's Experience** below. About 500m up this canyon, you'll come to a 2m-high wall; climb up, then turn right and scale the wall along one of several routes. This will put you on top of the Wingate. In the upper end of the 2-km-long South Fork, first veer left or north, then back to the right or south and up the uppermost ledges to get out. From there walk a little east of due south and into a minor drainage. Follow the map carefully. At the end of the drainage will be a short section of **man-made trail** up through oak brush, then around the corner to the south will be another **man-made** portion of a **horse trail**.

Once on top, head east over slickrock and **south** of a **dome rock** to the top of a **little butte** with the elevation of **5802T** (1768m). From there go down a little, then up into a sandy area. By the time you're near the + inside a circle marked 2-158 on *The Pinnacle quad*, you should be able to locate a 4WD track. Once on it, walk east back to your vehicle and retrieve your ropes at the Big Rappel.

Elevations North Fork trailhead, 1695m; drill hole site, Northeast Fork, 1902m; Main Fork TCBC parking, 1870m; parking near the Big Rappel, 1860m; lower end of South Fork's escape route, 1320m.

Time Needed North Fork, expect to take 5-7 hours round-trip. **Northeast Fork**, 2-3 hours, or less. To hike all the way down **TCBC** and return via the **South Fork**, 9-14 hours--but this will depend in part if you have two 4WD's, and how close you park to the head of the South Fork. Another option would be to make the Big Rappel, hike down past the K-W dropoffs, then return and jumar back up the Big Rap; which could be a little spooky for some. This might take 6-7 hours (?). Lots of options.

Water Take your own, but you will likely find scattered seeps in the middle parts of the lower TCBC.

Maps USGS or BLM maps Hanksville & San Rafael Desert (1:100,000) for driving; Robbers Roost Flat, Burr Point & The Pinnacle (1:24,000--7 1/2' quad) for hiking.

Flash Flood Danger High in the North Fork, less in NE Fork, little or no danger in the Main Fork TCBC.

Best Time to Hike Spring or fall, but all canyons can be hiked anytime the roads are dry and open.

Author's Experience After a brief scouting trip, the author and Scott Patterson rappelled into the **North Fork** and left a 60m rope. Below that, we pulled the rope below the near-keeper pothole & R2--big mistake! We got down to where the canyon turned west and decided to regress. The only possible place to climb out was about 100m below R2, which is an inclined wall about 30m high at about a 65° angle. Hooking up that, using a G-pick & grappling hooks, took the author 4 2/3 hours. After that, he rappelled back into the slot above, jumarred up the other side on the rope that was left, walked back to the cars and finally lowered a rope from the rim to Scott who jumarred up. A 12 hour day! The next day, the author did a better scouting trip and found they were about 200m short of getting through the slot!

Later on 9/14/2002, and with Steve Brezovec & Marty Glaubitz, went into the **North Fork**. We tied a rope to the bumper of one car, rappelled in and left a rope at R2. We wore wetsuits which kept us warm in the bottom and saved a lot of skin. Round-trip took a little less than 5 1/2 hours. Then yours truly rapped down again and retrieved the rope at R2, but lost lots of skin upclimbing to the long rope. Jumarring up to the car was easy except for the rough surface at the top which made it difficult to get the ascender up. It's recommended you walk down in and rap into the slot from the east side.

Years ago, he went down the **South Fork** from the Twin Corral Flat Road in 6 hours round-trip; in 2002, he did it again in 7 hours. After a couple of scouting trips, he, Byron Lemay and Tiffany Winget went directly to the Big Rappel, measured it at 75m, but for various reasons, didn't do the rappel, then came out the big horn sheep trail.

On 7/19/2003 with Terry Acomb & Jon Jasper, we did the **Northeast Fork** of TCBC in about 2 3/4 hours. In 3/2006, he walked down the **Burr Point Trail** on the west side of the Dirty Devil River and up both **Sams Mesa Box** and **TCBC** on a 4 day backpack. He made one camp at mouth of **South Fork**, then walked to the bottom of the Big Rappel, and back in 8 hours--that's 4 hours one way with a day-pack. On 4/27/2007, with Ryan Cornia & Cristina Amat we did the **Northeast Fork** in 2 3/4 hours. On that day we found some pools but stemmed over them--big floods change canyons!

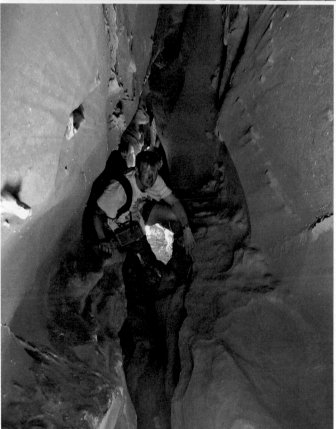

Above Left Steve Brezovec sliding through the tightest part of the **North Fork of Twin Corral Box Canyon**. This and the Northeast Fork are 2 of the better slots around. Both would rate **PG SLOTS**

Above Right Ryan Cornia & Cristina Amat crawling through the **Northeast Fork of TCBC**.

Left Jon Jasper going through the tunnel part of the **Northeast Fork of TCBC**.

Sams Mesa Box Canyon, Robbers Roost Country, Utah

Location & Access Sams Mesa Box Canyon is the most-southerly in the Robbers Roost Country (in this book). It's perhaps the deepest canyon in the area and is the first one south of Twin Corral Box Canyon.

To get there, drive along Highway 24 about halfway between Hanksville & I-70. Between mile posts 135 & 136, turn east & south on the **Maze Road** going in the direction of **The Maze District** in Canyonlands NP. When you get to **Burr Pass (50.9 kms/31.6 miles from Highway 24)** and the **signpost**, veer left and head in the direction of the Maze or Hans Flat Ranger Station. Drive across Robbers Roost Flat, then 13.5 kms (8.4 miles) from Burr Pass, turn right or south from the Maze Road onto a good graded road (64.7 kms/40.2 miles from Highway 24). After 1.6 kms (1 mile) you'll pass the head of **Twin Corral Box Canyon**. Continue south across **Twin Corral Flat**, over some low hills covered with piñon-juniper (P/J), then down to another flat grassy area. About 7.1 kms (4.4 miles) from the Maze Road, stop & park, perhaps next to a **lone cedar (juniper) tree** at **1914m** altitude. This will be the starting point when going down the main drainage of **Sams Mesa Box Canyon**.

Rating This one is rated: **3A III** (or **IV** if you go over R3, down to the spring, then exit from there).

Equipment RG, K&EP, 2 ropes at least 30m long (depending on the 1st slot, maybe one 50?), one 15m rope, W&R/QL, one Ibis hook, map & compass as always, and perhaps ascenders.

Route From your car, and with map & compass in hand, walk northwest across the flats. When you reach the P/J and the very shallow drainage, stay on the right or north side and locate a cattle trail. Later, the drainage deepens and veers to the left or west. As you're walking west, and as a minor drainage enters from the right or north, walk down to the left and to the bottom where the 2 forks meet. Immediately below that is where the bottom drops out and the **Upper Slot** begins.

At that point you have several choices. One would be to DC into the drainage about 40m, then you'll come to a very narrow slot & vertical dropoff. There are a couple of logs & roots where you could tie off one rope (or W&R/QL), then rappel **(R1)**; but **be warned**, this is a tight squeeze and for slender people only. Or, it could be a **tight DC**, but it does widen at the bottom. Lower packs separately, or hang it from your harness.

Or if this looks a bit too tight (this skinny writer who was alone, backed out of this one), you could UC back out, and rim-walk west about 60m on the north side of the slot. Locate a boulder or tree, attach a 50m rope (if you have a 60, take it), and rappel down on the other side of that really tight part.

Another option is to walk the canyon rim west for about 1 km, then head down & back east along an **east-west faultline side-canyon** to enter the main drainage. This is an easy walk-down route. From there you could head back up to see the end of the tight slot, or just head downcanyon.

Once in the canyon, you'll find it narrow but not a slot. The walking is easy with several escape routes. Then comes a **Lower Slot** which begins about 100m above where this main canyon meets an **east fork**. This is where the bottom drops out again, but it's much easier than it first appears. Get into the slot, then span it while climbing over a couple of log jams. Then DC 5m using lots of perfectly spaced foot & hand holds into the bottom. The last 40m is very fotogenic; one of the nicest short slots around. Right where the canyon opens, is an **11m** drop; do **R2** using a 15m rope & large **Ibis hook** attached to one of several natural anchors (make sure the hook doesn't get jammed into a crack or it'll be hard to flip off from below).

Below the Lower Slot, the canyon opens with the Kayenta ledges starting to appear at the bottom. Roughly 4 kms below R2, you'll come to **R3**, at the big Kayenta/Wingate **(K/W) dryfall**. Now at a point about 100 & 350m above this K/W dryfall, you can exit the canyon bottom to the north and climb out via a **man-made constructed trail**; or you can descend this 3rd and last rappel.

If you decide to do the entire technical section of Sams Mesa, then tie a short rope to a nearby tree or boulder, then attach a single 30m rope (or use one 50 and no sling) to that. You will first rap down about 10m to a ledge with a **sharp edge**, then another 15m to the top of a Wingate bench. At this point, bench-walk to the right or north along a big horn sheep trail. After about 350m, start

This scene is immediately above Rappel 2 in the the Lower Slot of **Sams Mesa Box Canyon**.

114

Map 22, Sams Mesa Box Canyon, Roost Country, Utah

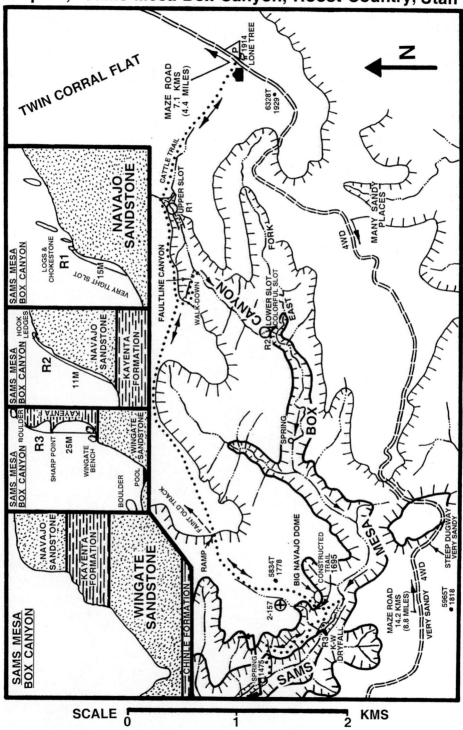

SCALE

0   1   2   KMS

walking down a talus slope to the slickrock bottom, bypassing another dryfall on the right or north. Below that it's boulder hopping for another 300m or so until you come to yet another **5m drop**. Below this is a **spring & pool** no matter how dry the drought may be. There is a DC'ing route to the right, but you may want to use a rope (about 250m below on the left is walk-down route to the green canyon bottom). Fill up on water there, then you could continue on down to the Dirty Devil River, if you like camping & backpacking.

However, for this book, let's return upcanyon and back to your car. From the **spring, walk east** angling up into the **1st minor side-canyon** along Wingate slickrock. As you near the Kayenta ledges, route-find either right or left to reach the upper part of this minor side-canyon. When you reach the 1st prominent bench, contour along big horn trails back upcanyon to the southeast. Counting the canyon you just climbed, you will pass through **3 short side-canyons** before reaching a point above R3 where you can DC to retrieve your rope (another option would be to jumar back up your rope, but place a towel, rag or pack on the sharp ledge going down; or during the act of ascending you'll nick the sheath of your rope for sure!)

After retrieving your rope, back up about 100m from the big K/W dryfall & R3, and walk up through the Kayenta benches east toward a **big Navajo dome**. As you walk up, first veer right a little, then left toward the bench west of the dome. Right at the corner is a crack in the rim and 2 places where some old sheep or cattle man piled up rocks & logs to create a trail. It's hard to imagine taking any kind of livestock up or down this trail because the crack is so narrow! As the author was within 75m of this trail coming down to it from above, he spooked-up 3 full-curl big horn sheep which scattered in all directions. One went down the chute, but a horse would have a hard time doing the same thing.

Once up through this short escape trail, head due north and walk between marks **2-157** in **a circle**, and a **Navajo dome rock** marked **5834T** (1778m) on *The Pinnacle quad*. After about 1 km, veer right or east and as you climb up, be looking for an easy route up through the uppermost Page SS/Carmel rim. Once on top of the Carmel, head northeast around one side-canyon, then east past the Upper Slot (pickup a rope if you left one there), then southeast back to your car. Don't try this without the proper map & compass.

Elevations Car-park, 1914m; big horn sheep & man-made escape trail, 1695m; spring, 1475m.

Time Needed If doing just the upper canyon without going over the K/W dryfall, about 5-7 hours. Or if you choose to rappel over the K/W dryfall and turn around at the spring, it may take 9-12 hours.

Water The only water may be at the spring downcanyon from the escape route.

Maps USGS or BLM maps Hanksville & San Rafael Desert (1:100,000) for driving; and The Pinnacle (1:24,000--7 1/2' quad) for hiking.

Flash Flood Danger Some danger, but only if a cloudburst occurs right above you while in the 2 slots.

Best Time to Hike Spring or fall, but it can be done any time the roads are dry, including summer or winter.

Author's Experience On one scouting trip, he drove his Chevy Tracker 4WD to a point just south of the K/W dryfall and was able to see the escape route or trail on the other side. Later that same day, he hiked in from the east & north, bench-walked the Kayenta north and found the route down through the Wingate to the spring. Round-trip took just over 6 hours.

On a second trip, he attempted to go down the Upper Slot, but with a 20 kg pack full of ropes, backed out, reentered via the faultline side-canyon, checked out the upper slot, then headed downcanyon. Later, he rigged up ropes and did R3, headed down to near the spring, returned and jumarred up his rope (nicking the sheath in the process!), and returned via the escape trail, all in 10 hours. If you're in a hurry and not taking fotos, you can do it faster than this.

This convoluted slot is not far above the 2nd Rappel in the Lower Slot of **Sams Mesa Box Canyon**.

Above Left After the big floods of 10/2006, this 4m DC & fast slide appeared in the middle of **Northeast Spur Fork**. Or you can rap from a knot & short rope on the right side.

Above Right This picture was taken in the **Corkscrew Slot** in the lower end of **Northeast Spur Fork**.

Left DC'ing just before the only real rappel in the **Northeast Spur Fork**. The yellow knee pads with straps worn by the author are not recommended. (All fotos on this page by Ryan Cornia)

Northeast Spur Fork & Big Spring Canyons, Robbers Roost & The Spur Country, Utah

Location & Access The **Northeast Spur Fork (NSF)**, and the **East, West & Northwest Forks** of **Big Spring Canyon (BSC)** are located southeast of the **Horseshoe Canyon Unit** of Canyonlands National Park & the head of Water Canyon, north of the **Hans Flat** or **Maze Ranger Station**, and in that part of the greater Robbers Roost Country known as **The Spur**.

To get there, drive along **Highway 24** about halfway between Hanksville & I-70. Between **mile posts 135 & 136**, turn south & east and using the **Maze Road**, head for the **Hans Flat Ranger Station**, 73.7 kms (45.8 miles) from the highway (see **Area Map**, page 65). On very rare occasions, the road to the ranger station is difficult to reach with a 2WD car due to **sand**, and **road grader & political problems**. If you're concerned, or if you want to check on road, weather or other information, call the rangers at 435-259-2652 between 8am & 5pm daily. You can buy maps there too. From the ranger station/visitor center, drive north on the occasionally-graded **Spur Road** signposted for *Horseshoe Canyon, 22 miles [35.4 kms]*. After driving **19 kms (11.8 miles)**, park on the right on a flat plain just before going down a hill. This is the parking place for the upper **East & West Forks** of BSC.

Or to reach what this writer calls **NSF**, drive to a point **20.1 kms (13 miles)** north of the ranger station to the turnoff to the **Deadman Trail**. You can park there at the **Upper Car-park** for easy access to the upper end of the drainage. Or, if you have a 2nd vehicle (preferably a 4WD to get past one bad place) you could save some walking by spotting it at one of the **Lower Car-parks**. To do that, drive toward the Deadman Trail, but after **3.4 kms (2.1 miles)**, pull off the road & park next to some cedar (juniper) trees. There are several entries/exits nearby and a 2nd parking place just to the west.

To reach the **Northwest Fork-BSC**, continue north on the **Spur Road** past the turnoff to **Deadman Trail**. Roughly 1 km (.6 mile) past the Deadman turnoff, park along the Spur Road about 500m SSW of elevation **5600T** (1707m) on the *Sugarloaf Butte quad*. This will be approximately 21.9 kms (13.6 miles) from the ranger station. Normally, cars driven with care can make it to this place.

Rating The lower part of the **Northeast Spur Fork**, and **West & Northwest Forks**: 3A II or III. The **East Fork** doesn't have any rappels, but does have some challenging DC's: **2A II.**

Equipment East Fork--DC'ing only. **West Fork**--RG, a 40m rope, G-pick & Ibis hook (or leave W&R/QL in the canyon!)--read more below! **Northwest Fork-BSC**--one 50m rope, Ibis hook (or build a deadman anchor at the shortest rap & leave W&R/QL!). **Northeast Spur Fork**, 50m rope (or a 30 & pull cord), and W&R/QL. **K&EP** for all tight slots.

Route Do the **East & West Forks** of **Big Spring Canyon** together. From where you park, route-find down off the flat-top mesa going northeast. After about 800m, you'll be in the **West Fork**; head down the drainage. Soon the canyon slots-up and you'll come to **Rappel 1**. We rapped about **8m** from a chokestone & Ibis hook, then flipped the hook off. About 30m beyond R1 is **R2**. This is about a **12m** rap. On the right are several cracks to place an Ibis hook, but a hole may have to be touched up with a G-pick so you can flip it off from below (flipping it didn't work for us!). Or you can leave a rope at least 20m long around a chokestone 7m from the dropoff, make the rappel, then pick it up on your way into the East Fork. Or leave 7m of W&R/QL on the chokestone and pull your rope. Below the rappels is some challenging DC'ing, then the canyon gradually opens.

From somewhere in the lower end of the West Fork, route-find south up between the 2 slots until you reach the head of **East Fork**. This slot has no rappels, but does have some pretty good DC'ing. If you're taking less-experienced hikers through, you may want to belay them in a place or two. Also, in each fork, there are several places you'll have to make short jumps, or help each other down 2m-dropoffs. You'll find some fairly tight slots, plus 3 bridges & one arch.

Northwest Fork-BSC From where you park, head east a few meters to the red Carmel rim, then down the white Navajo slickrock for about 150m to the bottom and the start of the slot. Almost immediately, you'll come to the **10m R1**. There are lots of holes and bumps there, so the best way down would be to rap off an Ibis hook. From there to the 2nd rap is about 750m; it includes 2 tight slots, 2 escape routes on the left or west, and a couple of dropoffs you may have to jump or help each other down.

At the top of **R2**, which is the beginning of a dark slot, is a large chokestone with W&R/QL which may have to be replaced. You'll have to rope-up, DC a bit, then rap at a low angle into a cavern, then while still on-rope, make the 2nd half of the rappel down into a dark hole. This is about a **20m rap**. At the bottom of R2, the canyon is very deep and impressive for about 200m. From there, simply walk down-canyon with an occasional DC until you reach the end of a shallow slot where there's a drop of 18m into the lower drainage. Beyond that, the canyon opens and is uninteresting, so turn left or northwest, and route-find up slickrock back to the top of the Navajo & Carmel rim. From there head southwest back to your car, perhaps along a faint old vehicle track.

Northeast Spur Fork From the **upper car-park**, and with map & compass in hand, walk southwest roughly 800m and enter the upper end of the canyon at one of several places. Observe this map and compare it to your *Sugarloaf Butte quad*. There are 3-4 entry places above the **circular feature**. This **upper slot** is very fotogenic in places. After about 1 km, the canyon opens for the first time and there's an entry/exit via a **steep gully** coming down from the north above.

After this first open place, you can walk for about 3 kms with alternating walk-through slots and wide open places. However, since the 100 year storm & floods of 10/2006, one dropoff has been created. DC this about 1m, then let it rip for a fast 3m slide--it's easier & safer than it first appears.

In the lower part of the canyon, there's another very fotogenic **corkscrew slot** that rivals Antelope Canyon. Because of the 10/2006 floods, this place was filled in with sand and much different than in 2003. There used to be a 3m drop just beyond a little **natural bridge**, but in 3/2007, it was a walk through. Further along you'll come to an open chamber which is actually a tunnel caused by a **collapsed wall**, then after 200m will be a steep **6-7m DC**. Some may want to be belayed. Immediately below that will be a huge boulder and a **23m rappel**. Rig W&R/QL around a small boulder if needed.

Beyond the rappel, the lower canyon opens and you may see tracks of cattle & wild burros. Roughly 1 km below the rappel will be a side-canyon coming in from the left or southeast; immediately across the canyon northwest of that will be a small indentation in the wall and a steep but easy climb out of the canyon. At the bottom of that route out and on the left is a rock art panel (maybe paleface made?); near the top will be some moki steps. Because of these steps, it's assumed this is an **old Indian trail** (maybe paleface moki steps too?). From there, route-find back to your car.

Elevations Car-parks, 1816m, 1719m, 1695m & 1745m; exits--Northwest Fork, 1585m, NSF, 1548m.
Time Needed Most can do the **West & East Forks BSC** in 4-5 hours; **Northwest Fork BSC** in 2 1/2-

Map 23, Northeast Spur Fork & Big Spring Canyons, Robbers Roost & The Spur Country, Utah

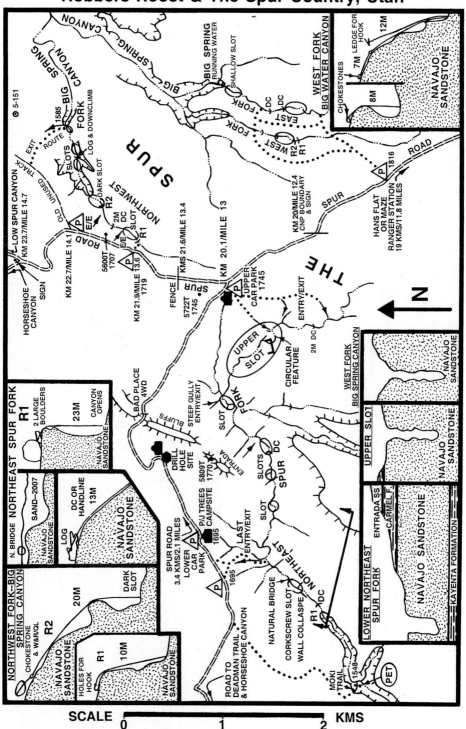

SCALE

0   1   2   KMS

119

-3 1/2 hours. The **NSF** should take 5-7 hours round-trip depending on whether you have one or 2 cars.
Water You'd have to route-find down to **Big Spring**, which runs for 100m year-round, so have water in your car & pack. **Note:** if you're low on water, you'll have to **buy it** in jugs at the Hans Flat or Maze Ranger Station because all their supply is hauled in by tanker truck--and that's not for sale!
Maps USGS or BLM maps San Rafael Desert & Hanksville (1:100,000) for driving; and Sugarloaf Butte (1:24,000--7 1/2' quad) for hiking.
Flash Flood Danger High in the deepest slots, but these are short.
Best Time to Hike Spring or fall, but it can be done anytime the roads are open and dry.
Author's Experience After a couple of scouting trips, Jon Jasper, Terry Acomb & the author, did the **West & East Forks BSC** in a little over 4 hours. Later the same day, we did the **Northwest Fork BSC** in about 2 1/2 hours. On one trip to the **Northeast Spur Fork**, and from the lower car-park, he entered at the lower corkscrew slot, left ropes at 3 dropoffs, found the moki trail exit, went down to Spur Fork, then jumarred back out, all in less than 5 hours. On 3/5/2007, with Ryan Cornia, we did it from the upper car-park in 4 1/2 hours, and with only 1 rappel.

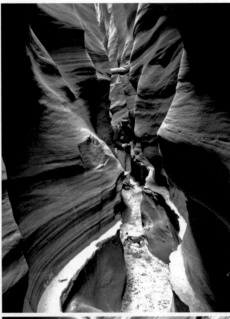

Above Left This is from one of the 2 upper forks of **Big Spring Canyon**--either East of West Fork (?).

Above Right A scene from the **Upper Slot** of **Northeast Spur Fork**. The white stuff on the small ledge is windblown sand.

Right Part of the **Corkscrew Slot** in the lower end of **Northeast Spur Fork**. This foto was taken in the morning hours, but it will be much more colorful if you get there in the afternoon with much more light getting down into the slot.

Above Left Jon Jasper DC'ing just above Rappel 3 in the **East Fork of Water Canyon**.

Above Right This is part of the **East Fork of Water Canyon** between R3 & R4.

Left From the top of Rappel 3 in the upper end of **East Fork of Water Canyon**. This is a 2-stage rap ending in a dark slot.

Water, Tidwell & Low Spur Canyons, Roost Roost & The Spur Country, Utah

Location & Access   These 3 canyons are located in the far northeastern corner of the greater Robbers Roost Country; locally that section is called **The Spur, or Low Spur**. This is all east of Hanksville and north of the Hans Flat or Maze Ranger Station. **Water Canyon** and it's 3 upper forks enter the detached **Horseshoe Canyon Unit of Canyonlands National Park** from the east and southeast. **Tidwell Canyon**, so named by this writer because it runs north from near the old **Phillips Oil Well**, later used by the **Tidwell family** of Green River as a line cabin when grazing cows in the area. **Low Spur Canyon** is located just east of lower Tidwell Canyon in the area known locally as the **Low Spur**.

To get there, drive along **Highway 24** about halfway between Hanksville and I-70. Between mile **posts 135 & 136**, turn south & east onto the **Maze Road** heading for the **Maze or Hans Flat Ranger Station**, a distance of 73.7 kms (45.8 miles). If you're there between about 8am & 5pm, you can get road or other information. From the ranger station, head north on **The Spur Road** signposted for *Horseshoe Canyon, 22 miles [35.4 kms]*. See **Area Map** page 65. This road is graded occasionally and is generally passable for cars, but it has a couple of short sandy or rough places. After driving 23.7 kms (14.7 miles) is **The Spur Junction**; turn left toward **Horseshoe Canyon**. At about Km 25.4/Mile 15.8, pull off the road to the left in the area between **Willow Spring** and elevation **5525T** (1684m) shown on this map & the *Sugarloaf Butte 7 1/2' quad*. This will be your parking place for **Water Canyon**.

If going to **Tidwell Canyon**, continue north to about Km 26.9/Mile 16.7. Park there where the road turns west and goes down a steep sandy dugway toward the old **Tidwell Ranch** and **Horseshoe Canyon**. For a history of the well & Tidwell Ranch, read this writer's other book, ***Hiking, Biking and Exploring Canyonlands National Park & Vicinity*** (out of print 2007, but may be republished later).

If you plan to completely descend either of these 2 drainages and return via Horseshoe Canyon, you can leave a 2nd 4WD vehicle on the rim of Horseshoe to be used as a shuttle back to your car on top. **Warning:** Don't drive down the **steep sandy dugway** unless you walk it first to see if your **4WD** can get back up; normally it's not a problem (In 2003, the author drove his Chevy Tracker from the Phillips Well/Tidwell Ranch to the rim of Horseshoe and found it to be one of the roughest roads he's ever been on). The sandy dugway can change quickly with any big windstorm!

Now back to **The Spur Junction**. To reach **Low Spur Canyon**, turn right or northeast and proceed downhill into the Low Spur Country. About **31.9 km (19.8 miles)** from the ranger station, park at the bottom of a steeper section of road, and just south of a **small butte** marked **5345T** (1629m).

Rating East & Southeast Forks of Water Canyon: 3A II (the big South Fork isn't interesting). If going down to the last rappel in **Tidwell Canyon**, and regressing the same way: 3A II or III. If going all the way through Tidwell, pulling your ropes, and walking up Horseshoe Canyon, then east up the old road passing the Phillips Well, will be 3B III or IV. If doing just the upper part of **Low Spur Canyon**, it rates only 2A II (3A for some), but to do the entire canyon down to Horseshoe, then up Horseshoe and back to the Tidwell Ranch, will be at least 3A IV--probably an overnight trip (?). Long hike. Best to go to the big rappel in each, see the best slots, and return the same way. Tidwell & Low Spur may be **PG SLOTS**.

Equipment Water Canyon, RG, one 50m rope, W&R/QL, K&EP and possibly a G-pick & Ibis hook. Or if leaving ropes and returning back upcanyon from the last 2 dropoffs, take 2 ropes--1 at least 20m & a 50m, plus ascenders. **Tidwell Canyon**, same as above, but if going all the way through and returning via Horseshoe Canyon, two 60m ropes (or one 60 & pull cord), W&R/QL's for 4 raps. If going down just to the big drop and returning, then 2 short ropes of about 15m and ascenders. **Low Spur Canyon**, 1 rope about 30m long, headlamp & K&EP, and maybe ascenders. Or if making the big rap and coming out Horseshoe, then two 60m ropes (or a 60m and pull cord), BK and lots of W&R/QL.

Route Water Canyon   From where you park, locate one of many walk-down routes over the rim to the west; one of these is an **old cow trail** marked with a cairn (unless some ranger has kicked it over!). With map & compass in hand, head WSW for either the East or Southeast Fork. In the upper part of **Southeast Fork**, you'll come to some potholes in an open drainage, then a drop of about **18m**. With no chokestones in sight, you could make a small hole with a G-pick and rap or HL down from an Ibis hook, or install a B&H, or perhaps build a deadman anchor. About halfway down is a resting place, then a vertical drop (this whole thing might even be DC'able (?). From the bottom of this single rappel, head down a narrow canyon with one challenging corkscrew DC, in which some may want to be belayed. This canyon is really not that good. Then walk down to the confluence with the **East Fork**. You could continue downcanyon from there; or exit east up an **escape route**. From the confluence, get upon the bench and route-find southeast, then eastward back to the head of East Fork. Or, you can see 99% of the Southeast Fork from the bottom end as a side-trip to the more interesting East Fork. From the confluence, walk/climb upcanyon to the bottom of the rappel mentioned above and return the same way.

East Fork   If doing this entire canyon, you'll first come to a **14m rap** from a **cedar (juniper) tree** on the right (install W&RQ/L, or leave one rope to be picked up later), then a short slot, followed by a 4m rap/HL from a hook & natural hole down to a flat place. From there, someone will have to install 1 B&H (or build a deadman) and **rap 18m** down a low angle slope. Below this it's open for a ways, then another short slot, and another open place with an easy escape route heading southeast.

After that 2nd open area **(you can enter the canyon at this point)** comes the best part of Water Canyon, a **125m-long slot** which is deep & dark at the bottom. You'll want the smallest pack possible as it's real tight in the first section. After that are 2 DC's; the 2nd of which looks difficult at first, but you can help each other down easily. Then comes a low-angle **2-stage 24m rappel** from **2 B&H** on the left above your head. This puts you into the deepest & darkest part of the canyon.

Below that, you may have to chimney across a pool or two, then just below the confluence of the **East & Southeast Forks**, will be a small tree on the left and 8m beyond, a **10m drop**. If your plans are to retreat back upcanyon, tie a rope to the tree and rappel (or leave W&R/QL and pull your rope). After that, walk 400m to the **last rappel** in the canyon. If you decide to return, tie off a rope to a tree on the left, rappel down about 25 vertical overhung meters, explore lower Water Canyon, then jumar back up and take ropes with (if you do this, place a shirt or something between your rope and the sharp edge so you don't nick your rope jumarring up).

Of if pulling all your ropes, walk down lower Water Canyon which is an interesting Navajo SS gorge. Once in Horseshoe Canyon, check out the 4 or 5 pictograph panels before walking downcanyon (north) to where the old **Phillips Oil Company road** crosses the canyon. Walk up this to the rim above and back to a 2nd 4WD, or road-walk back to your car.

Tidwell Canyon   From the car-park (1719m), you have 3 choices of entry. One is to walk eastward

Map 24, Water, Tidwell & Low Spur Canyons, Robbers Roost & The Spur Country, Utah

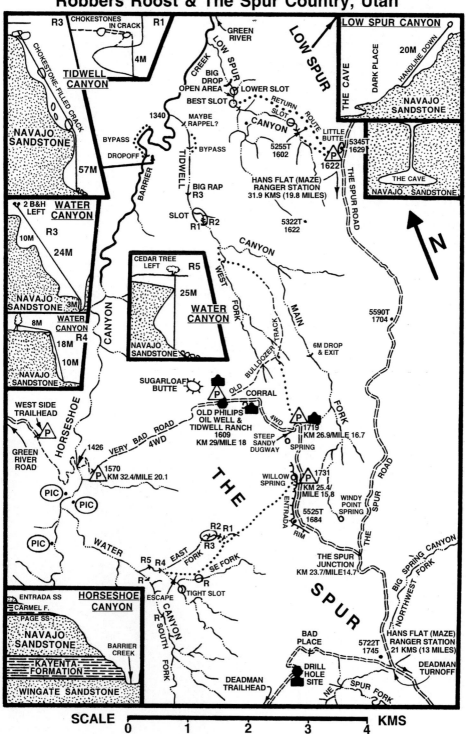

SCALE

0   1   2   3   4   KMS

into the **Main Fork** and head north downcanyon. It's easy walking in a shallow V-shaped drainage. Part way down will be a 6m drop, but you can bypass it on either side and reenter below. Further down is one short slot, then the very shallow **West Fork** enters on the left, your 2nd possible entry. You can also get to that point from the Tidwell Ranch, or directly from the car-park as shown.

From the confluence of the 2 forks, the canyon gets deeper and all of a sudden slots up. At that point is a partial wall collapse and a **4m rappel**, first down through a **tight crack**, then it opens bell-bottom style below. Rapping down from a small chokestone is quite easy, but if you're leaving ropes and ascending the canyon, jumaring back up this part is somewhat strenuous, especially for a beginner. One solution would be to place a B&H out as far as you can reach on the left (west) wall. This would allow you to jumar up just beyond the tightest part.

Below **R1**, the canyon opens momentarily, then you'll rappel **(R2)** down under one of several big chokestones. Install W&R/QL, or tie off a rope at least **10m long**. It's in this part, you may need a **headlamp**. Below R2, it's a dark walk for about 100m, the first part of which is barely shoulder width. This is one of the best slots around. Then it slowly opens and 650m below R2 comes the last rappel, a 1, 2 or 3 stage 57m low angle rap. Most people would stop there at the end of the slot and jumar back out; or you could use a single 60m to reach the bottom, explore down to Horseshoe Canyon, and retreat from there. Or, by pulling your ropes and using **two 60's (or a 60 & pull cord) at R3**, you could hike down lower Tidwell to Horseshoe Canyon. About halfway down lower Tidwell, which is characterized by Kayenta benches, veer right at a big drop and reenter the canyon bottom below. After 300-400m will be one rap of about **10m** from a chokestone. When you reach **Horseshoe Canyon**, which has a **year-round stream**, turn left or west, and hike upcanyon to the national park section and back to your car via the old Phillips Oil Well road heading east (if you do the walk through Horseshoe Canyon, expect a little bushwhacking and use an old cattle trail bypassing one dropoff). This trip will be 5.5 kms (3.4 miles) shorter if you can leave a 2nd 4WD on the rim of Horseshoe Canyon.

Low Spur Canyon From the **little butte** parking, head northwest and into any of the upper drainages. After about 3 kms, will be an 8m dryfall, skirt it on either side; 200m below that will be the beginning of the **best slot** around. After about 40m, you'll come to a couple of chokestones; tie a rope (at least 20m long) to one and HL down 2 steep pitches (we HL'ed back up, but a pair of jumars might be comforting. From there you'll need a headlamp for about 5m, then a tight squeeze & 1 1/2m squeeze-drop to the floor of a **virtual cave**. At that point, above you will be a half-meter wide slot going up to 25-30m, but at your level it widens to 10m quickly and is roughly 5m high. It ends 50m from where you drop in. This is one of the most unusual features this writer has seen. Getting back up the 1 1/2m squeeze-drop will be a challenge, so build a cairn for the last person to stand on.

From the **cave**, walk about 300m in a open canyon with trees, then a 2nd slot. Below that will be a 5m drop into a pool (?) which is easy to climb back out of. Just below that is a large boulder, possible pool, then the **big drop** to a wide open canyon below. If you want to continue, you'll have to rig W&R/QL around a chokestone, or perhaps install a couple of B&H's. You'll first DC a 45° 15m tight slot, then it drops for at least 40m (?). We didn't do this, but the canyon below looks like a walk-through to Horseshoe Canyon.

Elevations Car-parks: **Water Canyon**, 1731m; **Tidwell Canyon**, 1719m; Tidwell Ranch, 1609; **Low Spur Canyon**, 1622m; bottom of Horseshoe Canyon in the national park, 1426m; confluence of Horseshoe & Tidwell, about 1340m.

Time Needed To go down the **East Fork of Water Canyon** to the last rap and return will take 3-6 hours--but how long depends on exactly where you enter the upper end, and if you take in the **Southeast Fork** on your way out. Or, if you pull all your ropes and hike down to Horseshoe Canyon, then out via the Phillips Oil Well road to a 2nd 4WD on the rim, the time maybe be 5-6 hours. To see just the slot of upper **Tidwell** and regress, 4-6 hours; this is what's recommended. To just see the upper part of **Low Spur Canyon**, 3-6 hours; maybe 12-15 hours to make a loop of Horseshoe Canyon.

Water Take your own from a culinary source, but there may be some pools and/or seeps in lower Water Canyon (?), and always running water in Barrier Creek below or north of the national park boundary. Plan to treat or filter this creek water, because of the possible presence of beaver.

Maps USGS or BLM maps San Rafael Desert & Hanksville (1:100,000) for driving & orientation; Sugarloaf Butte, Keg Knoll, Bowknot Bend & Horsethief Canyon (1:24,000--7 1/2' quad) for hiking.

Flash Flood Danger Potentially high in the short tight slots of each canyon as discussed above.

Best Time to Hike Year-round, but late spring with long days is best, or the cooler fall season.

Author's Experience After several scouting trips to this region, the author and Warren Egbert parked at the 1719m car-park, were at R1 in **Tidwell Canyon** in 1 hour; went through the slot and rappelled down a single 60, then walked halfway down lower Tidwell before returning. Round-trip time was 7 1/4 hours. After 3 quick scouting trips into **Water Canyon's 3 forks**, and along with Jon Jasper & Terry Acomb, we bypassed the upper part of **East Fork** and entered at the beginning of the 125m slot. We installed 2 B&H, then proceeded to the confluence with the Southeast Fork, and returned from there in about 3 1/4 hours. Next morning, Jon and the author reentered Water Canyon at the confluence and got to the last or 5th possible rappel, and turned back. Round-trip was 2 2/3 hours (the author has been in lower Water Canyon on several occasions). Along with Ryan Cornia & Cristina Amat, we did **Low Spur Canyon** to the top of the last big drop in less than 5 1/2 hours.

Opposite Page Ryan & Cristina in the Lower Slot of **Low Spur Canyon**; **Above Left** Warren Egbert in **Tidwell Canyon's** nice slot; **Above Right** Part of the slot in **Tidwell Canyon**. **Left** This is part of **The Cave** in **Low Spur Canyon**; **Below** DC'ing in the upper part of the Best Slot in **Low Spur Canyon's PG SLOT.** (Ryan Cornia foto)

Clearwater & Easter Pasture Canyons, South Robbers Roost & North Lake Powell Country, Utah

Location & Access **Clearwater Canyon** drains into the upper end of Lake Powell/Cataract Canyon from the west and **Waterhole Flat**. To get there, it's best to come in from the southwest. First, make your way to the Hite area near the upper end of Lake Powell along **Highway 95**. Between the 2 bridges spanning the Colorado & Dirty Devil Rivers, and between **mile posts 46 & 47**, turn east onto the **Hite Road**. After .6 km (.4 mile) will be an information board with maps and other information about the Maze District of Canyonlands National Park & vicinity; and all their new restrictions.

From there, continue east & northeast along the **Hite Road**, which can be driven in most passenger cars, but beware, you'll have to drive slowly in places. It's best to take a higher clearance car. About 31 kms (19.3 miles) from Highway 95 turn right; and **44.6 kms (27.7 miles)** from Highway 95, will be a **cattle guard**, and a **sign** stating **camping permits required north of that point**. Much of the area ahead or north of you is still in the Glen Canyon National Recreation Area, but the NPS has hijacked the place, and now is charging $30 for camping. Needless to say, most of you will want to camp south of that cattle guard & sign.

If doing the entire main drainage of Clearwater, continue north, then east to **Km 47.5 /Mile 29.5** and park in the dry wash, or nearby. This will be your starting point. To save walking about 3 kms at the end of your hike, drive a 2nd vehicle east to **Km 48.3/Mile 30**, veer right and drive southeast to **Km 50.2/Mile 31.2** and park it at or near the old **Chaffin Camp**, with a nearby stock pond & corral. This will be the end of your hike (or the beginning and end of a hike if going down **Waterhole Canyon**, an upper tributary of Clearwater). The author has yet to do this. Can someone who has please send info (?).

Rating Clearwater: **3A IV or V**. If you decide to make Rappel 6 over the 86m? waterfall it may have a risk factor of **R (?)**. Most people will be reluctant to invest in ropes that long.

Equipment RG, two 50m ropes (or a 50 and a pull cord; or better still a 2nd 50 or 60 as an emergency backup) Or if you're planning to stay in the drainage and do the last rappel measured at 86m?, you'll need 2 ropes of nearly 100m or so and **leather gloves**), lots of W&R/QL's & biners, ascenders just in case. If doing the big rap, **2 radios** would help with communications.

Route From where you park, walk south downcanyon. After 20-25 minutes, and at a dropoff with cottonwood trees below, veer right above **Cottonwood Spring**, and walk down in on an old **cattle trail**. About 600m below that, will be **Rappel 1**. With two 60m ropes, you can place W&R/QL on a tree or rock and descend over 2 dropoffs going **straight down the stream course**; or use the W&R/QL on a tree on the left/north side and do 1 rappel of 40m to a point just below the 2 drops mentioned above.

From R1, you'll walk about 2 1/2 kms before coming to **R2**. This will be from W&R/QL next to a large chokestone and down about **8m**. Less than 100m below Rappel 2 will be a **HL** down about **4m**, then **R3** from a chokestone. This is again down about **8m** and into a **deep swimming pool** that will be there no matter how dry the drought. After the pool, the canyon opens up. Or, instead of doing R2 & R3 and swimming, you can bench-walk on the right about 50m beyond the pool and **rap 28m** from a large bush. In cool or cold weather, this would eliminate a cold swim and the need for a **wetsuit**.

After walking a little more than 1 km with possible intermittent pools, and 350m below where **Waterhole Canyon** enters on the left or north, you'll come to **Rappel 4**. Below this **10m** rap is a nice spring and the beginning of running water which flows as far as the big pool below the 86m rappel. At R4, rap from a **tree** on the left and down a wall made of **travertine**. About 300m below that, and just around the corner to the right, will be **R5**. There's W&R/QL around a small cedar tree and down **15m**.

After R5, and more walking along a very small stream, you'll come to a huge dropoff of 86m and **R6**, the **Big Rappel**. At this point you have 2 choices; rappel straight down the water course and through a very small waterfall much of the way landing next to a large swimming pool (walk around it), or stay on the bench to the right and make **2 shorter raps** with normal ropes. The advantage of doing this route is, it can be done with two 50m ropes (or one 50 & pull cord).

If going down the Big Rappel, locate some chokestones under a ledge immediately above the drop. There are other anchors in the area if needed. Our group left long W&R/QL on chokestones, and parts of them were there after 4 years. An important thing to remember here is, make sure the W&R/QL extends to the edge of a rounded cliff edge so you can pull the rope from below. It took us 10 minutes and 4 guys using ascenders to pull our ropes which measured 122m each! Having full-sized ascenders or jumars really helped grasping the rope. Also, make sure the last guy down uses an ATC so the ropes don't get twisted, then pull from the far side of the pool at the bottom. You may have to tie another rope to the pull-side to get a better angle.

If you don't want to buy ropes long enough for this big rap, then walk around the corner to the right about 300m and look for part of a rope around a small boulder. **Rap 6A** is about **40m** landing on an intermediate bench. From there walk to the right or south again about 100m, and look for W&R/QL on another boulder. This last rappel (**R6B**) will be about **25m** and you'll land on top of a talus slope. **Note:** The exact location of the Elephant Canyon & Honaker Trail Formations in the rappelling cross-sections are likely not in the proper place. The author has guessed some as to their location.

From the last rap, walk downcanyon to running water in lower Clearwater Canyon. From there, walk northeast along a talus slope or bench just above Lake Powell roughly 3 kms to the mouth of **Easter Pasture Canyon**. Climb up the dry stream channel. Near the middle are 2 pools under several trees, then near the top are springs, a little running water and an Anasazi ruin. In that area, route-find east up to a prominent bench, then bench-walk north into the upper canyon. In the upper part, head east, sometimes on old cattle trails, back to Chaffin's Camp and your 2nd car.

Elevations Car-park, 1634m; Chaffin's Camp, 1621m; Lake Powell/Colorado River HWM, 1128m.

Time Needed Going light with a fast 3 person crew, it's possible to do this entire trip in one long day. This means you'll want to do it in May or June with long days. Otherwise, pack as light as possible and do it in 2 easier days, making a bivouac in lower Clearwater with a permanent year-round stream.

Water In the deep pool below Rappel 3, then a trickle of running water from beneath R5 all the way down to the pool at the bottom of R6. Also in the lower 500m or so of Clearwater, and in pools & running water in the upper half of Easter Pasture Canyon.

Maps USGS or BLM maps Hite Crossing & Hanksville (1:100,000) for driving; Clearwater Canyon & Teapot Rock (1:24,000--7 1/2' quad) for hiking.

Flash Flood Danger Low, with only a couple of short slots below Rappel 1, and between R2 & R3.

Best Time to Hike May, with favorable temps and long days. In winter, or in March, April or October, you'd likely want to skip the swim. Summers are hot, but no need for a sleeping bag or wetsuit.

Map 25, Clearwater & Easter Pasture Canyons, South Robbers Roost & North Lake Powell Country, Utah

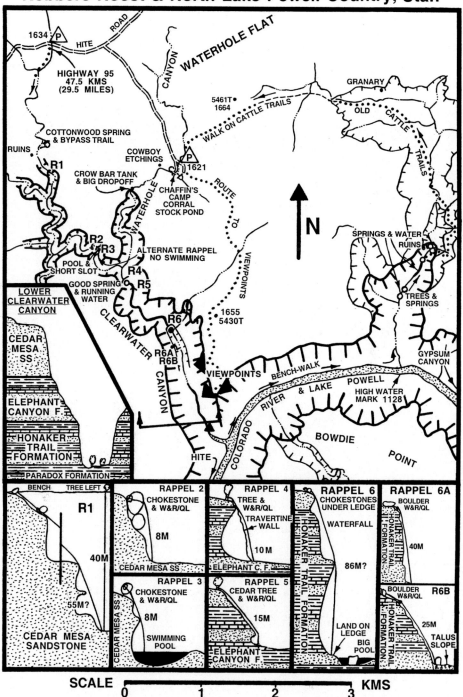

Author's Experience After Scott Patterson scouted the canyon from the rim, and the author went down Easter Pasture for the 5th time and walked above Lake Powell to the mouth of Clearwater on a scout trip, we and Steve Brezovec, Ryan Cornia, John Gilmore and Mark McCray started out with 2 days supplies. We had no problems and the hike was fun, but in the beginning, the new ropes Scott had were very fast and we were worried about a fast rap at the Big Dropoff. However, we dunked the ropes in water, then by being in an almost constant, but light spray, going down, the last rap was actually enjoyable, and not the panic scene one sees at the end of Heaps Canyon. We camped on lower Clearwater under an overhang on 5/24/2003 (9 1/2 hours walk-time), then with an early start, bench-walked to Easter Pasture which had many waterholes. We were back to a vehicle at Chaffin's Camp in 6 hours on Day 2. Total 2-day walk-time was 15 1/4 hours.

On a cool 5/10/2008, along with AJ Pastula, Jonas Fast & Stuart Paul, we camped just south of the **cattle guard**, and got an early start. The author bypassed R2 & R3, and did the longer side rap--while the others went for a cold swim! (They also got a rope stuck and lost 45 minutes!) We did the 2 shorter rappels instead of the big rap at the end, then high-tailed-it out. We took just over 13 hours.

Opposite Page Rappel 4 in the middle of **Clearwater Canyon**. You rappel over a cone of travertine left by spring water. This water flows to the bottom of R6.

Above Left Walking beside good drinking water flowing for 500m in lower **Clearwater Canyon**.

Above Right This is **Cataract Canyon** immediately around the corner (northeast) from the mouth of **Clearwater Canyon**. You walk this bench & talus slope just above the tamaracks about 3 kms to the mouth of **Easter Pasture Canyon**--just barely visible in the upper right-hand corner of this foto.

Left From near the top of Rappel 6A, looking southeast down into the lower end of **Clearwater Canyon**. In the lower right-hand part of this picture are the benches you rap to when making 2 shorter rappels as opposed to doing one long rap in the stream channel. In the distance is **Cataract Canyon**; that's where you turn left toward **Easter Pasture Canyon**.

Above Left The Big Rappel at the end of **Little Bull Canyon**. At the top of this big dropoff is a short DC or rappel, then the final plunge to the pool below.

Above Right Rappel 3, the next-to-last rap in the **North Fork of No Mans Canyon**. The drainage makes a 90° turn at the bottom of R3, then comes the final 33m rappel into the lower canyon.

Right Cristina Amat in the Best Slot in the lower end of the **West Fork of Lost Park Canyon**.

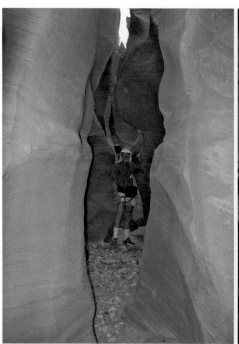

Above Left This is a 2004 foto of the **Corkscrew Slot** in the lower end of **Northeast Spur Fork**. This picture was taken with a flash, and it was in the afternoon with the sun shining down in to parts of the slot; that's the reason for the nice reds. The 10/2006 floods deposited a lot of sand in this part.

Above Right Warren Egbert jumarring up the nearly 60m dropoff in the lower end of **Tidwell Canyon**. This boulder filled crack & rappel is at the end of one of the better walk-through slots around.

Left Part of the deep upper section of **Three Canyon** below the only rappel in this part of the drainage.

Above Nat Smale crawls along a ledge just above the big drop into a dark hole in **Middle Fork/West Butler**. Just beyond is where you chimney down 10m.

Right Ryan Cornia stems over a water-filled pothole in the middle of **North Fork of Trail Canyon**.

Map 26a, Area Map--Poison Spring, Cedar Point, North Wash & Trachyte Creek Country, Utah

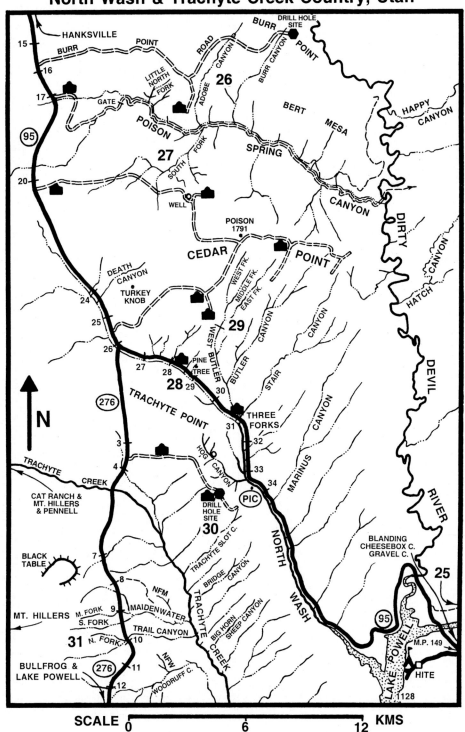

North Forks of Poison Spring Canyon, Cedar Point-
North Wash Country, Utah

Location & Access **Poison Spring Canyon** is located south of Hanksville and just east of Highway 95. It drains eastward from Mt. Ellen in the Henry Mountains into the Dirty Devil River. To get there, drive south from Hanksville on **Highway 95**; or north from Hite and the north end of Lake Powell. You could head east right at **mile post 17** and drive down Poison Spring Canyon and do what this writer calls **Little North Fork** (**LNF**) & **Adobe Swale Canyon** (**ASC**), but that's a rough cobblestone road and you'll need a least a HCV, and maybe 4WD.

The better way of access for all canyons, especially for people with cars, is via the **Burr Point Road** (**BPR**). To get on it, turn east between **mile posts 15 & 16** (see **Area Map**, page 133), and head toward **Burr Point**. To do **LNF**, stop at the road junction marked **5122T** (1561m) on the *Baking Skillet Knoll 7 1/2' quad*. This is 9.5 kms (5.9 miles) from Highway 95. If you have one car, park at this same junction and do **ASC**. Or you could leave one car at the end of the little side-road 1.8 kms (1.1 miles) from the BPR (this side-road is a little sandy, but any car can get there); and a second vehicle about 1 1/2 kms (1 mile) further up (eastward) the BPR from the junction marked 5122T (1561m). This puts you on the upper west side of ASC. If doing **Burr Canyon** (author's name, it starts at Burr Point), park near the head of the canyon and at or near another junction which is 16.7 kms (10.4 miles) from Highway 95.

Rating **LNF** & **ASC: 3A II** or **III**; no major problems in these 2 canyons which are relatively easy. **Burr Canyon: 3A III -X SLOT (or +R SLOT).** This one has a short tight slot, high-stemming and one big rap.

Equipment For all canyons--RG, one 60m rope (plus a 2nd 60, or 60m pull cord for **ASC & Burr**), K&EP, W&R/QL and compass & map, For **Burr**, include a R/QD/DC for high-stemming & small pack.

Route **Little North Fork** From the car-park, head southwest in the shallow drainage. It's a sandy wash for nearly 1 km, then it starts to dive into the Navajo (most likely the Page SS at this point?) Sandstone. Soon you'll turn northwest to find another little side-drainage coming down from the northeast. It has a short slot which for most is DC'able. If you want to try it, circle around and come down from the top end. After another km or so, the canyon drops quickly with yet another steep side-canyon coming in from the northeast. Just before that junction will be some DC'ing & a **16m rappel** from small chokestones in some holes in the drainage bottom. About 300m beyond this only rap will be a nice slot with DC's, then it opens slowly. After nearly another km will be a canyon junction. At that point, turn left and climb northeast up steep slickrock between the **LNF** or its **Northeastern Fork**. Climb the nose of this ridge between the 2 canyons and return to your car. The author has been down the NEF, but it is uninteresting; however, it does have a 7m rappel going under a big chokestone, then an interesting 2-part DC, but no slot, before you walk out the bottom end to the escape route just mentioned.

Adobe Swale Canyon For those with one car, you might as well park at the same place as if going into LNF. From there walk northeast along the BPR for about 1 1/2 kms, then head east into upper ASC. For about 2 kms, it's easy walking, then it slots up a little. Finally, you'll come to the **8m Rappel 1** with W&R/QL around a chokestone. Rap down, possibly into a pool (?). A very short distance below that will be the **3-part R2**. There you may find a **deadman** with W&R/QL as your anchor (better have **a BK & B&H's** in case its gone and there's no more stones to build another deadman anchor!). R2 has 3 drops (the last one may be into water?) totaling about 45m, but be careful with your rope getting hung up because its a long pull going around 2 corners. We had to jumar back up and try it a second time. Or, pull your knot past the 2nd corner as the last person raps down. Just below the end of R2 is **R3**. Someone has buried a large stone (?) in the sand as an anchor--which may or may not be there when you arrive. Rap **10m** into some good narrows with several DC's. About 450m beyond the last rap, the canyon opens quickly. Walk down this deep Navajo Sandstone gorge for about another 2 1/2 kms, then as the high walls fade away, head southwest following cattle trails into a minor canyon in the Kayenta Formation. Once on top of the Kayenta, circle around the rim of this short rounded drainage and first head northeast, then NNW up the nose of a slickrock ridge. Some cairns mark portions of an old **constructed horse trail**, which has been blasted out of the slickrock in places. Follow this trail as it zig zags up to the end of the side-road mentioned above at 1561m. Road-walk back to your car.

Burr Canyon From a point at or near the road junction & trailhead mentioned above, head southwest into the upper drainage. You'll find a couple of short shallow slots which can be bypassed easily, then the beginning of the **125m-long -X SLOT**. Right from the beginning, stay high, from 3-10m above the deck. High-stem to about the halfway point. There you'll have to do **an off width move** where the crack opens suddenly into a silo. Keep your body in the crack while worming your way down 5m. From the resting place at the bottom of the silo, it's up a little and moving about 10m horizontally where the slot has smooth vertical walls less than half meter wide. For this awkward width, move hands-to-elbows, knees-to-heels, and hang your pack next to your harness--not hanging down where it can get stuck. Right after that, it's down to the bottom with a little walking & DC'ing to the end of the slot. About 10m back from the **48m drop**, are 3 chokestones/anchors. If our rope is gone, attach new W&R/QL and rap down a crack (take measures to prevent your rope from being cut on one sharp edge near the top).

After the rappel, walk southwest down a broad, wide-open canyon (some running water, but with cows in winter) lined with Kayenta benches. After about 4 kms, and when you come to the **15m-high Wingate dryfall**, head back upcanyon about 200m and into a sizeable side-canyon coming down from the northwest. Walk up this for about 1 km, then at the 4800' contour mark on the *Burr Point quad*, veer left or west and walk up a sandslide to the next rim. From there, continue northwest up slickrock another km (aim for **5248T**/1600m), then head northeast cross-country back to the BPR and your car. Better have a map & compass for this last part.

Elevations Trailheads 1561m, 1561m, 1597m & 1625m; down to 1396m & 1399m at the low ends.

Time Needed LNF, 2 1/2--3 1/2 hours. **ASC** may take 4-6 hours round-trip. **Burr Canyon**, 5-7 hours.

Water Take your own water on all hikes. Poison Spring Canyon has seeps in a few places.

Maps USGS or BLM map Hanksville (1:100,000) for driving & orientation; Baking Skillet Knoll, Burr Point, Turkey Knob & Stair Canyon (1:24,000--7 1/2' quads) for hiking

Flash Flood Danger High in the short slots, but these are all short drainages.

Best Time to Hike Spring or fall, but LNF and Burr can be done any time. ASC may have pools & potholes, so stay out of it in winter--unless you're prepared with a wet/drysuit.

Author's Experience After spending 2 1/2 days scouting these canyons, he and Nat Smale did **ASC** in 4 1/3 hours using 2 vehicles. We also did **LNF** in 2 1/2 hours from the same car-park. On 3/4/2007, and with Ryan Cornia, Cristina Amat & Tom Tolboys, we did **Burr Canyon** in 5 1/4 hours.

Map 26, North Forks of Poison Spring Canyon, Cedar Point-North Wash Country, Utah

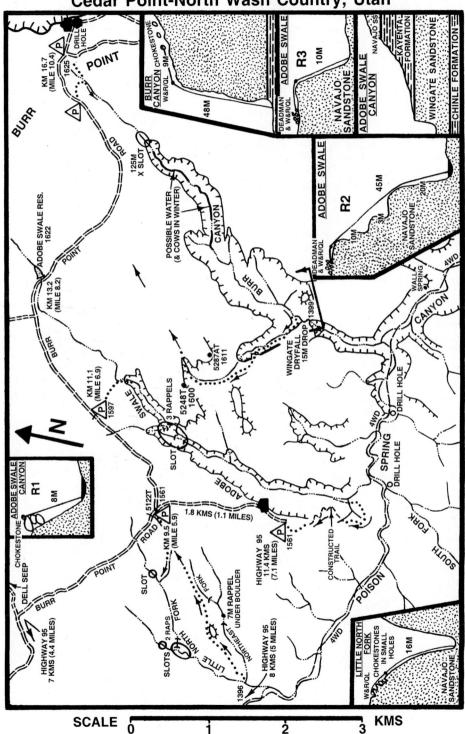

Top Left Tom Tolboys & Cristina Amat in the middle of **Burr Canyon Slot**. They're about 5m above the bottom at that point. **Top Right** Nat Smale in the middle of Rappel 2 in **Adobe Swale Canyon**. Don't get your ropes crossed here--45m is a long pull! In the distance in the upper right you can see W&R/QL buried in the sand; that's the beginning of Rappel 3. **Above** Nat getting ready for the 2nd rappel in the **Little North Fork of Poison Spring Canyon**. The anchor here is W&R/QL wedged into a crack with small chokestones.

Above Left Stefan Folias starting Rappel 3 into a dark hole in the **East Fork of South Fork Poison Spring Canyon**.

Above Right Nat DC'ing somewhere in the **West Fork of South Fork of Poison Spring Canyon**.

Left This is the end of the technical section of the **Middle Fork of South Fork of Poison Spring Canyon**; the bottom of Rappel 2. This last part is almost a DC for some (?).

South Fork, Poison Spring Canyon Slots, Utah

Location & Access Shown here are 3 short side-canyons, **East, Middle & West Forks** (AKA's: Constrychnine, Slideanide & Arscenic) at the upper end of the **South Fork of Poison Spring Canyon** (this is what Hanksville cowboys call it, but it's more like a Southwest Fork). This canyon is located about halfway between Hanksville & Hite, at the upper end of Lake Powell. To get there, drive south out of Hanksville on **Highway 95** (or west from Blanding and/or Hite on the same highway). Between **mile posts 20 & 21**, turn east onto a graded road. After 400m or so, you'll pass the **Lone Cedar Reservoir** with all the tamaracks. Continue east--don't turn north or south onto side-roads. After crossing a couple of upper tributaries of South Fork, you'll come to a **drill hole** with a fenced-in **water pump** powered by a **solar panel**. This is **9.2 kms (5.7 miles)** from Highway 95. You can camp there, but there may be noise from the pump, or cows milling about (winter months only). Continue east another 300m, turn left or north, then east and stop at another campsite which is **9.9 kms (6.1 miles)** from the highway.

Rating All 3 slots are probably **3A II**, but **Middle Fork** could be rated **R** if you DC instead of being belayed or rappel over 2 big tight dropoffs (?). All groups should have at least one experienced leader.

Equipment For each canyon--RG, W&R/QL's, leather gloves, K&EP, R/QD/DC for downclimbing with pack suspended from your harness. **Ropes**--East Fork, two 60m (or a 60 & pull cord); **Middle Fork**, one 60m; **West Fork**, one 60 to be left at the top & picked up later.

Route East Fork From either of the car-parks or trailheads, walk northward--you'll need a map & compass. In the East Fork, you could rappel from the very first dropoff which is **30m**, but you'll have to install a long piece of W&R/QL. Or better still, walk out to the point between East & Middle Forks, work your way down to the first bench (probably below the Page SS), then bench-walk back upcanyon 400m and descend steep slickrock to the drainage bottom and **Rappel 1**. This will likely be from a deadman, or tie W&R/QL around a boulder on the right. This will be **51m**; and puts you in the bottom of the slot. Right after that will be W&R/QL for the **14m R2** (experienced people may choose to DC this). Further along will be **R3**, which is a **21m rap** from a **single piton** on the left above. You may find a shallow pool at the bottom of this fairly dark slot. After a ways, you'll come to **R4**, which is a low angle HL of **6m** anchored by some cobblestones in sand. From there, it's a walk to the lower end of the combined East & Middle Forks. When you reach the main drainage marked **1414m** on the map, turn left or southwest, and walk up to the bottom end of the **West Fork**. Walk up this drainage about **100m**, then climb steep slickrock on your right or south side. Route-find southward up the nose of the ridge which is the south wall of West Fork. Return to your car, or do a second slot.

Middle Fork Right at the head of this slot is W&R/QL around a boulder; **R1** is down about **24m**. After a ways, you may see more W&R/QL for a **6m rap**, but most people are **DC'ing** this--your choice. Then a dropoff of about **17m**. Everybody seems to be **DC'ing** this--which the author did. It's about shoulder or half-meter in width, so it's chimneying *back to hands & knees* all the way down. Finally near the bottom will be the last rappel, **R2**, which may be a double rap (?). Find W&R/QL around a boulder, rap down a dark, slanted, twisting slot. After about **20m**, will be another chokestone with W&R/QL. Most people pull and retie their rope, then continue down the last **17m** or so. Changing anchors is good, because it may be difficult to pull your rope from the bottom in this convoluted slot. However, you could continue the first half of this rap another **5m** to a **flat place**, then maybe DC the rest of the way (?). If doing this, you may have to let go and slide the last 1 1/2m to the sandy bottom. From there walk out of the canyon and return the same way described under East Fork.

West Fork The best way to do this is to tie one **60m rope** at the top and make a **3 stage rappel**, then pick up your rope at the bottom of the hike. This way you won't have to lug it through the slot (as of the spring of 2005, there was W&R/QL around 2 other intermediate chokestones, so you could do it in 2 or 3 stages--but the first option seems much better). After this first rap at the head of West Fork, there is one slightly difficult DC--which some are rappelling, then other easy DC's and walking the rest of the way. It might be a good idea to take a short rope along just in case. Return the same way as described above and pick up your 60m rope.

Elevations Both trailheads, about 1573m.

Time Needed About half a day for each fork; and 2 slots in one day for most people.

Water Take your own, but there may be seep water in the lower end of Middle & East Forks.

Maps USGS or BLM map Hanksville (1:100,00) for driving & orientation; and Turkey Knob (1:24,000--7 1/2' quad) for hiking.

Flash Flood Danger Fairly high, but these drainages are very short, so it would take a cloudburst immediately above you to cause concern.

Best Time to Hike Spring or fall, but these great slots can be hiked year-round.

Author's Experience With Nat Smale & Stefan Folias, we did **East Fork** from the very top, rappelling 30m, then found an easy way in or out below that. The W&R/QL was pickup later. In East Fork we used a 50mm pull cord and a 2nd 60. After the East Fork, we exited, then left a 60 at the head of West Fork. The slot in **West Fork** might be the best. We returned to our camp after 8 2/3 hours. A few days later, the author returned & did **Middle Fork** solo in 4 1/2 hours.

The beginning of the rappel in the **West Fork** of the **South Fork of Poison Spring Canyon**.

Map 27, South Fork, Poison Spring Canyon Slots, Utah

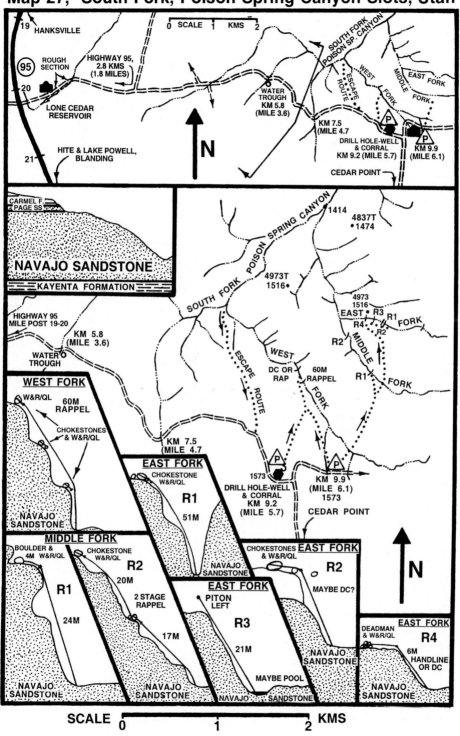

SCALE 0 1 KMS 2

HANKSVILLE

19

ROUGH SECTION
95
20

HIGHWAY 95, 2.8 KMS (1.8 MILES)

LONE CEDAR RESERVOIR

HITE & LAKE POWELL, BLANDING
21

SOUTH FORK POISON SP. CANYON

WEST FORK
MIDDLE FORK
EAST FORK
ESCAPE ROUTE

WATER TROUGH KM 5.8 (MILE 3.6)

KM 7.5 (MILE 4.7)

DRILL HOLE-WELL & CORRAL KM 9.2 (MILE 5.7)
KM 9.9 (MILE 6.1)

CEDAR POINT

CARMEL F. & PAGE SS

NAVAJO SANDSTONE

KAYENTA FORMATION

HIGHWAY 95 MILE POST 19-20

KM 5.8 (MILE 3.6)

WATER TROUGH

SOUTH FORK POISON SPRING CANYON

1414
4837T •1474
4973T 1516•
4973 1516
EAST • R3 R1
R4 R2
FORK
WEST
DC OR RAP
60M RAPPEL
FORK
R2
R1
MIDDLE FORK
ESCAPE ROUTE

KM 7.5 (MILE 4.7)
1573
DRILL HOLE-WELL & CORRAL KM 9.2 (MILE 5.7)
KM 9.9 (MILE 6.1)
1573
CEDAR POINT

N

WEST FORK
W&R/QL
60M RAPPEL
CHOKESTONES & W&R/QL
NAVAJO SANDSTONE

MIDDLE FORK
BOULDER & 4M W&R/QL
CHOKESTONE W&R/QL
R2
20M
2 STAGE RAPPEL
17M
R1
24M
NAVAJO SANDSTONE

EAST FORK
CHOKESTONE W&R/QL
R1
51M
NAVAJO SANDSTONE

EAST FORK
PITON LEFT
R3
21M
MAYBE POOL
NAVAJO SANDSTONE

CHOKESTONES & W&R/QL ## EAST FORK
R2
MAYBE DC?
NAVAJO SANDSTONE

EAST FORK
DEADMAN & W&R/QL
R4
6M HANDLINE OR DC
NAVAJO SANDSTONE

SCALE 0 1 2 KMS

NAVAJO SANDSTONE

North Wash Side-Canyons, Cedar Point Area, Utah

Location & Access **North Wash** begins as **Crescent Creek** high in the Bromide Basin of the Henry Mountains, then heads east and southeast to enter Lake Powell just south of the mouth of the Dirty Devil River. Most of North Wash flows beside Highway 95. The area covered here is roughly halfway between Hanksville in the north and Hite to the southeast. Along this stretch are 5 short slot canyons entering North Wash from the northeast. According to Alvin Robinson of Hanksville, only Death Canyon has a name, so all the others on this map are named according to the nearest mile post markers--or the AKA's mentioned below.

To get there, drive south out of Hanksville (mile post 0), or northwest from the Hite area (mile post 49) on Highway 95, until you're somewhere between **mile posts 24 & 29**. As you drive, observe mile post markers & canyons closely and match them to this map. Along this stretch are several pullouts for camping/parking, most of which are near the mouths of these canyons.

Rating These canyons are short & sweet with most rating a **3A I or II**. These are some of the tightest slots around with some moderate DC'ing and some rappelling. **Mile 28.1 Canyon's West Fork** is a **PG SLOT** that's pure fun & popular. **Mile 28.5 Canyon** might be **4A III X SLOT**. Read more below.

Equipment In general, and for each canyon, you'll need one rope of 30 to 50m (**read each hike**, two 50's for the **East Fork** of Mile 27.4 Canyon), RG, W&R/QL, K&EP, and as small a pack as possible. Read below for what is needed in **Mile 28.5 Canyon**--it's a difficult one with special equipment & skills needed.

Routes **Death Canyon** (named for some cows that got trapped in the upper canyon and choked to death) there is no place to park, so just pull off the highway at about **mile post 24.4**. Hike upcanyon about 1 1/2 kms until you come to the pool below the only rappel. From there, backtrack 50m, climb up the west-side wall to the rim, head upcanyon about 400m and reenter the drainage which is carved out of Entrada Sandstone. Head downcanyon to the very short slot above the pool. Wrapped around one chokestone should be some W&R/QL, so chimney over to it, tie in and make a **5m rap** to the bottom of the slot (you can **DC** this, but this may be your only anchor for the rappel). From there, and while still roped-up, walk downcanyon about 10m and make an **8m rappel** into waist-deep water (this is a spring). Use a **50m rope** for this. This last part may also be DC'able for some (?).

Mile 27.1 Canyon is a very short slot. Park on the highway and walk upcanyon. After 600m, you'll be in a shallow tight slot, then an 8m dryfall some can UC (or backtrack a little and bench-walk above it on the left). A short distance above that may be a pool. From there, backup 50m and climb east up to the rim on a talus slope. From there, walk 75m north and back to the main drainage. For safety, tie off a short rope and **HL/rap 10m**, then DC to the pool mentioned above. Return & retrieve your rope if necessary, and head back.

Mile 27.4 Canyon (AKA Shillelagh) has **2 forks**. To do the **West Fork**, park right at the mouth of the drainage and walk upcanyon. After 500m, and near the junction of the 2 forks, veer left or west, and climb steep slickrock to the Carmel rim or mesa top. Once on top, rim-walk north about 1 km and walk into the shallow drainage. Shortly you'll come to **Rappel 1**, which drops only 5m, but you'll have to tie into W&R/QL from a small boulder up-slope making the total rap about **18m**. Immediately below that is **R2** from a couple of chokestones. This is also an **18m rap**. Below that it's easy walking with some DC's to the bottom.

For the **East Fork**, you might walk up the main drainage and route a route to the rim to the east, but a guaranteed way is to park at the mouth of **Mile 27.6 Canyon**, walk 100m up the drainage and veer left and climb up the obvious slickrock ridge to the Carmel rim. Continue north between 2 canyons for about 1400m, then walk down into the bottom of the East Fork. After about 100m, you'll come to the **15m R1** from a small boulder. Just below that will be **R2** from a boulder on the left. This one drops about **32m**, which means you'll need **two 50's** (or a 50 & pull cord). The slot below is moderately good.

Mile 27.6 Canyon (AKA Blarney) Drive 100m up the short road to the mouth and park; walk upcanyon about 100m before veering left or north and onto slickrock. Climb up the obvious ridge to the Carmel rim, then continue rim-walking north on the west side of the canyon to the head of the **Main Fork** (this is the same route as if going to the **East Fork of Mile 27.4 Canyon**). Drop into the upper end to find a chokestone on the right with 4m of W&R/QL. This is the top of a **15m rap** which gets you down over the Page Sandstone cliff. As you get into the slot, you'll chimney down under a large chokestone, then after another 125m or so, you'll come to **R2 & W&R/QL** wrapped around a chokestone. This is a **10m rap** with an awkward start. Below that you'll find a good fotogenic slot for 350m, then it opens up.

The **Little East Fork--Mile 27.6 Canyon** is shorter, has some difficult UC or DC'ing, and not much of a slot. Climb upcanyon to one serious UC, get out of the bottom on the left going up steep slickrock, reenter above and head downcanyon. It's easier going downcanyon than up.

Mile 28.1 Canyon (AKA Leprechaun) has 2 main forks (and a little west fork this writer hasn't seen) both of which are about as tight as anyone can crawl through. The **West Fork** is one of the longest sustained tight slots anywhere and is the **best canyon on this map**. For both forks, take the smallest pack possible--maybe in cool weather, no pack at all! Just a camera, K&EP and one rope. Larger folks will have to chimney up a little.

To do the **West Fork**, walk upcanyon about 1 km, veer left and climb another km up steep slickrock between 2 forks to the Carmel rim. After rim-walking another km, veer right or east crossing one drainage to get into the main slot. You'll immediately come to **R1**. Do this **7m rap** from W&R/QL under a chokestone; *no need for a harness here*, just **body-wrap rappel** or **HL** down. After that it's DC'ing & squeezing until **Rappel or HL 2**; about **6m**. All together, the slot lasts for a little more than 1 km and will take 2-3 hours. Right where the West Fork meets the East Fork, the canyon opens a little with graffiti on the walls, then a deep & impressive narrows, and finally regular walking back to your car.

The **East Fork** is another tight one. Hike up the main drainage about 600m, then veer right or east at a little butte and route-find up Navajo slickrock. As you climb and head NNE, don't get out on the Carmel rim or mesa top; stay on the upper part of the Navajo just below what appears to be the Page Sandstone cliff. As you near the upper end of the canyon, walk into the slot and you should immediately find **R1**. This is from a small chokestone and you can **HL** down 5m. About 200m below that is **R2**; you can also **HL** down this one about 6m. About 500m below R2 is a pothole which will likely have water year-round (?). After another 800m or so, will be the **6m R3** which is an overhang. Best to use a harness or the Dulfersitz (body wrap) method on this one. Below that is a good DC, then the junction with the West Fork and the walk out.

Mile 28.5 Canyon (AKA Sandthrax) This very tight slot requires high-stemming & hard climbing, 1 off-width move. Equipment needed; two #5 camalots, two #6 camalots & one #4 Big Bro; and knowledge to use it.

Elevations Upper entries to these canyons are about 1525m altitude; Highway 95 about 1360m.

Time Needed From 1-2 hours for the shorter slots; up to about 5 hours for Mile 28.1's West Fork.

Water Always have water in your car for your trip; carry some into each canyon, at least in summer.

Maps USGS or BLM map Hanksville (1:100,000) for driving; Turkey Knob (1:24,000--7 1/2' quad) for hiking.

Flash Flood Danger These are very short slots so the only danger would be if a storm dumped right on you

Map 28, North Wash Side-Canyons, Cedar Point Area, Utah

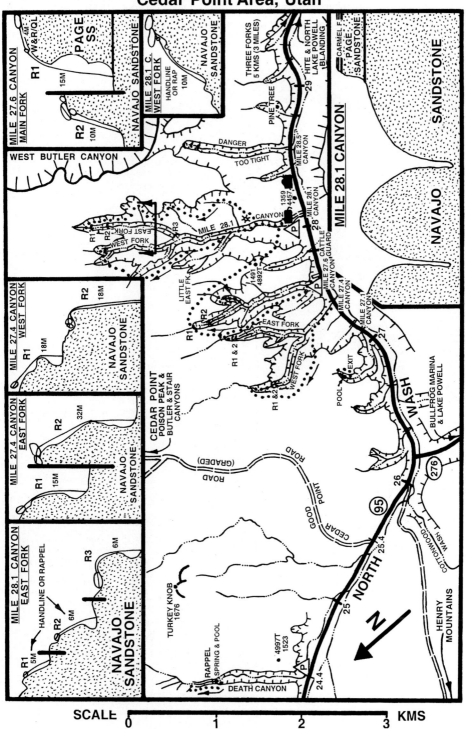

SCALE

0   1   2   3   KMS

or immediately upcanyon.

Best Time to Hike In the cooler half of the year so you'll be wearing more clothes--to save skin!

Author's Experience He explored **Death Canyon** twice in about 2 hours each. He and Steve Tyler did the **Main Fork** of **Mile 27.6 Canyon** in just over 3 hours; he did the **Little East Fork** in less than an hour; and the **West Fork of Mile 28.1** with Tyler in 4 2/3 hours. Later, the author & John Summerson did the **East Fork** of **Mile 28.1** in 3 1/3 hours, then **both forks of Mile 27.4** in 3 3/4 hours round-trip, all 3 on the same day. In 9/2007, and with Richard Patterson & Dan Logan, we revisited the **West Fork** of **Mile 28.1** with digital cameras; it took 3 4/5 hours (about 2 1/2 hours in the tight slot part).

Above Left Rappel 2 in the **East Fork of Mile 27.4 Canyon**.

Above Right Dan Logan & Richard Patterson chimneying over a pool in the upper end of **West Fork of Mile 28.1 Canyon**. Everyone who goes to North Wash scramblers down this slot, one of the best around.

Right Dan DC'ing one of many simple dropoffs in **West Fork of Mile 28.1 Canyon**.

Above Left The **West Fork of Mile 28.1 Canyon** is usually dry, but not on this trip.

Above Right Nat Smale is halfway down the 10m DC which is immediately after the shelf-crawl in the lower end of **Middle Fork/West Butler**.

Left This section of **Middle Fork/West Butler** is just below the shelf-crawl & 10m DC. If you're here around noon, you get a streak of sunlight at the bottom which enhanced the colors.

Upper Butler, West Butler & Stair Canyons, Cedar Point Area, Utah

Location & Access The canyons featured here drain the **southern slopes of Cedar Point**, which is the high ground forming a triangle between **Poison Spring Canyon** on the north, the **Dirty Devil River** on the east, and **North Wash & Highway 95** on the southwest. Shown here are the upper forks of Butler, West Butler and Stair Canyons, all of which have some of the tightest slots around.

To get there, drive south from Hanksville on **Highway 95**; or west from the Blanding area on the same State Road 95. To reach the bottom end of these canyons, make your way to a large parking and/or camping area just east of **mile post 31** at a place called **Three Forks**. You could leave one vehicle there where Butler & Stair meet North Wash. To reach & do the upper slots, drive to a point halfway between **mile posts 25 & 26**, and turn northeasterly. Drive this good graded road about 6.4 kms (4 miles) to a **3-way junction** marked **5268T** (1605m) on the *Turkey Knob quad*. From there, drive **southeast 600m** on an good, ungraded track to the **1585m** car-park (**park & camp**) if you'll be doing any of the **3 upper slots** of West Butler; then continue northeast 1.5 kms (.9 mile) and park a second car at 1637m which is closer to an entry point to the **West** and **Middle Forks**.

Or if doing the **East Fork** of West Butler, **Butler's main fork**, or **Stair Canyon**, continue northeast on this graded Cedar Point Road for a total distance of **13.6 kms (8.3 miles)** from Highway 95. This is another **3-way junction** just south of a high flattop point labeled *Poison, 5874* on the *Stair Canyon* quad; or 1791m on the *Hanksville* metric map. Park and/or camp just south of Poison Peak at 1762m or 1707m.

To do any part of **Stair Canyon**, head east from Poison Peak. 2WD cars can make it to Poison Peak, but east of there, the road gets **extremely sandy** and is for tough **4WD's only**. If you have the right vehicle drive (or walk) east 3.7 kms (2.3 miles) and turn south for another 800m (.5 mile). Stop & park.

Rating The 3 slots of **West Butler (all PG SLOTS)** and the upper main fork of **Butler**, should be rated **3A II or III**--if you choose to use the upper exits shown. To go all the way through Butler to Three Forks will be **3A III or IV. Upper Stair Canyon Slot: 3A II R SLOT**; but if going all the way down to Three Forks it's **3A III or IV R SLOT** (soon after rains all will be **3B**).

Equipment RG, one or two 50m ropes (check the longest rappel and other info for each canyon), W&R/QL, K&EP, a drybag/waterproof box for cameras, maybe a headlamp for the 3 slots of West Butler and R/QD/DC for Stair (right after floods in spring or fall, go prepared with a wet/drysuit and Ibis hook & G-pick).

Route West Fork/West Butler, park one car at a campsite at **1585m**, then from the **1637m** car-park, walk due east 500m to the east side of the upper end of the drainage; there are several ways in. From there, walk downcanyon about 1 km. This part is easy with some DC's, then it narrows quickly with steeper DC'ing. A short way past that, it falls away into a dark hole and a **2-stage 23m corkscrew rappel**. There should be a small chokestone with W&R/QL to rap from; if not, find a cobblestone to use as an anchor. Rappel, walk through one pothole, then comes a 2m drop into a 2nd pothole; jump or help each other. More DC'ing for 50m, then comes the **10m R2** into a **keeper-type pothole**. Best way down is to tie a rope & biner to an Ibis hook, place the hook over a very solid lens in the sandstone right at the beginning of R2. Rappel, then with a little slack, flip the rope to recover your hook.

Once in this pothole, 2 people should be able to help each other out (if you're alone you might need a hooking kit or pack toss). Once out of that, climb out of the shallow slot to the right or west onto a sloping bench. By doing this you avoid about 50m of meaningless potholes & short drops until you reach the last dryfall & **R3**. There is W&R/QL around a small buttress on the right above the big drop. From this 2m-long W&R/QL, rappel down about **8m** to the final drop, then **another 16m** to the bottom, perhaps with a shallow pool. If your 50m rope looks short, then perhaps add a little more W&R/QL to the sling to make this last partly-free rappel.

After R3, the drainage widens at the top of the Kayenta and it's a walkout downcanyon to the highway at Three Forks near mile post 31; **or better still**, walk 200m to find the Middle Fork coming in on the left; another km to East Fork--again on the left, then after another 500m turn right or west and climb up a steep faultline side-canyon. Near the top, turn right or north, climb out onto a bench, then route-find north & west up to the canyon rim. From there, and with compass & map in hand, walk northwest cross-country about 1 km to the car-park marked **1585m**, or the Cedar Point Road.

Middle Fork/West Butler (AKA Shenanigans) Park in the same place as if going into the West Fork, then walk about 1 1/2 kms a little north of due east. When you come to the upper end of Middle Fork, route-find down 2 levels or benches, then at the first dropoff into the slot, either **rap** down the watercourse chute from a chokestone; or **DC about 8m** south of the chute. This DC is a little tricky.

Once in this **1st slot** head downcanyon. After about 600m of DC'ing & chimneying, you'll come to an escape on the left or east, then you enter the **2nd slot**. This part has one deep dark twisting section, plus a dark tight right-leaning slot with daddy long leg spiders before the canyon opens for a 2nd time. Here you can escape to the east up a rockfall. To continue into the **3rd slot**, you must weigh roughly 80 kgs or less, otherwise you'll never make it through (big fellas might chimney over the tightest part)! If you're up to this last part here's what you'll find. This 3rd part slots up quickly again, with steep tight DC'ing, then it falls away into a dark hole. From there, **belly-crawl** along a ledge on the right or west for about 20m, then hop down 1m onto a chokestone. From there, DC 4m to a 2nd tier of chokestones. It's easier than it looks, but it's hard to see in the dark; take a headlamp. From there, chimney down another 5m, and continue downcanyon.

Next is a twisting **corkscrew slot**, then comes the **really tight part**. This section has vertical walls of about 10m (some can chimney up and over) and for about 30-40m it's only about 20-25 cms wide--gotta be skinny for this! It's here you can hardly turn your head. Halfway through this part is a 1m dropoff caused by a small chokestone. If you make it to that point, you can make it the rest of the way. As you're heading west, you'll come to an opening with some large boulders. Immediately after that, it slots-up again. Get right down into this crack; it's not quite as tight as before. Soon you'll come to more large chokestones and a drop. On top of the highest boulder will be W&R/QL for a **13m rappel** (might be DC'able?). With a 60m rope, some may be tempted to do this and R2 as one rap--but don't. Do this last part as **2 raps of 13m & 14m**. At the bottom of R1, walk a skinny ledge for 5m, then tread your **rope** of at least **30m** through W&R/QL hanging from a chokestone above and do **R2**. At the bottom of this will always be a pool of drinkable water. If you're very careful, you can walk/climb along tiny ledges on the left and save swimming about 3m. From this 2nd & last rappel, head downcanyon. Use the same exit as described above for the West Fork/West Butler.

East Fork/West Butler Best to drive to the 2nd 3-way junction marked **5780T** (1762m) immediately south of **Poison Peak**, then continue southwest down a sandy road for 1 km (.6 mile) to another junction (1707m). Park 200m back from where everyone is turning around and walk west on a track for 75m, then turn south on still another never-used track for about 750m and turn west (look at the map; there's an **alternate route** into the head of this slot by walking down the drainage from the 1707m parking place). After another 500m, bypass the upper part of East Fork which has 3 open drops in 60m, and make your way down in where you see a **minor seep with reeds** on the east side. You'll enter right where the slot begins, and about 300m south-

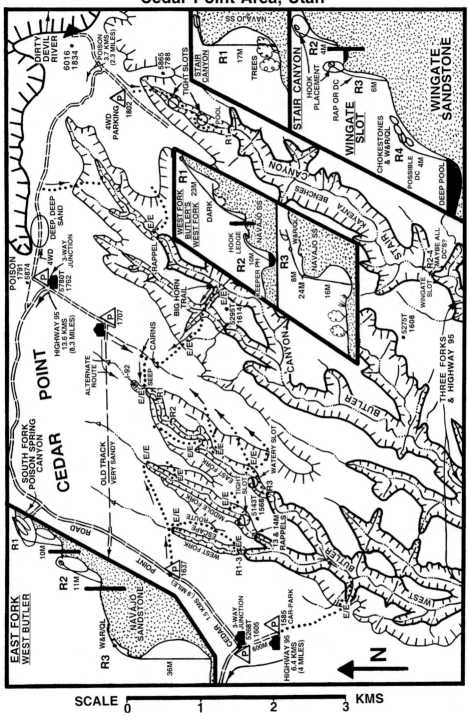

SCALE 0   1   2   3 KMS

west of where **I-92** is marked on the *Stair Canyon 7 1/2' quad.*

The **East Fork** has several slot sections with good DC's; below each slot is an escape route to the east. Halfway through the 1st slot is **Rappel 1**, a drop of about **10m** from chokestones and W&R/QL. Not far into the 2nd slot is **R2** from a small chokestone on the left side of a large boulder and down about **11m**. This 2nd slot is fairly deep & dark at one point

At the beginning of another slot, DC on the far left side of a shallow open slot avoiding several drops. Once down in, you'll find a slanted slot but no rappels. About 75m into the last **watery slot**, you'll come to a tight section. Chimney up 3m to a tree stump, stay high for about 15m, then slide down to the bottom into water. This 100m-long part always has water, so in cold/cooler weather, take a wetsuit. In 5/2008, the author found 2 waist-deep pools, then chimneyed down 2m into a short swimmer. Soon after that is the end of wading.

About 175m below the watery slot, will be **R3**, the last rappel in the East Fork. Here you'll have 2 choices; leave the canyon by going up steep slickrock either east or west and go back to your car; or find W&R/QL around a rock slab near the edge and rappel about **36m**. If you do this rappel, you then have a choice of going all the way downcanyon to **Three Forks** and **Highway 95**; or use the escape route as describe under the West & Middle Forks to get back to the Cedar Point Road and the **car-park** marked **1585m**. Most escape up **steep slickrock** to the east because the slot is much more exciting than a long hike in an open canyon.

Upper Main Fork of Butler Canyon For this one, it's best to camp/park at or near **Poison Peak** (or down the road south 1 km), then walk east on the road going to an **overlook** of the Dirty Devil River. After the deep sandy place, leave the track and head due south 800m and drop into upper Butler in one of several places. There are 2 dropoffs in the upper end, but you can walk around these into the bottom.

Once in the canyon, you'll first encounter a tight slot where you'll have to walk sideways and drag your pack behind. After the first tight part, the canyon opens a little, and it's like that the rest of the way. You'll have several pretty good DC's as well, but no rappels. A little over 1 km from where you enter are 2 dropoffs and an **escape route (E/E)** heading upslope southeast. Exit there if you don't want to rap into the lower canyon.

If you choose to proceed, then it may be possible for one person to belay everyone down the first dropoff, then that person can possibly DC with help from below (?). The author, being alone, made 2 small holes in the rock and rapped down from grappling hooks (Ibis hooks are better!), but it was hard dislodging the hooks from below. Best to take a partner and possibly DC (?) this. Or look for a natural anchor for an Ibis hook.

Immediately beyond this dropoff, will be the only real rappel in the canyon. You can DC about 3m to a chokestone and install some W&R/QL to rappel from; or possibly **rap off an Ibis hook**. Total drop is **11m**.

After the last rappel, walk downcanyon about 1 km to find a possible seep. Half a km below that, and if you want to use an **escape route** back to your car at Poison Peak, look to the west for a break in the canyon wall. Climb up along a **big horn sheep trail** into a faultline side-canyon which is just north of **5295T** (1614m) on the *Stair Canyon quad.* Walk 400m northwest through this little canyon, then cross a minor drainage and continue northwest up a hogsback ridge just southwest of the upper part of the faultline canyon. Once on top, continue northwest on the flats 500m to find a track which leads northeast back to Poison Peak.

Or you can continue down the main fork of Butler Canyon to **Highway 95** at **Three Forks** near mile post 31. It will take nearly 3 hours to reach the highway from the last rappel.

Stair Canyon Somewhere in the area of the **4WD parking** at **1802m**, veer left or southeast off the track and work your way into the upper end of Stair Canyon. About 400m west of **5865T** (1788m) are the first of **2 tight sections**. Here there are no rappels and ropes are not needed. It's all steep **DC'ing & high-stemming**. In places you'll have some stemming up to 8-9m above the bottom. If this part is too much, you can bypass both **R SLOTS** by walking in on the east side and route-find downcanyon.

Further along is the **2nd tight slot**. In this lower slot, you'll have to stem over some places 8-10m above the bottom. Don't forget to have shoes with good rubber and a R/QD/DC to hang your pack from a harness.

After the 2nd tight slot, you can route-find around both slots and return to your 4WD. Or, you can continue downcanyon. If you choose to do this and continue on to Three Fork, right after the slot, you'll DC into a possible pool, then 150m below that will be **Rappel 1** down off the lower Navajo and into the Kayenta part of the canyon. We installed W&R/QL on the left and landed in some small trees about **17m** below. Always install new W&R/QL because this one is exposed to the sun.

After R1, you'll walk in an open canyon as it cuts down through the Kayenta which always forms benches. After about 4 kms, you'll come to a couple of pretty good DC's which you can bypass on the right or west. After 75m or so, you can DC into the beginning of the last or **Wingate Slot**. There you'll DC into a possible pool then comes a possible rap from a large chokestone. We rapped **4m** using an Ibis hook. Immediately below that is another chokestone which you can **DC** about **6m**. Just around the corner from that are about 4 choke-stones just above a deep swimming pool; or a large mud hole! It's possible that some can **DC** this, especially if the water is deep, which would save a fall, but we installed W&R/QL and **rapped** down about **5m**, then swam out. From this last dropoff & pool, it's a 1 1/2 hour walk to **Highway 95** at **Three Forks**.

Elevations Car-parks for Butler are at 1585m, 1605m & 1637m, or 1707m south of Poison Peak; 1762m for Main Fork Butler & Stair Canyon; 1802m at 4WD parking to Stair Canyon; Three Forks is 1280m.

Time Needed East, West & Middle Forks of **West Butler**, 4-6 hours round-trip; a little more if you go on down to **Three Forks & Highway 95**. **Main Fork of Butler Canyon**, 4-6 hours, if you exit via the shortcut exit/escape; 2-3 hours more if you go down to Three Forks. All of **Stair Canyon** will take 7-10 hours and you'll need an extra car (and/or shuttle) at Three Forks (camp there and drive one car to the head of the canyon in the morning). **Upper Stair Canyon R SLOTS** only, 2-4 hours round-trip from the 4WD parking area.

Water Have plenty in your car, but there may be (?) drinking water below the Navajo slots where the canyons widen in all forks of Butler, especially West Butler, in the upper Kayenta Formation. Lower Butler probably has running water--and cows!, but Stair Canyon is dry most of the time, except for possible potholes.

Maps USGS or BLM map Hanksville & Hite Crossing (1:100,000) for driving; Turkey Knob & Stair Canyon (1:24,000--7 1/2' quads) for hiking.

Flash Flood Danger Fairly high danger exists in the slots, but all canyons are short, so a heavy rainstorm would have to pour down right on you, or just above, to cause any concern.

Best Time to Hike Spring or fall, but all can be done any time of year.

Author's Experience After several scouting trips, he and Nat Smale went down the **West Fork/West Butler** exiting via the faultline side-canyon and returning to the Cedar Point Road in 4 3/4 hours total. We finally made it all the way through the very fun **Middle Fork/West Butler** and out the faultline side-canyon in 5 2/3 hours. The author's solo **East Fork/West Butler** 5/2008 trip was finished in about 4 1/2 hours from the car-park marked 1707.

The author parked under a shade tree south of Poison Peak, then did the loop-hike of the upper end of the **Main Fork Butler Canyon** in about 5 3/4 hours. Steve Tyler joined in for **Stair Canyon**. We camped at Three Forks, left one car and drove to Poison Peak. The hike took 8 hours, but the upper slots were by-passed. Later, and with N. Smale, J. Keener & S. Folias, we did the upper slot in 2 1/2 hours.

Above Left Nat starts down the last rap in **Middle Fork/West Butler**. **Above Right** The pool at the end of the slot of **Middle Fork/West Butler**. **Left** The beginning of the **Watery Slot** in **East Fork/West Butler**. Notice the log/root in the middle? Chimney up & over to that, continue horizontally another 15m, then DC to shallow water at the bottom. **Below** Jay Keener & Stefan Folias high-stemming in the middle part of **Upper Stair Canyon**.

Hog & Trachyte Slots, and Bridge Canyon, North Wash & Trachyte Country, Utah

Location & Access Featured here are 3 nice slots in the upper end of **Hog Canyon**, and a couple of unnamed northeast **tributaries of Trachyte Creek**. These are located roughly halfway between Hanksville & Bullfrog. To get there, drive south out of Hanksville; or north from the Lake Powell & Hite area, on Highway 95. People with nice cars doing Hog Canyon could park along **North Wash** at the **Hog Springs Picnic Area** about halfway between **mile posts 33 & 34**. Or, best to park along a road on **Trachyte Point**. To get there, turn south from Highway 95 onto **Highway 276** at **mile post 26** heading toward Bullfrog (that junction is mile post 0). Drive south to a point about halfway between **mile posts 3 & 4** and turn east onto a pretty good graded road. Drive **uphill** 3.5 kms (2.2 miles) to the **Hog Canyon TH**, which you can do with a 2WD--even a car. Park at the crest of a hill & at elevation **5172T** (1576). Or to reach the **Trachyte Slot (TS)** continue southeast to the **drill hole site** at Km 7 (Mile 4.4), then an ungraded road to Km 8.8 (Mile 5.3). This is the head of what this writer is calling **Trachyte Slot** (AKA Witch's Cauldron). Ryan Cornia's website calls the bridge in Bridge Canyon, Tik-Tok Bridge

Rating Hog Canyon, each fork: **3A II PG SLOT**, but **East Fork** could rate **3A II R SLOT** if you DC one 14m vertical part rather than being belayed, or rig an anchor & rap. **Trachyte Slot: Upper Slot: 3B III +PG/-R SLOT**. The **Lower Slot** may be **+R/-X SLOT** because of a **4m-deep KPH** & medium to high-stemming in the slot below. **Bridge Canyon, 3A II**, with a long single rope rappel.

Equipment Hog: RG, one 50m rope, plus 1 shorter rope, K&EP & W&R/QL for all slots. **TS:** RG, two 50m ropes, Ibis hook & G-pick, R/QD/DC & ascenders. **Bridge:** one 60 & 30m ropes, gloves.

Route Hog Canyon If starting at the **picnic area**, walk upcanyon along a hiker's trail. It's a good trail up to the **Kayenta Falls**, then you may have to wander around looking for a way to avoid bushwhacking up to the area of **Hog Spring**. At that point look for the side-canyon coming down from the northwest. Scramble up this **steep drainage**. Once on top, rim-walk north to the head of the canyon and the fork of your choice. From the **Upper Trailhead**, simple rim-walk NNE to the head of the slots. Have a USGS map & compass in hand. The **steep drainage** just mentioned will be your **exit route**.

   West Fork Scramble down into the slot. Shortly you'll come to a **6m drop**. Some might DC this (?), but a small group can help each other. Or set an Ibis hook on a shelf to the left (make a small hole with the sharp point so it can't migrate), then DC-on-rope, or rappel. About 7-8m beyond that, is a boulder on the right with W&R/QL. Use this to make the **2-stage R2**. Or helping each other, you can do this without rappelling; or even HL down from an Ibis hook (lots of natural holes for this). For the next 300m or so, you'll be in a nice tight slot. You can more or less walk down this, but there are several shoulder-width sections & dropoffs to DC. Just before R3, will be 30m of chimneying or low-stemming 2-3m above the tight bottom. For this, hang your pack from the front of your harness with a R/QD/DC so you can do a *back-to-knees-&-hands DC*. Just below this is a large chokestone with W&R/QL for the **13m R3**. Below R3, it's an easier slot with possible shallow pools, and one more good DC, then the 800m-long slot ends. Walk downcanyon to the highway; or, climb up the **steep drainage** escape route and return to the **Upper Trailhead**.

   Middle Fork This might be the best of the 3 forks. There are 2 ways to do this one. If you go right down the drainage, you'll first have a shallow slot, then steep slickrock leading to a big drop. You can rap this but you'll need long ropes for a **42m drop** (start at W&R/QL under a ledge at the beginning of the steep slickrock). However, you can do it easier with fewer ropes if you walk down the **hogsback ridge** between the **West & Middle Forks**. You'll soon find a walk-down route heading left (east) into this drainage below the 42m rap. Further along, you'll come to the **2-part 18m R1**; then just below that, a partly free-rap **(R2)** of **16m**. So one 50m rope will do. After some walking you'll come to the best part; a short slot ending with a **dark & tight 9m vertical DC**. Some may want to be belayed, or rig an anchor for a rappel/HL. However, with K&EP, you can slither (DC) down this one--it's easier than it first appears. Halfway down will be a scallop on the west wall where you can rest, then facing east, continue down. At the very end it **bell-bottoms**, and you may have to let go and slide of 1 1/2m; or lower yourself from a toehold (?). Or the first person down can help others at the end (there could be a small pool right after floods). From there, the canyon opens quickly and you walk out and up the **escape route**.

   East Fork This entire slot drops very quickly right to a 22m rap at the end. Throughout, you'll chimney down a straight section, then a 90° left turn and after 3-5m the next drop. Getting started at the top of each drop can be scary for some, so the leader may have to do a belay. At one of these big drops the wall on the opposite side curves back under or bell-bottoms, making for a tricky first step--to do this lean to the right as you leap. This is a **risky move!**, and perhaps most will want to be belayed. After that it's *back-to-feet chimneying* down. Near the end is a **22m rap**, then walking out of the canyon. There may be water at the bottom of this fork from the nearby Hog Spring.

   Trachyte Slot From the end of the road, route-find south into the head of the prominent drainage. After about 2 kms, you'll pass (& bypass) 2 short insignificant slots. After another 2 kms or so, you'll come to the **Upper +PG/-R SLOT** lasting about 250m. Skinny folk can stay in the bottom most of the time, others may have to high-stem a little (or if it's wet and you want to keep feet dry).

   Right after the Upper SLOT opens will be a **drainage** coming in from the left or east, and opposite to the right or west will be an **escape route**. Remember this place, because after another 300m or so will be the **10m Rappel 1**. You may want to tie one of your long ropes onto a big rock about 17m back from the drop and leave it--to be picked up later (instead of dragging it through the lower slot or taking the time to build a deadman anchor closer by). Soon after this simple rap is the **Lower +R/-X SLOT** which begins on the northeast side of the **circle feature** marked **4696T** (1431m). Right away you'll DC to a big **4m-deep KPH**. To get 1 person past this--you might try stemming over the top (perhaps belayed at one point); or while on-rope, reach around the corner and make a pack toss down the exit side which is a steep slot; or get into it and swim across; or one person get in and do a kind of shoulder stand but with outstretched arms pushing a second person out. As life insurance, a **G-pick** could be used to make one small step, or clear mud from a critical point--but this would be your last choice.

   Just beyond the KPH, is a **steep DC** in a **smooth water-worn groove** and into a 3m-long **swimming pool**. After that it's moving laterally 2-8m above the deck in a 1/2m-wide smooth vertical slot. Looking back, this may have been easier if our packs had been attached using quick-draws to the sides of our harness instead of hanging down and getting hung-up below our feet. After that section are chokestones 7-8m above the bottom. It would be easier to DC this part after & below the chokestones, but this will be you're last anchor before the 2nd half of a **22m R2**. A really good (or stupid) climber might DC the 2nd part but the risks are high! After that the canyon opens (after a full-body span over a wa-

Map 30, Hog & Trachyte Slots, and Bridge Canyon, North Wash & Trachyte Country, Utah

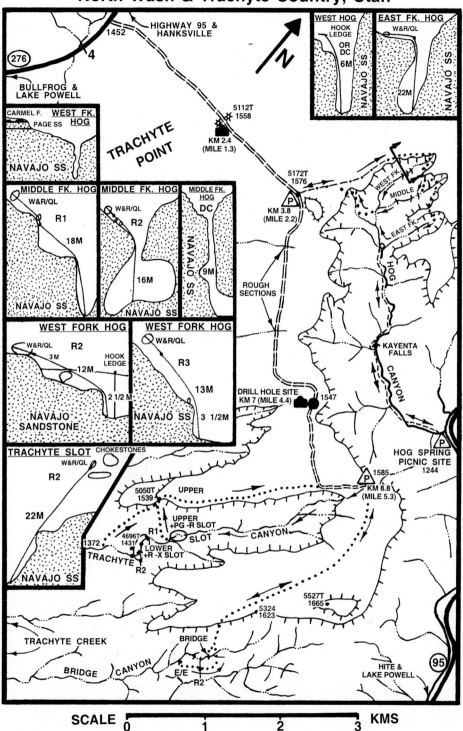

SCALE 0 1 2 3 KMS

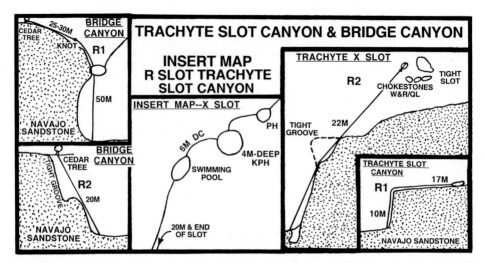

TRACHYTE SLOT CANYON & BRIDGE CANYON

BRIDGE CANYON

CEDAR TREE
25-30M
KNOT
R1
50M
NAVAJO SANDSTONE

INSERT MAP
R SLOT TRACHYTE
SLOT CANYON

INSERT MAP--X SLOT

TRACHYTE X SLOT

R2
CHOKESTONES W&R/QL
TIGHT SLOT

TIGHT GROOVE
22M

5M DC
PH
4M-DEEP KPH
SWIMMING POOL

TRACHYTE SLOT CANYON
R1
17M
10M
NAVAJO SANDSTONE

BRIDGE CANYON

CEDAR TREE
R2
20M
TIGHT GROOVE
NAVAJO SANDSTONE

20M & END OF SLOT

tery PH), then after 600m, exit north on the right. Pickup your rope at R1 if necessary, and return.

About 1.8 kms SSE of the circular feature marked **4696T** (1431m) and in the **1st drainage southeast of Trachyte Slot is Bridge Canyon** and a natural **bridge** some have rapped through. It's a little like Golden Cathedral, but this bridge originates from a single pothole, instead of 2 or 3. To do it, hang a 30m rope from a cedar tree on the north, then tie a 60 to that and rappel below the knot. In the canyon below that is a **tight 20m rappel** from another cedar tree (last guy down might DC this, with one risky step around a corner). Below that is another drop; tippy-toe along **skinny ledges** to the right or north. **Exit east 500m below the bridge.** South of this drainage are not-so-interesting canyons with potholes.
Elevations Upper Hog Trailhead, 1576m; Hog Springs Picnic Site, 1244m; Trailhead for Trachyte Slot, 1585m; canyon bottom exit altitude, 1372m.
Time Needed Half a day, or 4-6 hours for each of Hog Canyon's Slots; longer if starting at Hog Springs on Highway 95. For Trachyte Slot, about 6-8 hours.
Water Take your own, but there should be creek water at Hog Springs.
Maps USGS or BLM map Hite Crossing (1:100,000) for driving & orientation; and Black Table & Hite North (1:24,000--7 1/2' quads) for hiking.
Flash Flood Danger High for all slots, but a thunderstorm would have to dump right on you to cause concern; these are very short drainages. Trachyte Slot is longer and would be more dangerous.
Best Time to Hike Spring or fall, but you can do these hikes anytime of the year.
Author's Experience Starting from the Upper Trailhead, he did the **West Fork** solo in 3 3/4 hours. He DC'ed the 1st rappel on-rope using an Ibis hook. **Middle Fork** was done with J. Summerson in 4 3/4 hours from the Upper Trailhead. The **East Fork** was done with N. Smale, J. Keener & S. Folias. We left one car at the highway, then drove to, and started at, the Upper Trailhead--4 hours car to car.

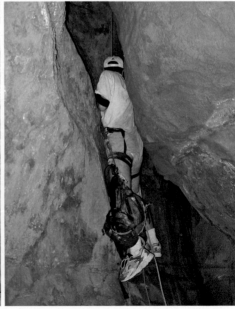

Opposite Page--Left The beginning of the tight 9m DC in the **Middle Fork of Hog Canyon**.

Opposite Page--Right John Summerson at the end of the 9m DC in the **Middle Fork of Hog Canyon**. It would be best for him to face the opposite direction doing this last part--face the east, instead of west like he's doing.

Above Left Ryan Cornia high-stemming through the **Upper +PG/-R SLOT** of **Trachyte Slot Canyon**.

Above Right The lower end of Rappel 2 at the end of the **Lower +R/-X SLOT** in **Trachyte Slot Canyon**. Some might DC this except for one short move in the Tight Groove.

Left Rappelling through the natural bridge for which **Bridge Canyon** is named. The person on top is roped-up and taking fotos of the rappeller below.

Maidenwater, Trail & Woodruff Canyon Slots,
Trachyte Country, Utah

Location & Access Discussed here are the **2 forks** of **upper Maidenwater**, plus the **North Forks** of Maidenwater; the **North Fork** of upper **Trail Canyon**; and **4 slots** entering **Woodruff Canyon**. These are located just east & west of Highway 276 between **mile posts 8 & 12**, and due east of Mt. Hillers, which is the central peak of the Henry Mountains.

To get there, drive south out of **Hanksville**; or northwest from **Blanding or Hite** on **Highway 95**. At **mile post 26**, turn south onto **Highway 276** (mile post 0), the paved road heading south toward Bullfrog Marina on Lake Powell. Drive 13 kms (8 miles) and stop at **mile post 8** and a little sandy track heading east toward the upper end of **Maidenwater's North Forks**. Cars should park (& camp) near the highway; 4WD's can go about 2.9 Kms (1.8 Miles) partly downhill, and park. Or stop at **mile post 9**, which is immediately next to the **Middle Fork of Maidenwater**. Or continue south to about mile 9.4 and park about 150m south of a **cattle guard**. This is where you park if going into the **South Fork of Maidenwater**. Or park & camp just north of **mile post 10** where UDOT stores gravel near where **Trail Canyon** crosses the highway.

If going into the **North Fork of Woodruff**, stop & park on the highway about halfway between mile posts 10 & 11. For the **3 South Forks of Woodruff**, stop just north of mile post 12 on the east side of the highway (or park & camp at the corral 125m west of mile post 12 and next to Woodruff Creek)

Rating North Fork Maidenwater (NFM): 2B I; Little NFM: 2A I R SLOT; Middle Fork Maidenwater (MFM): 3A II; South Fork Maidenwater (SFM): 3A II; North Fork Trail (NRT): 3B II PG SLOT; North Fork Woodruff (NFW): 3A II; Little Fork Woodruff (LFW): 3A I; Middle Fork Woodruff (MFW): 3A II PG SLOT; Pothole Fork Woodruff (AKA--Woody Canyon)(PFW): 4B R III.

Equipment NFM & LNFM, 15m rope & K&EP; **MFM**, RG, 30m rope, Ibis hook, W&R/QL, K&EP; **SFM**, RG, 15m rope, Ibis hook, W&R/QL, K&EP; **NFT**, RG, 40m rope, W&R/QL, K&EP; **NFW**, RG, 30m rope, W&R/QL, K&EP; **LFW**, RG, 20m rope, Ibis hook & G-pick, W&R/QL, K&EP; **MFW**, RG, 20m rope, W&R/QL, K&EP; **PFW**, 2x15m ropes, toss bags for KPH escapes, Ibis hook & G-pick, K&EP, wetsuits.

Route North Fork Maidenwater From **mile post 8**, walk/drive east on the sandy track to **Km 2.9/Mile 1.8**, then with map & compass in hand, head eastward around the head of upper NFM. Look for a route down through the Page SS (?) cliff, then route-find southwest into the slot, which begins with a slide into a long **pool** (see foto page 154). This is shallow slot with **3 fairly challenging DC's**. At the end of the slot, you'll find trees & brush, and after 100m you can turn left where the LNFM comes in on the left or northeast. Walk up steep slickrock and backtrack to your car; or climb up about 100m and enter the upper part of the **LNRM**. This slot has **4 very challenging 3-5m DC's** which border **R or X** in difficulty. For this you must have one excellent climber and a 15m rope to be used as a hand/safety line for the less-experienced. A bit spooky, but they're short DC's; or fast slides with hard landings! Afterwards, escape back to your car the same way as described above--no need to go downstream and up Trail Canyon.

Middle Fork Maidenwater From **near mile post 9**, climb west up the slope which is immediately south of the **Middle Fork**. To begin, rim-walk southwest with the canyon on your right. Later you'll be walking & route-finding west around the upper part of 2 short side-drainages. Soon, you get pretty close to a buttress to the west made of the reddish Entrada Sandstone and marked **5752T** (1753m) on the *Black Table* quad. From that area, gradually turn right or north around the upper end of the **2nd side-canyon**, and into the bottom of the upper Middle Fork.

At first the Middle Fork is wide open, then you can avoid a 15m dropoff by DC'ing just to the south and back into the main drainage. About 1/3 the way through the canyon are **3 dropoffs in about 40m**. For the **first 2 rappels**, the author, who was alone, rapped off an grappling hook (Ibis hooks are better), but you may want to leave W&R/QL around some chokestones (with a partner assist, this might be DC'ed?). For the last drop (R3) and with help from friends, you can likely help each other down the **2nd** and **3rd raps** (?). After these 3 dropoffs, you'll have about a dozen DC's from chokestones before you reach the highway.

South Fork Maidenwater From where you park, walk north 150m to the **cattle guard**, then climb and walk west along an easy slope on the north side of this biggest drainage of upper Maidenwater. After nearly 3 kms of walking along the contact point between the Navajo Sandstone (maybe the Page SS?) below, and the Carmel F. above, you'll arrive at the head of the canyon. About **150m** from where the dry creek bed drops in, and on the **south side**, are 2-3 places you can DC into the drainage.

After about 100m will be **R1**. We did that from an Ibis hook in a crack, but you might do it with a partner assist. After another 250m is **R2**. We again rappelled from an Ibis hook in a crack on the left; but don't let it get wedged in or you can't flip it off. You may find W&R/QL there; or maybe DC it (?).

About 400m further will be the **5m drop** at **R3** from a small chokestone & W&R/QL. In the 200m below that will be the deepest part of any canyon on this map--nice slot. Next will be a possible exit on the right, then the **5m R4** with W&R/QL. Below that will be one possible deep PH, but it should never be a KPH. After that you'll walk and DC for 1500m, then the **7m R5** (maybe DC) which is about 100m from the highway.

Upper Trail Canyon Trail has 3 main components west of Highway 276; the best one is the **North Fork**. From the **gravel storage parking**, walk northwest crossing the main drainage; after 100m get into the combined North & Middle Forks; veer right into North Fork. About 400m from the highway, you'll come to a PH where the canyon constricts. From there, route-find NNW along a maze of ridges separating the North and the Middle Forks. If you don't know how to read a map or use a compass, maybe this isn't the place for you!

Eventually you'll arrive at the head of North Fork which is about 1 km southeast of where you enter the South Fork of Maidenwater. Route-find down in on the south side of the upper slot and start DC'ing immediately. After about 700m you'll come to the **18m R1**. On our trip, the anchor for this rap was a pile of rocks, or deadman, with W&R/QL attached to the bottom stone--if it's gone, built a new one with W&R/QL, or do something else (?). Below that, the slot gets more narrow and after another 500m will be **R2**. You should see W&R/QL attached to a couple of chokestone for a drop of about **7m**. Below that are a couple of potholes, likely with water. You can avoid wading by climbing along a ledge to the right of the first, but the next one you may get your feet wet--unless you have a long reach. About 500m beyond R2 will be the narrowest part. At one point you'll have to UC about 2m to pass a narrow spot. After that, and just before the slot ends, will be a place where **only the skinniest folk** can pass; everyone else will have to chimney up a meter or two to get through. After that, walk out and back to your car.

There are also a couple of short slots with DC'ing & pools in the **Middle Fork of Trail Canyon**. Another hiking option would be to go down the **South Fork of Maidenwater** east of the highway all the way to **Trachyte Creek**, then south to **Trail Canyon** and back up to the highway. Lower Maidenwater may have a short slide into a pool, otherwise it's just a non-technical walk. It has running water, a few pools, and is fotogenic. **Lower Trail Canyon** is a **walkup** and a good exit from Trachyte Creek.

Map 31, Maidenwater, Trail & Woodruff Canyon Slots, Trachyte Country, Utah

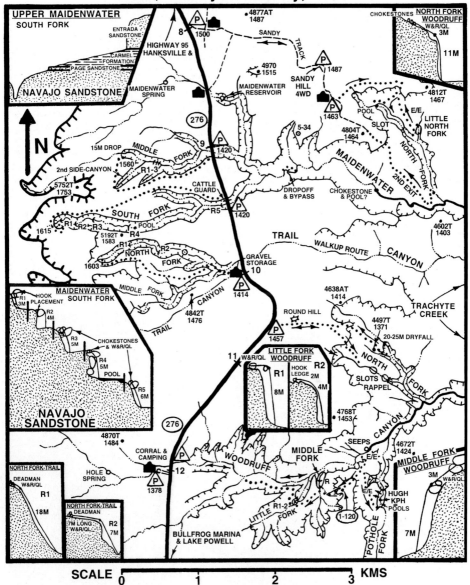

North Fork Woodruff Canyon From the car-park at 1457m, head ESE about 1 km, then look for one of several ways down through the Page SS (?) cliff. There are 2 slot in this canyon, with an escape/entry between the 2 sections. The upper slot has easy DC's, then an open area, and the lower slot. It has **2 interesting DC's**, one of which is a partner assist. At the end is an **11m dryfall**. **Rap** this by installing W&R/QL around a small chokestone (you may have to run back upcanyon to find one) about 4m back from the drop.

From the bottom of the only rappel, continue downcanyon 150m, turn left or east and walk up another tree & water-filled drainage. About 700m into this 2nd canyon, you'll come to a 20-25m. From there, back up about 75m and scramble up to the southwest. You'll soon come to a minor drainage; head northwest up this and route-find just east of the dome marked **4497T** (1371m). Leave this drainage by walking up a sandslide which covers the Page SS, and head back the same way to your car.

South Forks of Woodruff Canyon From near mile post 12, head east down Woodruff with map in hand. After 1 km, will be a large canyon coming in from the right or southwest. Right at that junction, look southeast at the slope with Navajo SS below, and Carmel F. above. To reach either the **LFW or MFW**, climb up toward

153

the top of the Navajo slickrock, then contour around the point and into the **LFW** just above the **40m-long slot**. Once there, DC 15m to the **1st rappel**, about **8m**. You'll have to install W&R/QL. Below that is easy DC'ing to the **2nd rap**. In the bottom of the stream course are 2 natural ledges 2m from the drop; we set an **Ibis hook** (a Spoonbill hook will also work) in one, and rapped **4m** to the bottom, and walked out.

Middle Fork Woodruff Use this same approach route, but continue east past the LFW, and to the beginning of MFW. Or if doing the 2 slots on the same day, walk out of LFW and into the main Woodruff Canyon. About 300m down Woodruff and on the right or south, will be the bottom end of the MFW; you might check-out the lower end first. From there, continue east another 700m and turn to the right or south and climb up on a sandy or alluvial bench, then route-find up a little hogsback to the rim. From there, again route-find along the top to the west & southwest past the circle on the USGS map marked *1-120*, and into the head of MFW.

At the beginning of MFW, which is where 2 upper forks meet, is a **slightly challenging DC** into the slot. At first stay 3-4m above the deck, then slowly work your way down to the bottom. Further along is a big rock-fall & rappel. We left a single strand of W&R/QL for a **7m rap**. After that are 2 pretty good DC's (one might be a 3m rap), then a tight slot to the edge of a deep pothole, but you lunge forward and grab the lip on the right side to escape; and keeping your feet dry. Then out the bottom of this **PG SLOT**.

Pothole Fork Woodruff From the bottom of LFW or MFW, head back up to the rim as described above, then after about 200m, instead of heading southwest to MFW, head south and a little southeast to reach the beginning of the main slot of PFW. DC in and immediately find big potholes in the first 75-80m. Some were dry for us, and they were no problem, then a drop of **3-4m** into a pothole; we rapped off an **Ibis hook** set in a **natural cup hole in the wall** to the left. Here you can do it without a harness by wrapping the rope around your body, then **HL down**.

Then **2 KPH**, so be ready to toss a bag filled with sand, or somebody's pack, over one pothole and down the other side. At that point, we found a bridge over 2 deep potholes; if water is real low, you can go under, but you'll need a rope & pack toss to get one person up & out the far side.

After those potholes, it's a short walk to another **huge KPH**. This one is so long, no one can possibly throw a pack or bag across it, so to get out, it must be full of water, or completely dry. We were there with water 1m down from the escape-side rim, and below that a 45° slippery slope under water. No way! So we climbed out the north wall bypassing the huge KPH; you'll need one good climber to do this (a G-pick might be life insurance at this point). After that, you'll find an escape from the canyon on the left or north, or HL back down into the canyon bottom (you'll need a **30m rope**, or a combination of PR's & webbing, to do this). From there, it's 100m in an easy slot filled with small potholes to a 2nd escape on the left or north. Then another 75m or so to the drop at the end of the slot. To do this, wedge your left elbow & forearm in the groove and DC backwards to some tiny steps at the bottom. Then walk out to the main canyon and back to your car. An interesting slot with special challenges.

Elevations Car-parks/trailheads, between 1378m & 1500m.

Time Needed Both slots of **NFM**, 3-5 hours (depending on where you park); **MFM**, 2 1/2-4 hours; **SFM**, 5-6 hours; **NFT**, 3-5 hours; **NFW**, 2 1/2-4 hours; **LFW**, 2-3 hours; **MFW**, 3-4 hours; **PFW**, 4-5 hours (all south forks of Woodruff can be done in one long day,--if you have a strong team for Pothole Fork).

Water There are few springs or safe-to-drink running water (except in lower Maidenwater & Trail) because cows graze the area from October to May, so take water in your car and pack.

Maps USGS or BLM map Hite Crossing (1:100,000) for driving; Black Table & Mt. Holmes (1:24,000--7 1/2' quads) for hiking.

Flash Flood Danger Higher risk right in the tight slots, but all these drainages are very short, so to be in great danger, a heavy storm would have to hit directly above you.

Best Time to Hike Spring or fall, but you can hike these anytime there's no snow or too much mud. Summer would be good for **PFW**, especially if the potholes are full.

Author's Experience On his 2nd trip to **MFM**, he did the hike suggested above in about 3 1/3 hours. Later, he and John Summerson did the **SFM** in 5 1/4 hours. **NFT** was done with Cristina Amat & Ryan Cornia in 3 1/3 hours. After scout trips, he & Ryan Cornia did both **NFM** in just under 3 hours; **NFW** in 2 3/4 hours; **LFW** in less than 2 hours; **MFW & PFW** together in 5 3/4 hours.

Opposite Page The first big pothole in the **North Fork of Maidenwater**.

Above Left One way out of a pothole is to stand on a pack, a log or pile of rocks. This is the first pothole in **Pothole Fork of Woodruff Canyon**.

Above Right This is the last rap in **Little Fork of Woodruff Canyon**. There are 2 natural cracks where an Ibis hook can be placed for the 4m rappel.

Left Ryan Cornia DC's the last dropoff in the lower end of **Pothole Fork**. There's a nice groove for the left arm, and solid hold for the right hand.

Meeks Mesa Slot, Capitol Reef National Park, Utah

Location & Access **Meeks Mesa Slot** (AKA Pandora's Box) is located in the northwest part of Capitol Reef National Park, just north of Highway 24, and east of Torrey. This canyon starts on top of **Meeks Mesa** and flows east into **upper Spring Canyon**. To get there, drive along **Highway 24** between Capitol Reef NP Visitor Center and Torrey. Just west of **mile post 73**, turn north into a flat area with a number of free campsites & P/J trees just west of the park boundary. Park & camp there.

Rating **3A or B IV +PG or -R SLOT**. Lots of high-stemming over tight slots with big rappel at the end.

Equipment RG, two 60m ropes (or one 60 & pull cord), R/QD/DC, W&R/QL, K&EP, smallest pack possible, leather gloves, PR, Ibis hook, 15m rope for short drops or belays & drybags for possible pools.

Route From the trailhead, locate and walk up an **old cattle trail** to the top of the obvious **flat bench** to the north. Follow this trail northwest roughly 3 kms, then as the trail is heading in a northerly direction, look for another trail turning southeast. Take this old **unused cattle trail** to the top of **Meeks Mesa**. Once on top, and with **map & compass** in hand, head northeast toward the upper **South Fork** (look back occasionally for landmarks because at the end of your hike, you'll return to the top of the same trail).

At the South Fork, drop down one bench and contour around to the head of the **North Fork**, which is apparently the better of the 2 drainages. Right at the start, you'll have 2 easy DC's, then the first of **about 7 rappels**. **R1** will be from W&R/QL on logs and down **21m**. The first person down should walk a short distance to R2 and make sure there are logs & W&R/QL, something to rap from. If not, you'd have to put in a B&H, or throw down some logs or something and build an anchor to rappel from. **R2** is a 2-stage rap of about **26m**. Just below that is what some have rigged-up as a rappel, but we **DC'ed 5m** instead.

If notes are correct (didn't have the usual 7 1/2' quad for this one!), you'll soon have some high-stemming, then some easy walking down to the confluence of the **2 upper forks**. Soon after that is **R3**, from W&R/QL & logs down **13m**. This is into a little dark hole, but Nat Smale DC'ed it; however it's risky! In the lower half of the drainage, you'll find short sections of tight slots where you have to chimney up (nearly 10m) and stem over, then open areas with short faultline side-canyons intersecting the route, then another tight slot. At what has been a rappel for some, we stemmed up & over a bell-bottom section and down the other side; then we came to what is likely a pool most of the time. With W&R/QL from a log, you can rap--**R4**; or use a hook from natural holes to **HL** down; or if you have good knees, let it rip & slide the last 2m. At that point the canyon opens, then closes again with another drop of **4m** and **R5**. You can HL over it; or help each other down using W&R/QL from above.

Finally at the end of the **Wingate SS slot**, the bottom drops out as the canyon opens with the Chinle clay beds below. There you'll do **Rappel 6**, an easy **17m rap** down to a second level, then the **45m R7** to the bottom. There will always be plenty of anchor material around. As usual in cases of long rappels, don't twist your ropes--have someone below separate the pull rope from the rappelling rope so you get everything down OK.

From the last rap, it's boulder hopping about 800m down to **Spring Canyon**. From there walk west upcanyon about 2 1/2 kms. Just above the **1st side-drainage** on the right (north) from where you en-

Rappel 4 into a pool in **Meeks Mesa Slot**.   High-stemming somewhere in **Meeks Mesa Slot**.

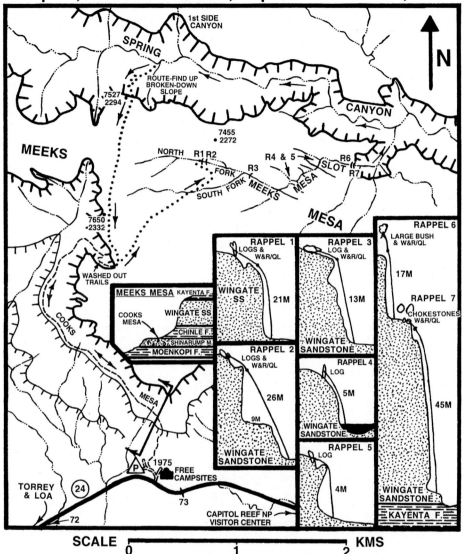

tered Spring Canyon, be looking for a broken-down slope to the left or south. Route-find up this slope with benches, boulders & P/J trees to the top, then with compass in hand, head in a magnetic south direction back to the top of the **cattle trail**. From there it's an easy fast walk back to your car.

Elevations Trailhead, 1975; top of Meeks Mesa, roughly 2300m.

Time Needed Depending on how many foto stops you make, and strength of your group, roughly 10-12 hours--or longer (take a headlamp just in case, and maybe a space blanket/bivvey sack too).

Water This is a dry canyon, so take plenty with you.

Maps USGS or BLM map Loa (1:100,000); Twin Rocks (1:24,000--7 1/2' quad) for hiking; the plastic Earthwalk Press chart, Hiking Map & Guide: Capitol Reef National Park (1:62,500) for hiking & driving.

Flash Flood Danger With higher altitudes & lots of slickrock, this could have high danger for short distances, but there are lots of open spaces too.

Best Time to Hike Spring/fall, June has longest days; higher altitudes make it possible in summer.

Author's Experience Along with N. Smale, J. Keener & S. Folias, we camped at the trailhead and with a fairly early start, did the hike in 11 hours.

Burro, Cottonwood & Fivemile Washes, Capitol Reef National Park, Utah

Location & Access These 3 canyons are located in the middle part of **Capitol Reef National Park**. To do either hike, first head to the park **visitor center** located between new **mile posts 79 & 80** on Highway 24. While there, buy a good map, ask about road conditions, and perhaps when the last rains fell in the area. As of 2007, no permit was needed to do these canyons, but it won't hurt to let someone know of your plans.

These canyons can be done using one car and making a loop hike from the bottom of each canyon; or with a shuttle (2 HCV's), spotting one vehicle at the lower east end of each canyon then drive to the other end and hike back to the first HCV. Either way, first drive **east** from the **visitor center** on Highway 24 in the direction of Hanksville. After 14.5 kms (9 miles) and between **mile posts 88 & 89**, turn right or south onto the **Bullfrog-Notom Road (BNR)**. Drive south to **Km13.8/Mile 8.6** where **Burro Wash** crosses the road; park there, or drive up the slightly sandy dry creek bed as far as 2.9 kms (1.8 miles) and park (2WD cars can usually make it that far). Or from Highway 24 drive south to **Km 15.9/Mile 9.9** at **Cottonwood Wash**. You can park in the shade about 50 & 250m up from the BNR; or with a HCV, drive up the dry wash about 1 1/2 kms and park (unless the NPS or BLM have blocked it off). Or from Highway 24 drive **17.9 kms (11.1 miles)** south to **Fivemile Wash** and park next to the BNR; or under the new bridge for shade. In 2007, the BNR was paved to Fivemile Wash. Park one car/HCV as suggested and hike from there; or leave one vehicle in one of these areas and head back to the park visitor center and the west side of Capitol Reef.

To reach the head of each canyon from the visitor center, drive southeast along the paved road going toward the **Fruita Campground, Grand Wash & Capitol Gorge**. That's where the pavement ends. Continue south on a good graded dirt road toward the **Sleeping Rainbow Ranch** on **Pleasant Creek**. From the visitor center to the crossing of Pleasant Creek is **17.6 kms (10.95 miles)**.

From Pleasant Creek, continue south on a never-graded road suitable only for HCV's during dry conditions. This track follows a dry creek bed. Drive to **Km 21.7/Mile 13.5** (from the visitor center) and you may see an old miner's track heading east, but which is now blocked off by a couple of posts. Park there if you're heading for **Burro Wash**. Or continue south to **Km 22.9/Mile 14.2** and be looking for another old blocked-off miner's track veering left. Park in that area if going to **Cottonwood Wash**. That point will be about 700m due west of high point *6930T (2112)* on the *Golden Throne* quad. Or if you're going to **Fivemile Wash**, continue south up the road another **150m** or so, and when the road climbs out of the dry wash to the right or west, park and begin hiking from there. For any last minute information call the visitor center at Tele. 435-425-3791; or see their website at **nps.gov/care**.

Rating Each canyon: **3B III**; Fivemile may be the longer hike of the group--but not by much. All canyons are relatively short, starting at the top of the Reef and draining east.

Equipment RG (Ropes; **Burro** a full length 60m; **Cottonwood**, one 50m; **Fivemile** one 15m), K&EP, W&R/QL, PR, if going solo an Ibis hook to HL down short drops, and drybag/waterproof case for cameras. In spring/fall/cooler weather a wet/drysuit--unless the rangers tell you the canyons are dry, which is seldom.

Route These 3 canyons are located close together and are very similar in appearance. Each canyon passes through the Wingate Sandstone, Kayenta Formation, Navajo Sandstone and ending in the Carmel Formation. For the most part, the best narrows/slots and the technical sections are in the Navajo Sandstone.

Burro Wash Starting on the east side with one car, from where you park 2.9 kms from the BNR, leave the drainage bottom and walk up the slope heading southwest with map & compass in hand. Route-find to and/over the top of elevations 5898T (1798m), 6305T (1922m) & 6655T (2028m), then route-find south & west and eventually into an upper south fork of Burro. Once in that drainage, head north into the upper main part of Burro in the Kayenta section. Read more below.

Or if you're doing **Burro** with a **shuttle**, walk east on the old track to near the **old mining prospect** shown on the *7 1/2' quad*, then once on top of the Shinarump bench, veer left or north and continue up a shallow drainage until you're just west of the point marked **6660T** (2030m). From there turn right or east, climb up the steep Chinle clay slope with Wingate boulders and drop down into the head of Burro. The first 350m or so will be walking through brush (but no bushwhacking), then you'll come to a nice slot in the Wingate with 3 pretty good DC's (2 B&H were removed from the first dropoff in 2002, so you'll now have to DC it, rap off some W&R/QL, or have the leader belay beginners). After the Wingate, the canyon opens and you'll come to the Kayenta and several potholes you can walk around, then the Navajo Sandstone appears.

About 100m past a prominent side-canyon coming in from the south (where the 2 routes meet) will be a **10m dropoff**. In the past some have rappelled here, but you can climb out and walk around this on the right or south side. After another km or so, you'll come to a good slot and **Rappel 1**, which is down about **8m** from a **single B&H** on a ledge to the right. There are several deep pools in this section, but you can span most of them. Just below R1 is a DC & steep slide into a possible pool.

After that good slot, you'll have walk for about 1500m, then the bottom drops out (**Note:** at that point you can backup 100m, climb up on the bench or hogsback to the west, walk north, then east, and finally down into the canyon **below R2 & R3**. Let's call this the **Bypass Route**). There you'll find a possible **KPH** and a large chokestone. You can rappel off an Ibis hook on the right, or build a deadman anchor, and try going through that pothole by using a shoulder stand or pack toss. Or, immediately above that, and on a ledge to the right, is a **tree** 15-20 cms thick, and W&R/QL. Perhaps reinstall W&R/QL, then **rappel 11m** to just below the KPH. About 10m beyond that will be **R3**. This is from 2 B&H's on the left with the total drop of about **29m**. The first 1/3 the way down is along a low angle slope to a small pool; then the last part is first a groove, then a free rap into a **round chamber** which is very fotogenic. This may be the best part of Burro Wash. After the chamber is about 100m of tight slot, then a **pool** which may be a swimmer. After that are several easy DC's, 2 more fotogenic walk-through slots, and nearly 2 kms of walking to your car (or another 2-3 kms to the BNR).

Cottonwood Wash If starting on the east side with 1 car, walk or drive west up the dry wash 1 1/2 kms to the end of the vehicle track. From there continue upcanyon about 300m and look for an exit to the right or north. Once out of the Cottonwood drainage, and with map & compass in hand, head due west toward elevation **5979T** (1821m). Continue west to **6655T** (2028m), then route-find south & west toward an upper divide between Burro & Cottonwood. From there head due south in an upper north fork of Cottonwood until you're in the main drainage. Read the rest of this description below.

Or if doing **Cottonwood** with a **shuttle**, from where you park, walk southeast. Look for an **old miner's track** heading for the top of a bench formed by the **Shinarump Member** of the Chinle Formation. After about 1 km, you should be just west of elevation **6421** (1957m). From there, simply head east downcanyon passing through the Wingate and Kayenta Formations (the 1st route mentioned above comes in here). About 300m into the Navajo, you'll come to the **1st of up to 4 rappels/DC's**. Look for W&R/QL around a small tree on the right--if the W&R/QL is gone, rig another one. Below that will be a nice slot for about 1 km. Expect a

Map 33, Burro, Cottonwood & Fivemile Washes, Capitol Reef National Park, Utah

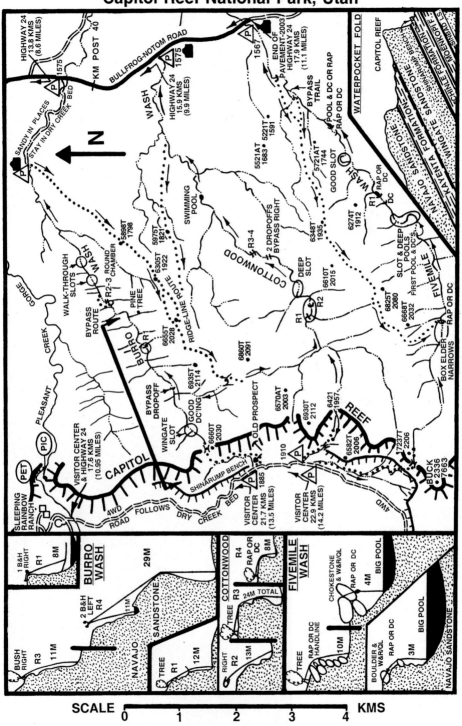

SCALE 0 1 2 3 4 KMS

few pools and pretty good DC's. About 250m below Rap 1 will be **R2** from an outcropping & small chokestones on the right. Nearly 2 kms below R2 will be 2 dropoffs you can skirt around on the right. About 500m below that will be **R3 & R4** one after the other.

To do R3, you'll need a 50m rope attached to W&R/QL tied to a small tree. Even through the tree is about 10m from the dropoff, pulling the rope from below was no problem for us. Just below that is R4 which can be done by wrapping a short rope around limbs of a small tree; or DC 3m with a short jump at the bottom. About 2 1/2 kms below R4 you may find a long narrow pool to swim, then the canyon slowly opens up.

Fivemile Wash From the Fivemile bridge on the BNR, walk upcanyon. After about 2 kms, route-find out of the canyon to the right or north and head west just north of the minor drainage & elevation **5721AT** (1744m). Continue west over the top of **6348T** (1935m), southwest past contour altitude **6600** (2012m), zig zag west north of **6825T** (2080m), then southwest north of **6668T** (2038m), & finally route-find south to the bottom of Fivemile just east of contour altitude **6200** (1890m) and the **box elder narrows**. From there head downcanyon. Read more below.

Or if doing **Fivemile** from the top end with a **shuttle**; from where you park, walk south up the dry wash about 1 km. In that area you should see an old miner's track climbing out of the dry creek bed on the left. This should be approximately 500m southwest of elevation **6582T** (2006m). Once on this track, follow it south with some switchbacks for about 1 km to where it vanishes high on the slope about 500m north of *Buck Point* marked **2336 & 7653**. From there, route-find south up the steep Chinle clay slope to the rim, which will be about 300m west of **7237T** (2206m). Once in the upper part of Fivemile, head east downcanyon.

About 2 1/2 kms into the canyon, you'll come to the Wingate narrows which is filled with **box elder trees.** After 500m in the narrows will be a **10m dropoff** which you can **HL or rappel** down using a tree as an anchor (some **DC** this--or help each other). From there (just below this is where the other route enters from the left or north), and for the next 3 kms or so, it's easy walking, then the canyon **slots-ups** and you'll have to DC into **pools** with possible swimming. In this section, the canyon takes a twisting turn to the north & northwest. About 1 km below that will be the real **R1** from a large chokestone into a possible swimming pool. After another km, you'll come to some pretty good narrows about a km long. At the very end of this last slot will be some tight DC'ing and another short rap, or DC & short slide/jump into another swimming pool. Just beyond that is a wading pool, then the canyon opens and it's an easy walk to the road.

Elevations Upper trailheads, 1885m & about 1910m; end of all canyons about 1570m.
Time Needed Doing the full length of Burro and Cottonwood, about 5-7 hours each; Fivemile, 7-9 hours. If doing either drainage from the bottom end with just one car, plan on 9-12 hours round-trip.
Water Take your own drinking water, but there will be pothole water most of the time (more after rains).
Maps USGS or BLM map Loa (1:100,000) for driving & orientation; Golden Throne, Bear Canyon, Notom & Sandy Creek Benches (1:24,000--7 1/2' quad) for hiking; or the plastic Earthwalk Press chart, Hiking Map & Guide: Capitol Reef National Park (1:62,500) which is best for driving.
Flash Flood Danger High in the short slot sections, low to medium otherwise. These are all short drainages, so for you to be in great danger, a storm would have to dump right on you, or just upcanyon.
Best Time to Hike June to mid-September; in spring or fall you'll need a wet/drysuit.

Author's Experience With Andrew Fitzgerald, we left the author's old VW Rabbit on the BNR and drove around to the heads of Cottonwood & Fivemile on consecutive days, then did the hikes. Walk-time for **Cottonwood**, 5 1/2 hours; **Fivemile**, just over 7 hours. The car-shuttle took about 1 1/4 hours each way. These hikes were done on 10/15 & 10/16/2001 and although the weather was fine, we had to swim through several ice cold pools in each canyon. We definitely would have enjoyed the hikes more, especially Fivemile, if a wet/drysuit had been taken. We both had mild hypothermia at the end!

On 10/10/2002, and with James Heller, we spotted one vehicle on lower **Burro**, drove the loop to the beginning of the hike, and did the canyon with wet/drysuits in just over 7 hours. However, we had bad information and didn't have enough rope to make the last rap. So, we used the **Bypass Route** and got into the round chamber from below to checkout that part. Still later on 5/25/2004, and with John Summerson, we rehiked **Burro** using 2 SUV's. Hike time, 5 hours.

On 6/23/2006, with John Summerson, we parked near the BNR and hiked up the ridge between Burro & **Cottonwood**, got into the canyon as suggested above, and did the hike in 9 2/3 hours. Next day, the author parked under the new bridge on the BNR, walked up the ridge route described above and into **Fivemile**. Hike-time was again 9 2/3 hours.

Opposite Page The first DC in the upper end of **Burro Wash.** This is the Wingate Sandstone section.

Above Left The last 29m rappel into a nice round chamber in **Burro Wash**.

Above Right John Summerson doing Rappel 2 in **Cottonwood Wash**. This is DC'able & a fast slide for some.

Left This pool is below what for some might be the last rappel in **Cottonwood Wash**. Others can DC this one. Often times this pool is dry.

Sevenmile Canyon Slots, Ticaboo M. & C. Lake Powell Area, Utah

Location & Access **Sevenmile Canyon**, its **Big & Little East Forks**, and **2 forks or branches of Seven-mile East** are located about 10 kms southeast of **Mt. Ellsworth**, which is the southern-most peak in the **Henry Mountains**. They're also about 20 kms northeast of Bullfrog Marina, and they drain south from Ticaboo Mesa into Lake Powell.

To get there, drive south from Hanksville on **Highway 95**. At **mile post 26** (26 miles south of Hanksville), turn right or south onto **Highway 276**. From that junction, drive 41 kms (25.5 miles) south to between **mile posts 25 & 26**, and turn left or east onto a graded dirt road. This usual 2WD/HCV (4WD recommended) road heads east toward **Ticaboo Mesa** (see Area Map). Drive **11.5 kms (7.15 miles)** to a **junction** marked **5050T** (1539m) on the *Ticaboo Mesa* quad. This is right on top of a north-south oriented hogsback hill. From there, turn right and head south about 2.3 kms (1.4 miles) and park at 1475m. This place will be just before the road begins to go downhill into a sandier area. You can take a 2WD to this point, but beyond that, better have a 4WD. This will be a good place to start if doing the **Main Fork of Sevenmile Canyon**.

To reach the **Little & Big East Forks of Sevenmile**, and the **2 branches** of **Sevenmile East**, drive south from **junction 5050T** again, but after 250m, veer left; at Km 12.2 (Mile 7.6), turn right; at Km 13.9/Mile 8.6, go straight--not left; at Km 14.7/Mile 9.1 is a **stock pond**; at Km 15.5 (Mile 9.6) stop at a very shallow drainage, the starting point if going to either the **Little** or **Big East Forks**; at Km 15.9 (Mile 9.9), stop & park at another very shallow drainage (you may see a cairn on the right?). This is the best place to park if going into the **West Branch** of **Sevenmile East** (AKA Good Day Jim). Or, continue southeast another 500m to Km 16.4/Mile 10.2 and just beyond a **NPS boundary sign**, park on the left. This is where you start down the **East Branch** of **Sevenmile East** (AKA Hard Day Harvey).

Rating The **Main Fork of Sevenmile: 3A III** in normal conditions. It has 6 possible rappels/HL's, but a good climber can DC & UC all but 1 or 2. The **Big East Fork: 3A III** has 2 raps, and a good short slot. **Little East Fork: 3A II** has 2 raps, but isn't very interesting. **West Branch of Sevenmile East: 3A R III**, it's easy except for one big KPH--so be prepared for a pothole escape. **East Branch of Sevenmile East: 3B III +PG or -R SLOT**, lots of tight spots & DC's.

Equipment RG, one 50m rope, and a shorter one for each, W&R/QL, K&EP, R/QD/DC (for **East Branch**), PR for everyone, a couple of toss bags for **West Branch**. If going to the lake and coming out at Exit 2 in Sev-enmile, take 2 short ropes to leave in place for UC'ing, just in case. In **both branches** of Sevenmile East, take a wet/drysuit in cooler weather.

Route Main Fork of Sevenmile From the parking at 1475m, walk southeast about 750m to the bottom of the drainage. Continue south downcanyon. After roughly 1 km will be the **2-part Rappel 1**; rappel once or twice; or help each other down the 1st part, then belay some members at the second--or DC it. About 400m below that will be **R2** from W&R/QL on small chokestones on the right or left of a big boulder and down **5m**.

About 200m below R2 is **R3**. There is W&R/QL around a chokestone, but a good climber can DC, then UC to return--but not if there's mud at the bottom. After about another km is **R4** from the left or east side of a house-size boulder. There is W&R/QL, but some might DC it--with a helping hand from a friend below.

Between R3 & R4 will be **Exit 1** to the west. About 500m below R4 will be **Exit 2**, a diagonal route run-ning up to the northwest. About 1 km below this 2nd escape will be 2 more dropoffs, **DC5 & DC6**. Anyone reading this book can DC both of these then chimney back up; but maybe leave a short rope at each. Below the 6th dropoff, the canyon begins to open, then you'll find several places with large boulders which you'll have to crawl around or under. After that, you'll find trails made by Lake Powell boaters, some pools, then pos-sibly some running water down to the lake. Return the same way with an **escape via Exit 2**.

Little East Fork From the **Km 15.5/Mile 9.6 parking**, walk due west with map & compass in hand to the high point **4715T** (1437m) and route-find in as shown. **R1, 3m**, is from on top of a big boulder (leave a rope if coming back up) then it's easy walking to the end with an **8m DC** or difficult **DC** (a 15m rope left on W&R/QL will allow an easy HL back up). About 5m from the end is the Main Fork. Not a great canyon.

Big East Fork From the same parking as for Little East Fork, walk southwest down the drainage. At the first dropoff (Page SS?), veer right and rim-walk about 200m, then scramble down a talus slope to the top of the Navajo Slickrock. Enter the drainage bottom and soon you'll come to **R1** about **7m** under a boulder. Then easy walking down to the **16m R2**. After that, skirt right or left around a **10m drop**; then on the left is an **es-cape** to the south (we didn't go up, but it sure looked like it would go from the bottom). Finally you'll come to a **nice 100m slot** with a couple of interesting DC's; then the Main Fork. Get out by going back upcanyon to the **escape** mentioned above, or head up the Main Fork and escape to the west. Our group in 5/208 actu-ally UC'ed the entire Main Fork by tossing up a short rope with Ibis hook on the end to get over R4 & R2. In the process, we saw several possible escapes on the east side (?). Later we came down the Little East Fork, and returned from there, all in an 8 hour day.

West Branch/Sevenmile East From the parking place, head southwest in the very shallow drainage. After about 1 km, DC what may be a Page SS dropoff on the right, and continue in a wide open valley. Gradually you encounter slots and after another km, will be **R1** into a **KPH** about **3m deep**. To get over/out of this, the best option is to do a shoulder stand to get one person out, then he/she can help others out and down the rest of the low angle **18m rap** (see picture on page 164) Or, throw out a couple of **toss bags or packs** filled with sand, attached to the end of a rope, then pull one person out. If the KPH is 4/5's full of water, you may have problems. In this case throw something over before starting the rappel. The tallest person may have to go underwater and push the smallest body out first.

About 200m below the KPH, are **raps 2 & 3** of **15m & 7m**. Further along, will be a huge **near-KPH**, but you can climb out if it's dry; or better still, walk around this one on a **bench** to the **right or west**. Near the bot-tom of West Branch is **R4**. Look up to the right 10m to see **W&R/QL** from an **outcrop & bush**, then rap **17m** from the long W&R/QL. About 100m below R4 will be a steep **exit/escape** up to the right or west. Along this **shortcut exit** are 2 steep sections for which the group leader may have to belay others over rotten rock. Or, continue downcanyon for another 250m to the **confluence** of the **East & West Branches**. About 125m below this will be the **final rap** (7m over a large boulder) for both forks. About 1 km below that will be a big **15m dropoff** into the lower canyon with cottonwoods & a **spring**. At that point stay on the bench to the right. Fur-ther along, you can get down to the dry wash, or stay on the bench. About 1 km below the **spring**, and fol-lowing this map carefully, route-find west, then north up though the Navajo SS bluffs. Further north and just below the Page SS/Carmel rim, continue north on an emerging hikers trail back to the DC you did when you first entered the canyon, and back to your car.

East Branch/Sevenmile East From the parking place & NPS boundary sign, walk southwest in the shal-low drainage. After 400m walk down the **10m R1** from a boulder. The next 1 1/2 kms will be easy canyoneering, then things get tight and in the lower 40% of this canyon it's a challenging slot. About 1 km into this part is a **rockfall** with one **huge boulder**; you can either crawl down under it, or rap from a shelf on its far side. Below

Map 34, Sevenmile Canyon Slots, Ticaboo Mesa & Central Lake Powell Area, Utah

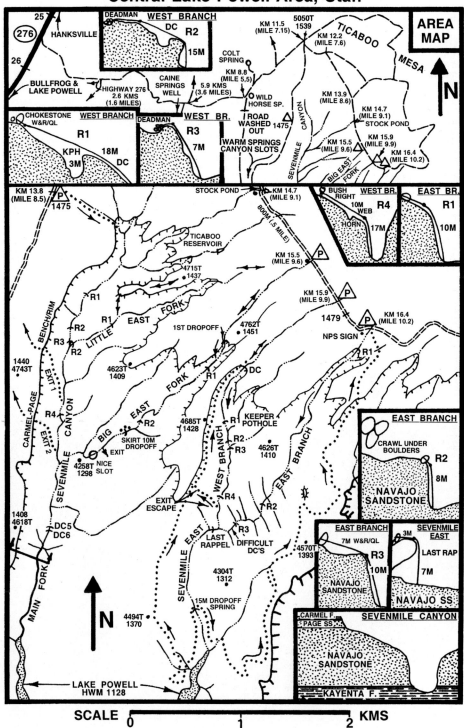

SCALE 0 _____ 1 _____ 2 KMS

this boulder is the real **R2** down about **10m**. About 500m below R2 will be a small boulder with W&R/QL and the **10m R3**. Below this is where things get slow, but more interesting, challenging & darker. For about the last 250m above the **confluence** with the **West Branch**, you'll have at least 2 challenging DC's and the tightest slot in the area--**no fat guys allowed!** Near the end, may be for some, a **3m rap**, but we DC'ed a wide chute. Below that are a couple of near-KPH's, then the confluence. This is the best, most fotogenic & most challenging canyon on this map.

At that point you could head up the **West Fork** and leave via the **shortcut exit/escape**, but you'll have a couple of strenuous UC's getting there. Or, just head downcanyon as described above and exit to the west and back to the road & car. Or, following this map carefully use another route out on the **east side** of the **East Branch**. Either route involves some route-finding with **map & compass** across Navajo slickrock domes.
Elevations All trailheads, about 1480m; high water mark (HWM) Lake Powell, 1128m.
Time Needed Depending on where you enter the **main fork of Sevenmile**, and/or exit, it could take from 6-9 hours to the lake & back. For **West Branch/Sevenmile East** using the shortcut exit, 4-5 hours; via the longer exit, 6-7 hours. **East Branch/Sevenmile East** should take 7-10 hours, maybe more for some.
Water Take plenty in your vehicle, but there is good running water just below the Navajo-Kayenta contact in the **lower main fork**. Possibly some water at the spring (below 15m drop) in Sevenmile East.
Maps USGS or BLM map Hite Crossing (1:100,000) for driving & orientation; Ticaboo Mesa (1:24,000--7 1/2' quad) for hiking.
Flash Flood Danger Moderate to high danger in the slot sections (**East Branch** has highest danger!), otherwise there are many places to run to & hide in the event of a flood.
Best Time to Hike Spring or fall, but you can hike here anytime. If there is water or mud in the potholes or low places, it will be more difficult to climb out, or back up, with mud on your shoes.
Author's Experience The author has been up from the lake on 3 occasions in the **Main Fork** of **Sevenmile**. Years ago and for this book, he headed down from the top while leaving 3 ropes. He stopped at Rap 4, then returned gathering his ropes, but scouted out Exit 1 along the way.

Later the same day, he rim-walked south and found Exit 2, got back into the lower canyon, walked up to Rap 4, then went down to the lake, leaving 2 ropes at DC5 & DC6, just in case since he was alone. On the return from the lake, he managed to UC the last 2 places without ropes, climbed up Exit 2, and walked to his car using a sandy road part way. Total walk-time for the day was 10 1/4 hours.

Later on 7/29 & 30/2006, he and Ryan Cornia & Cristina Amat, did the **West Branch** in the afternoon (we did North Fork of Trail that morning) using the shortcut exit in 4 1/3 hours. It was a dry hike that day. Next day we did the **East Branch** (the author returned via the east-side route and scouted the 2 short drainages east of the East Branch along the way. His time was 9 hours round-trip) and got wet feet in shallow pools.

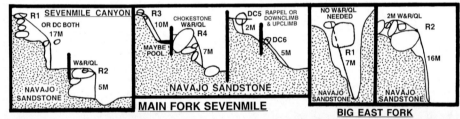

164

Opposite Page Cristina Amat & Ryan Cornia do a shoulder stand out of the **KPH** in the **West Branch Sevenmile East**.

Opposite Page Cristina starting R2 in the **West Branch Sevenmile East**.

Above Left Ryan DC'ing in the **East Branch Sevenmile East**.

Above Right On our trip there were several wading pools in the **East Branch Sevenmile East**.

Left AJ Pastula & Jonas Fast in the nice slot in the lower end of **Big East Fork of Sevenmile Canyon**.

Warm Springs Creek/Canyon Slots, Central Lake Powell, Utah

Location & Access This chapter includes 6 short slots (some with silly names--for lack of anything better!) all of which enter the lower end of **Warm Springs Creek/Canyon** next to **Lake Powell**. This area is southeast of **Mt. Ellsworth** in the southern Henry Mountains, east of **Highway 276**, in between **Sevenmile & Smith Fork Canyons** and northeast of **Bullfrog Marina** on Lake Powell.

To get there, drive along **Highway 276** between **Highway 95** in the north, and **Bullfrog Marina** in the south. Between **mile posts 25 & 26**, turn east onto a graded dirt road. After 5.9 miles (3.6 miles) is a junction--go straight east (don't turn south toward Cane Spring Well); from that junction on, best to have a **4WD** vehicle with a little ground clearance; at Km 9 (Mile 5.6) is another junction near **Wild Horse Spring**, head northeast (don't turn right or south because in 2008, there was a big washout not far away); at **Km 11.5 (Mile 7.15)** and on top of a big north/south hogsback hill, turn right or south (instead of heading NE toward Ticaboo Tank); after 250m is a road going east & south toward the 2 slots of Sevenmile East, but you continue south instead; at **Km 17.2 (Mile 10.7)**, make a hard left turn to the south (this is where the road you're on begins to turn west, then north again); drive south then southeast along the **high ground between lower Sevenmile & Warm Springs Canyons**. The end of the road is about 21.4 kms (13.3 miles) from Highway 276. However, from that point, drive back along the road about 650m (.4 mile) and park & camp (**beware**: it's on a ridge and can be a windy place!).

Rating Upper Narrows of Warm Springs Creek (UNWSC): 2A I; Lower WSC: 3B II; Fun Slot: 2A I; KPH Fork: 2B R I; Best Slot: 4A II -X SLOT; Potholes Fork: 4B II R SLOT (?).

Equipment K&EP in all. **UNWSC:** DC/UC'ing over boulders only; **LWSC:** RG, 15m rope, ascenders. **Fun Slot:** nothing but camera; **KPH Fork:** wetsuit and equipment for getting out of KPH's; **Best Slot:** RG, rim crew, 60 & 30m ropes & ascenders; **Potholes Fork:** RG, rim crew & 100m rope for extraction from end of slot, ascenders, wetsuits and KPH gear. Read more below under each slot.

Route From the parking at 1385m, walk south with map & compass in hand and look for an easy walk over the first rim (probably Page SS?). Once on to the Navajo slickrock, you should be in the upper drainage of the **Best Slot**. Walk southwest to the various slots.

Or, about 800m back up the road northwest from the parking/camping place, you could, with map & compass, walk WSW and into the uppermost end of the main **UNWSC**. Route-find over the Page SS rim and into the upper narrows. This part of the canyon has a number of large boulders/chokestone to get over, but we UC'ed the whole thing. It should be a little easier going down. Further along it opens with easy walking for 300-400m.

After the open area comes a nice slot that gradually tightens & deepens. This **Lower WSC** has several DC/UC's where a partner assist will help (no solo hikers here). Near the bottom is a **7m drop**, so unless you're planning to rappel into a boat on Lake Powell, better leave a short rope there. Below that, you'll pass the end of **KPH Fork** as it comes down a tight crack with an off-width move into a pool, then the big 60-70m? drop into the lake. Best to forget that and retreat the same way.

We had no idea what to expect in **Fun Slot**, so 2 of us started at the lower end (bypassing one short section because of water) and worked our way upcanyon staying on the bottom almost all the time; it's tight & shallow for 300m. Leave everything behind but K&EP & camera. We also didn't try going through a short upper section that could prove interesting.

If you're planning to go down **KPH Fork**, then leave a rope at the **7m drop** in **LWSC first** (this is UC'able, but a rope makes it a lot easier), exit LWSC, route-find around to the east and enter KPH Fork. We missed this one, but word is there's 1 KPH where you'll need to use a toss bag/pack toss to get out, one stretch of good DC'ing and an off width move at the end. From the bottom, head up **LWSC** and out.

Here's what's recommended for the 400m-long **Best Slot**. First, rim-walk the west side, plus circle around the entire canyon and look at it from the opposite side of Warm Springs Canyon & Lake Powell. This will give you a better look at the very end of the slot, but few if any would want to make the 2 raps into the lake. So here's what we think is the **best option**. Near the end of the slot on the west side is a little slickrock valley with a small shallow **pothole** in the middle. With rocks we left behind, rebuild a **deadman anchor** in the pothole. From there and using a 60m rope, rappel in and checkout the lower 50m or so. There you'll find a difficult DC, and a drop into a widening silo. For this you'd need to install a B&H, or be belayed by someone. Just below that, a rim crew could drop a rope (a 60m might do it?) from a shelf, and you could jumar up. Or you could rap on down to the lake, but who wants to do that?

Instead, at the end of a 60m rope tied to the deadman anchor, leave one pack with water, rappelling harness and ascenders (one or 2 of each). Then walk back to the head of the slot. There you'll find some slabs of rocks in 2 places where you anchor and leave a rope (a 15 or 30m rope will do). DC or rap using a body rap/Dulfersitz method. After about 60m of easy DC'ing, you'll come to a **7m drop**. Most people will want to be belayed when DC'ing this tight groove and into a widening silo. Or someone could install 1 B&H for this trickiest part in the canyon. If someone is belaying others, that person could return and stay on the rim for any emergency.

From the 7m drop, head downcanyon with nothing but a camera & K&EP. There are no potholes and the few silos are non-factors. You can walk on the bottom in a few places, but mostly you'll be off the deck 3-8m, sometimes a little more. In places you'll be squeezing through tight places high up using hands-to-elbows--this is the reason you'll want only a camera--no harness or pack (a rim crew is strongly recommended). There are a fair number of ledges and footholds, so it's not a true X SLOT (DC'ing the 7m drop would make it an X SLOT however). At the end of the slot, jumar up & out.

Potholes Fork We didn't do this one, but it's got 3 big KPH's after a rappel at the beginning. You'll need at least one 100m rope, or several shorter ones to hang down at the end (finding an anchor may be a problem, so have a BK & at least 1 B&H ready). This slot is less than 200m long ending with a wider part & big boulders before a rappel of 50m (?) into the lake. Expect big KPH's & some high-stemming. Scout it from both sides, and from across the bay first. If you do it, please send some feedback. Or contact Steve Ramrus, he has recently done it/or will shortly (?), **groups.yahoo.com/group/canyons/**.

Elevations Trailhead parking, about 1385m; Lake Powell at HWM, 1128m.

Time Needed With a long weekend you should be able to do all or most of these slots. Once at the slots; **UNWSC**--1 hour; **LWSC**--2-3 hours round-trip; **Fun Slot**--1-2 hours; **KPH Fork**--2-3 hours round-trip; **Best Slot**--3 hours; **Potholes Fork**--3 hours (?).

Water In potholes only, so take your own.

Maps USGS or BLM map Hite Crossing (1:100,000) for driving; Ticaboo Mesa, Knowles Canyon & Lost Spring (1:24,000--7 1/2' quads) for hiking.

Flash Flood Danger High in all slots, but they're all short drainages. Expect a little water anytime.

Best Time to Hike April, May, late September and early October. In cool weather take wetsuits.

Map 35, Warm Springs Creek/Canyon Slots, Central Lake Powell Area, Utah

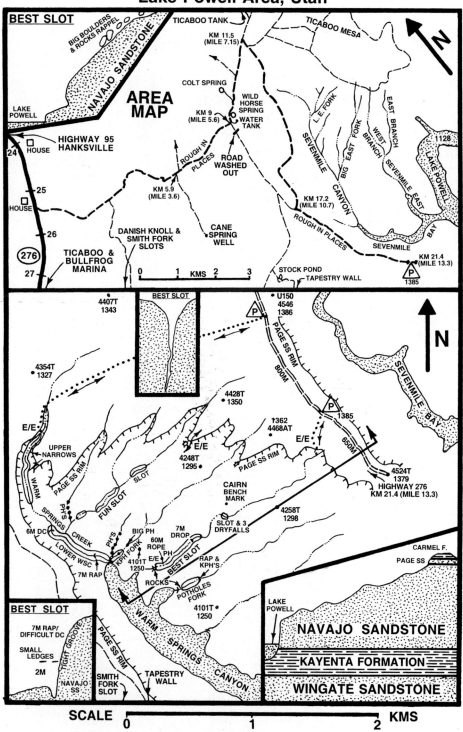

Author's Experience After the usual scout trip, the author along with Jon Jasper & Tim Barnhart, looked into all slots and from all angles, went down upper Best Slot to the 7m drop, went down LWSC & back, up most of Fun Slot, then up & out via UNWSC; 8 1/2 hours. On Day 2, we finally made the deadman anchor, then JJ & MRK went down into the lower end. We checked out the lower short section, then UC'ed to the 7m drop. The author returned to pick up discarded clothes & wetsuit in the canyon, while JJ jumarred up a 100m rope at the 7m drop, then returned, all in 6 2/3 hours.

Above Left Jon Jasper & Tim Barnhart in the **Lower Slot of Warm Springs Creek**.

Above Right Fixing a handline for the 7m rappel in the **Lower Warm Springs Creek**. (Tim Barnhart foto)

Right Entering the upper end of the **Best Slot**. About 60m below this entry is the 7m drop.

Above Left There are lots of tight squeezes like this 4-5m above the bottom in the **Best Slot**. At this point it's mostly hands-to-elbows, knees-to-heels that keeps you up & going. Best not to have a pack here--just K&EP and camera. (Jon Jasper foto)

Above Right Jay Keener starting down the last big DC near the end of **Smith Fork Slot**.

Left Jay (Stefan Folias behind) is high-stemming in part of the more sculptured sections in the upper end of the **Smith Fork Slot**.

169

Smith Fork Slots, Central Lake Powell Area, Utah

Location & Access Featured here is at least one challenging **X SLOT** which flows west from the west side & top of **Tapestry Wall** (rising from **Lake Powell** on the east side) and into the middle part of **Smith Fork.** Let's call it **Smith Fork Slot (SFS)**(AKA Psychological Damage). This is on the west side of Lake Powell, south of Mt. Ellsworth of the Henry Mtns., and NNE of Bullfrog Marina.

To get there, drive along **Highway 276** between the Henry Mtns. (and/or Highway 95), and Bullfrog. Right at **mile post 32**, turn east (or between mp 33 & 34) & soon cross **Hansen Creek** at Km 1.4/Mile .9. Continue northeast to **Danish Knoll** (Km 4.5/Mile 2.8) and turn east. At Km 5.3 (Mile 3.3) turn right uphill and head SSE following this map. Finally at about Km 9 (Mile 5.6) stop in a shallow valley where a handful of others have parked before (can someone please erect a cairn?). If you have a boat, start at Bullfrog and cruise up Smith Fork Inlet. This slot enters Smith Fork right at the HWM (1128m).

Rating **4A III +R or -X SLOT.** Best part is a 700m-long section. The rating comes from high-stemming across a **silo**, followed by a 10m vertical DC, then a near-vertical 10m DC in a chute at the end.

Equipment Small pack, PR, harness & R/QD/DC for suspending pack while high-stemming (consider depositing some packs at the bottom end of slot for a quicker & easier climb), **K&EP**, shoes with good soles & a 15m rope just in case.

Route From your car, walk ENE down a shallow wash aiming for the **Page SS point** shown. From the end of that point, veer right or south and route-find down. Continue east down steep slickrock to the bottom of **Smith Fork**. Walk downstream to the mouth of **SFS** and check out the bottom end first. Then following this map, climb up the point between the **main Smith Fork** and **SFS** & another drainage (this one has some challenging slots higher up). Zig zag eastward and enter the slot about 400m south of elevation **4002T** (1220m), which is about 700m from the slots' end (east of this is another slot with potholes). The first 1/3 of the way is a nice introduction to high-stemming; maybe 5-7m above the bottom. About 40% into the slot is an escape (or entry) to the north. Beyond that, the stemming gets higher. About 50m before the end of the slot is the crux of this hike. It's a widening, roundish, vertical opening people are calling a **silo**. Some have stemmed over the top, but that's about 15m above the deck; and risky! The best way past this is to **wiggle through** at about **mid-level**. There you'll find a wider part for your body (even for pretty big guys), plus a narrow place below for your feet to foot-jam sideways. At that point, hand drag your pack behind you, but leave it jammed in behind you while stretching across the silo for some pretty good hand & foot holds. With one person across to a resting place, relay packs using a PR. Your departure point is tight & awkward, then you fall foreword catching a ledge, that's why it's **best done without a pack**. Beyond the silo, stay high for 20m, then chimney straight down about 10m, then worm your way 10m down the last crack or chute to the bottom; see picture below. There you may find a **pool** (?); that's why you check out that part first before doing the slot. In cooler weather

& with a pool, a **wetsuit** would save skin and body warmth. Also, consider leaving all or some packs there to be picked up after the slot because there are no rappels.

South of SFS are several short slots in drainages where they drop into Lake Powell (see map). These may prove interesting for an explorer. There's another short slot north of SFS.

Elevations Trailhead, 1250m; end of slot, 1130m (which is just above the HWM of Lake Powell).

Time Needed About 6-9 hours round-trip (maybe 3 hours, or more, in the slot).

Water Take your own, but there will always be some water (with trees & willows) in lower Smith Fork.

Maps USGS or BLM map Hite Crossing (1:100,000) for driving; Bullfrog (1:24,000--7 1/2' quad) for hiking.

Flash Flood Danger Moderately high in the slot, but it's a short drainage to the east.

Best Time to Hike Spring or fall, but summer isn't all that bad. Best not to do this right after a flood because you'll likely have wet feet-- which increases the risk while high-stemming.

Author's Experience With N. Smale, J. Keener & S. Folias, we did this hike on 5/19/2006 in 7 1/3 hours. We were in the slot about 3 hours, and had lunch in the nice cavern at the bottom end.

The last big 10m DC in a chute at the end of **Smith Fork Slot**. Immediately above that is the silo & 10m vertical DC to the top of this chute. Much of the time you'll find a pool of water at this point.

Map 36, Smith Fork Slots, Central Lake Powell Area, Utah

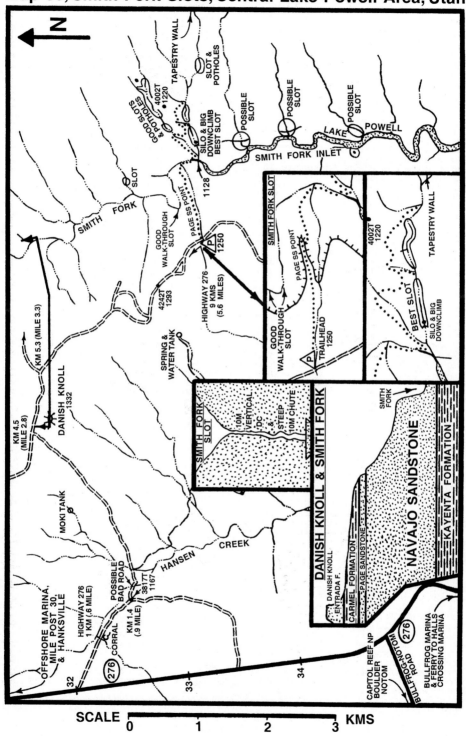

Horse Tanks, Fortknocker, Short, Lower Long & White Canyons, and the Black Hole, Utah

Location & Access Shown here are several tributaries to **lower White Canyon**, including the best hike of all, the **Black Hole**. The part of White Canyon shown here is located about halfway between Hite on Lake Powell, and Fry Canyon Lodge (closed in 2007). To get to the **Black Hole**, drive along Highway 95 between the Hite & Natural Bridges NM, and park near the new **mile post 57** on the north side near an information board. Leave a 2nd car at one of 2 parking places near **mp 55**. A great hike for those beginning technical canyoneering. For **Long Canyon**, park where a road crosses White Canyon between **miles posts 64 & 65**.

For **Short Canyon**, leave the highway just northwest of **mile post 61** and at a place called **Duckett Crossing**, turn west onto a big open flat, then following tracks, double-back and drive under Highway 95 through a cement culvert and into the bottom of White Canyon on **Road #2081**. Continue east up to the east rim of the drainage, then at Km 1.1/Mile .7, turn right onto **Road #43**. Follow it east to Km 2.7/Mile 1.7 and park. This will be almost even with a vertical cave in the canyon wall to your left.

For **Fortknocker**, leave the highway at **Duckett Crossing** near **mile post 61** and drive along **Road #2081**. There are many places you can park & enter the canyon. To get to **Horse Tanks Canyon (HTC)** (this canyon has no name, but it's near a place on the map labeled **"Horse Tanks"**) and **Little HTC**, leave Highway 95 near **mile post 53**, head east on **Road #2731**, then southeast on **Road #2081**. Follow the map closely and head for one of 2 possible entry points near the head of HTC, or the stock pond next to Little HTC .

Rating Horse Tanks: 3B II; Fortknocker: 2A II; Short: 2B I; Lower Long: 2A II; Black Hole: 2B III.
Equipment Horse Tanks, RG, 50m rope, W&R/QL. Read below for the others.

Route Black Hole This is a challenge for most people, especially beginners, and a great adventure for all. To do this hike successfully, you'll need **a short 15m rope**, drybags, plus a **wet/drysuit** for spring or fall hikes (nice in summer too). Before doing this hike, check at the Hite store (the marina was gone in 2003 due to the drought/low water) to see if there have been storms upcanyon at Natural Bridges NM the day or night before. In September, 1996, one girl was drowned in this slot. It had rained hard the night before upcanyon near Bridges and the flood reached the hikers at the worst possible time--right in the Black Hole. Start hiking between 10 & 11am so you arrive in the darkest part between 1 & 2pm. This will give you warmer temperatures for swimming, and more light for better fotos.

From the car-park & information board near **mile post 57**, walk 300m north on a trail to a little side-canyon, then turn right and head eastward down to the bottom. From there walk downcanyon. This is a dry gorge, except for the Black Hole (there are usually other pools above and below it). Along the way you'll have a number of easy downclimbs, plus 4 long swimming pools--3 close together, so you'll need to drybag cameras, lunch, maps, etc. Put the drybag inside your pack and swim with your pack on. To float better, include an empty water bottle or two in your pack. Until 2004, you had to swim most of a 125m-long section (3 separate pools) in very cold water, then wade other potholes below that. It's best to go with several people and help less-experienced members of your group down into the first part of the Black Hole.

In the fall of 2003, the flood from a 100-year storm roared down this canyon and left **2 huge log jams;** one immediately above the Black Hole, the other between what used to be the 2nd & 3rd swimming pools. In 6/2004, we crawled under both log jams and swam through lots of **log soup**. Then after the biggest floods of all in 10/2006, the log jams were gone. Expect changes with every big flood.

About 3 kms below the last long swimming pool, pay attention to your map and look for cairns & foot prints marking the beginning of 2 routes out as shown on the map. The first is up a little side-drainage labeled **Exit Canyon** on the map (**more difficult**); the next is 200m below that and **easier**. Follow cairns as you zig zag up through ledges. Both exits are on your left or south side.

The upper part of **Horse Tanks Canyon** has some narrows, dropoffs and pools that may interest those who prefer to rappel through canyons. But there are also a couple of easy routes into the upper end as shown. Most of this canyon is just another Cedar Mesa Sandstone drainage, then as you approach White Canyon just above the normal entry for the Black Hole hike, you'll come to 3 dryfalls with pools below. You can actually bypass the first 2 on the east, then comes the interesting part. You can slide down in just above the last dryfall, then you have to rappel into a colorful chamber (from a chokestone anchor to the bottom may be 30m? Take W&R/Q). From there it's scrambling down through boulders into White Canyon. Or, 50m below this rappel, scramble up to the east and return to upper Horse Tank Canyon; or exit completely and walk to **Road #2651** and **Horse Tanks**.

Little Horse Tank Canyon is the least interesting hike here. There are a number of dryfalls/dropoffs you can bypass on one side of the other, then a big 35m dropoff as it nears White Canyon. Return the same way.

Fortknocker is a walk-through canyon that is rather shallow. It has some narrow places and there are a number of entries/exits. The author did this one in the early 1990's, and didn't consider it all that great.

Short Canyon, take K&EP, a 15m rope and a wetsuit in spring or fall. For this short & sweet slot, walk from where you parked 50m east to the canyon rim and route-find down in over 3 ledges/rims to the beginning of the 125m-long slot. DC/slide down about 50m below the first drop, which is 6m. At the end of the slot will be a 25m-long pool with deep wading. After that continue down to White Canyon and look right or west to see some pretty good **rock art**--petroglyphs. From there, you could continue down-canyon to **Duckett Crossing** and road-walk back to your car; or **better still**, walk back upcanyon 200m from the rock art panel and climb steep slickrock on the west side. This takes you back to the bench you were on when you dropped into the slot. Work your way up to the final rim and back to your car.

Long Canyon has some nice narrows & slots. To enter the bottom end of the canyon, start from the side-road between mile posts 64 & 65, then walk down White and up into the bottom end of Long. There are a couple of upclimbs in the area of the **tight slot** where some hikers may need a little help, otherwise most people can make it up to about where the **narrows & 1 upclimb** are shown on the map. If you're a strong hiker, you might make it up to the bottom of the **4th Rappel** (for more information on doing the full length of Long Canyon, see the next map). Return the same way to your car.

Elevations Mile post 57, 1404m; mile post 55, 1295m; Lake Powell at high water mark (HWM), 1128m.
Time Needed Black Hole, in at mile post 57, out at mp 55; 5-7 hours. Plan on all day for **Long Canyon**. **Short Canyon**, 1-1 1/2 hours, if taking the short-cut back to your car. **Fortknocker**, all day or half a day. **Horse Tanks Canyon**, 5-7 hours round-trip.
Water These are dry canyons except for some permanent potholes, so take your own drinking water.
Maps USGS or BLM map Hite Crossing (1:100,000) for driving; Copper Point, Indian Head Pass, Jacobs Chair & part of Mancos Mesa NE (1:24,000--7 1/2' quads) for hiking.
Flash Flood Danger High in the **Black Hole** & parts of **Long**, otherwise moderate to high. Read above.
Best Time to Hike For the Black Hole, hot summer weather with temps 35°C or higher--preferably over

Map 37, Horse Tanks, Fortknocker, Short, Lower Long & White Canyons, and the Black Hole, Utah

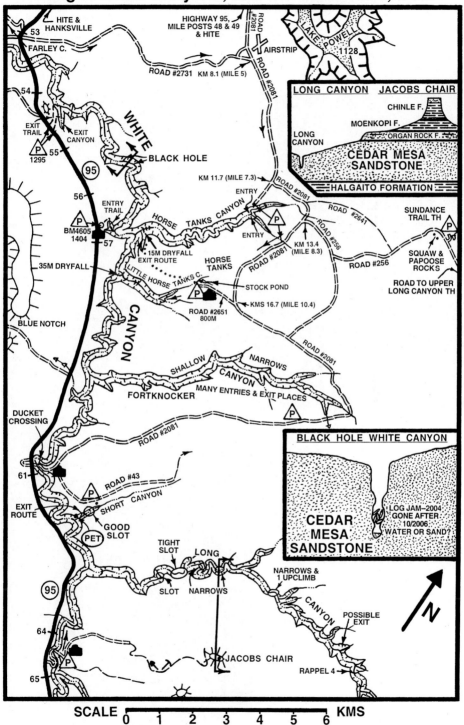

HITE & HANKSVILLE

HIGHWAY 95, MILE POSTS 48 & 49 & HITE

ROAD #2081

LAKE POWELL

53

FARLEY C.

AIRSTRIP

ROAD #2731 KM 8.1 (MILE 5)

1128

ROAD #2081

54

LONG CANYON   JACOBS CHAIR

CHINLE F.

WHITE

MOENKOPI F.

LONG CANYON

ORGAN ROCK F.

CEDAR MESA SANDSTONE

EXIT TRAIL

EXIT CANYON

P 1295

55

BLACK HOLE

HALGAITO FORMATION

95

KM 11.7 (MILE 7.3)

ROAD #2081

56

ENTRY

ROAD #2641

SUNDANCE TRAIL TH

ENTRY TRAIL

HORSE TANKS CANYON

P

P

P

BM4605 1404

57

ENTRY

KM 13.4 (MILE 8.3)

ROAD #256

SQUAW & PAPOOSE ROCKS

15M DRYFALL EXIT ROUTE

ROAD #2081

ROAD #256

ROAD TO UPPER LONG CANYON TH

35M DRYFALL

HORSE TANKS

LITTLE HORSE TANKS C.

STOCK POND

CANYON

P

BLUE NOTCH

ROAD #2651 800M

KMS 16.7 (MILE 10.4)

ROAD #2081

SHALLOW

NARROWS

CANYON

FORTKNOCKER

MANY ENTRIES & EXIT PLACES

DUCKET CROSSING

P

ROAD #2081

BLACK HOLE   WHITE CANYON

CEDAR MESA SANDSTONE

61

P

ROAD #43

SHORT CANYON

LOG JAM–2004 GONE AFTER 10/2006 WATER OR SAND?

EXIT ROUTE

GOOD SLOT

PET

TIGHT SLOT

LONG

SLOT

NARROWS

NARROWS & 1 UPCLIMB

CEDAR MESA SANDSTONE

95

CANYON

POSSIBLE EXIT

N

64

JACOBS CHAIR

P

65

RAPPEL 4

SCALE 0 1 2 3 4 5 6 KMS

40°C. Black Hole water is frigid & hypothermia a real possibility. If you have a wet/drysuit, take it for more pleasant swimming--even in the hottest weather. One hiker recorded the Black Hole water temperature at 11°C (52°F) in summer! Spring or fall for the other canyons.

Author's Experience It was 41°C (106°F) the day the author first went through the **Black Hole** in 1985, but he still got a touch of hypothermia (uncontrollable shivering). On a 4th trip, he returned to get better fotos and it was 34°C (94°F) outside. He thought he was going to freeze to death, and hardly got any pictures! That was on 5/25/1994. In 1996, he and Steve Tyler arrived at the darkest part in early afternoon, thus things weren't so cold, and they got better fotos because of better light in the slot. On his 3rd, 4th and 5th trips, he did it in 5-5 1/2 hours each, then used a mountain bike or car to get back to mp 57. In 6/2004, again with Tyler, we did it in 6 hours, starting at 11am, but we took lots of digital fotos which slowed us down. On that trip, the author wore a wetsuit and didn't shiver at all. Nice trip. With Scott Patterson, we scouted for about an hour first, then did **Short Canyon** in 1 hour 11min making an exit up the steep slickrock.

Above Left Steve Tyler wading through *log soup* in the **Black Hole** in 6/2004. The 10/2006 floods took all the logs out.

Above Right Swimming through logs in the **Black Hole** in 6/2004. Logs are all gone now. (Steve Tyler foto)

Right Just below the 2nd big log jam in the **Black Hole** in 6/2004.

Above Left You'll likely have at least 1 pool to wade in **Short Canyon**.

Above Right Nat Smale rappelling off a deadman anchor at R4 in **Long Canyon**. With determination, you can upclimb Long Canyon to this point from the bottom end.

Left Some nice fotogenic narrows & slots exist in the lower end of **Long Canyon**.

Long, Lower Gravel & Cowboy, Middle White Canyon, Utah

Location & Access Long & the lower end of **Gravel Canyon** (including **Cowboy Canyon**) are located southeast of Hite in the middle part of White Canyon. They both drain into White below the lodge at Fry Canyon (closed in 2007). To reach the bottom end of each, drive along **Highway 95** between Natural Bridges NM and Hite. Between **mile posts 64 & 65**, turn north onto a side-road at a place called **Gravel Crossing**; this is just downstream from where Gravel enters White Canyon. From this site, you can hike into the lower ends of **Long & Gravel**; or camp there and leave one car while you circle around to the north and/or east to a second car-park if doing the entire length of either canyon.

If you have just one car, and want to visit parts of the **lower end** of **Gravel, and Cowboy Canyons**, then continue down into White Canyon and cross to the other side on an old mining road heading toward **Jacob's Chair**. About **1.1 kms (.7 mile)** from the highway, turn right or **east** onto an ungraded track. This track ends about **3.5 kms (2.2 miles)** from the highway in a **sandy area**. From there you'll see a **cow trail** continuing east. Pack & camp in that area at the **Cow Trail Trailhead**.

To get to **Long Canyon's upper end**, drive northwest from Gravel Crossing on Highway 95. Just south of the road to **Farley Canyon** and near **mile post 53**, turn north onto **Road #2731**. You'll need the *Hite Crossing metric map*, and the **previous map** in this book on the **Black Hole**, to do this drive. First head northeastward along the same route as if going to the **Sundance Trail** in Dark Canyon. From the **Sundance Trailhead** turnoff near **Squaw & Papoose Rocks (18.6 kms/11.5 miles** from Highway 95 and mile post 53), continue southeast for another 15 kms (9.3 miles) and park near the **6418AT** (1956m) mark on the *Black Steer Canyon quad*. This is about **33.6 kms (20.8 miles)** from **Highway 95**. In normal conditions, this graded road is good for most cars, but something with HC is recommended just in case.

Rating Long: 3A or B III or IV; Gravel: 3B IV or VI. Cowboy: 3B III -PG SLOT.
Equipment RG, 60m rope (or one 30 & pull cord), 15m rope for Cowboy, K&EP, W&R/QL, wet/drysuits in spring/fall (also in summer after floods, or any time in **Gravel & Cowboy**), and maybe an Ibis hook.
Route **Long Canyon** For a pretty good day-hike, you can walk down (northward) **White Canyon** then up Long. There are several pretty good slots, some perhaps with water. About halfway up the drainage is one fairly difficult upclimb, but if you can get above that, you can go up to the **last rappel, R4**.

Or, with 2 cars, start at the upper car-park and head southwest into the short drainage. On the right or west side is a break in the cliffs where you can climb up. On the ridge top is an old mining road. From there look almost due south to locate another break in the cliff 1 km away. This break is 1300m due east of the elevation mark **5875T** (1791m) on *The Cheesebox quad*. To reach this site, walk southeast up the road about 600m, then route-find south to the descent place and head west to **5875T** (1791m). Right there at the end of a little north fork is **Rappel 1**. We rigged W&R/QL from a bush and rapped **26m** in 2 stages (you may find a pool?). You can rap from other places as well.

Then it's downcanyon nearly 2 kms to **R2**, a rap of about **13m** from a bush on the right. Just below that, we made a **5m rap** from an **Ibis hook** set in a natural hole for **R3**. After another 2 kms will be a narrowing of the canyon and **R4**, the **last rappel**. We rigged W&R/QL to a couple of large cobble stones set in a hole and rapped down about **8m** into a shallow pool. You could also rap from an Ibis hook, but make a small hole with the hook's tip (or G-pick) so it can't migrate.

Immediately below R4, we found footprints from someone who had walked up from White Canyon--but that's a long day-hike from Gravel Crossing! From R4, it's mostly easy walking (with a DC or 2) all the way to White Canyon, then nearly 5 kms upstream to Gravel Crossing and your other car. Along the way are several nice sections of narrows & slots.

Lower Gravel Canyon For now, let's start on the righthand side of this map; this is where you finish a loop-hike of the upper part of Gravel and exit. **See next map**. From that exit, there are several places with potholes, and 2 more high ledge with Anasazi ruins as shown. Soon you'll pass through boulders, then a tunnel made by a wall collapse. Not long after a major side-canyon comes in from the east, and a tight bend, you'll come to a dryfall. This makes a rappel of about 8-10m, but everybody is walking along a **bypass trail** on the **north** side. The bypass lasts about 1 1/2 kms, with a couple of places where you can return to the bottom of the canyon. In the area where another side-drainage comes in from the north, is a nice slot with potholes & no doubt filled with water. If you want to go through it, leave your big pack somewhere, do the slot, then retrieve it afterwards via the bypass trail.

About 2 kms further along, you'll come to the **1st mandatory rappel** in this lower section of Gravel. You'll find a slot filled with DC's & potholes, and likely a rappel into a swimming pool, but everyone seems to be walking along one side or the other for about 100m and rapping about 12-15m. If you do this, **install you're own W&R/QL**, nothing was there in 6/2007. However, if you don't want to do this rap, or if you want to start here and do just this lower part, there is an E/E back **upcanyon 500m** from this 1st Rap. This is up a steep slickrock slope, thus the name, **Slickrock E/E**. You could walk from the Cow Trail TH, east along the cow trail shown, around the head of **Cowboy Canyon**, and enter Gravel here--or exit and do the reverse. Lots of choices.

Further along is the mouth of **Cowboy Canyon**. You can walk up it for about 600m to find water and a couple of 2 dryfalls which you can climb around (more later). Further along in Gravel, you'll come to **R2** in this lower canyon. You could rap from a cedar tree (with a very long W&R/QL) near a natural bridge at the beginning of a **60m-long subway-like slot**, and next to a big pool. Last person down--don't cross the ropes or you'll have a hard pull! Do this with about 10m of W&R/QL and 2 ropes about 20m long (or one 50m). Or you could bench-walk on the right or west side along a trail. About 400-500m beyond the subway-like slot, you can either locate a cedar tree or boulder to rap 15-20m from; **install your own W&R/QL**. Or you can escape straight up the canyon wall. This is easy, but there is a little climbing up through a section that resembles **Swiss Cheese** (with big holes). From there you could walk to the Cow Trail TH overland; or continue downcanyon. Or enter at this point.

If you stay in the canyon bottom, you'll come to a dryfall and a groove or narrows and bypass. You can bench-walk around this on either side. Near the end of Gravel, is a 100m-long narrows which will surely have water and wading/swimming all the time. Better have a **wetsuit** for this (except perhaps in the hottest part of summer). Another 2 kms or so will put you at Gravel Crossing and your car.

Cowboy Canyon Perhaps the best part of this canyon complex is Cowboy Canyon. Here's a good all-day hike. Begin at the **Cow Trail Parking**, walk east along a pretty good trail; head one minor drainage, then about 2/3's of the way up Cowboy from Gravel, route-find into the bottom. There you'll come to a nice slot beginning with a near-vertical **6m DC** in a 1/2m-wide groove. Rim-walk the east side of the slot for about 60m, then DC & slide to the bottom. It's easy. In this 600m slot, you'll find 1 DC & drop into what will likely be a swimming pool (belay one person down first to test the depth); then

Map 38, Long, Lower Gravel & Cowboy Canyons, Middle White Canyon Area, Utah

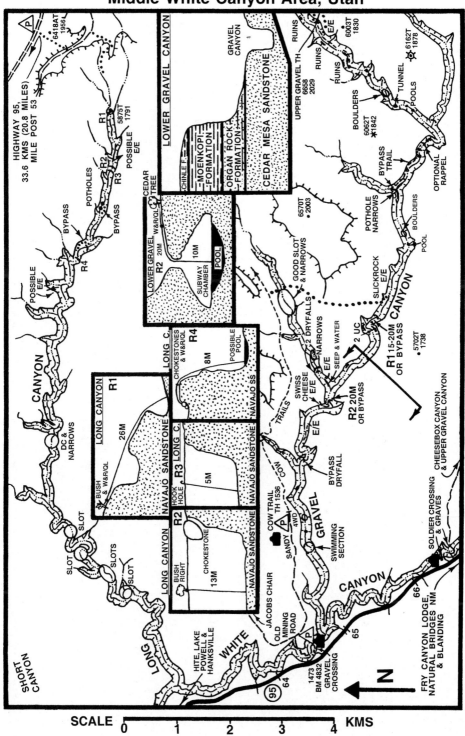

SCALE 0 1 2 3 4 KMS

another DC & slide into a 2nd pool at the end of the tight part. Finally the canyon widens a little and at the end of that section is another DC & jump into another possible pool. Further along is another narrow section, then 2 dryfalls, one after the other. At the first, climb over some boulders to the right or west and DC a steep tricky pitch to 2 logs set upright--some may want to be belayed down this **6m drop**. At the 2nd dryfall, head to the right or west side again and walk down to the next level, then continue down to Gravel. The lower part of Cowboy will likely have good drinking water.

Elevations Upper car-park for Long Canyon, 1956m; Cow Trail Trailhead, 1536m; Upper Parking for Gravel Canyon, 2029m; Gravel Crossing, 1473m.

Time Needed Long Canyon--to do the full length, 9-13 hours, plus nearly 1 1/2 hours each way for the car shuttle. **Gravel Canyon**--full length one-way with 2 cars, 10-13 hours (?). **Cowboy Canyon** and lower Gravel, 6-9 hours if starting at the Cow Trail TH. Or do the full length of Gravel in 2-3 days--if you enjoy camping.

Water Take your own on day-hikes, in lower Cowboy, and use water tablets or filter for pothole water.

Maps USGS or BLM map Hite Crossing (1:100,000) for driving; and Black Steer Canyon, Indian Head Pass, Mancos Mesa NE, The Cheesebox & Jacobs Chair (1:24,000--7 1/2' quads) for hiking.

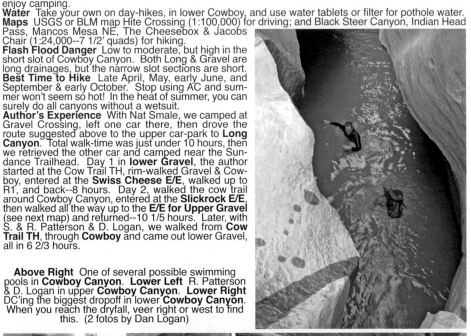

Flash Flood Danger Low to moderate, but high in the short slot of Cowboy Canyon. Both Long & Gravel are long drainages, but the narrow slot sections are short.

Best Time to Hike Late April, May, early June, and September & early October. Stop using AC and summer won't seem so hot! In the heat of summer, you can surely do all canyons without a wetsuit.

Author's Experience With Nat Smale, we camped at Gravel Crossing, left one car there, then drove the route suggested above to the upper car-park to **Long Canyon**. Total walk-time was just under 10 hours, then we retrieved the other car and camped near the Sundance Trailhead. Day 1 in **lower Gravel**, the author started at the Cow Trail TH, rim-walked Gravel & Cowboy, entered at the **Swiss Cheese E/E**, walked up to R1, and back--8 hours. Day 2, walked the cow trail around Cowboy Canyon, entered at the **Slickrock E/E**, then walked all the way up to the **E/E for Upper Gravel** (see next map) and returned--10 1/5 hours. Later, with S. & R. Patterson & D. Logan, we walked from **Cow Trail TH**, through **Cowboy** and came out lower Gravel, all in 6 2/3 hours.

Above Right One of several possible swimming pools in **Cowboy Canyon**. **Lower Left** R. Patterson & D. Logan in upper **Cowboy Canyon**. **Lower Right** DC'ing the biggest dropoff in lower **Cowboy Canyon**. When you reach the dryfall, veer right or west to find this. (2 fotos by Dan Logan)

Above Left From the bottom of the only rappel in **Fry Canyon** looking down at the swimming pool that's always full of cold water. **Above Right** The swimming pool in Fry Canyon. The bottom of the rappel is in the upper middle part of this picture. **Bottom** Scott & Richard Patterson have just completed Rappel 2 in the middle part of **Gravel Canyon**. There's a big pool just to the left of this picture. This is a nice fotogenic cave-like chamber, so don't miss it. (Dan Logan foto)

Fry, Upper Gravel, Cheesebox and Hideout Canyons, Upper White Canyon Area, Utah

Location & Access The canyons featured here drain into the **upper part** of **White Canyon** which is about halfway between Blanding & Hite and immediately north of Highway 95. This immediate area extends west from **Natural Bridges National Monument** downstream past **Fry Canyon Lodge** (closed in 2007) to about mile post 65. Access is easy and you're never far from pavement & Highway 95.

To hike **Fry Canyon**, park immediately north of the bridge about halfway between **mile posts 71 & 72** and about 800m east of the Fry Canyon Lodge. If doing the **East Fork of Cheesebox** or **Hideout Canyon**, proceed to a point just west of **new mile post 75**, turn northward and drive about 500m on a good dirt track and park/camp on the rim of White Canyon opposite the mouth of Cheesebox at **1720m**.

If going to the **West Fork of Cheesebox Canyon**, drive to a point between **mile posts 66 & 67** near **Soldier's Grave**, and look for **Road #2271** going down into White Canyon and up the other side. This is called **Soldier Crossing**. Read its history in the author's other book, *Boater's Guide to Lake Powell*. Drive this once-graded **Cheesebox Road #2271** for about **13.2 kms (8.2 miles)** while following your USGS map carefully. At about that point, the road has begun to head northeast. Park one car (or maybe a mtn. bike) in the area marked **1795m**. This will be near where you may want to exit Cheesebox before it enters White Canyon.

With a 2nd car (or your only car), drive northeast on the same **Cheesebox Road** for a total of about **17.5 kms (10.9 miles)** from Highway 95 and park at **1980m**. By starting there, you can enter the West Fork via a little side-canyon just as the good slot begins. Or you could drive further northeast to a side-road heading east **(19.2 kms/11.9 miles** from Highway 95). That junction is marked **6658** (2029m). You could walk that rough road a ways (or use a 4WD) and enter the upper part of the canyon, but that part is not so interesting.

To hike **Gravel Canyon**, you could get into the bottom end of the drainage by driving along Highway 95 to between **mile posts 64 & 65**. Locate the road running down into White Canyon to **Gravel Crossing** and up the other side. By parking at Gravel Crossing, you can walk up White Canyon 500m and enter lower Gravel.

Or to enter **Upper Gravel**, drive to between **mile posts 66 & 67** and leave Highway 95 at Soldier's Grave, Soldier's Crossing and Cheesebox Road #2271--same road as if going to the West Fork of Cheesebox as described above. Follow the Cheesebox Road for **19.2 kms (11.9 miles)** and park at the junction marked **6658** (2029m). This is very near the southwestern end of **Ram Butte**. From that car-park, the middle part of Gravel Canyon is only about 800m to the west. Cars driven with care can usually make it to that point.

Rating Fry Canyon: **3B** I or **II**; **Upper Gravel: 3B III** or **VI**; **Cheesebox--both forks: 3B III** (**IV** for some); **Hideout Canyon: 3B III** or **IV**. Attractions in all these canyons, except for the West Fork of Cheesebox, are Anasazi ruins as shown on the maps. They also have some fotogenic narrows, slots and cold swims.

Equipment RG, one 50m and a 15m rope (one 15m rope for Gravel), PR, W&R/QL, K&EP, wet/drysuit much of the time, drybags or plastic camera cases for all drainages, and an Ibis hook for making short rappels.

Route Fry Canyon First walk upcanyon (east) from the bridge about 125m and enter; in the first 100m may be a small pool or two. Head downcanyon in the shallow drainage. Walking is easy for 3 kms then you come to the only rappel in the canyon. This is about 300m above or upcanyon from some **Anasazi ruins** (access from the rim only) and about 1 km before Fry enters White Canyon.

At the rap station will be 2 B&H's on the right about 3-4m from the rim. From the B&H to the bottom is about **14m**. You'll rappel into a dry pothole, then slide & DC about 6m to the edge of a large 8m-wide pothole that always has water and is always a swim. Below that will likely be some shallow wading for another 75m, then the slot opens some, and you'll see the ruins just ahead & above in an alcove facing south.

To get out of Fry, continue downstream 600m to **White Canyon**, then walk up White about **500m** and be looking for a way up to the rim on the right or south side. There is now a trail/route marked with cairns.

Once on top and out of White Canyon, and if you want to see the ruins, veer right and walk to the rim of lower Fry. Locate the bend above the ruins, then rim-walk back to the north for about 150m until you came to a slight break in the wall. Route-find down very carefully while veering to the left or south. You'll find steep slickrock with half a dozen **moki steps** in the steepest part. Once safely down on a little bench, walk south (upcanyon) to the ruins. Retrace you steps from the Fry Canyon Ruins, then head southeast cross-country back to your car near the bridge.

Hideout and **East Fork of Cheesebox Canyons** From the end of the road opposite the mouth of Cheesebox Canyon between mile posts 74 & 75, is a recently-created hiker's trail zig zagging down into White Canyon. To start, head east over the first little rim, then the track circles around to the west on lower benches. Just follow the most-used trail & cairns; it's getting better every year. Once down in, if you walk directly into the bottom end of Cheesebox, you'll soon come to a pool & dryfall. You cannot go upcanyon from there.

To actually get into the lower end of Cheesebox and/or on the route to Hideout or East Fork, walk to the mouth of lower Cheesebox and on the east side, climb up a crack & log to get onto the next bench. Once there, you could turn left and enter lower Cheesebox; but instead, veer right and route-find eastward up through several benches while heading for the feature called **The Cheesebox**. Head due east on the north side of The Cheesebox, then gradually veer northeast along the northwest rim of **Hideout Canyon**.

If you're interested in **Anasazi Ruins**, you could explore one group located about 750m ESE of the elevation marker **6248T** (1907m). You'll need one rope to HL down/up over one ledge. You can only get to these ruins from the rim (if you're a good climber and have 2 partners, you might climb up from the bottom?). Further along, and immediately east of **6560AT** (1999m), is a short side-canyon running southeast into upper Hideout. Route-find down this little drainage to a prominent rim, turn left or north, walk about 200m, and **rappel 12m** from W&R/QL on 2 big rocks (100m above this is where Hideout drops into its own canyon).

Once in Hideout, walk downcanyon. When you reach a point about 500m upcanyon from the above-mentioned ruins, and just below a big pothole, you'll come to more ruins on the northwest (right) side of the canyon.

About 60% of the way through Hideout, you'll come to a **short slot** with a **5m DC** into a **pool**, and 2 **rappels** in the space of **100m** or so. You'll first DC into a pool, then 75m below that is the **10m R2** from W&R/QL between 2 big chokestones. About 25m below that is **R3**. This is from a chokestone with 3m of W&R/QL, then DC'ing a chute into a pothole with year-round water at the same level. Immediately below that is a 2nd drop of about **10m**. Use one **50m rope** (or a 30 & pull cord) to do this short section of about 20m from the same anchor & W&R/QL. **Warning:** The author soloed this canyon on 6/10/2007. In this last pool, he swam across with the pull cord attached to his harness in a bag, and the rappelling rope in hand. But as he was trying to climb out while swimming, the rope kept pulling him backwards toward the anchor. The ledge at water level was covered with mud and slippery. He had a few near-panic moments before climbing out because it was a cool day and he wasn't waring a wetsuit. This is not a KPH, but it's best to have one person swim across unencumbered, get out, then have a second person throw him/her the rope.

From R3, it's walking again. Near the bottom end of Hideout, you'll come to an exit about 400m upstream

Map 39A, Fry Canyon, Upper White Canyon Area, Utah

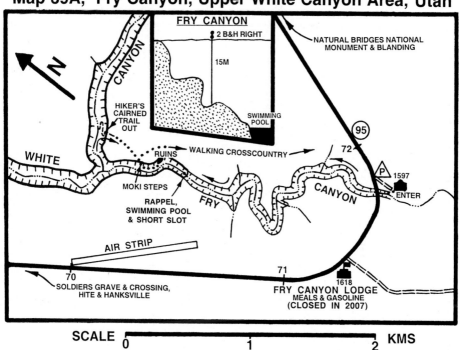

Map 39B, Area Map--Upper White Canyon Area, Utah

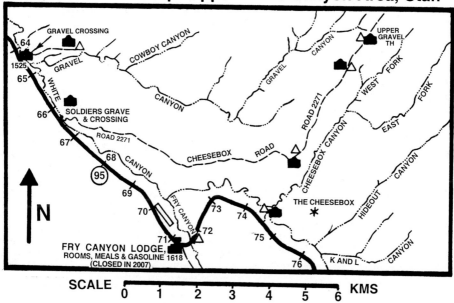

Map 39C, Upper Gravel, Cheesebox & Hideout Canyons, Upper White Canyon Area, Utah

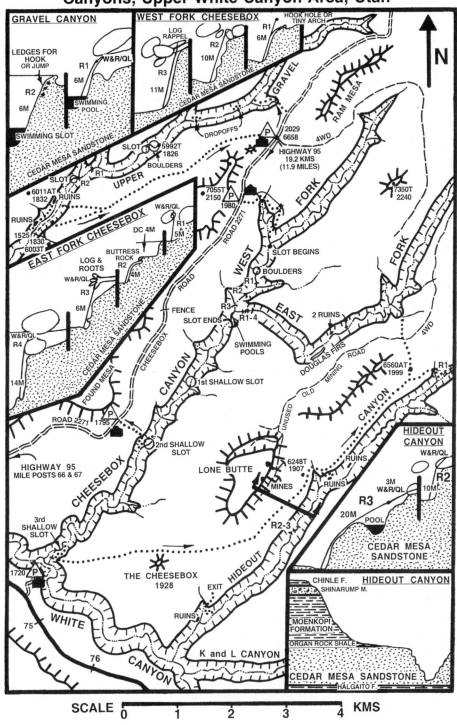

GRAVEL CANYON

WEST FORK CHEESEBOX

HOOK HOLE OR TINY ARCH

LEDGES FOR HOOK OR JUMP

LOG RAPPEL

R1 6M

R1 6M

R2 10M

W&R/QL

R1 6M

N

R2 6M

R3 11M

RAM MESA

SWIMMING POOL

CEDAR MESA SANDSTONE

GRAVEL

SWIMMING SLOT

CEDAR MESA SANDSTONE

DROPOFFS

P 2029 6658

4WD

SLOT 5992T 1826

BOULDERS

HIGHWAY 95 19.2 KMS (11.9 MILES)

7350T 2240

SLOT R2 R1

UPPER

7055T 2150

P 1980

FORK

6011AT 1832 RUINS

W&R/QL

RUINS 1525 1830 6003T

EAST FORK CHEESEBOX

DC 4M

R1 5M

ROAD 2271

WEST

SLOT BEGINS

FORK

BUTTRESS ROCK R2 4M

BOULDERS

LOG & ROOTS

W&R/QL

R1

R3 6M

CEDAR MESA SANDSTONE

ROAD

R2

R3

EAST

2 RUINS

4WD

FENCE

SLOT ENDS

R1-4

W&R/QL R4

FOUND MESA

CHEESEBOX

CANYON

SWIMMING POOLS

DOUGLAS FIRS

OLD MINING ROAD

6560AT 1999

R1

14M

1st SHALLOW SLOT

UNUSED

CANYON

RUINS

HIDEOUT CANYON

ROAD 2271 1795

P

2nd SHALLOW SLOT

LONE BUTTE

6248T 1907

RUINS

W&R/QL

R2

HIGHWAY 95 MILE POSTS 66 & 67

CHEESEBOX

MINES

RUINS

3M W&R/QL

R3

10M

3rd SHALLOW SLOT

R2-3

20M

POOL

CEDAR MESA SANDSTONE

1720 P

THE CHEESEBOX 1928

EXIT

HIDEOUT

CHINLE F.

SHINARUMP M.

HIDEOUT CANYON

75

RUINS

MOENKOPI FORMATION

WHITE

K and L CANYON

ORGAN ROCK SHALE

76

CANYON

CEDAR MESA SANDSTONE

HALGAITO F.

SCALE   0   1   2   3   4   KMS

from a side-canyon on the right or west; this is a route-find up through several ledges. Just below the side-drainage are more uninteresting ruins on the right (west) as shown on the map. From there, it's another km to White Canyon, then another 4 kms or so downstream to the trail leading back to your vehicle.

East Fork of Cheesebox Canyon To do the East Fork, you can get to the upper part via an old 4WD track which begins at the road junction marked **6658** (2029m), but the author is unfamiliar with it. You certainly could park one car at the junction, walk the track, drop down into the canyon, and eventually exit lower Cheesebox to the car-park marked 1795m.

Or here's another option that worked for us. Park and/or camp near mile posts 74 & 75 and walk the exact route as if going to Hideout Canyon; but head north from the elevation marker **6560AT** (1999m) and over a low divide into a short side-canyon, then into the East Fork as shown. It's easy walking downcanyon.

About 1 1/2 km from where you enter East Fork, look to your right or west from a couple of big potholes to see a very well-preserved **ruin under a ledge**. This one is covered with mud and has a round hole in the front. About 500m past that, is a group of **ruins** under an alcove. Not many people have seen either of these.

In the first km below the 2nd ruins are **douglas fir trees** on the left, then the big 90° bend in the canyon. About 500m above the junction of the East & West Forks of Cheesebox, is the beginning of the only **technical part** of the **East Fork**. R1 is from W&R/QL down about **5m**. R2 is from a rock jutting out in the drainage bottom and down **4m**. Immediately below that is the **6m R3** from a tree stump with roots. To do **R4**, climb under a large boulder covering the lower slot, and rap **14m** from W&R/QL around another smaller chokestone. From there, it's 200m to the junction with the West Fork of Cheesebox.

Once in the combined Cheesebox Canyon, you'll soon come to a long wading (maybe swimming?) pool, then shortly after, another long swimming pool. Immediately below that will be a 3m DC, probably with the help of webbing from a small chokestone into a short swimmer. You may have to do a beached-whale move out of this one, but the author got out twice alone with no problem. Just below this 3rd swimming pool, the canyon opens up into nice deep gorge. If you're hiking up Cheesebox from mp 75, this is as far you go.

About 2 kms below where the canyon opens up, you'll come to the **1st shallow slot** where your choice will be a wade/swim of about 100m; or you can jump back & forth several times above to avoid this shallow slot. If you stay above the slot, get on the east side to reenter the canyon bottom below.

Less than 2 kms below that will be a **2nd shallow slot** which requires a swim of up to 75m. If this doesn't appeal to you, backtrack about 150m and climb up to the west. Walk south or downcanyon on an intermediate bench until you're above the lower end of the swimmer, then DC to another bench and finally route-find to the bottom on one of 2 routes. At this point, you can go up to another bench and head back north, then climb west through another bench and **exit Cheesebox** completely. More on this exit in the **West Fork part**.

Below the 2nd shallow slot is more boring walking, but about 400m before the end of Cheesebox, you'll have another choice. Most people veer left as the canyon turns west and walk along the lowest bench. Finally, they DC over the log route mentioned above, then walk the trail back to their car near mp 75. Or you can DC into this **3rd shallow slot** and end up rappelling into the pool which is located about 150m up lower Cheesebox from White Canyon. The author still hasn't done that part.

West Fork of Cheesebox Canyon To see only the best parts of the West Fork, stop at the **1980m** car-park which is 17.5 kms (10.9 miles) from Highway 95. From there, walk due east directly toward the butte in the distance marked **7350T** (2240m) on *The Cheesebox quad*. Skirt the head of the obvious minor side-canyon and rim-walk south avoiding another short drainage. As you near an overlook of the confluence of Cheesebox and the north-south running side-canyon, veer right or west and route-find down into the lower end of the side-canyon about 300m above the junction. Once at the bottom, head south for the confluence where the 1st of several slots begins. In that 1st slot, you'll have one slide-in pothole which always has water, so if you're taking a wet/drysuit, put it on there.

Less than 1 km below the confluence are several large boulders you'll have to crawl under, then DC. About 400m below that will be the **1st rappel**. Right in the bottom of the drainage is a tiny bridge that some have rapped from, but the author didn't think he could pull his rope through it from below, so he used his G-pick to make a small hole on the left for his grappling hook and rapped from that (Ibis hooks are better, since you can usually use an natural hole). Some have apparently DC'ed this part.

About 500m below R1 is **R2**. For this one, wrap your rope around an upturned pointed rock and rap **10m**. No need to clutter the place with W&R/QL; unless things have changed. This also might be DC'able? After another 400m will be **R3** which was from a log (build a new anchor if necessary), and down about **11m**. Someone told the author you can DC a little ways past this rap (?). From there it's 200m to the junction of the East & West Forks of Cheesebox. Below that are the 3 swimmers, then the 2 shallow swimming slots. If you have another car at the place marked 1795m, when you reach the 2nd shallow slot, regress 150m and climb up on the next bench above to the west and route-find up to the rim, then walk west 650m to the road.

Upper Gravel Canyon The best place to start is at the road junction marked **6658** (2029m) which is 19.2 kms (11.9 miles) from Highway 95. From there, road-walk northeast about 6-7 minutes or 600m, turn left and head WNW for another 600m and route-find down into the upper end of a little side-canyon of Gravel. From there walk downcanyon.

In the first 1 1/2 kms of Gravel, you'll walk on a bench to the right to get around 2 dropoffs (or rappel/HL if you like). The 2nd drop would be down through a pretty potty-type bridge, formerly a deep pothole.

When you get to the area just west of elevation **5992T** (1826m), you'll find several large boulders to navigate through then it slots up with many foto opts. This goes on for another 1 1/2 kms then you come to **R1** on the left. There you'll find W&R/QL underneath a large boulder. Rap down about **6m** and likely into a swimming pool (big floods can change this!). From there, you'll find short slots with wading or swimming for 300m, then a **6m dryfall** and perhaps **R2** into another swimming slot. Around the beginning of this drop, both above and just under the lip, are a number of natural holes you can attach an Ibis hook to rap from. Another way to the water would be to belay at least one person down to test water depth, then if deep enough, the last person can jump/slide in.

Below R2 are many short swims or wading in tight slots, which somewhat resembles what you find in Neon or South Fork of Choprock Canyons. Finally, when you reach the area between **6011AT & 6003T**, the canyon opens some and you'll find cottonwood trees spread out in the next 150m. At that point, climb up on the bench to the west and head back to the north. In a space of about 600m you'll see 2 groups of **Anasazi Ruins** (a **3rd ruins** is 750m below the 2nd and on the highest ledge). The **1st ruins** further north are extensive and will take a pair of climbers helping each other to get up. However, your best view is from the bench below.

For those who want to do this **upper canyon** only, this is also where you'll leave the canyon and return to your car. To do that, and while viewing the ruins, look to the east side of the canyon for at least 2 ways up to the rim. The route the author used went up to the east rim about halfway between the 2 sets of ruins. Once on top, head generally ENE while observing your USGS topo map carefully. In about an hour, you'll be back to your car.

See the **previous map** if you're interested in seeing the rest of Gravel Canyon below this part. With 2 cars and a very early start in late spring/early summer with 16 hour days, it's possible to do the entire length of Gravel in one long day.

Elevations Trailhead for Fry Canyon, 1597m; Fry Canyon Lodge, 1618m; trailhead for Hideout & East Fork of Cheesebox Canyons near mile post 75, 1720m; trailheads for West Fork of Cheesebox & Gravel Canyons, about 1795m, 1980m & 2029m; exit place of Gravel Canyon on this mapped area, about 1525m.

Time Needed The hike through **Fry Canyon**, about 3 hours; **Hideout** and **East Fork of Cheesebox**, about 10-12 hours each using the routes discussed above; **West Fork of Cheesebox**, 7-8 hours; **Upper Gravel**, 8-10 hours; **all of Gravel** (one day), 11-14 hours (?-take a headlamp), and not much time to rest or take fotos!

Water Carry plenty in your car at all times. At Natural Bridges National Monument. There are no springs around but there is always lots of pothole water, which should be treated or filtered.

Maps USGS or BLM map Hite Crossing (1:100,000) for driving; Jacobs Chair & The Cheesebox (1:24,000--7 1/2' quad) for hiking; or Natural Bridges (1:62,500) if available.

Flash Flood Danger In each canyon the danger level alternates between high and low, depending on how narrow the slot is.

Best Time to Hike Late spring or early summer, or early fall. During these times wet/drysuits are required. In the heat of summer, the swimming parts might be welcome, but cold without wet/drysuits. You'll enjoy the swims better with a light-weight wetsuit. Outside the slots it's pretty hot in summer, but do-able.

Author's Experience He has been into the lower end of Cheesebox on several occasions, but finally with Byron Lemay & Scott Patterson, did Hideout Canyon in 10 1/3 hours; and East Fork of Cheesebox in 10 1/2 hours; both from the mile post 75 trailhead. Next day, the author did the West Fork of Cheesebox alone from the trailhead marked 1980m, and road-walked back to his car in 7 hours round-trip.

Later in cool October, and using a drysuit in the swimming parts, he did upper Gravel alone in 8 3/4 hours. He rapped off an Ibis hook at R2, and both raps were done with a 15m rope. To get an idea of the full length of Gravel, read the author's experiences in lower Gravel under the previous map.

Above Right The short slot just below what could be **R1** in middle **Gravel Canyon**. **Below Left** Immediately below what could be *Optional Rappel* in **Gravel Canyon**. **Below Right** From the top of the last rappel in **Hideout Canyon** looking down a short drop into a swimming pool. On the far side is the 2nd half of the 20m rappel. Your anchor is a small boulder with W&R/QL

Above Left Preparing Rappel 2 in **Cheesebox Canyon**. No need for W&R/QL, just loop your rope over a small boulder (unless things have changed because of floods--take W&R/QL anyway). **Above Right** A nice mud-covered Anasazi granary in the middle of the **East Fork of Cheesebox Canyon**. **Bottom** In 6/2007, the author rigged these boulders to be the 1st rappelling station in upper **Hideout Canyon**. This is from the side of the canyon and down about 12m.

185

Lehi, Anasazi & Moepitz Canyon Slots, South Lake Powell Country, Utah-Arizona

Location & Access These 3 canyons are located between **Lake Powell** and the northwest slopes of **Navajo Mountain** in extreme southern Utah. If you have a boat, coming in via the lake might be the best way. If you do come by boat, start at Bullfrog or Halls Crossing Marina in the upper end of the lake and head southwest. Or start at the new Antelope Point Marina northeast of Page, Arizona, and head ENE. If starting there, you can refuel at Dangling Rope Marina, before docking in a short unnamed inlet this writer is calling **Little Oak Canyon**. This is just southeast of Oak Canyon Bay, and not far north of Rainbow Bridge National Monument. This is where you'd dock if going into **Moepitz & Anasazi**. If doing just **Lehi** Canyon, boat to **Nasja Canyon Inlet** in the lower end of the **San Juan River Arm** of Lake Powell, and hike from there. See the author's other book, *Boater's Guide to Lake Powell* for more information on boating.

To get there by road, you'll first need a hiking permit from the Navajo Tribal Parks (see *Introduction to the Navajo Nation*) and an Arizona state highway map, then head for either **Page** near the Glen Canyon Dam, and drive east on **State Highway 98**; or if coming from the east, drive southwest from **Kayenta** on **US Highway 160**. About 35 minutes out of Kayenta, turn north on Highway 98 in the direction of Page. About halfway between Page and Kayenta on Highway 98, and between **mile posts 349 & 350**, turn north onto **Navajo Nation Road 16**, more commonly known as the **Navajo Mountain Road.**

Near **mile post 5** is the **Inscription Chapter House** and the **last gas station**--fill up there as the old Navajo Mtn. Trading Post was closed in 1990 (Utah Historical Quarterly, Spring, 2000). Continue north from Highway 98 for a total of **50.7 kms (31.5 miles)** to **Navajo Junction** with **Navajo Mountain** directly in front of you. Turn left to hike to Rainbow Bridge; but instead turn right or north and drive to **Km 59.4 (Mile 36.9)** and turn left toward the old trading post which now the **Navajo Mtn. Chapter House**. You can get information there on weekdays (or at the Inscription Chapter House). From the chapter house, head north to **Km 65 (Mile 40.4)** and **Rainbow City** with a large **school**. From there, continue north about 1500m to find a steep dugway running down to the west. From that point you'll need a HCV/4WD, to reach the end of the road near Cha Canyon. From Rainbow City to **Cha Canyon Trailhead** is 6.9 kms (4.3 miles). You can park & camp there at the northeastern end of the North Rainbow (Bridge) Trail.

Rating To do all of **Moepitz** from Little Oak Creek Inlet: **3B III**. Moepitz maybe the easiest of the group, but this could change depending on lake levels. To do **Anasazi** from Little Oak Creek Inlet: **4B VI -R SLOT**. Doing **Lehi** from Nasja Inlet: **3B IV**. Ratings will change depending on whether you go down to about the High Water Mark (HWM) and return; or whether you have a boat waiting at the head of each inlet and you continue to the lake itself. If you don't have a boat and will be coming by land, you have no choice but to descend each canyon and return the same way. To do all 3 canyons from the land route will take 6-8 days (?). A little less by boat; or if one person knows the access routes, then all 3 can likely be completed in 5-6 days from any marina (?). Planning the logistics for these canyons is a nightmare.

Equipment RG, K&EP, ascenders, one 50m rope, and depending on tactics, several short ropes 10-15m long, W&R/QL, Ibis hook & G-pick, drybags, compass & the 7 1/2' quads suggested, and a wet/drysuit for all hikes except in the heat of summer.

Route Moepitz One route out of **Little Oak Creek Inlet** is to walk upcanyon on hiker's trails. Just above the HWM is a little waterfall and just above that a **hogsback** ridge coming down from the right or south. Climb this slickrock ridge using **moki steps** in the steeper parts. After 250m, you'll see cairns marking the way as the route veers southeast. After roughly 1 km, you'll be in a flat area. From there, turn left and head northeast. After about another 1 1/2 kms, you'll be heading due north at about where the constructed **Navajo Trail** is shown on this map. It's too difficult to describe every turn, bend, DC, etc., so follow this map carefully and match it to the 7 1/2' quad. Further along, you'll head north and a little east down the drainage from the dome marked **4321T** (1317m) to the bottom of Moepitz.

Another route out of Little Oak Creek Inlet begins about halfway upbay. Look for a gentle sloping slickrock ridge coming down from the northeast with steps cut in the middle part. Once above this ridge, head north in a drainage, then generally northeast until you can walk down more steps cut for sheep & goats. This allows you to enter Moepitz 500m north or downcanyon from the previous route mentioned above. **If you don't know how to read a map or compass, don't go into this area--take up another hobby instead!**

When you reach Moepitz drainage, head south upcanyon. After about 3 kms, you'll come to a very impressive **gooseneck bend & big undercut**, one of the best around. Above that are some spectacular cliffs & domes and after another km, you'll reach the head of the canyon. The upper part of Moepitz has lots of water, water-loving plants, but no beaver or livestock. This means the water should be drinkable-as is (?).

Or head down Moepitz to find the technical part and soon a rappel of **6m** into a swimmer. You can construct a **deadman** for an anchor (leaving W&R/QL as trash); or install a piton or B&H; or make a hole on one side and rap off an Ibis hook. From this rap down to the HWM is about 1 km and includes perhaps a dozen swimming pools and a couple of pretty good DC's. If you're planning to regress upcanyon, best to tie off short ropes at several places. At the HWM are a couple of pools with nothing to tie ropes to; if going all the way through, simply jump in & go. Or if plans are to regress back upcanyon, build or make a small hole and leave an Ibis. Four hikers went through this canyon to the lake in 5/2003, but this writer and Byron Lemay went only to the HWM and returned. The remaining short part doesn't look like anything serious, but if the lake is really low, there will likely be some rappels (will someone send an update?).

Anasazi is likely best reached by way of the lake and Little Oak Creek Inlet and Moepitz Canyon. However, to get from Moepitz into Anasazi, there is no easy way to describe the route--**so everyone should have a topo map & compass in hand and follow this map carefully every step of the way.** From inside Moepitz, and the pothole & **Navajo inscription**, go upcanyon about **350m** and look for a walkup route east out of the drainage. From there, head generally ESE for about 1 km, then general south along the east side of the biggest dome around. From there it's south to the top of a tiny **flat gravel-topped mesa**, then circle around the dome marked **4683T** (1427m). Head NNE down the drainage, then at the big dropoff, circle left to a 2nd drainage and walk down to the bottom near the **cave camp**. Just upcanyon is **likely water (?)**.

Head downcanyon. Soon you'll come to **Rappel 1**. Rig **7m** of W&R/QL around small boulders and **rap 15m** into a swimming pool. Then it's mostly easy walking for about 2 kms. Immediately east of the dome marked **4259T** (1298m) will be the **15m R2** with **1 piton & 1 B&H (Beware:** one group has threatened to remove these!) on the right shelf and into another swimmer. This could be a DC, slide & jump (?). About **500m below** that will be an **easy escape** on the left or west; 200m beyond that will be a **DC** into a **big pool**. Immediately below that will be your **last escape** on the left which is **600m above the HWM**. This place has **moki steps** and is a pretty good chimney job straight up, then over a slot. Above that, help each other over one boulder, then an easy walk down into Moepitz. If you don't have a boat waiting below, this is where you leave

Map 40, Lehi, Anasazi & Moepitz Canyon Slots, South Lake Powell Country, Utah-Arizona

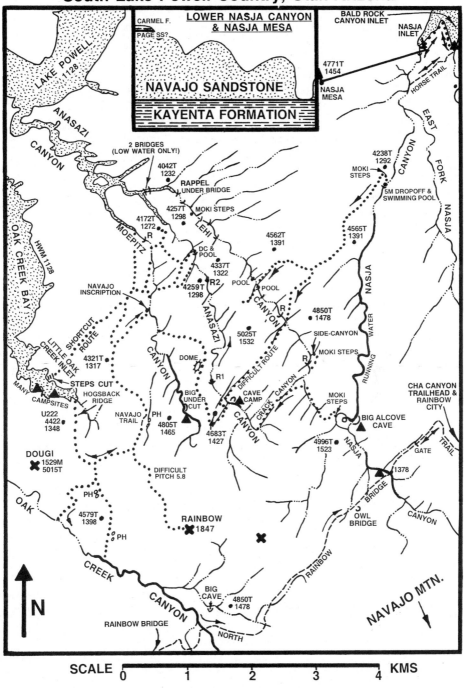

LOWER NASJA CANYON & NASJA MESA

CARMEL F.
PAGE SS?

NAVAJO SANDSTONE

KAYENTA FORMATION

LAKE POWELL 1128

ANASAZI CANYON

OAK CREEK BAY

HWM 1128

MOEPITZ CANYON

2 BRIDGES (LOW WATER ONLY!)

4042T 1232

RAPPEL UNDER BRIDGE

4257T 1298   MOKI STEPS

4172T 1272   R

LEHI

DC & POOL

4337T 1322

R2

4259T 1298

NAVAJO INSCRIPTION

ANASAZI CANYON

POOL

POOL

CANYON

R

4562T 1391

4565T 1391

BALD ROCK CANYON INLET

NASJA INLET

4771T 1454

NASJA MESA

HORSE TRAIL

EAST FORK NASJA

4238T 1292

MOKI STEPS

5M DROPOFF & SWIMMING POOL

SHORTCUT ROUTE

LITTLE OAK CREEK INLET

4321T 1317

STEPS CUT

HOGSBACK RIDGE

MANY CAMPSITES

U222 4422 1348

NAVAJO TRAIL

DOME

BIG UNDER CUT

5025T 1532

R1

DIFFICULT ROUTE

CAVE CAMP

PH

4805T 1465

4683T 1427

CRACK CANYON

CANYON

R

4850T 1478

SIDE-CANYON

MOKI STEPS

NASJA

RUNNING WATER

MOKI STEPS

BIG ALCOVE CAVE

4996T 1523

CHA CANYON TRAILHEAD & RAINBOW CITY

TRAIL

GATE

1378

DOUGI
1529M
5015T

OAK

CREEK

CANYON

DIFFICULT PITCH 5.8

PH

4579T 1398

PH

RAINBOW
1847

BIG CAVE

4850T 1478

RAINBOW BRIDGE

NORTH

RAINBOW

OWL BRIDGE

BRIDGE

OWL CANYON

NAVAJO MTN.

N

SCALE   0   1   2   3   4   KMS

187

the canyon.

If you have the luxury of a boat waiting in the Anasazi Inlet, and if you have 3-4 experienced canyoneers, then continue downcanyon. About 200m below the last exit, is a tight slot and some good DC'ing. This is where we stopped. But others have stated that as you near the lake, and close to the HWM, you'll have a 20m near-vertical DC/HL?, then at least one tight slot where you'll have to do some **high-stemming**. What you find exactlly will depend on the lake levels; if the lake is real low, expect lots more slot canyoneering. But it's been done a few times. Contact Steve Ramrus' website: **http://groups.yahoo.com/group/canyons/** for more info on the last little ways in this slot. From what this author has seen from his boat in Anasazi Inlet, it'll be a nice tight passage & DC'ing. If the lake level is low, and you get down to the lake, you'll have some strenuous UC'ing to get back. As of **7/24/2007**, with the lake water at **1099.79m (3608.18 ft.)**, there was an 8m waterfall at the very end. Jump this, if you have a boat waiting.

Another option. On your first hike in Anasazi, do the **upper part** of the canyon and exit as discussed above at the **moki steps**. On another day, head out of Little Oak Creek Inlet via the **cut steps** in the middle of the bay and head straight for the same exit. Enter there and head downcanyon. You can probably get into Anasazi in about 1 1/2 hours (?), then a shorter trip to the head of Anasazi Inlet. But again, this all depends on lake levels.

Lehi & Anasazi Canyons by Road Walk west from the Cha Canyon Trailhead on the very good **North Rainbow (Bridge) Trail** about 9 kms to find Nasja Creek, a sometimes year-round (?) flowing creek. Walk downstream about 1 1/2 kms to the **big alcove cave** on the east side--a good place to camp, but treat all water in the area because of Navajo livestock.

You can also get to this point from the **Hogsback Ridge** route out of **Little Oak Canyon Inlet**. From the route suggested above, instead of heading north to Moepitz, walk south over one pass just east of the dome marked **4579T** (1398m) and into **Oak Creek Canyon**. From there walk upcanyon to the **North Rainbow Trail,** then follow it northeast to **Nasja Creek** and to the alcove cave mentioned above.

From the **big alcove cave**, walk due west past springs & willows to the west side of the abandoned meander. There on the east-facing near-vertical wall, are some **moki steps** behind some cedar (juniper) trees. You won't believe they're there until you're up close! These were made with a large miner's pick of some kind.

These large moki steps angle up to the right or north, then when the slope eases, heads west to a flat bench. From there walk southwest roughly 60m to find a near vertical wall maybe **10m high** with more moki steps. For this, you may need to pull your big pack up behind you with a PR or another short rope. From there, continue west around a bluff, then northwest down into **Lehi** Canyon. About 1 km from the moki steps, you can turn left and head southwest up a **crack canyon** into the upper end of **Anasazi**. Or continue down Lehi another 600m, avoiding potholes & a dropoff by route-finding down a little **side-canyon** with more **moki steps** on the east, as shown. At the bottom of that dropoff should always be water. About 1 km beyond this, will be your entry if coming by boat from Nasja Canyon.

If just doing **Lehi Canyon**, perhaps the best starting point would be from **Nasja Canyon Inlet**. If water levels in the lake are low, you will find a sandy campsite at the head of Nasja Inlet; otherwise camp around the corner in **Bald Rock Canyon Inlet**.

You can walk from Bald Rock Inlet into Nasja via a **horse trail**, or along the narrow Kayenta Bench between the two as shown. Otherwise, in the middle of Nasja Inlet, climb up the talus slope to the south and using a **short rope**--or natural handholds--climb to the top of the Kayenta, then bench-walk upcanyon. After about 3 kms, you'll come to a **pool**, and just beyond, a 5m dryfall. From the pool, turn around and after 75m, walk north up a short drainage between the main canyon wall and a separate fin. After 150m, and at a little divide, climb west up a steep slope using **moki steps**. We did it both ways with large packs, but consider using a G-pick to enlarge about 3 steps, especially for a return trip down. At the top of the steps, continue north, then west and turn southwest just north of the hill marked **4238T** (1292m). Further along, walk along the west side of a little dome labeled **4565T** (1391m). From there climb up on top of another mini mesa and down into Lehi using the most southerly of 2 routes shown; the one between elevations **5025T** (1532m) & **4850T** (1478m).

Once in Lehi, you can camp there and walk upcanyon 1 km to get drinking water. If plans are to go down to or near the lake and regress, then it's recommended you camp there and finish the hike the next day.

On our trip, we went down to the HWM leaving **9 ropes** at dropoffs to ensure we got back up with a minimum of effort. But if you're going one way to the lake, then you can chimney, slide or jump all dropoffs except for a **7m rap** from a small chokestone in a crack directly under a bridge. Just below that is what could be a difficult pothole to get out of, then after another 100m, the HWM. If lake levels are lower than they were in early 8/2008, then you'll **walk (swim) under 2 bridges** immediately above where Lehi meets Anasazi. What you'll find between these 2 points is anyone's guess, but this entire canyon has been hiked or explored.

Elevations Lake Powell at the HWM, 1128m; moki steps in Nasja Canyon leading to Lehi, 1353m.

Time Needed To explore all 3 of these canyons from one boat in Little Oak Creek or Nasja Inlet will likely take 5-6 days (?), maybe longer. To do the same trip from the Cha Canyon Trailhead will take at least a week (?). But this will depend on the strength of your group and whether you're going one-way to the lake, or returning upcanyon.

If you can scout out and know these routes ahead of time, then Moepitz and Anasazi can be completed in one day from Little Oak Creek Inlet with a boat waiting in the lake below each (Anasazi will be the longest & most difficult of the 2). The same for Lehi from Nasja Inlet, if you have a 2nd boat waiting--otherwise it could take 2 or 3 days from Nasja.

Water Moepitz--running water in the upper canyon, potholes in the lower half; Anasazi--minor seeps or potholes throughout; Lehi---lots of pothole water throughout.

Maps USGS or BLM maps Navajo Mountain & Kayenta (1:100,000) for driving, boating & orientation; Nasja Mesa, Wilson Mesa, Navajo Begay & Rainbow Bridge (1:24,000--7 1/2' quads) for hiking.

Flash Flood Danger High in many places. These are short drainages, but the countryside is almost 100% slickrock, so any storm of consequence will send water down these canyons quickly.

Best Time to Hike Later spring or early summer with the longest days of the year. Early summer can be very hot for hiking, but swimming potholes would be best then.

Author's Experience After several scouting trips and getting information from Herb Taylor and Nat Smale, Byron Lemay and the author boated from Bullfrog Marina, had a quick look into the inlets of all 3, then camped at the head of **Nasja Inlet** (6/9-12/2003). Early the next morning, we headed for **Lehi**. Upon arrival, we left all camping gear and headed downcanyon. We left 9 ropes for an easier return, but without wet/drysuits, got scrapped up pretty bad. Getting to the HWM took 5 1/3 hours; back to our big packs another 3 2/3 hours. Total hike-time for the day was 12 3/4 hours.

From **Little Oak Creek Inlet**, we walked via the **Hogsback Ridge into Moepitz**, then into **Anasazi** in about 3 1/4 hours. We walked down Anasazi to less than 600m from the HWM, and used the exit route to return to our boat, while scouting the short-cut route out of Little Oak Creek Inlet. Total walk-time for that hike was 11 3/4 hours. The next day, we left Little Oak Creek via the **short-cut route** and entered lower **Moepitz.** We rapped off a pile of stones & W&R/QL left there by the previous group (that webbing will be washed away and deposited downcanyon somewhere so why not install 1 B&H or make a small hook hole). From there, we walked & swam down to the HWM and returned using an escape exit 600m below the only rappel. Total walk-time for the day was 8 hours.

Above Left Byron Lemay makes a rappel under the natural bridge in the lower part of **Lehi Canyon**. This bridge is just above the HWM of Lake Powell.

Above Right Rappel 1 in the upper end of **Anasazi Canyon**. There's always a pool and a swim below this rap.

Left This is the only rappel in **Moepitz Canyon** above the HWM of Lake Powell. In 6/2003, we used a deadman anchor for this rap. However, you can climb up and around this dryfall on the left or southwest side.

Above Left DC'ing into a pothole right at the junction of **KPH Slot & the Lower Warm Springs Creek**. Just to the right is a tight crack and an off-width move into this pool. Just behind the cameraman is the end of both slots and a big drop into Lake Powell. (Jon Jasper foto)

Above Right High-stemming in the middle of the **Best Slot**, one of several short & sweet slots in the **Warm Springs Creek** drainage. (Jon Jasper foto)

Right Spanning a long pool of water in the middle part of **Sevenmile Canyon**.

Above Left Brian Olliver climbing up the Squirting Waterfall in the middle part of **West Canyon**. We got there from Lake Powell.

Above Right Brian sliding down the Slippery Slide in the **Upper Slot of West Canyon**. This is a drop of about 5m, but it's an easy DC/slide.

Left Scott Patterson in one of several deep pools in **Cowboy Canyon**, which is a major tributary of **Gravel Canyon**. This side-drainage is the best part of Gravel Canyon.

West Canyon, South Lake Powell Country, Navajo Nation, Arizona

Location & Access Featured here is perhaps the **best all-around canyon hike** on the Colorado Plateau. It's **West Canyon**, located about halfway between Navajo Mountain and Page on the south side of Lake Powell. This hike is also covered in the author's other books, **Boater's Guide to Lake Powell** and the **Non-Technical Canyon Hiking Guide to the Colorado Plateau.**

It's recommended you reach West Canyon by boat, but it can be reached by **road & car**, preferably a HCV or 4WD. To do that, have the *Kayenta* metric map in hand, then drive along **Highway 98** about halfway between **Page & Kayenta**. Near mile posts 337, 338 & 339 are 2 well-used roads heading north. About 3 kms north of the highway, and at the power lines, these 2 roads meet to form one heading north. Continue north. About halfway to Navajo Creek will be a junction--the main most-used road runs left or northwest toward **Tse Esgizii Rock** and Starting Water Wash, but don't take that one. Instead, veer right or north passing 2 nearby houses. There are many side-tracks along the way but continue on the most-used. Further on, you'll drop down off the clay beds of the Carmel Formation, but before reaching Navajo Creek, there will be a couple of steep & rocky places which 2WD vans & pickups use all the time. Any HCV can make it to the bottom of **Navajo Creek** and back out under normal conditions, but some may want to use a mtn. bike or 4WD from the top of the dugway. The author went down in his VW Rabbit Diesel once but only after walking & clearing the way of rocks first. From Navajo Creek, make your way northwesterly on the road running toward John Yazzilo's home, but stay along Jayi Creek until you reach the end of the track. Park somewhere.

To avoid any hassles using this driving route, pick up a **Navajo Nation hiking permit** at the visitor center at Cameron, Tribal Park Headquarters in Window Rock, or at the Antelope Canyon Tribal Park entry gate east of Page between **mile posts 299 & 300** on Highway 98. That office is used to collect fees for hikers in Antelope Canyon and is normally open daily during the warmer 8-9 months of the year.

Or if going by **boat**, start your trip at the **new Antelope Point Marina** which is just east and a little north of Page. If you don't have your own boat, you can rent one at Stateline Marina; or for a better deal ask about renting one in Page, or drive northwest from the dam to Big Water on the Utah side of the line. You can rent a boat there too. You'll pay a pretty hefty price to rent a boat, but if shared by 3-5 people, it's not too bad.

This trip can be done in **2 long days** with a **fast boat**, but a 3rd or 4th day may be needed. If renting, you may want to get the boat in the afternoon, so you can get an early start the next morning; or begin in the late afternoon and camp at the trailhead the first night. If you're shooting for 2 days, then the first day will be the shorter of the two, so you could start from Antelope Point early, and hike up West to a camp. Day 2 would be to do the upper slot, and return to the boat, then boat back at Antelope Point just before dark. If you're driving from Phoenix or Salt Lake with your own boat, plan to have a **4 day weekend** at your disposal.

If boating, be sure to have a good map of the entire lake and a compass; even for the author, turning south toward the right canyon from the main channel is still confusing. Once in the West Canyon Inlet, boat to the head of navigation. Your landing site varies with the lake level. If arriving late in the afternoon, you can normally find a place to camp right where you tie down. During the middle of the day, many bikini-clad, barefoot day-hikers will be docking in the same place, so tie up accordingly.

Rating The **Upper Slot** has one simple rappel/HL, but some pretty good DC'ing and several potholes which, under the right/wrong conditions, could be a problem getting out of. It's rated: **3B VI PG SLOT**. Even though this canyon is technically easy, it may rate best on the Colorado Plateau for scenery and enjoyment. West has tight, dark slots, deep, deep narrows, running water, swimming in icy pools, fotogenic waterfalls and huge undercuts 75-100m deep.

Equipment One 15m rope (maybe an extra short rope to leave behind), K&EP, drybags for everything because of some likely swims, headlamp for upper slot, and no less than 3 people. If going in April, May, late September or October, take a wet/drysuit. Read more below.

Route If coming by **road**, you'll have to use your map & compass and head up the North Fork of Jayi Canyon to the divide at 1525m, then route-find into the head of West Canyon. Probably best to stay on the east side of the drainage near the Navajo (Page SS)-Carmel contact, then enter the upper slot from the east as shown on this map. If you don't know how to read a map or use a compass, best to stay away from this place!

Boaters From where you dock, simply walk upcanyon in the small stream with sandy bottom. You'll eventually come to a short slot, then a **swimming pool**, which can be a deep wade or a swim. If you're not quite ready to swim with a big pack, you could backtrack 350m and climb up to the south to begin the **Bypass Route** shown. There's now a faint trail marked with cairns, but it's best to pack everything in drybags, place them inside your pack, then float packs across this, and perhaps other pools. In 8/2007, there were about a dozen swimming pools up to the Upper Slot. On previous trips there were none. Go prepared.

Above this first slot, it's normally a dry open canyon. If you find running water there, that means it has rained & flooded recently, and perhaps not the best time to make this hike. The canyon up to the old log hogan site isn't so scenic, but you will pass 3 water/dryfalls. You may need to use a rope to get big packs upon a bench at the first 2 water/dryfalls.

At the site formerly occupied by a **log hogan** (it was washed away in 1999), you can hike out of the canyon to the west on a sheep trail. Above the hogan site is the 4th water/dryfall, then the canyon begins to narrow. After another 3 kms you'll come to a big undercut, then some deep, dark narrows where you'll have to carry your pack on your head, or swim with it on. Then the **Squirting Waterfall** (5th) and another big undercut and **Double Falls (6th)**. There should always be running water in this part of the canyon which is the most-fotogenic part (also, the best map of this canyon is now in the author's Lake Powell Guide, 5th Ed.).

Higher upcanyon, you'll pass a rincon or abandoned meander on the east, then the canyon opens a little with trees and campsites. About 1 km above the rincon will be an old sheep & goat trail out to the west (but not readily visible and is washed-out), and a side-canyon heading northeast which formerly had another **sheep & goat trail**, but which is now a hiker's path. Right at the mouth of this little canyon is a bench and a good safe campsite at 1263m. Be sure to carry the *West Canyon Creek 7 1/2' quad* in your hand so you know exactly where you are at all times. If you lose track of locations, you'll eventually come to a nice slot and the slit in the wall with a swimming pool. This marks the bottom end of the **Upper West Canyon Slot**. If you get that far, go back about 1 km to the campsite and exit to the east.

If you have time on the first afternoon, you could swim up through the **9 swimming pools**; or explore the trail to the east skirting the slot. This is getting to be a pretty good trail in places; in other places it's marked with stone cairns. The only way you can get through this exciting upper slot, is by walking up and around, then dropping in at the top and walking downcanyon. Follow the landmarks and cairns on this map to and across the cactus patch, then walk into the slot. You'll do this upper slot on the 2nd day.

Once into this 750m-long **Upper Slot**, you'll find one pothole after another in the upper half. And you'll wear out your pants sliding from one to another--K&EP are required! About 1/3 the way through, you'll come to the **Slippery Slide**, which resembles a water park slide (just wide enough for your rear end) down into a dark pot-

Map 41, West Canyon, South Lake Powell Country, Navajo Nation, Arizona

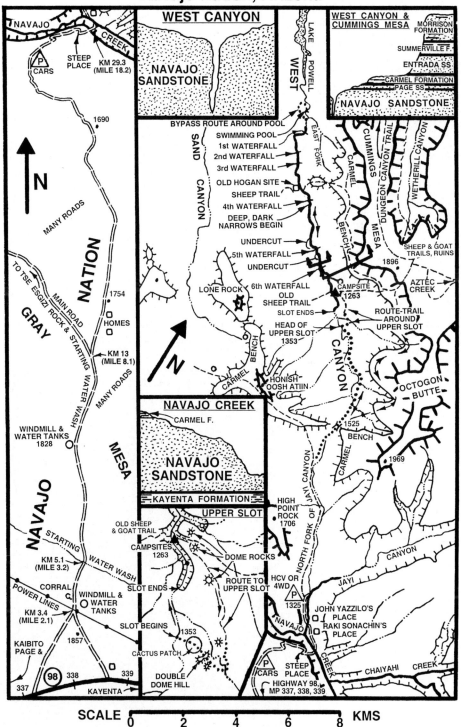

WEST CANYON

NAVAJO SANDSTONE

WEST CANYON & CUMMINGS MESA

MORRISON FORMATION
SUMMERVILLE F.
ENTRADA SS
CARMEL FORMATION
PAGE SS
NAVAJO SANDSTONE

NAVAJO CREEK
STEEP PLACE
KM 29.3 (MILE 18.2)
P CARS

1690

N

MANY ROADS

TO TSE ESGIZI ROCK & STARTING WATER WASH

GRAY NATION

MAIN ROAD

1754
HOMES
KM 13 (MILE 8.1)

WINDMILL & WATER TANKS 1828

MESA

MANY ROADS

N

NAVAJO

STARTING WATER WASH

KM 5.1 (MILE 3.2)

CORRAL
POWER LINES
KM 3.4 (MILE 2.1)
WINDMILL & WATER TANKS
1857

KAIBITO PAGE &

98 338 339
337
KAYENTA

SAND CANYON

BYPASS ROUTE AROUND POOL
SWIMMING POOL
1st WATERFALL
2nd WATERFALL
3rd WATERFALL
OLD HOGAN SITE
SHEEP TRAIL
4th WATERFALL
DEEP, DARK NARROWS BEGIN
UNDERCUT
5th WATERFALL
UNDERCUT
LONE ROCK 6th WATERFALL
OLD SHEEP TRAIL
SLOT ENDS
HEAD OF UPPER SLOT 1353

CARMEL BENCH

HONISH-OOSH ATIIN

NAVAJO CREEK
CARMEL F.

NAVAJO SANDSTONE

KAYENTA FORMATION

UPPER SLOT
OLD SHEEP & GOAT TRAIL
CAMPSITES 1263
DOME ROCKS
SLOT ENDS
ROUTE TO UPPER SLOT
SLOT BEGINS
1353
CACTUS PATCH
DOUBLE DOME HILL

HIGH POINT ROCK 1706

NAVAJO CREEK

HCV OR 4WD
P 1325

P CARS
STEEP PLACE
HIGHWAY 98, MP 337, 338, 339

LAKE POWELL
WEST

EAST FORK

CARMEL BENCH

CUMMINGS MESA

DUNGEON CANYON TRAIL

WETHERILL CANYON

1896

SHEEP & GOAT TRAILS, RUINS

AZTEC CREEK

CAMPSITE 1263

ROUTE-TRAIL AROUND UPPER SLOT

CANYON

OCTOGON BUTTE

1525
BENCH

CARMEL

1969

NORTH FORK OF JAYI CANYON

JAYI CANYON

JOHN YAZZILO'S PLACE
RAKI SONACHIN'S PLACE

CHAIYAHI CREEK

SCALE
0   2   4   6   8   KMS

hole. This is where you'll need a headlamp. At the bottom, you'll land on cobblestones. Be careful, don't sprain an ankle, and lookout for snakes!

Further down you'll come to a 3m-drop into a **KPH**. At the top are 2 B&H on the left and a rope to HL down (if you have a better one, tie knots in it and leave behind as a HL). This is the only dropoff of consequence in the canyon. Once into this pothole, you may find a chain hanging down from 2 B&H on the rim which you can use to climb out. If it's gone, push and/or pull each other out using a shoulder stand. Just below this KPH are other potholes--maybe help each other out; plus one good DC, which now has 1 B&H & rope.

Further down it gets easier, but deeper & darker. Then you'll come to some chokestones where you could **DC then swim or wade underneath (best option)**; or use a rope to get down over the front of this dropoff of 4m; or to the right on a bench and on the wall are **2 B&H** you can rappel from--or HL down 10m (in 8/2007, we found a hanging rope with knots, so we swam). Right after that, you'll have to slide down (there's now a **B&H & short rope)** into the 2nd of 9 icy swimming pools and the beginning of running water (you can get up to this point from the bottom of the slot). This is also where you'll want to put everything in drybags inside your pack so you can swim with hands free and your pack on. At one point, a pool is barely shoulder wide and slanted. After going through the slot, pickup your camp and head back to the boat; or out to your car.

Elevations High Water Mark on Lake Powell, 1128m; campsite, 1263m; head of upper slot, about 1353m.

Time Needed It's possible to day-hike from the lake up to the bottom end of the slot, and return, but that could take 10-12 hours or more. It took the author and 2 friends, 5 hours to walk from their boat up to the campsite with big packs. If you don't get lost, it'll take an hour to go up and around to the beginning of the slot, 2 1/2 hours through the upper slot, and another 5 hours or more back to your boat. Should take 2 very long days from Wahweap **(starting at Antelope Point Marina is now best)**, or a more leisurely 3-4 days. If coming by road, plan on 4 full days to do both the upper slot and the walk down to the lake.

Water Besides the lake, there's running water from the 2nd of the 9 swimming pools down to about the old log hogan site; then below the Swimming Pool down to the lake. Water should be drinkable as is in the upper canyon because there are no more sheep & goats in that area (?).

Maps USGS or BLM maps Kayenta (for driving) & Glen Canyon Dam (1:100,000); West Canyon Creek (1:24,000--7 1/2' quad) for hiking (if hiking in from the road better have Tse Esgizii and possibly Chaiyahi Rim SW); and any map of Lake Powell if going by boat.

Flash Flood Danger This canyon is rather short, but there's lots of barren slickrock in the upper part of the drainage, so runoff is quick. While in the upper slot or in the various short narrow sections further down-canyon, flash flood danger is high.

Best Time to Hike June through about mid-July, or before the first of the monsoon floods. Wait maybe 2-3 weeks after a flood has gone through before hiking, because floods create deep quicksand which makes hiking much more difficult. For an update on the last flood, call Dangling Rope Marina, Tele. 928-645-2969.

Author's Experience The author has been there on 6 trips from the lake, and once years ago in a helicopter as a guest of Eberhard Schmilinsky, who may have helped the first people through West Canyon in 1992. Once he day-hiked up to and swam through the 8 pools, but was stopped there. Later, and from the lake, he went with Steve Tyler and his son Mike. We camped next to the boat the first night, then in 5 hours were at the campsite. The next morning we did the slot, returned to the boat, and got back to Stateline at 7 pm. Long day! In August, 1998, the author and brother Gail, started in the morning and ran into lots of quicksand, some nearly waist deep. A km above the log hogan, Gail walked into a swimming pool that wasn't there before, and drowned a camera. We returned from there. That trip was in a 14 ft. boat with a 10 hp motor which consumed a total of 21 liters (5 1/2 gallons) of fuel from Stateline.

On 5/12 & 5/13/2001, he hiked alone upcanyon and attempted the upper slot on Day 2, but weather was threatening, so he missed the best part. Further downcanyon, it rained hard and he had to run ahead of the flood and under 4 waterfalls about 1 km down to the old hogan site. From there, a fast walk kept him ahead of the flood. About 4 days prior to that trip, Nat Smale and 2 friends came in by road and spent 4 days seeing all the canyon down to the lake.

On 7/20-21/2007, he and Brian Olliver did this hike in 2 days from Antelope Point Marina using 5 1/2 gallons of gas in a 14 ft. boat & 25 hp motor. They had everything is dry bags/kegs, and had to swim at least a dozen pools, most of which weren't there before. Look for Brian's filmed documentary on canyoneering in the next year or two. There are more B&H in the canyon now, but please leave them in place-- most people doing this hike are not skilled canyoneers! Safety comes first, right? Right.

Opposite Page If coming from the north & Lake Powell, this is the **1st swimming pool** you come to in **West Canyon**.

Above Left Brian Olliver inside the **KPH** in the **Upper Slot of West Canyon**. In his left hand is a chain attached to 2 B&H's above; the other end is attached to 1 bolt at shoulder level. This 2nd bolt keeps the chain inside the pothole.

Above Right This is the 1st pool you come to in the lower end of the **Upper Slot of West Canyon**. You can DC and swim through this pool; HL down on the rope shown; or rappel from 2 B&H's just to the left out of sight.

Left Deep narrows in the area just to the south of the 6th waterfall in the middle part of **West Canyon**.

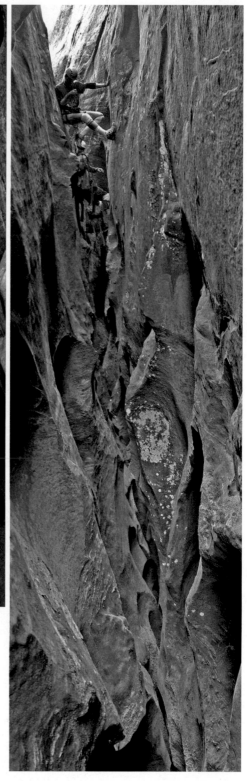

Both of these pictures were taken in the **Middle Fork** of the **King Mesa Slots**. Both were taken by Jon Jasper who spliced or stitched 2 digital images together using special software.

As you can see **X or XX SLOTS** are not recommended for the average canyoneering. If you slip on the moss here, and fall close to 15m, it would be death or serious injury. The local Sheriff and NPS rescue crews won't be happy fetching your body. Nor will the rescue/body removal be cheap or fast!

Map 42A, Area Map--Escalante River Drainage, Utah

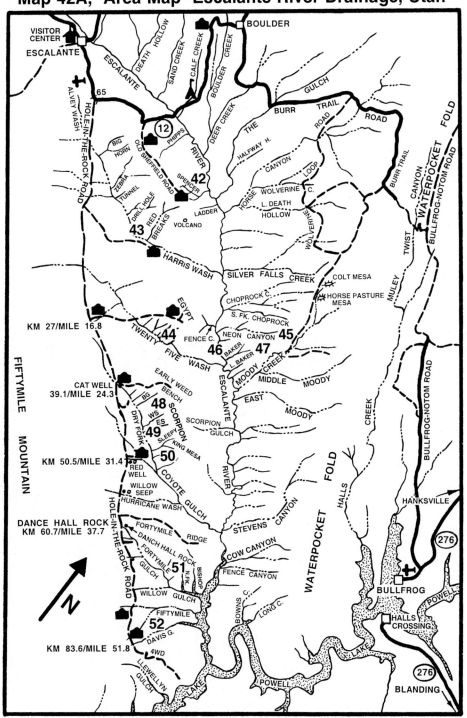

Spencer Canyon, Escalante River Drainage, Utah

Location & Access This canyon is unnamed but since it's immediately east of **Big Spencer Flat**, the author will call it **Spencer Canyon**--for lack of a better name. However, because of the way its shaped, some new-comers to Escalante are calling it Metro, State & Main, or Center & Main--or something (?). On topo maps, it looks like a schematic drawing of a big city's metro or subway system.

To get there, drive along Highway 12 between the towns of Boulder & Escalante. Right at new **mile post 70**, turn south onto the **Old Sheffield Road**. Head south, then east across the upper parts of **Big Horn Canyon** and into the middle of **Big Spencer Flat**. See **Area Map page 197**. After 9.4 kms (5.8 miles) the main graded road turns south and runs to an old drill hole site; but instead continue east on a less-used ungraded track. Soon you'll come to sandy places, so if you have a 2WD car, stop at a **little slickrock flat** 10.8 kms (6.7 miles) from the highway. Or with an AWD/4WD continue east to a **bigger slickrock flat** about 50m across. This is **11.4 kms (7.1 miles)** from Highway 12. Beyond this slickrock flat, it starts downhill and there are sandunes & **very deep sand** fit for tough 4WD's only! In winter, or when the sand is wet, a 2WD can go further, but if the sand is dry, even 4WD's could be in deep trouble! Best to park before getting stuck--and walk a little further.

Rating Spencer Canyon: **3B II or III**, there will likely be wading all the time in the 100m slot.

Equipment RG, a 50m rope, K&EP, W&R/QL's, possibly a wet/drysuit in cooler weather.

Route From either **slickrock parking**, walk east on the same track for roughly 1 km. North of that, are 2 little drainages heading into the upper west end of the **Main Fork** of Spencer Canyon. We'll call them **Forks 1 & 2**. If you go down into **Fork 1**, you'll eventually come to a **2-part dropoff**. In 2006, there was a deadman & W&R/QL in the sand for a 30m rappel, but you can walk around this on the left or north. Next is **dropoff 2**. There was also a deadman & W&R/QL there for an 11m rap, but you can easily walk around this one to the north--so what's the point of rappelling? About 30m further on, will be another deadman anchor in sand & W&R/QL. This will be **Rappel 1** and it's down about **18m**. At the bottom of that will be a 4th anchor in the sand and **R2**. This drops about **20m**. About 20m from the bottom of R2, the canyon slots-up immediately and drops about **12m**. Let's stop here for a moment.

If you don't want to bother with 2 raps (that part of the canyon isn't so interesting), you could reach this point by coming down **Fork 2**. It has several DC's and a fotogenic tight slot, but no raps before arriving at the last **12m dropoff**. A confident canyoneering can DC this meter-wide slot. However, it can be a little intimidating for some, so your group leader may want to belay some, then DC himself. You could also rig W&R/WL and rappel; or tie a rope to a chokestone or log, rappel, then pick it up later.

At the bottom will likely be a narrow **waist-deep pool**, then a longer & wider one that may be a **swimmer**. After that is perhaps a 3rd pool, then the dark **100m slot** ends and it begins to widen. Below the technical slot are 2 kms of narrows. In this part there are 3 pools; wade or swim these, or climb around each on ledges. When you finally come to the **South Fork** (marked 1573m), turn right or south and after 200m, turn right or west into another little canyon we'll call **Fork 3**. This one is also narrow and a little dark with a couple of upclimbs. At the top, get out to the south just after a big boulder, and walk southeast until you come to a little slickrock valley the author calls **Ponderosa Pine Valley** (with 18 ponderosas). From there walk west eventually coming to the sandy road and your car.

Here's another option for a long day-hike, or break it down into 2 separate hikes. At the junction with the **South Fork** (1573m), head north 2 kms to the Escalante River. In the last 350m or so, you'll have one steep DC, a short, deep pool (swim, wade or stem over), then wading through the beginning of a spring & running water before the river. Return the same way.

Elevations Car-park, 1755m; confluence of the Main & South Forks, 1573m; Escalante River, 1509m.

Time Needed Down Fork 2, Main Fork and back up Fork 3 should take 3-4 hours. Or if going down to the Escalante River, maybe 6-8 hours.

Water Start with your own, but you'll find it in potholes, the

This is the beginning of the dark slot in the upper end of **Spencer Canyon**. On this day the author tied off a rope, rappelled and picked it later. But for some this is a DC.

198

Map 42, Spencer Canyon, Escalante River Drainage, Utah

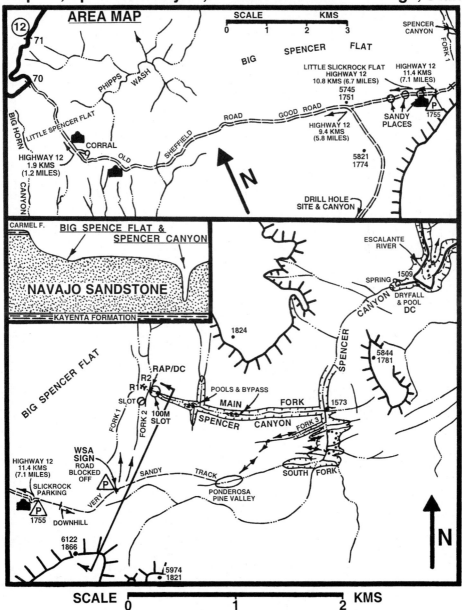

spring in lower Spencer, and the Escalante River (treat or filter).

Maps USGS or BLM map Escalante (1:100,000) for driving; Tenmile Flat & Red Breaks (1:24,000--7 1/2' quads) for hiking; and the plastic Trails Illustrated/National Geographic map, Canyons of the Escalante (1:70,500) for driving--but it's not much good for hiking in this section.

Flash Flood Danger High in the 100m slot; moderate to high in the rest of the Main Fork.

Main Attractions Deep dark slots, and a relatively unknown canyon.

Best Time to Hike Spring, summer (not so good, pretty warm) or fall, or in winter warm & dry spells.

Author's Experience After a scout trip, he stopped at the big slickrock parking and headed down Fork 2, left a rope at the 12m drop (he was alone), went down to the spring above the Escalante River, and back to his car in under 7 hours. That afternoon, he went down Fork 1 and picked up his rope, and returned in less than 3 hours.

Drill Hole & Red Breaks Canyons (& The Volcano), Escalante River Drainage, Utah

Location & Access Shown on this map is **Red Breaks Canyon**, and its 2 most interesting drainages; the **Big West Fork** and the **Upper West Fork**. Also, an unnamed canyon this writer is calling **Drill Hole Canyon**, because it starts right at an old drill hole site (plus there are 2 more drill holes on either side) and drains into Harris Wash. There are 2 ways into this area; for **lower Red Breaks**, head southeast out of Escalante on Highway 12. Between mile posts 64 & 65, turn south onto the **Hole-in-the-Rock Road**. Drive southeast until you see the sign *Harris Wash Trailhead, 7 miles [11 kms]* on the left. That junction is 17.2 kms (10.7 miles) from Highway 12. Drive east on this usually-good occasionally-graded road for 10 kms (6.1 miles) to the **Harris Wash & Red Breaks Canyon Trailhead (HWRBCTH)** where you'll find many good shaded campsites in cottonwood & Russian olive trees, and tall tamaracks.

For **Drill Hole Canyon**, and a **shortcut route** to upper **Red Breaks Canyon**, continue eastward on Highway 12 toward Boulder; right at **mile post 70** turn right or south onto the good graded **Old Sheffield Road (OSR)**. Head south, then east for **9.4 kms (5.8 miles)** to a **Junction**. From there head south on the **graded road**. Stop & park at the drill hole site 14.7 kms (8.5 miles) from Highway 12.

To reach an interesting hole-in-the-rock feature some locals are calling **The Volcano** (another name is Washing Machine), which is made of Navajo Sandstone, continue east on the **ungraded part of the OSR** from the **Junction**. Cars (2WD's) should stop at a **little slickrock flat 10.8 kms (6.7 miles)** from the highway; 4WD's stop at a **bigger slickrock flat** at Km 11.4 (Mile 7.1). Beyond that it's deep sand! Park/camp at either parking place.

You can get closer to **The Volcano** from the **HWRBCTH** with a **4WD**. From Harris Wash, head north on the obvious road and drive about 4.5 kms (3 miles). Stop and park at the **2nd set of rock salt yellow buckets**. Beyond that is a sandy sea and fit for winter 4WD driving only!

Rating Both forks of **Red Breaks**: **2A II** (for some **3A II**, depending on skill levels). Getting through these slots requires some chimneying, but no real rappels. **Drill Hole Canyon: 3A II**. Pretty simple.

Equipment Red Breaks: one 15m rope & Ibis hook, and K&EP for sure. **Drill Hole**: RG, two 50m ropes (or a 50 & pull cord), W&R/QL & K&EP. Read more below.

Route Red Breaks Canyon From **Harris Wash**, walk north into & up the main fork which has some fairly good narrows in the lower half. Halfway up will be the lower end of the **Big West Fork**, which you can't enter from the bottom because of a 15m dryfall. To enter it, walk back downcanyon about 200m, and climb up to the 1st bench on the west side, then head up Big West Fork. Soon it begins to slot-up. This lasts for 1 km and you'll have to UC over 7-8 chokestones, only one of which is a little strenuous. This is one of the better walk-through slots around. Be there at mid-day for colorful fotos.

Higher up, when the slot ends and the drainage opens, climb up to the right or east, walk over a ridge, and down into the middle part of the **Upper West Fork**; observe map carefully At the first drop, DC along the left side and into the slot. Not far beyond that will be a rockfall. If you're skinny, you can barely squeeze down through one of several holes to get into the slot below; or using K&EP, chimney down about 6m using the *knees to back* method. After that it's basically a walk-through slot, but in some places it's so tight, you'll have to walk sideways. Take the smallest pack possible.

Finally after about 600m, will be a drop of about 5m. You can DC half of this, then it's a fast slide often times into a pool. The leader can belay everyone down, then DC & slide; or with an Ibis hook, place it behind one of many small ledges at the top and everyone HL down.

Alternate Route to Red Breaks Start at the **Drill Hole** and head almost due east and around the south side of the obvious **butte** labeled **6136** (1870m). Once on top of the ridge, veer northeast and look for a trail used by guides & clients. This new trail heads in the direction of the NE corner of Section 16 marked **5821** (1774m). South of that, this trail-of-sorts heads east and into **Big West Fork**. From there, route-find in & out to the east and into **Upper West Fork**. Go down this fork, then up Big West Fork and back to the Drill Hole using the same route mentioned above.

Drill Hole Canyon From the drill hole, head northwest a little, then route-find due west and down along a minor drainage and into the main canyon. You can walk right down the shallow canyon, or along an old washed-out road which ends at another drill hole site on the east side. You can enter the canyon there as well. Soon you bypass 2 dryfalls on the west, then reenter the canyon just before the **1st Rappel** (this one can be bypassed). Simplest way to do this is to tie a 50m rope to a big rock or tree to the far right, make the rappel, then retrieve the rope at the end of your trip.

Or, rim-walk the west side bypassing this first dryfall/rappel and look for an entry 400m further downcanyon. From there, walk another 400m to a big dryfall, but turn left or east and scramble down a steep crack. At the bottom is an easy **9m vertical DC 60cms wide**. Checkout a little East Fork at that point, then walk through nice narrows. About 600m below the crack are 2 dropoffs into possible pools; DC the first on the right, then **rap from a hook** placed on the left at the second. A group might be able to help each other down this one (?). About 100m below R2 will be the last drop in the canyon, **R3**. On the far right will be the author's deadman anchor with W&R/QL. Rap about **10m**.

From the last rappel, the canyon opens, and after another 500m, you can turn right and rim-walk northward back to your car. Or, you can continue downcanyon into another narrow section and windup in Harris Wash. From there, choose one of several return routes. For more information on other nearby non-technical canyons, such as **Big Horn, Tunnel & Zebra Slots**, see the author's other book, *Non-Technical Canyon Guide to the Colorado Plateau.*

The Volcano While in the area, you might checkout an interesting site (non-technical) near the top of a big Navajo SS dome. Observe the map and walk in from the **HWRBCTH** side & the **rock salt yellow buckets** along **The V Road**, or **Big Spencer Flat** parking places. It's faster & shorter via **The V Road** (if the Monument BLM haven't blocked-off the road). Once there, head for the eastern side of this big natural hole and DC by way of **sawed-out moki steps**. Inside is sand, plus a feature that looks like a lava dome inside a volcanic crater. Best fotos can be taken from the western rim. The next dome north of The Volcano has some interesting foto opts with yellow & brown iron deposits. Interesting place.

Elevations Trailheads 1512m, 1755m, 1737m & 1609m; dome behind The Volcano, 1782m.

Time Needed Red Breaks: from the bottom up Big West Fork and down Upper West Fork, about 5-8 hours (Alternate Route, about 3-5 hours?). **Drill Hole Canyon**: 4-5 hours. **The Volcano**: 4-5 hours from the **Big Spencer Flat trailhead**; maybe 2-3 hours from the 2nd rock salt buckets on **The V Road**.

Water There's no reliable water source around, so take plenty in your car and pack.

Maps USGS or BLM map Escalante (1:100,000) for driving & orientation; Red Breaks & Tenmile Flat (1:24,000--7 1/2' quads) for hiking; and the plastic Trails Illustrated/National Geographic map, Canyons

Map 43, Drill Hole & Red Breaks Canyons (& The Volcano), Escalante River Drainage, Utah

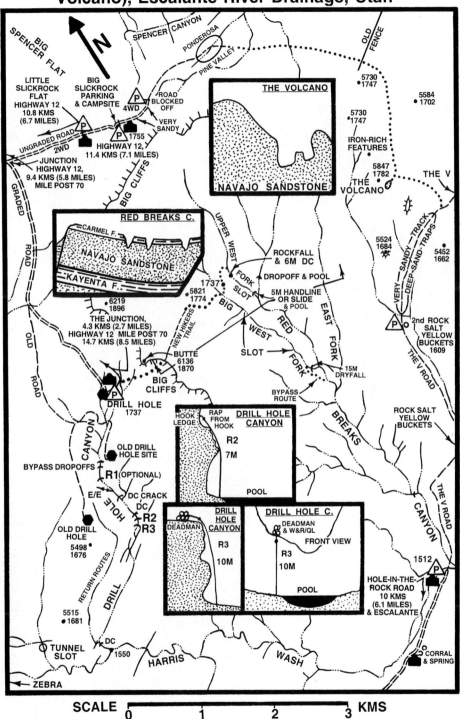

SCALE 0   1   2   3   KMS

of the Escalante (1:70,500) for driving & orientation, but not so good for hiking in this region.
Best Time to Hike Spring or fall, but summer is fine too--just turn off the AC and acclimatize to warmer temps. In summer, if you have to wade or swim, you can do it without a wet/drysuit. Warm dry spells in winter can be nice too.
Author's Experience After a number of scout trips, he finally walked up Big West and down Upper West Forks of **Red Breaks Canyon** in 5 1/2 hours. He scouted the **Alternate Route** to Red Breaks in 2 hours. He did **Drill Hole Canyon** in 5 3/4 hours solo, but had no idea what was there. He did **The Volcano** trip in less than 4 hours from the Spencer Flat TH. He has walked & mapped **The V Road.**

Above Left This is what some call **The Volcano**--but other names are used. It looks like water at the bottom, but it's sand except for a small puddle on the far left side. In that same area you can DC on steps cut in the wall.

Above Right In **Drill Hole Canyon**, this is where you DC about 9m in a crack. You'll need K&EP for this.

Right In **Drill Hole Canyon**, this is where you rap **(R2)** off a natural ledge with an Ibis hook and down about 7m. Some have used a partner assist at this point, others have slid & jumped. If water in the pool is deep, and it does have a sandy bottom, jumping is a good option. The yellow knee pads bought at a hardware store are fine except for the buckles; they keep coming off. Best to use **wrestler's knee pads**.

Above Both fotos were taken in the Big West Fork of **Red Breaks Canyon** around mid-day; that's when the sun is shinning down into the slot, and that brings out the colors. But don't shoot at a place where the sun is shinning directly into the slot--either back up a little, or move around a bend.
Below At the end of the **Upper West Fork of Red Breaks Canyon**, is this 5m HL or slide. One way to do it safely is to rig an Ibis hook to one of several ledges, then HL down.

Egypt Slots, Escalante River Drainage, Utah

Location & Access These 5 slot canyons are located southeast of Escalante and in between the **Hole-in-the-Rock Road (HRR)** and the **Escalante River**. All drain south from **Egypt Bench** into Twentyfive Mile Wash. To get there, drive southeast out of Escalante on **Highway 12**. Between **mile posts 64 & 65**, turn southeast onto the HRR and drive about **27 kms (16.8 miles)** to where you'll see a sign on the left or east near the county line reading, *Egypt--10 miles (16 kms)*. Drive east on this **Egypt Bench Road**, which can be done in most passenger cars (**Note:** The last part of this drive is slowly getting rougher).

To reach the parking place for **Egypt 0** (not a great hike), drive **9.9 kms (6.1 miles)** from the **HRR**. Park anywhere you can (or go to **10.2 kms/6.6 miles** to the 2nd of 2 upper forks). For **Egypt 1**, drive to **Km 10.9 (Mile 6.8)** and park in the big wash bottom. For **Egypt 2**, drive a total of **12.9 kms (8 miles)** from the HRR. At that point where the road crosses slickrock and the upper dry creek bed of E2, you'll see a huge dropoff to the east. Immediately south of the dry wash is a good place to camp/park in the cedars (junipers), but the NPS doesn't seem to want you there for some reason (?). Park/camp just to the north.

To reach the parking place for **Egypt 3 & 4**, drive a total of **14.8 kms (9.2 miles)** from the HRR. There you should see a minor dry wash, a large juniper tree on the right, and 60m back to the west, the best landmark around--a **shinny metal culvert** under the road. This is your parking place for both E3 & E4. Incidentally, the popular Egypt Bench Trailhead is 16.3 kms (10.1 miles) from the HRR.

Rating Egypt 0: 2A II; Egypt 1: 2A II; Egypt 2: 3A II or III (maybe a B); Egypt 3: 2A II PG SLOT, or 3B III PG SLOT if doing the lower technical section; Egypt 4: 4A IV X SLOT, this is the most challenging slot in these parts. Egypt 4 has no rappels, just high-stemming over a deep slot for 1 1/2 kms.

Equipment Egypt 0--K&EP for all, & short rope; Egypt 1--a 15m rope; Egypt 2--RG, one 100m rope (or a 50 if using the 2nd & **recommended** entry point) and another 20m (or longer) rope, plus W&R/QL; E3--one 50m rope, but only if doing the one & only rap/HL at the end; E4--small pack, R/QD/DC and KE&AP.

Route Egypt 0 Walk south down along either of the 2 upper forks. There's nothing of interest until you get below where to 2 forks meet, then there's short shallow V-shaped narrows, and a 12-14m drop into an open canyon. You could rap from a chokestone 5m above the drop, but it's really uninteresting.

Egypt 1 Just a few meters from where you park, and on the right, you can **DC 6m** into the beginning of the canyon. For the next km, you'll have 5-6 challenging DC's, or fast slides. At one 4m drop, the leader can belay everyone, then they can help him/her down from below. If you're alone, you can place an **Ibis hook** on a natural ledge and HL down. It's a hardrock landing if you jump/slide. After you leave the narrows and the canyon opens, get out on the right and walk around the cliffs to the west. Within 500m of the drainage bottom will be 2 steep slickrock routes up to the top of the bench. Rim-walk back to your car.

Egypt 2 Walk around the head of the vertical beginning of this slot (east side may be best) and decide which cedar (juniper) tree is the best anchor, then rappel down a single 100m rope (it's been measured at about 80m from the rim to the very bottom). Because it's a single rope rappel, use gloves & a variable speed ATC, and wrap the rope around your back if speed to too fast. Or, use a locking beaner on your brake-hand-side leg loop; or use a munter hitch. Leave the rope there and pick it up when you return (perhaps leave a note telling others to leave the rope where it is!). Now if you don't have a 100, then walk along the rim southeast about 200m and locate a steep slickrock route down to the south. At a point about 150m from the head of the canyon, look for some bushes & large rocks. Tie off a 50 or 60m rope and leave it there to be picked up later, then **rap** about **25-30m**.

Once in the bottom, head downcanyon, which involves easy walking along with some DC's. About 700m from the very head of the canyon is **R2**, a drop of between **5 & 8m**. Rap off a small chokestone with W&R/QL. About 200m below that is an entry/exit to the right or west. About 1 km below the exit is **R3**, about **5m**. You may have to rig W&R/QL around a small chokestone to do this one. Further down, you can exit either right or left and route-find back to the road.

Egypt 3 From where you park just east of the **metal culvert**, walk south along an emerging trail. After 5 minutes, you'll see a very deep canyon on your right or west, and another drainage coming in on your left. Route-find south over the edge of the Carmel rim and down steep Navajo slickrock. At the bottom, veer right or west and route-find into the main drainage. The upper part of this canyon is narrow and has interlocking potholes, one of which may be a little difficult for a short person to get out of. Further along, the canyon really slots-up and in many places you'll have to remove your pack and walk sideways. At one point you'll have to squeeze through a 20 cm (8") wide opening. Near the bottom end are 2 open areas where you can exit and return along the ridge between Egypt 3 & 4.

Beyond the last open area is the final slot with one short rappel/HL. As you walk down this slot, you'll first notice 2 tiny bridges, then 5m away, a **3rd small bridge**. Not far below that is a short dropoff & DC and the **1st big pothole**. Immediately after that is a **7m dryfall**. To get over that, you'll need a 50m rope (or a couple of shorter ones). Tying your rope to the 3rd bridge will get you over the 7m rap. Pick up your rope on the way back. See the cross-section.

After the 7m dropoff (no need for a harness; use a body wrap method, or just QL down) is a short series of interlocking potholes, then the **2nd pothole**. Once into the 2nd pothole, one person may have a hard time getting out of it if it's muddy, so this part of the canyon should be done with at least 2 people. Shortly after that, you'll drop into the last section which will always be a **35m wade** (or swim?). Once out of the slot, continue downcanyon 200m, turn the corner to the right or west, and look for a route back up to the bench above. Walk northeast along the west side of the slot to the open area, drop into the drainage, retrieve your rope, then follow the slickrock route running north between E3 & E4 back to the trailhead.

Egypt 4. Start out by heading east from the trailhead and basically follow the route down that you use to exit E3. Follow this route on a topo map carefully. Once into the head of E4, simple follow it downcanyon. It slowly turns into a very tight slot and you'll have to chimney or stem above. After a ways, you'll likely come to 2 small pools; you might stem or climb around these (?). Then comes the hardest part where you'll have to chimney up 6-7m or more above the bottom to pass through. In this part you'll be doing the *back-to-hands & knees move* while moving horizontally. Then comes a large log jammed up high. If you can make it to that point, or just beyond, you can make it the rest of the way. Otherwise turn back. If you slip and slide to the bottom, the Sheriff won't be happy pulling you out!

After that first very difficult section, the walls become more straight instead of convoluted, and you can chimney or stem doing the *back-to-feet method* for most of the rest of the way. One third the way through you can walk along the bottom a short distance, but then you'll have to chimney/**climb straight up 10m** or so to pass another very tight section, then more high-stemming (**Attention:** For this slot it's important to have good shoes and some kind of padding around your lower back to save a lot of rubbing and abrasion).

About 9/10's of the way to the **dome rock** on the west labeled **5201T** (1585m) on the *Egypt quad*, will be an exit on the left of east. You'll likely want to leave the canyon at that point, as the next 250m or so won't be

Map 44, Egypt Slots, Escalante River Drainage, Utah

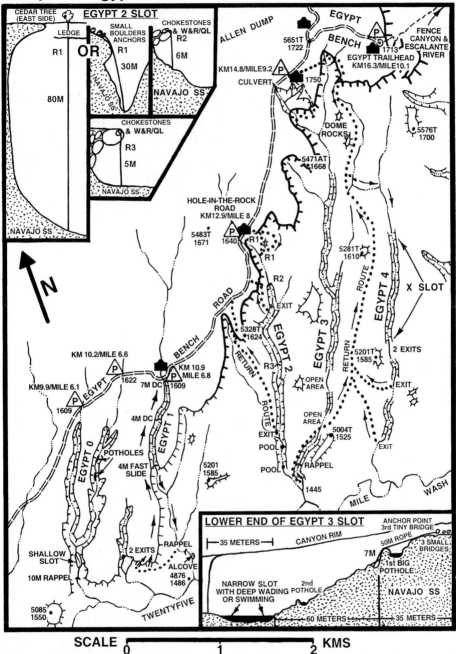

SCALE 0 — 1 — 2 KMS

the same challenge as the first part. You'll also likely be sore and ready for a break. Immediately east of 5201T will be another easy exit to the east. Below that is another short & quite easy slot, some of which will likely have water (chimney over the pools to stay dry). About 500m due south of 5201T is an Entry/Exit to either side; and just below that is a 3m drop but there's nothing to see below that.

Elevations Trailheads--Egypt 0 & 1, 1609m & 1622m; Egypt 2, 1640m; E3 & E4, 1750m; bottom end of Egypt 2 & 3 slots, 1445m.

Time Needed About 2 hours for **E0**; 3-4 hours for **E1**; 5-8 hours for **E2 & E3**, depending on skills, fitness and group size. For **E4**, plan on 7-9 hours.

Water None around, so always carry plenty of water in your car and in your pack (3-4 liters in summer!).

Maps USGS or BLM map Escalante (1:100,000) for driving; Egypt & Sunset Flat (1:24,000--7 1/2' quad) for hiking; and the plastic Trails Illustrated/National Geographic map, Canyons of the Escalante (1:70,500) which is best for driving, and can be used for hiking as well.

Flash Flood Danger These are short drainages, so all you have to worry about is a heavy thunderstorm immediately above you. But this is slickrock country, so any rain will put running water in the slots.

Best Time to Hike Spring or fall when the roads are dry. For the last section in E3, and because of deep wading, June through early-September; or use a wet-suit in cooler weather. Winter, or at least cooler weather, might be best for E4, as you'll be up & above all or most of the watery potholes; and you'll also be wearing more clothes, which means less skin abrasions. Anytime for E1, but it can be muddy after rains.

Author's Experience He has hiked down E3 on several occasions while car-camping on Egypt Bench. With John Summerson, we scouted out the entire length of E4, but missed 1 1/2 kms of the most difficult part. Later that day, we headed west crossing over E3 and rim-walked north up along the east side of E2. We entered E2 about midway by leaving a 30m rope attached to a bush and HL'ed down 25m. From there we hiked up to R2 and down to R3, before returning to the rim & cars in a total time of 8 1/2 hours.

Later, the author walked east from E2 and down to a point 150m downcanyon and tied a rope to several large rocks and rappelled to the bottom. From there he went upcanyon, then down. He left E2 on the west side and made his way back to his car in a total round-trip time of 2 hours. Afterwards he retrieved his rope.

In 2004, and with Jon Jasper & Tim Barnhart, we got about 100m into the difficult part of E4, but weren't quite prepared so we turned back. On 5/27/2005, along with Nat Smale, we did the entire length of E4 in 7 1/4 hours round-trip, but were actually in the main slot about 4 hours (about 5 hours in the slot if counting the last 2 short sections). After a scout trip along the rim in 2007, he went down E1, and came up parts of E0, all in 3 1/4 hours.

Above Jon Jasper in **Egypt 4**. This is the pool that's there all the time. The hard part begins right after that.

Right Typical scene in **E-4**. We're up 8-10m, but there's resting places throughout.

Above Left A common scene in **Egypt 1 Slot**.

Above Right There are 3-4 DC's in **E1**, but a couple are fast slides with hardrock landings. The author used an Ibis hook from a natural ledge, to get down this 4m drop. With at least 3 people, and hopefully a tall person or 2, you can do it without the hook.

Left John Summerson wading in near-chest-deep water in the 35m-long pool at the end of E3.

South Fork of Choprock & Neon Canyons, Escalante River Drainage, Utah

Location & Access Neon and the **Golden Cathedral**, a canyon that's unnamed on all USGS maps, and **South Fork of Choprock (SFC)**, are located in about the middle of the Escalante River drainage southeast of the town of Escalante. They're both on the east side of the river and in between the Moody Canyons to the southeast, and the rest of the Choprock Canyon complex to the northwest.

To do the **full length** of either canyon will normally take 2 days, but that means **lugging a large pack** through each--**this is definitely not recommended!** Depending on which way you exit, both may involve some kind of car shuttle (or perhaps a mtn. bike in the case of SFC). Or you can do **shorter day-hikes** through each from the Egypt Bench Trailhead and/or from a camp at or near the **Escalante River.**

Here's the best option for both canyons. From the NPS/BLM/Forest Service **visitor center** on the west side of Escalante, drive east through town and southeast on **Highway 12**. Between **mile posts 64 & 65** veer right and continue southeast on the graded & graveled **Hole-in-the-Rock Road** for about 27 kms (16.8 miles). At that point will be a sign on the left reading, *Egypt 10 miles [16 kms]*. Drive this graded road to the **Egypt Bench Trailhead**, a distance of **16.3 kms (10.1 miles)**. Park there, which for most people will be the beginning & end of their hike into both canyons.

Or if you choose to carry a large pack through the entire length of either canyon, then backtrack to **Highway 12** and proceed east to the town of **Boulder**. From there continue east on the paved **Burr Trail Road** for roughly **45.7 kms (28.4 miles)**. There you'll see a road veering right or south and a sign, stating among other distances, *Moody Creek Trailhead--20.4 miles [33 kms]*. Drive south on this graded road which is good for cars in normal conditions. Along the way, you'll pass the beginning of Silver Falls Creek, then after 22.9 kms (14.4 miles), stop at a potential parking place & exit just west of **Horse Pasture Mesa** marked **1630m**. It's possible to exit the Choprock canyons there if starting at the **Moody Canyon TH**. To get there, continue south along Moody Creek and stop where the road makes a big switchback and climbs up on a bench; this is **31.1 km (19.3 miles)** from the **Burr Trail Road**. This is the **Moody Canyon TH** at 1490m altitude. Park there if you insist on doing the full length of each canyon. The drive between trailheads takes roughly 3 1/2 hours, which means you'll likely look for an alternative.

Rating Neon: **3B V** or **VI** (a shortened 1 day version of Neon rates **3B IV**, a day-hike from Egypt), but this could rise to **R** if the big KPH in lower Neon just above the Golden Cathedral is low on water. **SFC: 4B IV;** some people believe **SFC** should be **4B R IV -R SLOT**, at least under the worst-case conditions (but this is not just a tight slot canyon with continuous high-stemming, only places with good DC's). April 13, 2004, 2 BYU students tried SFC in skimpy wetsuits without arms or legs, became hypothermic and drowned. At times, the SFC can be full of **water & logs**, making it difficult to drag a large pack through. Day-hike it with small pack!

Equipment Both canyons: RG, one 60m rope (or a 30 & pull cord), a 15m rope for short raps, W&R/QL, K&EP, R/QD/DC, drybags, wet/drysuit, water purification tablets (or light weight filter for rancid pothole water if camping) and perhaps a Ibis hook. In late spring/early fall, take neoprene gloves for swimming.

Route Neon From **Egypt Bench TH**, walk east over the rim on a good trail which goes down steep slick-rock then through sand into **Fence Canyon** and the **Escalante River**. From there, and with map in hand, take the most-used trail downriver and into the first drainage on the east side of the cottonwood bottoms. This is lower Neon Canyon, and you can walk upstream about 1 km to see the **Golden Cathedral** from the bottom.

Or, to see the canyon above and to rappel down into the Golden Cathedral, regress back downstream to within 200m of the river, then walk north up an old cattle trail on to one of the benches above. Just follow this good hiker's trail. Soon the trail heads upcanyon to the northeast along one of several benches. Most people now go upcanyon about 2 kms to a prominent drainage coming in from the left called **North Fork**. Walk up this 300-400m or so, then get into the drainage and head downcanyon. Soon you'll come to a DC (if you have a wetsuit, put it in there) into a tight slot. Not far below you'll come to the **main Neon drainage.**

Now if you enter Neon there, you can do it in one long day from Egypt. Or, camp somewhere near the mouth of the canyon and do it as an easy day-hike. Or, if you're camping nearby and want to see more of upper Neon, continue bench-walking northeast along the north side of the entrenched drainage. Some have entered about where R3 is located. From there you'll have lots of DC's into pools, swimming or wading, walking for a short distance, then another DC.

From North Fork, head downcanyon with alternating pools & swimming and a few DC's. About 500m above the Golden Cathedral, the canyon opens some and you may see **moki steps** on the right. Not far below that is a steep DC into a slot with more swimming. Just after that used to be 1 B&H on the left and R6, but it wasn't there on the author's last trip. More possible swimming, then around a corner you'll come to a **big potential KPH.** If it's full of water you won't know it's there. When water is moderately low, we rapped in from a **mini arch** on the wall behind and had no problem getting out because of ledges under water. But we were apparently standing on an overhang. Others couldn't get out when water was a lot lower as it was in the spring of 2008. Some tried a shoulder stand, but the guy under water wasn't touching the bottom then! They turned around and UC'ed back out missing the rappel into the Golden Cathedral. Take a **hooking kit & G-pick** just in case. Canyons change with every flood, so go prepared for a worse-case situation.

Then the **last rappel**, which is from the top of the **Golden Cathedral**. You may find 6m of W&R/QL and a **22m rap** down through one of **2 huge natural bridges** (formerly bottomless potholes) into a large pool about 1m deep (?). Most of this is a free-rappel, so it's easier if you tether your pack to the front of your harness and let it hang down. Mid-day or earlier should give you the best foto opportunity (in mid or late afternoon, the sun shines down into part of gorge making a high contrast scene). This is one of the premier foto stops on the Colorado Plateau. From the Golden Cathedral, it's back to the Escalante River, upstream to Fence Canyon and west to the Egypt Bench Trailhead.

South Fork Choprock Canyon To do the SFC, walk down from Egypt through Fence Canyon to the Escalante, then north 2 kms to the mouth of **Choprock Canyon**. Walk up Choprock about 200m, to find a campsite on the right or south next to the wall. The standard way of doing SFC is to camp there, or somewhere in the area, then with day-packs, walk a little further upcanyon to find an **old cattle trail** climbing upon the first bench. Walk south about 1 km, then route-find due east up through ledges. After 500m, veer northeast until you're on top of the **1st little mesa** just north of elevation **5381T** (1640m); then ENE to a **2nd little mesa** with an altitude of **5776T** (1761m). From that high point, continue easterly nearly 1 more km, then route-find north down into the middle/upper part of SFC between what could be R1 & R2.

Head downcanyon. Along the way will be some DC's, a **4m R2** into a pool, then DC'ing 1 slot which will garner at least an **R SLOT** rating, then the **15m R3** into the lower middle part of SFC. This puts you into a green canyon with running water or pools for another 2 kms or so. In this section will be lots of trees for a ways, then a long stretch of subway-like narrows. Still further will be **R4** (which maybe misplaced on the map) and

Map 45, South Fork of Choprock & Neon Canyons, Escalante River Drainage, Utah

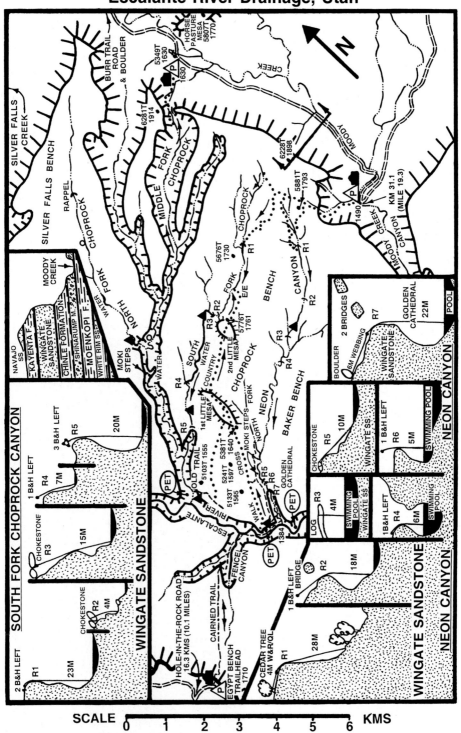

SCALE

0 1 2 3 4 5 6 KMS

down **7m**. Just above that will be a steep section where you may want to throw in a short piece of W&R/QL behind a chokestone for the last person to HL down.

Below R4, the narrows and slot seem endless. This part is similar to Neon in that you'll walk a little, then comes a narrow slot and a drop/DC into water, then swimming or wading. Some of these short slots are very narrow and getting a large pack through is very tiring. This part of the canyon is, or can be, **full of logs** and other floating debris and all too often you'll have to swim or wade through a real mess--log soup! However, on our last trip, it was mostly dry, and wetsuits really weren't needed once we got into the lower part.

In the last 600m or so, there may be 300m of DC'ing, swimming, etc., etc., then what could be a 100m swim/wade, and finally 200m of regular walking. At the end is **R6**, which is a **22m drop** landing next to a clear pool. From there, it's trees, grass and walking about 1 km to the main Choprock Canyon, then after another 1 1/2 kms, you'll arrive at your campsite.

If you want to do the full length of either canyon, match this map to the USGS quad, and walk the routes shown from the **Moody Creek Trailhead**. For either canyon, you could return to the Moody Road via the Main Fork and/or **Middle Fork of Choprock Canyon**. See map. Or you could do the full length of either canyon from Egypt. Do that by walking along **Choprock Bench** to the head of either canyon, then walk right down the chute of each. But it's highly recommended you camp somewhere near the mouth of either canyon, then with **smaller day-packs**, do the trips suggested above as **day-hikes**.

Elevations Egypt Bench Trailhead, 1710m; Moody Creek Trailhead, 1490m; Escalante River, 1384m.

Time Needed Neon: from Egypt, 8-12 hours. **SFC**: plan on camping in lower Choprock Canyon somewhere, then 7-10 hours for the day-hike. Come out at end of Day 2, or on Day 3?

Water Carry plenty in your car. There should always be some pothole water in both canyons; take a filter. **SFC** has running water in several places, and there's always a pool of pretty good water at the base of the 3 higher rappels. Also good running water just north and south of North Fork in the main Choprock Canyon. Also in the Escalante River & Fence Canyon (filter both).

Maps USGS or BLM map Escalante (1:100,000) for driving; Egypt, Silver Falls Bench & Horse Pasture Mesa (1:24,000--7 1/2' quad) for hiking; or the plastic Trails Illustrated/National Geographic map, Canyons of the Escalante (1:70,500), best for driving.

Flash Flood Danger Overall, moderate to high danger with a mixture of intermittent narrows, slots & swimming pools and occasional wide places in both canyons. SFC is the more difficult, strenuous and with a slightly higher degree of danger. Both canyons have lots of slickrock above in the Wingate and Kayenta benches, so any significant rain will run off into the canyons in a hurry.

Best Time to Hike Late spring through early fall. Even in the heat of summer better take a wetsuit in either canyon because of all the likely swimming. Late spring/early summer would give you more daylight hike-time.

Author's Experience Nat Smale and the author met and camped at Egypt on 10/3/2001, then with an early start drove one car 3 1/2 hours to the Moody Creek Trailhead. Once in the upper end of **Neon**, time was lost not knowing if we should go right down the drainage; or bench-walk on the north side. We camped 300m above R3 with a walk-time on Day 1 of just over 5 hours. Day 2, we had a cold splash early and soon found

the author's brand new Pelican camera case wasn't waterproof--which drowned a 2-day-old $400 Pentax--but his point & shoot backup got a few fotos. It took about 6 1/2 hours to reach the bottom of the Golden Cathedral (R7), then nearly 3 1/2 more hours to make it back to Egypt. Total 2-day walk-time was 14 1/2 hours, then driving part-way to Moody before camping after sunset. Counting drive-time, the trip took about 2 1/4 days.

For **SFC**, Nat and the author met near Horse Pasture Mesa and car-camped. Next morning (4/27/2002) we drove one car down to the Moody Creek Trailhead and walked in. We struggled for 8 3/4 hours before camping--lots of water & logs. Day 2, made it to the last rappel in 2 hours, then the long walk through the Main Fork, then the Middle Fork of Choprock back to one car at Horse Pasture Mesa. Walk-time for Day 2 was 7 1/3 hours. Total walk-time for the 2 days was about 16 hours.

The following 2 trips are what is recommended. Along with Nat & Cihan Bilginsoy, we left Egypt in the late pm and camped near the mouth of **Choprock Canyon**. Next day, 5/21/2004, we walked the route suggested above and dropped into upper **SFC** in 2 hours. We wore wetsuits, but only had some shallow wading. It took 8 2/3 hours to get back to camp. Nice trip under ideal conditions.

On 8/2/2004, along with Jon Jasper & Tim Barnhart, we left cars at Egypt, walked down to the Escalante and up the trail into the **North Fork of Neon** (2 3/4 hours). Went down to the top of the **Golden Cathedral** in less than 2 hours. After the GC, walked back up to Egypt, all in just over 8 hours (1 hour, 8 minutes from the river to Egypt with small day-packs).

This scene is below Rappel 4 in the **South Fork of Choprock Canyon**. On this trip the canyon was quite dry or with pools and low water--but there's always wading or swimming pools around.

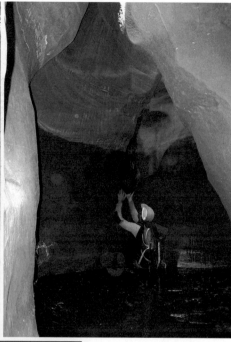

All pictures on this page are from the **South Fork of Choprock Canyon**.

Above Left You'll have lots of DC's like this throughout the canyon. Nat Smale & Cihan Bilginsoy.

Above Right Lots of DC's into pools as well.

Left The last rappel of about 20m from 3 B&H's is into a shallow pool.
(Cihan Bilginsoy foto)

Above Left At the bottom of the rappel into the **Golden Cathedral in Neon Canyon. Above Right** Tim Barnhart stands on an underwater ledge in the **big KPH** just above the rappel into the **Golden Cathedral.** When this KPH is low on water, it's almost impossible to get out even with a shoulder stand. A pack toss is useless to some eyewitnesses. Consider taking a hooking kit & G-pick. Above and behind this KPH is an arch you can tie a rope to; that will allow you to jumar out if needed.

Bottom This foto of the **Escalante River** was taken from the bench just above the mouth of Neon Canyon by Stefan Folias on **10/5/2006**. You're looking northwest and Fence Canyon can be seen in the middle background. Some have described this as the 200 year storm & flood. It was estimated the river ran at 40,000 cfs. For 10/2006, and after a couple of big storms, Lake Powell rose 2.07m.

A wide angle view of what some call **The Volcano**. It's located just east of the upper end of **Red Breaks Canyon**. It's not a volcano of course, but it has the shape of one. It sits near the top of a big Navajo Sandstone dome. (This foto was taken by Reinhard Kirchner, who disappeared in the Little Colorado River Gorge on April 1, 2007. He has not been seen since, and is presumed dead.)

Looking up at the rappel down through the **Golden Cathedral** at the end of the technical section of **Neon Canyon**. From this angle, it appears there are 3 openings, all the remains of former potholes.

213

Baker & Ringtail Canyons, Escalante River Drainage, Utah

Location & Access **Ringtail and West & East Baker Canyons** are located in about the middle of the Escalante River drainage, and immediately southeast of Neon Canyon and the **Golden Cathedral**. To get there from the NPS/BLM/Forest Service Visitor Center on the west side of Escalante, drive east through town and southeast on **Highway 12** for 8 kms (5 miles) to between mile posts 64 & 65, then veer right and continue southeast on the graded & graveled **Hole-in-the-Rock Road**. After about 27 kms (16.8 miles), and next to the county line, will be a sign on the left reading, *Egypt 10 miles [16 kms]*. Drive this graded road east to the Egypt Bench Trailhead, a distance of 16.3 kms (10.1 miles). In dry conditions, cars can make it, but there's a rough place or two near the TH. Camp in that area.

Rating After a half-day's backpacking trip, **Ringtail: 3B II PG SLOT; West Baker: 3B III; East Baker, 4B IV R SLOT.**

Equipment **Ringtail:** one 15m rope & Ibis hook, headlamp, K&EP and perhaps a wet/drysuit in the cooler season (shouldn't be needed in summer heat). **West & East Baker:** RG, a 50m rope & maybe a shorter 15, W&R/QL, K&EP, wet/drysuit in cooler weather, maybe an Ibis hook. For **East Baker**, add a BK and 1 B&H for 1 serious DC, so people don't have to continually build a deadman--or risk injury.

Route For all three, head east from Egypt Bench TH over the rim and down an old slickrock horse trail to the bench below, then northeast along a cairned trail to Fence Canyon. Continue east down Fence between its 2 forks. At the Escalante River, wade south for 50m to a trail, then follow it & topo map down-river past the mouth of Neon. About 500m south of Neon is a trail running up on a bench which leads to the upper part of **Ringtail**. From a campsite along the river at or near Ringtail, you can do all three canyons. Or you could camp at another place further downstream near the cowboy signatures & pet-roglyphs and get up on the same bench for an entry to either fork of **Baker**. You can also do **East Baker** from the **Little Baker Slots**; see next map for access.

Ringtail Use the **1st cattle trail** shown to get upon the **first bench** above the river, then follow it eastward to **another level**, then cross a short drainage before heading southeast on the slickrock bench which is the top of the Wingate Sandstone. When you come to the entrenched Ringtail, turn left or northeast until you can drop in. Soon you'll come to a pretty good DC, then it opens a little before get-ting real narrow again. Moments later, you'll come to what some have DC'ed, but for most hikers, it's the only possible rappel in the canyon. You can **rap/HL** off an **Ibis hook** from a ledge on the left and down about **5m**. This will get you over a deep pothole 1 person may have trouble getting out of alone (with several people, DC into the pothole and help each other out avoiding a rappel). If doing this, RG is not needed, but a short rope is still a good idea. Below that are 2 pretty good DC's into another pot-hole, then a tight slot which gradually gets deeper & darker; take 1 headlamp for this. Near the bottom, you'll likely do some wading.

Baker Canyon About 3 1/2 kms below the mouth of Ringtail, and 2 kms upriver from the end of Baker, is a **2nd cattle trail** running upon the first bench above the river. At its beginning are petro-glyphs & cowboy signatures. Once on the first bench, head east for roughly 400m until you come to an-other constructed **trail** up to the next level. From that bench, route-find up to a 3rd level which is the top of the Wingate, then bench-walk eastward to Baker and northeast along its rim. About 1 1/2 kms from the river will be **the confluence** of the **East & West Forks**. From there, continue north about 1 km before dropping into a little side-canyon just west of elevation marked **5201T** (1585m) on the *Scorpion Gulch quad*. This will get you into the **West Fork**. We didn't explorer above that, but there could be more slot that way (?). If you're going to the more difficult **East Fork**, route-find out of the West Fork and up along the East Fork above or below the confluence to one of several routes in.

West Baker About 400m into the West Fork will be your first rappel (you could jump about 5m into a deep pool, but checkout the depth first!). Look for **1 B&H** on the right-side wall behind a bush. If ever a bolt chopper should take that out, rap off a little ledge using an **Ibis hook**, a drop of 7m. Not far below that is **R2** from a **B&H** on the left and down **7m**. About 100m below that, the canyon is filled with large boulders. **R3** is from another **B&H** on the right and down past big boulders **11m**. **Rappels 4, 5 & 6** are within about 125m of **the confluence**. **R4 & R5** are down **6m & 7m** from a single **B&H** on the left, while **R6** is from a log and down **3m**. Just beyond that is the confluence of the East & West Forks. This section is reminiscent of The Subway in Zion, or the confluence of the Paria River & Buckskin Gulch, making it one of the nicest scenes anywhere.

East Baker Route-find in from the west above or below the little side-canyon shown. We accessed it from the Little Baker Slots to the south & east and entered from just below the little side-canyon shown. The hike is easy at first, then there are tight sections where you'll have to chimney up to 6-7m using a R/QD/DC at times. These hard parts are don't last long, and there are short walks or squeezing in be-tween--**take the smallest pack possible, and have good K&EP's. This canyon is for fit hikers only!** Finally you'll come to **R1** from **1 B&H** on the left. Rap 7m to a level place, but stay on-rope and continue down a tight part **11m** through a bell-bottom section. Then more high-stemming & slow move-ment a short distance to **R2** from a **log**. Not far away is a drop of **3m** into a dark hole. We built a **dead-man anchor** and rapped, but previous hikers have been belayed, then the last skinny guy goes feet first through the bottom of a really tight spot and is helped down on the other side by a pyramid of several hikers (or for the benefit of smaller groups, maybe add 1 B&H?). From there it's easy walking to the last rap in East Baker **E4**; an **8m rap** from **1 B&H** on the right. This puts you into a wider canyon about 125m upcanyon from **the confluence**.

From the confluence, head downcanyon. Really nice scenery lasts for about 300m, then after an-other 800m used to be a 3m rap from a log--but it was gone on 4/3/2008. That must have been where we swam through a pothole--in April with no wetsuits! Soon after that will be the **last rappel** from **2 B&H** & down **20m** into the lower end of the canyon. This is a partly a free rap. At the bottom is a pool with good looking water, then to the Escalante, a distance of 500m. Hikers regularly walk up to the bottom of this rappel on a trail.

Elevations Egypt Bench Trailhead, 1710m; end of Baker Canyon at the Escalante River, 1362m.

Time Needed Reasonably fit hikers could do Ringtail in one long day from Egypt TH, but to do either fork of Baker is too long for a day-hike. Backpack in for a long weekend (3 days) and do 2 or 3 from a camp somewhere on the Escalante.

Water Running water in Fence and the Escalante (purify both), in potholes in all canyons, and a pool and a little good running water in lower Baker.

Maps USGS or BLM map Escalante (1:100,000) for driving & orientation; Egypt & Scorpion Gulch (1:24,000--7 1/2' quads) for hiking; and the plastic Trails Illustrated/National Geographic map, Canyons of the Escalante (1:70,500), for hiking, driving & orientation.

Map 46, Baker & Ringtail Canyons, Escalante River, Utah

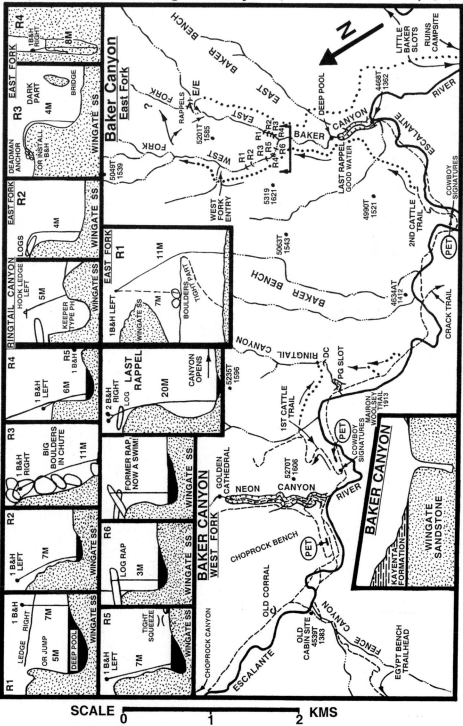

Flash Flood Danger High in places in Ringtail & East Baker, moderate to high in West Baker.
Best Time to Hike Spring and fall, but with potholes & likely water in both, summers are possible too.
Author's Experience After a scouting trip from Egypt, and along with Nat Smale & Cihan Bilginsoy, we first did the **South Fork of Choprock** from a camp near the end of the main Choprock, then that afternoon, headed downriver and camped near the mouth of Ringtail. Next morning, we started early and walked to & through **West Baker**. From our camp, it took 8 1/2 hours round-trip. Next morning we did **Ringtail** in 2 3/4 hours, then high-tailed-it to Egypt. From the river to the TH took 1 1/2 hours with large packs.

    With Jon Jasper & Tim Barnhart, we backpacked in from Early Weed Bench TH in 2 3/4 hours and stopped at the Ruins Campsite (see next map). Next day we did **East Baker** in just under 8 hours.

Above Left There are several big deep pools in the **West Fork of Baker Canyon**, so go prepared with a wetsuit most of the time.

Above Right This is the confluence of the **East & West Forks of Baker Canyon**, and one of the nicer scenes around.

Right The lower part of **Ringtail Canyon** which is similar in shape as Tunnel Slot (near the end of Drill Hole Canyon). This is a very dark place, but a flash & Photoshop help take the deep shadows out.

Above Left This is Rappel 1 in the lower end of **East Fork of Baker Canyon**. The anchor is 1 old B&H, but you have to angle down to one side to avoid a very tight part of the slot immediately below the anchor.

Above Right In the **East Fork of Baker Canyon**, you'll have to chimney up as high as 7-8m, or more, in several places. In some places it's moving laterally hands-to-elbows, knees-to-heels style. Sometimes it better to hang your pack down below your feet, but in other cases it seems best to keep it thy-high on one side of your rappelling harness. The author likes using a **quik draw** for this. (Jon Jasper foto)

Left Cihan Bilginsoy doing the **final rap** in **Baker Canyon**. At the bottom is a pool and good drinking water, then a short hike to the Escalante River.

Little Baker Slots, Escalante River Drainage, Utah

Location & Access Shown here are 3 unnamed short slots draining into the Escalante River from the **East Baker Bench**. This is a triangle-shaped area in between Moody Creek, Baker Canyon and the Escalante. To get there, drive southeast out of Escalante on Highway 12. Between mile posts 64 & 65 veer right and continue southeast on the graded & graveled **Hole-in-the-Rock Road**. After about **39.1 kms (24.3 miles)**, and **just before Cat Well**, will be a sign on the left or east reading, *Early Weed Bench, 6 miles*. Drive this sometimes-graded road northeast to the **Early Weed Bench TH**, a distance of 9 kms (5.6 miles). In dry conditions, cars can make it, but there's a rough place or two along the way (if driving a car, better have a shovel handy). Camp at or near the Cat Well, or the trailhead.

Rating After half a day of backpacking, **West Slot: 3A II; Middle Slot: 3B III -PG SLOT; East Slot: 3B III.** There could be a good slot higher upcanyon, but we didn't have the time to explore it.

Equipment For each canyon, RG, W&R/QL, K&EP, and perhaps a wetsuit in cold weather (?). **West Slot:** 60m rope & long pull cord, 15m rope; **Middle Slot:** 60 &15m ropes, BK & 2 B&H (or G-pick & Ibis hook); **East Slot:** 30m rope & pull cord, or a 60, and a 15m rope.

Route From the end of the road on **Early Weed Bench**, head **ENE** just south of the Navajo dome marked **5524T** (1684m) on the *Egypt quad*. After passing an old **brush corral**, continue ENE cross-country. To reach the **Escalante River**, you must have the topo maps or quads listed below, and a compass--and know how to use both. Along the way, you'll pass just north of a butte marked **5132** (1564m) and altitude marker **5109T** (1557m), then route-find down to the river zig zag fashion. Cross the river where you can, then make your way nearly to the crude Anasazi ruins, then double back to the west on a bench where you can camp near a cave--let's call this **Ruins Campsite**.

After depositing your big pack near the cave, walk west for 200m and look for some **moki steps** up steep slickrock to get upon a low bench, climb another more prominent bench, then head east via the top of the Wingate which forms a flat bench below the Kayenta benches. The first drainage you come to is the **West Slot**; enter about 500m before its end and just above **Rappel 1**. At that point you could tie a rope at least **17m** long to a large bush, go to the last rap, then return; or install W&R/QL, pull your rope and continue. Near the end are 3 DC's, one is through a bridge, the last is in a tight slot, and maybe a short walk in water, then a chokestone anchor for a **big rap** of **40-45m** (?) to the river bottom. Walk back along the river to camp; or backtrack from the last drop and continue to Middle Slot.

Middle Slot Enter this one at, or 100m below, the **circular feature** shown. It's 400m of pretty easy walking, then it dives and you'll likely have to **rap/DC 6m** from a chokestone (if coming back up, leave a short rope). About 60m away is **R2**, a drop of about 10m. About 10m above that drop were 2 large stones we used to tie a rope (because we returned); if these have been washed away, you'll have to roundup more, or install a B&H. Just below R2, the canyon slots up nicely with 6-7 PH, none of which should be KPH's (in cool weather, take a wetsuit). At the end of this section is a dropoff. You could rap from an Ibis hook, but you'd have to make a hole with a G-pick; or install **1 B&H** for the **5-6m R3**. After R3, the canyon opens up with a flat bottom. After about 300m of easy walking, you'll find 3 big PH, likely KPH's, making a big drop. But you can avoid this difficult section by getting up on the bench to the right and making a **rappel** of about **20-25m** from 2 small boulders. Take a 60m rope to be sure, plus lots of W&R/QL. At the bottom of **R4**, walk north and make an easy DC off the last bench down to the river. Walk back to camp from there.

East Slot If you choose not to make the last rap in the Middle Slot and have left ropes, UC & jumar back, and leave the canyon at the circular feature. Rim walk to the East Slot. We entered & explored from about the middle of the canyon down, but it appeared the upper part has a good slot (?). The bottom half of East Slot has a couple of short, tight slots with 2 rappels and DC's & at least 1 KPH, but with entries & exits on both sides. You can actually route-find around all these more interesting sections, and DC to the river at the end if you choose. From there, you can walk back along the river to camp, or route-find back up to the upper end of the slot.

Elevations Early Weed Bench TH, 1669m; Escalante River, about 1375m.

Time Needed With a big pack, plan on about 2 1/2 hours to reach the river. This means fast hikers might be able to do 1 slot in a long day from the TH; however, to get all 3, better plan to spend at least 1 night (maybe 2?) camping on the bench near the Anasazi Ruins (or anywhere). If you plan to camp 2 nights, you might also grab the East Fork of Baker on the same trip (it's an **R SLOT**).

Water Start with some, then get it from the Escalante River (but filter or purify first), or from potholes.

Maps USGS or BLM map Escalante (1:100,000) for driving; Egypt & Scorpion Gulch (1:24,000--7 1/2' quads) for hiking; and the plastic Trails Illustrated/National Geographic map, Canyons of the Escalante (1:70,500) which is best for driving, and can be used for hiking as well.

Flash Flood Danger High in the slots, but these are very short drainages.

Best Time to Hike Before or after spring flooding in the Escalante River, or fall, but summer too.

Author's Experience After a scout trip, the author, Jon Jasper & Tim Barnhart backpacked to the Ruins Campsite in 2 3/4 hours, then spent nearly 7 hours in the West & Middle Slots--more than a 10 hour day. Day 2 was spent going through the East Fork of Baker, just under 8 hours. On Day 3, we explored the lower half of East Slot in 5 2/3 hours, then backpacked another 2 1/3 hours returning to the TH.

The **Moki Steps** just east of the **Ruins Campsite**, which is a good base when hiking to the **Little Baker Slots**.

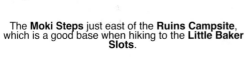

Map 47, Little Baker Slots, Escalante River D., Utah

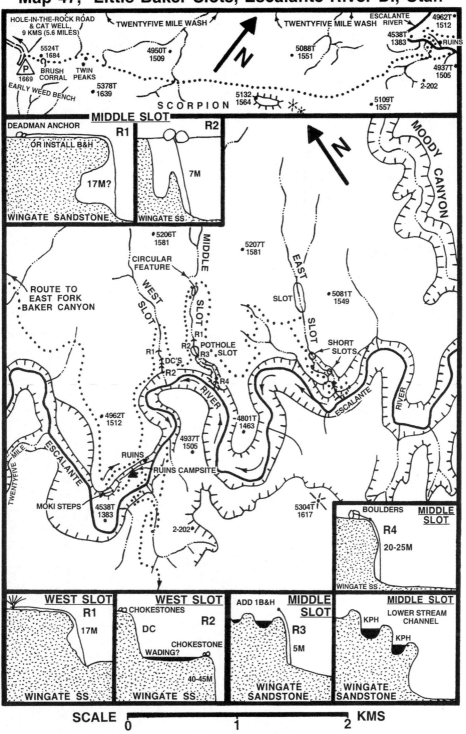

HOLE-IN-THE-ROCK ROAD & CAT WELL, 9 KMS (5.6 MILES)

TWENTYFIVE MILE WASH

TWENTYFIVE MILE WASH

ESCALANTE RIVER

4962T 1512

4538T 1383

RUINS

5524T 1684

4950T 1509

5088T 1551

4937T 1505

2-202

P 1669

BRUSH CORRAL

TWIN PEAKS

EARLY WEED BENCH

5378T 1639

S C O R P I O N

5132 1564

5109T 1557

N

MIDDLE SLOT

R1 DEADMAN ANCHOR OR INSTALL B&H

17M?

WINGATE SANDSTONE

R2

7M

WINGATE SS.

N

MOODY CANYON

5206T 1581

5207T 1581

CIRCULAR FEATURE

MIDDLE SLOT

WEST SLOT

EAST SLOT

5081T 1549

ROUTE TO EAST FORK BAKER CANYON

SLOT

R1

R2 POTHOLE SLOT R3

SHORT SLOTS

R1

DC'S

R2

R4

RIVER

ESCALANTE RIVER

4962T 1512

4937T 1505

4801T 1463

TWENTYFIVE MILE

ESCALANTE

RUINS

RUINS CAMPSITE

MOKI STEPS

4538T 1383

2-202

5304T 1617

BOULDERS

MIDDLE SLOT

R4

20-25M

WINGATE SS.

WEST SLOT

R1

17M

WINGATE SS

WEST SLOT

R2

CHOKESTONES

DC

CHOKESTONE

WADING?

40-45M

WINGATE SS.

MIDDLE SLOT

ADD 1B&H

R3

5M

WINGATE SANDSTONE

MIDDLE SLOT

LOWER STREAM CHANNEL

KPH

KPH

WINGATE SANDSTONE

SCALE

0 1 2 KMS

Above Left This natural bridge is in the middle part of the **West Slot**. (Jon Jasper foto)

Above Right Tim Barnhart in the middle part of **Middle Slot**. This Middle Slot is the best of the **3 Little Baker Slots**.

Right This little dropoff in the **Middle Slot** is either a rappel, HL or a pretty good DC. (Jon Jasper foto)

Above Left Taking the high road over a pothole to keep feet dry and out of the mud. This is in the upper part of **Middle Slot**.
(Tim Barnhart foto)

Above Right From the rim looking down at the end of the slot of **Middle Slot**. That dropoff you see would be Rappel 3.

Left Tim Barnhart DC's what could be Rappel 1 in the upper part of **Middle Slot**.

Brimstone Gulch, Little, Sandslide & Tightest Slots, Escalante River Drainage, Utah

Location & Access **Brimstone Gulch**, and other nearby slots (some for tourists) enter the **Dry Fork of Coyote Gulch** from the north just east of **Cat Well**. To get there, drive southeast out of Escalante on Highway 12. Between mile posts 64 & 65, turn southeast onto the **Hole-in-the-Rock Road (HRR)** and drive 38.8 kms (24.1 miles). There, just before **Cat Well**, you'll see a sign reading, *Early Weed Bench, 6 miles [10 kms]*. To do **loop-hikes from the upper end of any of these canyons**, turn left or northeast and drive 9 kms (5.6 miles) to the last trailhead parking on **Early Weed Bench**. Camp anywhere along that road which is sometimes good for cars--but take a shovel just in case.

To reach the bottom end of all these canyons and the **normal starting point for all**, continue southeast from Cat Well to **Km 42.7/Mile 26.5** and veer left at the sign reading: *Dry Fork Trailhead, Road #252*. From there, and following this map carefully, drive another 2.7 kms (1.7 miles) on a well-used road and park at the trailhead at 1512m altitude. This is the recommended starting place for **Brimstone & Little Canyon** slots. Or, to reach **Sandslide Canyon or Tightest Slot**, drive 1.1 kms (.7 mile) from the HRR, then continue east instead of turning north toward the Dry Fork TH. Drive to the end of this HC/4WD track (3.8 kms/2.4 miles) and park at the **1451m TH**.

Rating **Brimstone, Upper Slot: 3B III; Lower Slot: 2B III R SLOT; Little Canyon: 2A III +R or -X SLOT; Sandslide Canyon: 2A (or 3A for 1 possible rap) II; Tightest Slot: 2A IV X or XX SLOT (?).**

Equipment **Brimstone's Upper Slot**: 15m rope & Ibis hook; **Lower Slot**: nothing but skinny bodies! **Little Canyon**: KE&AP (leave packs at the top or bottom). **Sandslide Canyon**: nothing, except if you want to do the non-obligatory rap at the end, then RG, 50m rope, W&R/QL. **Tightest Slot**: KE&AP, rim crew, at least two 60m ropes, and jumars/ascenders.

Route **Brimstone** From the **Dry Fork TH**, walk north over the rim on a cairned trail to the bottom of Dry Fork. From there, it's about 150m downcanyon to the beginning of **Peek-a-boo Gulch**; another 800m to the end of **Spooky**. These are basically walk-through slots with many beginner hikers each day.

This book concentrates on **Brimstone Gulch** which is a step up in challenge & excitement from the 2 slots just mentioned. To enter the **Lower Slot** from the bottom, continue southeast down **Dry Fork**. You'll pass through some **nice narrows** with one possible chokestone DC (it was a walk-around in 5/2007, but often times a pool exists just below). About 650m below these narrows turn left into the **lower end of Brimstone**. Walk upcanyon 2 kms to the deep slot. Soon it will be dark, with water likely in a 20m-long low part. After that it lightens a little with easy walking. Soon you'll come to a **subway-like narrows**, then it really gets tight, and only a stick man/woman can get through staying on the bottom. Everyone else will have to chimney up 1-2m, maybe 15m (it's in this part you don't want a pack, so if you have one stash it, then give it a try, and pick your pack up later). This straight-walled section goes on for maybe 30m, then you come to a twisting convoluted slot. Soon after that is a **little east fork** coming in on the right or east; you can exit there and return to Dry Fork, and your pack if you left it somewhere. Or checkout **Little Canyon's X SLOT** near the mouth of Brimstone.

To do the entire Brimstone Canyon, rim-walk cross-country from the lower end; or route-find in from the **Early Weed Bench TH**, but starting at the bottom is best with a lot less driving on bad roads. Just below the 2 upper forks is a good place to enter. After 600m is the beginning of the **Upper Slot**, a tight walk. Below that are several exits, then after 1 km, is a **dryfall** and for most a **HL** down **6m**. There are several natural ledges for an **Ibis hook**; or maybe the best climber can belay everyone, then **DC--with some risk!** The author was alone and happy to have an Ibis hook & rope! Below that is a **tight slot & water**--you can chimney/stem over this for 35m. About 150m below that, and just below a **little west fork**, is another **drop** of 3m into a **pool**. If the water is deep, a fast slide or jump will work; otherwise, an Ibis hook from a natural ledge will allow an easy HL down. About 1 km below this 2nd drop, is the **little east fork** mentioned above, and the **Lower Slot**. The easiest way through it is with nothing but a camera. Best to leave pack and everything at the east fork, then retrieve it later--the distance isn't far.

Little Canyon This tiny canyon and 200m-long slot is short & sweet, but it's **all high-stemming** with lots of knees-to-feet, or knees-to-back, and some feet-to-back moves. It's rated somewhere between an **R & X SLOT** (flaky walls might make it an +R or -X SLOT?). To start, go into lower Brimstone about 600m, veer right or east and climb the wall to leave the canyon. From there, walk south to an entry to Little Canyon. Leave packs behind to be picked up later. Right from the beginning, you'll have to go high so be prepared with all the available body armor. At the end, retrieve packs, go through the lower slot (with one 2m jump, or bypass on the west) and into Dry Fork and back to your car.

Sandslide Canyon From the **1451m TH**, walk northeast 200m, then eastward to enter **Dry Fork** at one of 2 places with short fences & a gate. Head up Dry Fork and into lower Sandslide. Exit on the left via **the sandslide**, and rim-walk to the upper end of a shallow slot. Enter at one of many places and head downcanyon in a convoluted slot; part of this might rate a **PG SLOT**. Right at the end is a bridge & 2m away, a 20m drop. On 3/19/2008, there was no W&R/QL or rope anchors, so this rap may be a virgin, and a good place to try a **retrievable sling**. Or backtrack, exit, and return to your car.

Tightest Slot Best to start at the **1451m TH**. Get to the head of the slot from either side. The upper 3rd is incredibly tight; narrowing to 2 cms! There's no way to enter there, so walk south on the east side about 100m then try DC'ing when it begins to open. However, you'll be faced with a **big silo** right at the git-go. That put an end of our dream of a quick descent. We examined the slot carefully in its 125m-long course and decided that you'll need a rim crew with two 60m ropes to be tied to cedar trees on the west side and hung down in. That way, someone can drop food, water, ascenders, or can rap down in, jumar out, or get over a silo. Because of all the silos (it may be a lot easier once you get down in), this looks like a very difficult canyon with lots of high-stemming and ups & downs (?). So go prepared; maybe do a little scout trip yourself beforehand. You can walk/crawl into the bottom end which has a nice sculptured look. It may be best to start at the bottom then retreat the same way.

Elevations Cat Well, 1483m; Dry Fk. TH, 1512m; Lower Brimstone, 1400m; Early Weed TH., 1670m.

Time Needed You can see the best parts of **Spooky & Peek-a-boo** in half a day. Doing the entire **Brimstone** may take 7-10 hours; just the **Lower Slot**, 5-7 hours (?). **Little Canyon** alone, from the trailhead 5-6 hours (?). **Sandslide & Tightest** slots together 6-8 hours or more (?).

Water Take plenty in your car & pack (especially in summer). Don't expect to find water at Cat Well.

Maps USGS or BLM map Smoky Mountain (1:100,000) for driving & orientation; Big Hollow Wash (mostly) & Egypt (upper Brimstone only)(1:24,000--7 1/2' quads) for hiking; or the plastic Trails Illustrated/National Geographic map, Canyons of the Escalante (1:70,500) for driving.

Flash Flood Danger High in the tight parts, especially in the Lower Slot of Brimstone.

Map 48, Brimstone Gulch, Little, Sandslide & Tightest Slots, Escalante River Drainage, Utah

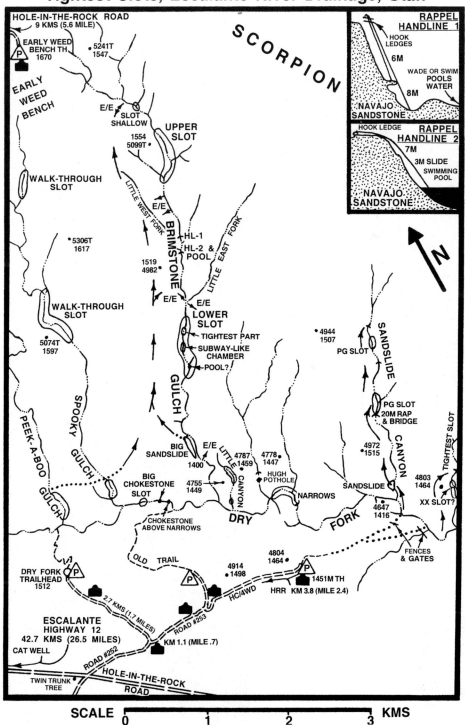

RAPPEL HANDLINE 1

HOOK LEDGES

6M

WADE OR SWIM POOLS WATER

8M

NAVAJO SANDSTONE

HOOK LEDGE

RAPPEL HANDLINE 2

7M

3M SLIDE

SWIMMING POOL

NAVAJO SANDSTONE

HOLE-IN-THE-ROCK ROAD
9 KMS (5.6 MILE)

EARLY WEED BENCH TH 1670

P

5241T 1547

EARLY WEED BENCH

SCORPION

E/E SLOT SHALLOW

UPPER SLOT

1554 5099T

WALK-THROUGH SLOT

E/E

LITTLE WEST FORK

BRIMSTONE

LITTLE EAST FORK

HL-1
HL-2 & POOL

5306T 1617

1519 4982

WALK-THROUGH SLOT

E/E

E/E

LOWER SLOT

TIGHTEST PART

5074T 1597

SUBWAY-LIKE CHAMBER

POOL?

4944 1507

SANDSLIDE

PG SLOT

GULCH

SPOOKY GULCH

PG SLOT 20M RAP & BRIDGE

4972 1515

CANYON

TIGHTEST SLOT

PEEK-A-BOO

BIG SANDSLIDE 1400

E/E

LITTLE

4787 1459

4778 1447

HUGH POTHOLE

4803 1464

BIG CHOKESTONE SLOT

4755 1449

SANDSLIDE

XX SLOT?

GULCH

CANYON

NARROWS

4647 1416

CHOKESTONE ABOVE NARROWS

DRY

FORK

OLD TRAIL

4914 1498

4804 1464

FENCES & GATES

DRY FORK TRAILHEAD 1512

P

P

P

1451M TH

2.7 KMS (1.7 MILES)

ROAD #253

HC/4WD

HRR KM 3.8 (MILE 2.4)

ESCALANTE HIGHWAY 12
42.7 KMS (26.5 MILES)

CAT WELL

KM 1.1 (MILE .7)

ROAD #252

TWIN TRUNK TREE

HOLE-IN-THE-ROCK ROAD

N

SCALE   0   1   2   3   KMS

223

Best Time to Hike Spring through fall, but it's cool in these slots even in summer. Maybe winter too?
Author's Experience In 5/2007, he walked the west rim of **Brimstone** from Dry Gulch and entered where the 2 upper forks meet. With a pack (and too much stuff), he tried going through the Lower Slot--much better with no pack! It got too tight, so he exited and walked the east rim south to **Little Canyon**. Round-trip from his car, 9 1/4 hours. Next day he tried Brimstone from the bottom end with no pack, but failed; he couldn't even turn his head at one point! Tried the 200m-long Little Canyon slot again, but it's not good for solo hikers--6 1/2 hours. On 3/18 & 19/2008, he checked out **Sandslide, Tightest & Pothole** (just east of Tightest) on short scout trips for about 3 hours each day. In 4/2008, along with Jon Jasper & Tim Barnhart, we explored **Tightest Slot**, but didn't finish it. Later that afternoon we did **Little Canyon**. Round-trip took 4 3/4 hours.

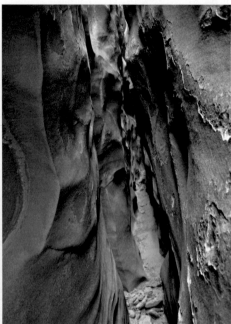

Above Left In the middle part of **Brimstone Gulch** is this pool; it's labeled **HL-2** on the map, and often requires a cold swim.

Above Right Part of the Upper Slot in upper **Brimstone Gulch**. This is an easy walkthrough.

Right This sculptured slot is just above the Subway-like Chamber in the lower end of **Brimstone Gulch**. Just ahead is the tightest part that few can squeeze through. Most have to chimney up a few meters.

Above Left Starting through **Little Canyon** 6-8m above the bottom. (Tim Barnhart foto)

Above Right From the opposite rim, looking at the end of **Sandslide Canyon**. Can you see the bridge about 2m back from the dropoff?

Left Another scene in **Little Canyon**. In some places you'll chimney up to about 10m above the bottom. (Jon Jasper foto)

Scorpion Slots, Escalante River Drainage, Utah

Location & Access The 2 main canyons on this map are unnamed, but since they drain south from a region known as **Scorpion**, we'll call them **Scorpion East & West (SE & SW)**. There are also 3 minor drainages which are mentioned below. To get there, drive southeast out of Escalante on Highway 12. Between mile posts 64 & 65, turn south onto the **Hole-in-the-Rock Road**. Drive about 50.6 kms (31.4 miles), and turn left or east at the sign pointing to **Redwell Trailhead**. Drive east 1.7 kms (1.1 miles) and stop at the large parking area. Any car can make it there in normal conditions, and you can camp/bivouac at the trailhead.

Rating Scorpion West: 4A V XX SLOT (Perhaps 2 separate day trips to complete). The most difficult & challenging slot canyon this writer has seen--**you've been warned!** Scorpion East: **3A III PG SLOT** (with perhaps one very short **R SLOT** section).

Equipment Scorpion East: RG, 1 rope at least 30m long (or one 15m & equally-long pull cord), W&R/QL, K&EP, smallest pack possible & PR, G-pick & Ibis hook, R/QD/DC, plus map & compass in hand for both slots. **Scorpion West**: Same as above, but add a headlamp, K&EP, a bolt kit (?) and a crew on the rim with long ropes which might make an easier escape/rescue.

Route For **both canyons** and from the **Redwell TH**, instead of walking east on the trail into Coyote Gulch, head due north crossing the lower part of **Big Hollow Wash**. After that, walk **ENE** over a low divide and into a small side-drainage of **Dry Fork of Coyote Gulch (DR-CG)** right where the "G" is in "Gulch" on the *Big Hollow 7 1/2' quad*. That will also be right at the bottom end of **Scorpion East**. To do this canyon, turn left or northwest right at the mouth of East, and rim-walk north. After about 4 kms, you'll come to 2 branches near the head of the drainage. The easiest of the 2 is the eastern fork. You can circle around the head of the West Branch to enter it. The **East Branch** has about 500m of shallow slot, part of which you can bypass. This is only moderately interesting.

The **West Branch** is better. It starts with a **3-4m drop** into a pothole. There's no anchor/deadman building material nearby, but the rim of a little pothole just above has a square edge and is perfect for using an Ibis hook. From there it's only 20m to a **2nd dropoff/rappel** of about **5-6m**. It's a little too high for a partner assist, and to use a hook you'd have to knockout a hole somewhere with a G-pick. We saw no building material within reach. A B&H is also possible (we climbed up to that point from the bottom and missed the 20m between the 2 dryfalls). Below these 2 dropoffs there's some **medium high, but easy stemming** and **2 silos** to chimney over (**PG SLOT**) which makes this section good training for future **X SLOTS**.

Below the confluence of the **2 upper branches** is a short shallow slot, then an open area with at least 2 escapes to the west. Further along is a deep pothole with clear drinking water, then another shallow pothole slot with a couple of pools you can get in & out of easily; however, you can walk around this part on the left or east. Finally more open canyon for 200m, then just below where you see a **lone ponderosa pine** to the east, is the final **75m slot & rappel** into the lower canyon. That slot is moderately strenuous (maybe an R SLOT?) with low-stemming & several potholes. In about the middle of this last slot, look for several softball-sized cobblestones to be placed in a groove as an anchor for a rappel. Think ahead, because there's a deep pothole just before the end. To get out of that, you'll need a partner assist; and going back upcanyon will not be easy. Or if you prefer, and for safety sake, it may be best to install 1 B&H for the **10m rappel**. After that it's about 3 kms of easy walking down to DF-CG.

Scorpion West For this canyon, use the same approach discussed above, then head northwest up DF-CG from the mouth of **SE**. When you're about halfway between a **little pothole drainage** and the **mouth of SW**, look for a trail up to a bench north (with a cairn on the rim). The author calls this the **water-to-pasture cow trail**; it's a cow-made path that in places has a hoof-worn groove in slickrock. Follow it north to the beginning of the slot & head downcanyon. Or better still, **rim-walk** the west side of the canyon for a better look at what you're in for.

Enter the canyon just west of elevation **4794** (1469m). This is where the **Upper Slot** begins and it lasts about 150m. This is an **X SLOT**. After that is the **1st Walking Section** which is about 100m long with an escape route to the west. Remember this place.

After that is the **Middle Slot**. We--Ryan Cornia, Jon Jasper and MRK rate it **XX SLOT**, and **feel it's impossible to do safely--unless you use hardware of various kinds. No one should attempt this unless they've scouted the Middle Slot from both ends; and only if you have a tough-assed group of climbers!** Here's a description and our experience.

The **Middle Slot** starts with a 40m-long straight wall section with medium-high stemming, then it's up & down the other side of that very tight part. Then it's down to the dark bottom with 2 potholes (they should never be too deep). It was there MRK slipped and fell backwards about 1m hitting his head on the wall; it was then we all put on helmets! It was not a serious thing, but with all the sweat, blood was running down his face all day long.

From the potholes, we chimneyed up as high as 15m with smooth, green, mossy, slippery walls. It was in this section Ryan lost both knee pads--the kind used by skateboarders with velcro straps--not the best kind. At the end of that section, we chimney up and around some very tight places, but then we would have had to climb up another 10m or so. It was there we decided it wasn't worth the risks, and turned back. As we did, Ryan somehow dropped his pack which is still at the bottom.

The hardest part of the retreat was getting back up the part with 2 potholes. We ended up hacking a hole in the wall with a G-pick (life insurance) for an Ibis hook, which gave the last man the right angle to climb up a very tight place on a rope with several loops for feet. This would have been **impossible without the hook & G-pick!** We finally made it back to the 1st walking section and escaped to the west. We ate what food we had and walked in the dark until 11:30pm, then laid down in the sand for the night after a 16 hour day. Dead tired & no sleep. This experience is good reason to always carry an **emergency bivouac sack**--MRK had one, and invited anyone else to join--but all declined! Next morning we were back at the cars in 40 minutes. Ryan's knees looked like hamburger!

Back to the canyon. Below the Middle Slot is the **2nd Walking Section** of about 175m (part of that is a shallow slot), then the **Lower Slot** which is another **X SLOT** for about 100m or less. The last 80-90m is a PG SLOT with no rap at the end. From there it's another 1 1/2 kms of sandy walking to DF-CG, then backtracking to your car.

Here are some alternate ideas on how to do this canyon. Do the Upper Slot and exit at the 1st walking section. Perhaps at another time, rap or HL down from a cedar tree into the **2nd Walking Section** and scout the end of the Middle Slot and complete the Lower Slot and retrieve ropes. At another time perhaps return to the Middle Slot, but maybe have a rim crew above as life insurance. We don't know what was around the corner from where we stopped, but to get up, then back down into the slot from

Map 49, Scorpion Slots, Escalante River Drainage, Utah

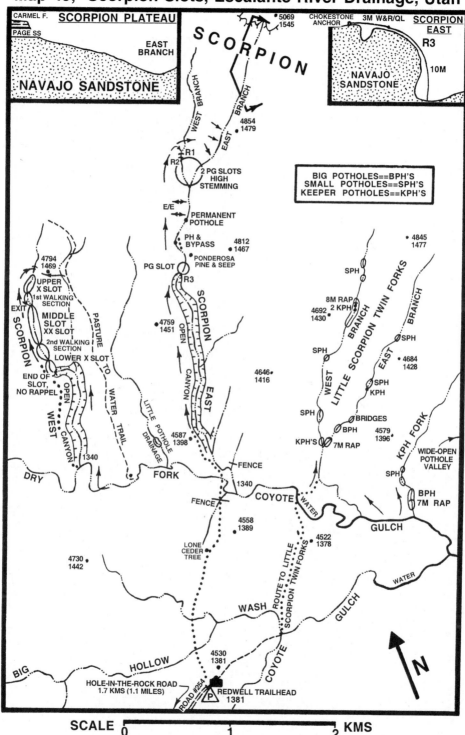

so high above, might take a G-pick & hooks, a bong and B&H's (?).

Just to the east of these 2 main slots are 3 open drainages with not a lot to talk about except for some KPH's. The author is calling them **Scorpion Twin Forks (East & West Branches)** and **Keeper Pothole Fork**. In the **East Branch of STF**, there's a short section with half a dozen little arches & bridges; and a couple of sections with what looks like KPH's and a drop or two. Same with the **West Branch**; lots of short PH sections and walk-arounds.

The **KPH Fork** has about 400m of V-shaped canyon lined with KPH's with 1 drop or dryfall of about 7m in the middle part. To get through that section you'd need the gear necessary to get in & out of KPH's, and make an anchor for a short rap. Also, you'd want a crew of no less than 3 so you can get in & out safely.

Elevations Redwell TH, 1381m; upper ends of all slots, about 1475m; end of SE & SW, about 1340m.

Time Needed For **Scorpion East**, 5-7 hours; **Scorpion West**, long time, maybe 2 days (?). Best to do this one in stages as suggested above.

Water These are dry canyons (except for a few potholes), so bring your own from a culinary source.

Maps USGS or BLM maps Escalante & Smoky Mountain (1:100,000); Big Hollow Wash & King Mesa (1:24,000--7 1/2' quads) for hiking; and the plastic Trails Illustrated/National Geographic map, Canyons of the Escalante (1:70,500) for driving & orientation.

Flash Flood Danger Moderate to high in the tight slots, non-existent in the open canyons.

Best Time to Hike Cooler weather in spring with long days.

Author's Experience After a scouting trip in 9/2006, then with Jon Jasper & Ryan Cornia, we did all parts of **Scorpion East** except for 20m in the upper East Branch. Round-trip, just under 6 hours. A 2nd scout trip to **Scorpion West** took the author 7 3/4 hours. Our trip on 9/1/2007 has been described above. In 4/2008, he explored the 3 little open drainages from along the route passing the 4522 (1378m) dome.

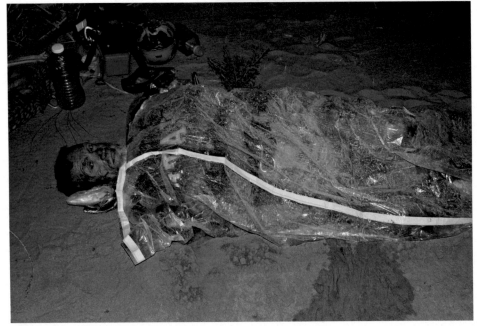

Opposite Page Above MRK & Ryan Cornia in the **Upper X Slot of Scorpion West**. (Jon Jasper foto)

Opposite Page Below MRK trying to sleep in a mylar emergency bag after escaping **Scorpion West's Middle XX Slot**. After 16 hours of walking, we finally just laid down in the sand and spent the night--notice the bloody face. (Jon Jasper foto)

Above Left Jon Jasper & Ryan in the **Upper X Slot of Scorpion West**. Lots of slippery moss in some places.

Above Right JJ & RC in the upper **West Branch of Scorpion East**. We're up 5-6m, but the sides flange out at the top, so this one seems extra safe.

Left JJ making the last rap, Rappel 3 in **Scorpion East**. This is the one not far below the lone ponderosa pine tree.

King Mesa Slots, Escalante River Drainage, Utah

Location & Access Featured here are several short side-canyons flowing south from **King Mesa** into the **Dry Fork** of **Coyote Gulch**. This is all just upcanyon from where **Hurricane Wash** enters. Only 2 seem to have official names; that's **Sleepy Hollow (SH)**(the upper end--AKA-Big Tony), and **Long Branch (LB)**, a northeast tributary of Sleepy Hollow. The author refers to the other 2 main drainages as **Middle Fork** (AKA-Don't Do It/DDI) & **East Fork** (AKA-Pintac). All 4 are **X SLOT's,** meaning there's lots of **high-stemming or chimneying, along with difficult UC/DC'ing.** Also included are 3 easier drainages just to the west and on the left side of this map. Some call them **Raven & Headless Hen (HH) Canyons or Slots.** Also, **Wide-Open Pothole Valley** (WOPV) which hardly deserves recognition.

To get there, drive southeast out of Escalante on Highway 12. Between mile posts 64 & 65, turn south onto the **Hole-in-the-Rock Road**; after about 54.3 kms (33.7 miles), turn left or east at the sign pointing to **Chimney Rock**. Follow this map closely as you drive to, and beyond, Chimney Rock. Part of this road is sandy, but 2WD's do it all the time--just keep moving fast in the sandy places. After 6 kms (3.7 miles) you'll be at the trailhead (about 300m northwest of **4715** (1437m) on the *King Mesa* quad).

Rating For SH, if going down the main fork: **Upper Slot 4B IV X SLOT**; or if going down the **Pothole Fork** route: **4B III R SLOT**. LB is very long & tiresome: **4B IV XX SLOT** (best to do it in 2 days). **Middle Fork: 4B IV X SLOT**. This slot has one of the longest sustained high-stemming sections of any canyon, plus the last 850m is *swamp time!* The **East Fork: 4A III X SLOT**; or if including the very difficult upper slot maybe: **4A IV XX SLOT**. **Raven** has 4 slot sections with easy exits between and ranging from **PG** to **X SLOTS.** **HH** has an upper V-shaped **technical section** filled with many **big KPH's** and should rate at least: **3B R III**; or if water is low in the KPH's: **4B R III**. Take a big group through this one, and people who have KPH escape skills!

Equipment For each canyon, RG, one 60m rope (or a 30 & pull cord), KE&AP, smallest pack possible, PR, plus R/QD/DC. As always, take a good supply of W&R/QL--lots for **LB.** If doing the **Pothole Fork** into **SH & HH**, add a 15m rope, G-pick & Ibis hook. For **Lower Slot** of SH, 1 headlamp per/person (this is now standard for all tough slots). In spring or fall, take a wet/drysuit into **SH & Raven** (year-round in **LB & HH**). Drybags for all, and equipment for exiting (lots of tossbags & extra rope) **KPH's** in **LB, Raven & HH**. Also, and for life insurance only, take a **G-pick** into any drainage with lots of big KPH's.

Route For all canyons, from the trailhead, walk northeast down the sandy slope aiming for a minor drainage which is about 300m east of elevation marker **4312** (1314m) on the *King Mesa quad.* Near the bottom, walk down just west of 2 beehive-like rocks. That little dry wash enters Coyote Gulch about 1 km upstream from the mouth of Sleepy Hollow. Or, if going to the western part of this map, there is a possible shortcut route from the trailhead running northwest (the author hasn't used it, but it looks good).

For **WOPV, HH & Raven**, turn left or west on the flat bench marked **4312** (1314m) and enter Coyote Gulch where you can; walk upstream passing 3 impressive undercuts. Just beyond the 3rd UC, exit upon another bench and follow the dry creek bed into the lower ends of HC & Raven. Walk north between the 2 canyons, and choose one to come down. Or walk further up Coyote Gulch, exit to the north and walk along side WOPV. At the last minute, the author was informed of another possible shortcut route to these 3 canyons by walking northwest from the trailhead. Route-find into Coyote Gulch.

Wide Open Pothole Valley On the far west side of this map is a minor drainage that's filled with what appears to be KPH's. Only problem is, it's an open valley and you can walk out at many places, so going right down the chute seems meaningless.

HH is filled with **big KPH's**--depending on water levels--so go well-prepared. In the upper end is a cottonwood tree right in the bottom of the drainage; enter there and head downstream in the V-shaped drainage. Lots of DC'ing with partner assists, and short raps in the middle of the technical section; especially if water levels are low. Going with a big group of 4-5 people is recommended--or use an Ibis hook (and a G-pick to make small holes) to HL from. The lower part is a more open drainage.

Raven Most people end up in the lower part which has 2 short slots, but with potholes--go prepared for a pack/bag toss & partner assists to get out of a pool or two. It doesn't look too difficult but the author didn't quite complete the lower section. There are also 2 challenging slot sections higher in the drainage. These rate at least R SLOTS with high-stemming. Strong groups can do most of Raven & HC in one long day.

Sleepy Hollow Once into Coyote Gulch, cross the small stream and immediately climb upon an alluvial bench on the left or northwest. From that point, route-find due north along the west side of SH while taking a peek down in occasional. Your last rappel will be ENE of **4751** (1448m) about 400m. Eventually you'll come to the head of the slot.

There are no rappels in the **Upper Slot**--instead just a lot of very high-stemming from 5-12m or more, above the tight bottom. To begin, you'll enter a shallow slot and do some easy low-stemming. After 150m or so, you'll come to a big boulder that has broken off the wall. Worm your way down through a slot on the right to find a short open area, then it's climb time. You'll have to UC at least 10m. After that, and for about 200m, you'll be high-stemming. At one point you'll have to cross a silo which is not good news for short people! Put your hands behind your back for added reach. Beyond that, you slowly go down nearer the bottom to just above where the **Pothole Fork** enters on the east.

Or, if you're out for some fun without putting your life on the line, then circle around to the east side of upper SH (or exit the Upper Slot at the big boulder), and walk south to the beginning of **Pothole Fork.** Getting around or through the upper potholes is easy, then there are 2 short dropoffs. We used a G-pick to make a small hole at the top of each, then rapped/HL'ed **6 & 7m** from an Ibis hook. Once into the main drainage, you can walk up into the lower end of Upper Slot about 150m.

From the mouth of Pothole Fork, walk downcanyon about 75m to find a large boulder with W&R/QL around a smaller chokestone. **Rappel** about **16m**. Below that, crawl under a rockfall with boulders, then just after that is the last major hurdle. First DC a couple of meters, then chimney 5m across a void to a chokestone & DC about 6m--but it is tight, **maybe 20 cms wide!** Or you could chimney out another 3m to a second chokestone, then DC in a slightly wider place.

From the bottom of this slightly spooky place, climb up a rockfall, then chimney down the backside into the dark **Lower Slot.** For about 100m you'll need a **headlamp** and will walk sideways dragging your pack. Near the end of the dark section, you'll be in water and will have to chimney up 2m--unless you're a stickman! After 50m of this, it opens some and you walk normal. In this last part of the slot are a couple of DC's, with one short swim likely. Immediately below this pool, the canyon opens wide. Locate some W&R/QL under a boulder on the right side and rappel about **14m** into the **lower part** of Sleepy Hollow. From there to Coyote Gulch, you'll find good running water, trees & campsites, an emerging hiker's trail and several large undercuts or alcoves.

Map 50, King Mesa Slots, Escalante River Drainage, Utah

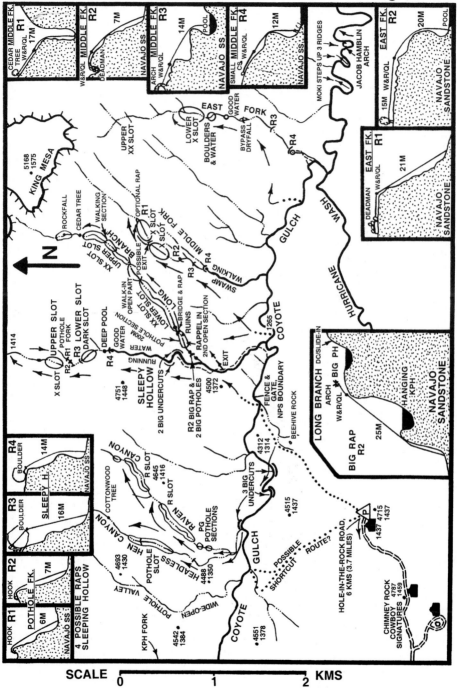

SCALE 0   1   2 KMS

At the bottom of the canyon, it gets a little brushy, so leave (or enter) via a little pass just west or up from the mouth of the drainage. Once into Coyote Gulch, walk upstream a short distance and backtrack to your car near Chimney Rock.

Long Branch The author hasn't completed LB yet, but Nat Smale & Jeff Webb have and here is Nat's account--with some editing by the author after his scout trip: *Got up at 5am on Saturday, 10/7/2007, and left camp at about 6am with headlamps. We headed straight to the rim of LB, arriving at the **walk-in open part** between the 2 main slot sections where we stashed water & gear. This is at the end of the **upper slot**. We then went to the head of the slot and dropped in. It was difficult from the git-go with a hard upclimb of an off-width crack. It was mostly high-stemming with some difficult downclimbs, and a little walking. Just before the **walk-in open part,** we had a difficult KPH that we dealt with a pack toss (and a swim).*

*The **1st narrows** took us just over **2 hours**, and we felt pretty good about our time since the **lower slot** was only slightly longer. In the **walk-in open part**, we rested and warmed up in the sun for half an hour, then got started on the **lower slot**. This part turned out to be very long, strenuous and difficult, taking us **4 1/2 hours**; we never touched the ground for the first 4. There were long sections with lots of silos and hard ups & downs. Other sections had parallel walls without footholds and very insecure. The entire stretch involved total concentration. At one point we arrived at a chamber with emerald green, moss covered walls. We were already pretty high up, but at the end of this section it narrowed, and gave us a very hard, slippery upclimb, with feet slipping on the moss. Later we rested on a ledge about 15m up. After the ledge was a wide silo all the way to the ground. We gave each other horizontal belays across this. It wasn't that hard but was a bit spooky. Finally the walls started getting lower, and we eventually came to our **1st rappel** into an open riparian area with trees, willows and grass, the **2nd open section**. Our only anchor was a **natural bridge** in the bottom of the slot about 25m from the dryfall. We rigged about 25m of webbing and did a rappel of about 13m.*

*We found something amazing in this **2nd open section**: an Anasazi granary [The only place the Anasazi could have gotten into this open section was downcanyon about 80m from the granary on the left or south. At one time there was windblown sand and fallen rocks from the wall at that point, but a big storm must have sent water over the ledge which washed away much of the debris. That has since left a 13m rappel 11m back from a cedar tree--a 24m rope is needed to rappel into that open section]. We headed downcanyon but were confronted with a 200m-long **pothole section**. We did some hard climbing around the first couple [or just get in and swim], swam through the rest, with a number of difficult climbs out and one pack toss out of a KPH. We finally arrived at the last **2 potholes and the big drop**. There was a downclimb into one pothole, followed by a 7m drop into a **big water-filled pothole**. This turned out to be a difficult (5.8 or 5.9) downclimb, wide-stemming, shoulders vs toes, into the pothole [or slide into it if water is deep enough?]. Just around the corner to the right from this big pothole we found a small perfectly placed **natural arch** to anchor the final **big rap [R2]**. The rappel was about 25m, but there is a hanging **KPH** halfway down. We tossed the rappel rope with a pack on the end so we could pull ourselves out of this pothole.*

*We got back to the car after about 12 1/2 hours; 9 hours in the slots. We were totally whooped. We both thought it was a great slot, the best on King Mesa. A little harder than upper Pintac [East Fork, King Mesa], though we were better prepared this time, and had cooler weather (temps in the 50's F) and more water. Because of all of the potholes, LB has more varied problems. The high-stemming was also more varied and perhaps more continuously difficult. We cut it a bit close in terms of time, doing it so late in the year, but it payed off with cooler weather. Before going, I found some **Asic wrestling knee pads**. These were awesome. I put them on at the beginning and never had to adjust them in the slot!*

By the sound of Nat's description, LB might be the most difficult slot in this book! If you try it, here's some recommendations: Do it in **2 days**; do the **1st narrows** and the **pothole section** on **Day 1**--this means rappelling into the **2nd open section** from the south side then doing the potholes & big rap. Before doing this last pothole part, get a good look at it from a ledge directly above (see foto). On **Day 2**, do the **2nd narrows** and jumar up from a rope placed beforehand and out from the **2nd open section**. If potholes have lots of water, do it with body armor, a wetsuit. Late April, early May might be best time to hike this because of longer days (it's recommended you take a wetsuit even in summer, but if potholes are dry or with little water, stash the wetsuit to be picked up later; a wetsuit is too hot in summer unless you're swimming a lot). Checkout the pothole water situation from the rim at the bottom end in 2 places before committing.

Middle Fork From where you enter Coyote Gulch, walk east downcanyon past a fence & gate to a point about **125m** beyond the mouth of **Sleep Hollow**. The walls are low at that point, marked 1265m on this map, so climb left or north up a little ridge. Once on top, route-find northeast to where you can overlook the lower end of Middle Fork (Use this route to do Long Branch as well. Or enter lower Sleepy Hollow and exit up a wall on the east side directly east of elevation 4500/1372m). Walk along the rim for an occasional look into the slot. Not far beyond a possible exit is a short section of **potholes & short drops**. Since this is a long & difficult canyon, most people are rappelling into the slot just below those potholes. **Rap 17m** from a cedar (juniper) tree in a parallel crack to the general flow of the other nearby canyons.

Immediately after R1, the canyon tightens up and you'll have to start high-stemming up to about 10m or so. This goes on for about **1 1/2 hours**, then you'll come to where the right side wall gives way and an expert climber can **escape**. Just below that area, we found several large scallops or cups in the canyon wall--and had lunch. From there it's about another **1 1/2 hours** of high-stemming, then a little walking to **R2**. You could rap from an Ibis hook here, or from a **deadman** placed in a crack and down about **7m**. Then more medium-high-stemming with intermittent short walks down to **R3**. This will likely be from a long piece of W&R/QL tied to a little arch in the drainage above. R3 is down about **14m** and over 2 drops and into 2 potholes; the second one will likely have some water. In this last section are a couple of **partner assist DC's**. A short distance beyond R3 will be **R4**, the last rap in the canyon. R4 is from a pothole, and getting into that requires an off-width move down a crack, then a short hop. On our trip, there was a small softball sized chokestone in a crack with W&R/QL attached. It won't be there forever, so your choices are; have the lead person shout back for someone above to bring a cobble stone down as there is no building material in that last pothole; or just place a **B&H** for both convenience & **safety**. This last rap is down a tight crack to a ledge, then past an overhang of about **12m**. At that point you'll be out of the slot and into the **lower canyon**.

From the last rap there is running water all the way down to Coyote Gulch. However, this apparently was never grazed by cattle, because it's full of water loving plants, namely joint grass, sprinkled with occasional poison ivy. Take a pair of long pants for this. It took us 50 minutes to walk about 850m to Coyote Gulch. This is one reason no one does this canyon a second time.

East Fork Using the *King Mesa 7 1/2' quad*, walk down Coyote Gulch to the mouth of East Fork and turn left or north and checkout the lower 600m or so; then backtrack up Coyote about 500m above where East Fork enters. Climb north up steep slickrock to the bench above, turn northeast and rimwalk the west side for looks down in. Higher up, you'll see where the slot ends marking the beginning of the lower walk-through canyon. About 600m above that you can enter what we'll call the **Lower X Slot**. However, there's another section above that starting on King Mesa, but Nat Smale insists that was perhaps the most difficult climbing he has ever done. Most people will be perfectly happy to do just the lower part of this drainage--which by itself is still an **X SLOT!**

From where you enter, you'll be high-stemming right from the git-go. You'll also find lots of **moss and lichens on the walls**, so make every move & foot hold count. It's not that difficult, but don't become complacent! The high-stemming here will also get up to about 10m but it only lasts for about 1 1/4 hours, then you move down to the open bottom where a large boulder is situated. After that, and for about 300m or more, you'll find large boulders in the middle of the slickrock drainage, but with a **deep undercut on the left**. This is really a pretty & impressive canyon. There will also be small amounts of good **drinking water** trickling along in this part. No animals here except birds, so it's safe to drink.

A little lower you'll come to a dropoff; wade through the brush on the left and scramble/walk down-- watch out for poison ivy. After that will be **Rappel 1**. Look to the right or west about 20m to find a nice solid **deadman anchor** (we built it out of the way of flood waters, so you know it's good!) with W&R/QL stretching out to the start of a low angle drop of about **21m**. There is sometimes water in the section below. About 400m beyond R1 is **R2**. Here you'll (maybe?) find a small tree with 15m of W&R/QL stretching out to just above the last drop into the lower canyon. This rap will be about 20m landing next to a pool. Hikers come up to this point. Below that is a trail down to Coyote Gulch (there's more good water in this short section as well). Retrace your steps back to the trailhead. Of the 3 main King Mesa Slots, this one is the shortest, but it has the longest approach.

Elevations Trailhead, 1437m; entry to Sleepy Hollow slot, 1414m; end of Sleepy Hollow, 1265m.

Time Needed HH alone, maybe 5-8 hours; but more if KPH's are low on water. **Raven** alone, 5-12 hours, but you can break it down into various parts--not necessarily doing the entire length. If doing the **Upper Slot** of **Sleep Hollow**, perhaps 9-11 (?), but if you enter via **Pothole Fork**, 8-10 hours. **Long Branch**, 12-15 hours; or a couple of 7-9 hour days. **Middle Fork**, 9-11 hours. **East Fork**, 6-8 hours.

Water In potholes in the slots, then running water in or near the lower end of each canyon.

Maps USGS or BLM maps Escalante & Smoky Mountain (1:100,000); King Mesa (124,000--7 1/2' quad) for hiking; and the plastic Trails Illustrated/National Geographic map, Canyons of the Escalante (1:70,500) for driving & orientation.

Flash Flood Danger High throughout the slots, but these drainages are short, so a big storm would have to dump on the south side of King Mesa for there to be a flood. The lower canyons are safe.

Best Time to Hike Spring or fall, but because of the potholes and likely swims in Fox, Sleepy Hollow & Long Branch, better have a wetsuit in cooler weather. Wetsuits are good in X SLOT's, because you'll loose less skin! But if wearing the full length models, you have to hike in cooler weather.

Author's Experience After some scouting trips, and along with Jon Jasper & Tim Barnhart in 2004, we entered the **Upper Slot** of **Sleep Hollow** after 1 1/3 hours. We went to just below the big boulder and gave up--then exited and went down **Pothole Fork**, then downcanyon. By the time we put our headlamps on, there was thunder overhead. We tore through the 100m dark slot in record time, but the rest of the trip was a joy. Round-trip took us 12 hours, but we were scouting & wandering around. Later, in 4/2008, the author scouted, and did short sections of **HH & Raven**, and explored **Wide-Open Slot Valley** along with a couple of other nearby drainages.

In 9/2006, with Jon Jasper & Ryan Cornia, we first did the **East Fork** in 6 1/4 hours round-trip. The next day, we did the **Middle Fork** in just under 9 hours. In 10/2007, Nat Smale & Jeff Webb did **Long Branch** in 12 1/2 hours. In 4/2008, the author scouted the entire length of Long Branch.

Getting a cool drink just as the **East Fork of King Mesa** begins to widen.

233

Above Left High-stemming in the **East Fork** of the **King Mesa Slots**. Depending on your style of chimneying and the width of the slot, a foam pad of some kind around your rear end & lower back saves a lot abrasion in these kinds of X SLOT's. (Ryan Cornia foto)

Above Right Jon Jasper high-stemming in the **East Fork** of the **King Mesa Slots**. Avoid the mossy places if you can--it can be slippery!

Right Doing Rappel 2 in the **Middle Fork** of the **King Mesa Slots.** (Ryan Cornia foto)

Above Left Ryan & Jon high-stemming in the **Middle Fork** of the **King Mesa Slots.**

Above Right Jon in the **Middle Fork** of the **King Mesa Slots.**

Left Tim Barnhart starting to rap off an Ibis hook in the **Pothole Fork** of **Upper Sleepy Hollow.**

Above Left Jon Jasper starting Rappel 3 near the beginning of the **Dark Slot in Upper Sleepy Hollow**. **Above Right** Starting into the **Dark Slot** where headlamps are required. We quick-timed-it through this part because there was thunder overhead and rain was falling!

This is the last rap & potholes in **Long Branch**. Look closely and you can see some W&R/QL tied to a small natural arch just to the left of the biggest pothole in the middle of this foto. From that rappel station, throw a pack over the hanging KPH to the bottom. That way you rap into the KPH and pull yourself out with your rappelling rope. Then finish the rappel.

Above Left This is the Anasazi (?) granary located in the **2nd Open Section of Long Branch**. The only way to this site is to come down the canyon and make the 1st rappel seen in the background; or rappel in from the south side from a cedar tree--which the author did.

Above Right These potholes in **Long Branch** are located just above the granary shown on the left. Look closely, and you might see W&R/QL wrapped around the bridge in the middle of the foto and just above the biggest pothole.

Left If going to **Raven** or other nearby slots draining into **Coyote Gulch**, you'll pass this big impressive undercut on the way. This is the most-easterly of the 3 shown on the map.

North Forks of Fortymile Gulch & Willow Creek, Escalante River Drainage, Utah

Location & Access The canyons on this map are located south of **Fortymile Ridge & Road**, and just north of (and drain into) **Fortymile Gulch** and **Willow Gulch & Creek**. The author calls these (from west to east) **Little Fortymile Gulch (LF)**, **North Fork Fortymile Gulch (NFF)**, **Short Fork (SF)**, **North Fork Willow Creek (NFWC)**, and finally **Bishop Canyon (BC)**.

To get there, drive southeast out of Escalante on Highway 12. Between **mile posts 64 & 65**, turn southeast onto the **Hole-in-the-Rock Road (HRR)**. Drive **59.4 kms (36.9 miles)** and turn left or east onto **Fortymile Ridge Road (FRR)**. Continue east. At **Km 7.1/Mile 4.4** is a **metal water tank** on a hill to the left or north; at **Km 8.4/Mile 5.2** is a **bend (Bend TH)** in the track--park/camp there to reach NFF, SF, NFWC & perhaps LF & BC. Or from the **bend**, head northeast (this part is very sandy in places and is best to have a 4WD/AWD) to **Km 11.1/Mile 6.9** and the **Fortymile Ridge Trailhead (FRTH)**; this is where most people park if going to Coyote Gulch (and from the metal water tank). The FRTH is one possible starting point if going to **BC**.

If going to **LF**, continue down the HRR to **Km 65/Mile 40.4** and park at the **Carcass Wash Boy Scout Monument**; or drive a little further and park on **Sooner Wash** at **Km 66.8/Mile 41.5**.

Rating LF: **2A II**; NFF: **2A** (maybe **3A** for some?) **III PG SLOT**; SF: **3A III +PG SLOT**; East Fork/NFWC: **3A III X SLOT** (if doing 1 of 3 upper forks and jumarring back up), or **IV** if going down one slot, pulling ropes and walking out the bottom end; **BC's Upper Slot: 4A IV XX SLOT**. Or bypass the XX SLOT and do 2 long raps to the beginning of **Lower BC** and walk out the bottom: **3A IV PG SLOT** with a boat pickup (or if the lake is really low, you could walk/wade/swim (?), then walk up Willow Creek).

Equipment LF: K&EP; NFF: K&EP & 10m PR; SF: RG, K&EP, 60m rope and W&R/QL (maybe bolt kit?); **NFWC**: to rap down each fork, then jumar back up, read the description below for each fork and the required ropes; **BC XX SLOT**: KE&AP, ascenders, headlamp, and a pre-hung 50m rope to jumar up from the end of the XX SLOT. If bypassing the XX SLOT, and heading straight for **Lower Bishop**: you'll need one 50 or 60m rope & pull cord to do 2 long raps (and a boat pickup at the end--or ascenders to return up the same way) and W&R/QL. Or, leave two 50's in place and jumar back up.

Route Little Fortymile To reach LF, walk east downcanyon from the Boy Scout monument on Carcass Wash (or Sooner Water). Bypass one drop on the right and enter **Fortymile Gulch (FG)**. After about 600m (see map), leave FG on the left or north, and walk northeast another 900m to enter the upper end of LF. There's easy DC'ing and shallow slots in 2 upper forks--a good place for beginners to start. Walk down through the canyon to FG, turn right and return. Or to do 2 in one day, start down to FG and immediately turn left or north to exit at one of 2 **escape routes** shown. Once out, walk north to the upper end of NFFG and descend it returning to the Boy Scout monument (or Sooner Water) via FG.

North Fork Fortymile Gulch Get to the head of this canyon by first going through LF as just described; or forget LF, and head directly there from the monument in **Carcass**, or **Sooner Wash**. You can also get to this canyon from the north and the **Bend TH** on the **FRR**--the distance from either TH is about the same. From the Bend TH, walk south on a **very sandy track** (it can only be driven in winter when the sand is wet/frozen); after 800m is a remote met. station & fence, after another 400m a T junction. From there head southeast just to the right of a low butte labeled **4687** (1429m); then SSE to the head of the canyon.

Once there, you'll find easy but steep DC'ing in the 75m-long **Upper Slot**, then an open drainage, and the **Lower Slot**. If you're going all the way through the Lower Slot & canyon, then slide/DC in at the top--if returning, tie a 10m rope to a large bush, so you can HL out (a partner assist should eliminate the need for a rope). At the very end of this shallow Lower Slot, there's a **9m DC**, then another 5m of easy DC'ing that might jack the rating up to PG SLOT. Your group leader could belay beginners, then DC this last place. Once in the lower canyon, you'll find trees and maybe water. Halfway down are old worn out **moki steps** going up to the west and an exit; or walk out the bottom and back to your vehicle via the escape route shown, or up FG to your starting trailhead there.

Short Fork Best done from the **Bend TH** on Fortymile Ridge. Follow the **very sandy track** south & east to the **winter water trough**. From there, route-find south along the west side of the **Carmel Mesa** and the little butte marked **4526** (1379m). Continue south into the main drainage of Short Fork. Eventually you'll find a PG SLOT about 400m long with lots of low-stemming. Just before the end is a **3m rap**; we did it with an **Ibis hook** placed in a **natural hole**. Then comes an open section and a 75m-long sculptured potholed slot that may rate an **R SLOT**, because of the best move (?). At the end of that is a big dryfall; if you want to continue down to FG, you'll likely have to install 2 B&H and do a rappel of around 50m (?), but take a 60m rope & pull cord to be sure. Most people would want to return the same way.

From the bottom of the big rap, walk 50m, and look for (or build) a deadman anchor for a 15m (?) rappel down to the bottom of FG and water. If you've left ropes, jumar back up; if you've pulled your ropes, walk upstream to the mouth of NFFG, exit there (or at the moki steps) and head north back to your car at the Bend TH.

North Forks Willow Creek Get to the upper end via the **Bend TH**. Walk south on the very sandy track to a winter watering trough & feed ground. The vehicle track ends near the top of the Carmel Mesa (covered with cheat grass) marked **4664** (1422m). From there, walk up along a cow trail to the south, then route-find southeast into the upper forks of NFWC.

West Fork Right at the top is a chokestone. If going into the lower canyon and jumarring back up, tie one 60m rope there and make **Rappel 1** with a short walk in between 2 parts. Walk 60m to an easy DC & maybe pool, then tie a rope to a chokestone for the **23m R2**. Walk another 100m, then the **10m R3**--but there's nothing to rap from; seems logical to install **1 B&H** to one side. From there, which is at the junction with the **Middle Fork**, walk downcanyon 450m to the last drop & **R4**. Rig up W&R/QL on the right or west side; or if returning the same way, tie a rope to a boulder for a **53m rap**. Part of that is a free-hanging rappel, so take precautions for what could be a fast ride!

Once into the lower canyon, you can jumar back out the same way, or continue downcanyon to **Willow Creek**. This will be a fairly long walk and in places involves a little bushwhacking in the lower end of **NFWC** because of running water & water loving plants. Once at Willow Creek, head west up to **Broken Bow Arch** and the **Willow Bench TH**; or up **Fortymile Gulch** to a second car; or exit **NFF** and return north cross-country to the **Bend TH**. Long walk! Best to jumar back out.

Middle Fork Only 250m from the start of the West Fork, is the beginning of what this writer is calling Middle Fork. There are several small boulders you can tie a rope to; or install W&R/QL and pull your

Map 51, North Forks of Fortymile Gulch & Willow Creek, Escalante River Drainage, Utah

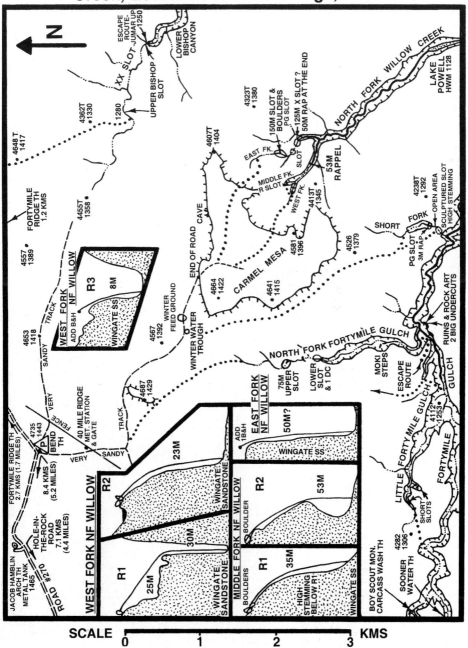

rope. About 30m of rope will get you down to a leveler place just before a moderately steep **R SLOT**. At the bottom of that 60m section you'll walk another 50m to some bushes which are at the beginning of another DC in a nice slot. At the end of that is the junction with the **West Fork** and its **R3**. The rest has been discussed.

The **East Fork** first has a PG SLOT, then a wall collapse & boulders in the middle where you'll need help from a friend and/or Ibis hook to make an **off-width DC move**. After that is a nice slot then open country, another short PG SHOT, and finally a section that must be an R or X SLOT (?). The author was alone and it looked like another X SLOT, so he turned back at the head of a 75m-long slot. What is known now is, right at the end is a rappel of what appears to be 40-45m, but take a 60m rope & pull cord to be sure. Better take a BK and 2 B&H's too. Return the same way jumarring up any fixed ropes; or take a long hike down NFWC and up Willow Creek to the HRR.

Bishop Canyon To get started from the **FRTH**, set your compass somewhere between southeast and ESE, and start walking. First aim for the high point marked **4668, then 4645T and 4648** (1417m). From that area continue south along a ridge of sorts to **4362T** (1330m). The **upper XX SLOT** begins about 550m south of that altitude mark. If you have a car, you can also get there from the **Bend TH**. From there, walk east on a very sandy track heading for **4455T** (1358m). From there continue southeast to the **Upper XX SLOT**.

This hike has several options. First, you can go down the **Upper Slot** which is rated **XX SLOT** (the author still hasn't done it, but has info from others). To do this, first with a **50m rope**, lower a pack or two which should include ascenders, water and other stuff, from the B&H's on the rim at the lower end of the XX SLOT. To get out from the end of this slot, you'll have to jumar up and out of the canyon.

Now the Upper XX SLOT. At first it's shallow, then it deepens & tightens and you'll have to do high-stemming. One reliable source says, there are 3 parts to this slot; *an upper R-SLOT, a middle X-SLOT, and a more difficult XX SLOT where everyone needs a headlamp and all the body armor possible. This XX SLOT has a long series of silos that we had to go down part way into and then climb (hard) back up, all in the 60-100 foot [18 to 30m] exposure range, 7 hours of stemming. Not for the faint of heart. A rim crew is highly recommended.*

Or if the XX SLOT is too big for you--and it will be for most--you can rim-walk to the lower end of the Upper Slot, **rappel 40m**, then scramble & DC a short steep 50-75m **R SLOT,** tie-off a second rope and rappel **40m** to the bottom. There you'll find running water in an **unusual 270° circular alcove** which marks the beginning of Bishop Canyon. If you're planning to return the same way by jumarring up, protect your rope against one sharp edge--yours truly nicked a brand new 60m rope by not doing so.

Here's another option if the lake is real low. Install W&R/QL & make the 2 raps pulling your ropes, then walk down Lower BC and swim/wade/walk up Willow Creek & Gulch--best to have a boat waiting, but that makes it a logistical nightmare! Check the level of the lake ahead of time at the visitor center (Tele. 435-826-5499) in Escalante.

Elevations Trailheads, 1329m, 1317m, 1280m, 1443m & 1426m; beginning of the XX SLOT of BC, 1280m; start of first rappel, about 1250m; end of raps in BC, 1165m; HWM Lake Powell, 1128m.

Time Needed From the monument in Carcass, through **LF & NFFG** and back, about 5-7 hours; descending one fork in upper **NFWC** and jumarring out and back to the Bend TH, 7-8 hours?; the upper XX SLOT of **BC**, about 11-13 hours from either trailhead.

Water Take your own. Water flows in lower NFFG (?), FG, WC, NFWC & BC--but none in the slots.

Maps USGS or BLM maps Escalante, Smoky Mountain & Navajo Mountain (1:100,000); Sooner Bench, King Mesa, Stevens Canyon South & Davis Gulch (1:24,000--7 1/2' quads) for hiking; and the plastic Trails Illustrated/National Geographic map, Canyons of the Escalante (1:70,500) for driving.

Flash Flood Danger High in any slot, low in the lower canyons.

Best Time to Hike Spring or fall, but if in the shade of a slot, it's not too hot even in summer (but you'll need lots of water then!). Or during some warm spells in winter.

Author's Experience In **BC**, he and Jon Jasper & Tim Barnhart rapped down all the way into the lower canyon, then ascended and went into the lower end of the XX SLOT about 40m and stopped. Round-trip, 7 3/4 hours. The author explored **LF & NFFG**, and the upper ends of **NFWC** in 9 1/2 hours from Carcass. He probed **Short Fork** from the Bend TH in 4 1/4 hours. A scout trip from the Bend TH to **NFWC's upper forks** took 5 3/4 hour; another scout trip into the **West**, then **Middle Forks** of NFWC from the Bend TH, took 8 1/4 hours. He descended **NFFG** solo from the Bend TH, and explored all exits in 5 hours.

In 4/2008, along with Jon Jasper & Tim Barnhart, we went from the Bend TH down the **Middle Fork of NFWC** to the 53m rappel, then walked to the bottom end of the **East Fork** for a close look at the last rap there, and returned. In the afternoon, we walked to **Short Fork** and the top of the big dryfall and returned; 9 1/4 hours for the day.

Jumarring up the last stage at the head of the **Middle Fork** of upper **North Fork of Willow Creek**. One way to see just the upper end of North Fork of Willow Creek is to rappel in, then jumar back out instead of walking all the way around.

240

Above Left This is the end of the short PG SLOT section of the **North Fork of Fortymile Gulch**. In it is a 9m vertical DC, then another 5m of easy DC'ing.

Above Right Near the end of the PG Slot in the upper part of what this writer calls **Short Fork**. (Tim Barnhart foto)

Left Jon Jasper observing the very end of the best part of **Short Fork** as it makes a big drop to the shelf below. We didn't do this rappel, but it seems to be at least 40m, or more. If you try it, go prepared. You may have to install 2 B&H's.

Above Left This is the big 53m Rappel in the West Fork of upper **North Fork of Willow Creek**. Tim Barnhart made this rappel, then jumarred back up rather than taking the long way around.

Above Right The author jumars up one of the 2 raps in the upper **West Fork of Middle Fork of Willow Creek.**

Right Tim and Jon Jasper have just rappelled to the bottom of upper **Bishop Canyon**, and are in the process of UC'ing back. Notice several B&H's in this steep slot. We didn't think the B&H were necessary, but strung a rope to them anyway.

Above Left Jon waits at the beginning of the last rap into **Lower Bishop Canyon**. The anchor for this rappel is about 5 B&H's!

Above Right Tim coming down to the beginning of the last rappel into **Lower Bishop Canyon.** He is more or less DC'ing, but having a rope handy makes it a bit safer.

Left Jon starting the last rappel into **Lower Bishop Canyon**. Below is good drinking water, and a nice hike down to Lake Powell.

Davis Gulch, Escalante River Drainage, Utah

Location & Access **Davis Gulch** is located in the lower end of the Escalante River drainage and near the end of the Hole-in-the-Rock Road. Drive southeast out of Escalante on Highway 12. Between mile posts 64 & 65 veer right or south onto the **Hole-in-the-Rock Road**. Drive southeast on the HRR about 83.4 kms (51.8 miles). This will be the Davis Gulch Trailhead. Nearby landmarks are: 1 km back up the road is Fiftymile Point; and 300m back will be a **fence & cattle guard**. The HRR is very good to as far as Egypt Bench turnoff (27 kms/16.8 miles) and the county line, then it slowly deteriorates with a mostly washboard surface the further you drive. After floods, and before the road graders arrive, you'll need a HCV for sure. The worst affected places are where the dry washes cross the HRR.

Rating For some: **2B III -PG SLOT**; but perhaps: **3B III PG SLOT** for beginners. The slot could well rate an **A** after a long dry spell, but has running water in the lower canyon. It has 6 pretty good DC's, but some of these may be rappels for beginners; or some people may want to be belayed down 2 or 3 drops.

Equipment K&EP & one 15m rope; or ascenders & 4 short ropes if returning the same way via the slot.

Route From the road, walk northward right down the main drainage. In about 200m you'll enter the **Upper Slot** which has 2 dropoffs created by chokestones. The group leader may have to help less-experienced hikers down with the rope, but most can chimney down easily. After about 200m of slot, the canyon opens for about 400m and you can walk out on either side.

Next comes the **Lower Slot** covering a distance of about 800m. This one gets rather narrow & fotogenic and there are 4 more dropoffs, all ranging from 3-5m. You may find a small pool at the bottom of each. Finally, you'll slide down the last dropoff (6th) and into a pothole with cobblestones, then the canyon opens up. After another 500m or so, look left and up along a natural ramp you'll see some **paleface-made moki steps** leading up to the rim. It looks climbable, but a little exposed!

Further along you'll find an 85m-deep undercut, a year-round stream, cottonwood trees, **Bement Arch**, beaver dams, a rock art panel, and the **Davis Stock Trail** with steps cut in slickrock; your only way out of the lower canyon. To avoid bushwhacking below Bement Arch, stay on hiker's trails on either side of the willow bottoms. Right at the bottom of the Davis Stock Trail, you may still see some logs & small corral; nearby on the south-facing wall are **petroglyphs**.

To return, walk up the Davis Stock Trail to the rim, then walk parallel to the canyon back to your car.

Elevations Trailhead, about 1300m; Lake Powell's high water mark (HWM), 1128m.

Time Needed It's only 7-8 kms down to the Davis Stock Trail (which was chopped out of the slickrock with a pick), but it's slow going in the slot and below the arch, so count on an all-day hike; 6-9 hours round-trip.

Water Have plenty in your car and in your pack. There are beaver in Davis, so take water directly from a spring source high in the canyon, or treat all water nearer the lake.

Maps USGS or BLM maps Escalante, Smoky Mountain & Navajo Mountain (1:100,000) for driving; Davis Gulch (1:24,000--7 1/2' quad) for hiking; or the plastic Trails Illustrated/National Geographic map, Canyons of the Escalante (1:70,500) which is best for driving.

Flash Flood Danger High in both slots, but no danger in the rest of the canyon.

Best Time to Hike Spring or fall. Summer is OK in the canyon, but a hot walk back along the rim.

Author's Experience He has hiked up Davis from the lake several times, then in 1998 soloed the entire canyon from top to bottom returning via the stock trail. Round-trip, 6 2/3 hours. On 10/17/2002, he took 4 short ropes and left one at each of the last 4 dropoffs. He walked down to the moki step and returned using ascenders. Round-trip 3 3/4 hours.

This is the bottom of the 5th rappel, DC or HL in **Davis Gulch**. If you're going all the way through and come out the old Davis Stock Trail, then slide down all of these short dropoffs. If returning back up the **Lower Slot**, leave a rope at each upclimb.

Map 52, Davis Gulch, Escalante River Drainage, Utah

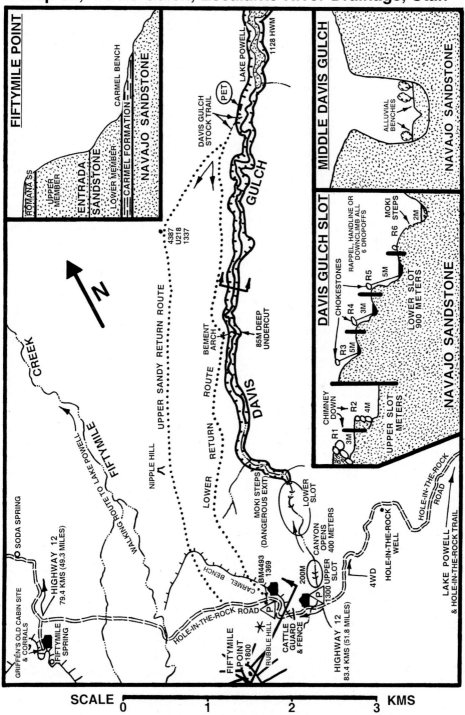

245

Stone Donkey Canyon, Paria River Country, Utah

Location & Access **Stone Donkey** (unnamed on USGS maps, but a nearby pinnacle has the shape of a donkey--thus the name) is an upper west-side tributary to **Hackberry Canyon**, which in turn is a tributary to the upper Paria River. Stone Donkey Canyon is located southeast of Bryce Canyon National Park, SSE of Cannonville, and almost directly south of Kodachrome Basin (KB) State Park. It's also in between the Upper Paria to the west, and **Hackberry Canyon** to the east, as well as due north of the **old Pahreah** ghost town.

To get there from the south and Highway 89, drive to between mile posts 17 & 18, and turn north onto the **Cottonwood Canyon/Wash Road**. After driving 61 kms (38.3 miles) on a well-maintained, but dirt & clay summertime-only road, you'll come to the turnoff to KB. Now, if you're coming from the north and Bryce Canyon, which is the recommended approach route, drive south through Tropic to Cannonville, then turn south onto the paved KB Road. From Cannonville to the turnoff to KB is roughly 11.4 kms (7.1 miles).

From where the paved KB Road turns north toward the state park (Km & Mile 0), drive east on the Cottonwood Canyon Road running toward Grosvenor Arch & Highway 89. After 1.3 km (.8 mile) you'll come to a low hill called Watson Ridge, and a road running SSW along the top. At Km 3.7/Mile 2.3 from the **KB Turnoff**, you'll cross **Rock Springs Creek**, then **Wallace Ott's Corral** after another 160m. From there continue SSE up a minor drainage. Near the top of the canyon is a road junction (Km 5.5/Mile 3.4). Straight ahead is a steep rough section, and is fit for 4WD/HCV's only; to the right is a better road--but you should have a 4WD. Best to turn right, climb the **steep hill** then once on top of **Rock Springs Bench**, pass through a **gate**, and make 2 left turns and be heading in a northeast direction. Finally you'll turn south at Km 8.9/Mile 5.5 on to a less-used ungraded road. Follow this deeply rutted track almost due south to where it ends at **2073m** altitude. This will be **Km14.3/Mile 8.9** from the **KB Turnoff**.

The author once got to the end of this track in his old VW Rabbit & newer Golf, but since then, heavy rains have made the road even worse and now you'll need some kind of HCV, or better still, a 4WD/HCV.

An alternate to driving this complicated route & bad road, is to drive north from Highway 89 on the **Cottonwood Wash Road 23 kms (14.2 miles)** and park opposite the mouth of **Hackberry Canyon**.

Rating 3B III PG SLOT. After a long hike you come to a tight slot with one or more rappels totaling about 40m, then steep & tight DC'ing in a 250m-long slot where you'll need a headlamp in one section.

Equipment RG, one 60m & a 15m rope, K&EP, W&R/QL, PR, one headlamp per/person, and of course a compass & map, ascenders, etc.--as always!

Route From the end of the **Rock Creek Bench Road**, and with the two 7 1/2' quads listed below in hand, route-find due south about 1 km to the southern tip of Rock Springs Point as shown. About 200m before the end of the point, look for a cairn on the left (east) marking a trail-of-sorts angling down over the rim heading southeast. This is **Ott's old horse trail** but is only visible in the upper part; further along, head straight down the sandy slope. Consider caching a bottle of water at the bottom of this steep slope for the return hike.

At the bottom of the big cliff & sandy slope, and with map & compass in hand, route-find SSE. After about 2 kms, you'll pass west of the peak labeled here **Little Mollies Nipple at 6825** (2080m) and immediately under the elevation marked **6444** (1964m) on the *Slickrock Bench quad*. Continue south along a hogsback ridge, but eventually veer southeast and drop down into the upper part of Stone Donkey. If you can't read a topo map, you're in the wrong place!

Once into upper Stone Donkey, walk about 4 kms with scattered ponderosa pines until you're almost exactly due south of Cottonwood Peak marked **2022m** (6634). When you reach the dropoff, you'll have 3 choices of how to get started. **(1)** Tie a 15m rope to a small bush, backed up with an Ibis hook under a small ledge, then a 60m rope to the end of that. From there you can HL or rap down over 2 dropoffs, then rappel over a chokestone and down another 20-25m to a walk-out pool. **(2)** You can help each other down into the slot, set up W&R/QL on one of several chokestones, then rappel down to the pool just mentioned. This way you'd be pulling your ropes and dragging then down the remainder of a tight slot.

(3) Or, look southeast from the head of the slot to a cedar (juniper) tree which is about 30m from the edge of the slot. From it you can rappel in immediately below the pool mentioned above. To do that, first tie a short rope of about 15m to the tree, then attach the 60. Rap in from there, but tether your pack to the front of your harness because you'll have about 30m of free rappel to reach the bottom. Better leave leather gloves too, or if your hands gets too hot, pull the rope around your back to your other hand to cause more friction. Leave the ropes in place to be picked up on the way back to your vehicle.

Once at the bottom and just beyond the pool, remove your rappelling harness and tie it to the rope to be hauled up later. From there, you'll have a short walk, then comes several interesting DC's in a tight slot. This is where you'll want the smallest pack possible. Later, you'll come to a **dark slot** where a **headlamp** is required. This goes on for about 35-40m, then deep wading or a short swim. A short distance below that, the canyon opens quickly. This entire slot is only about 250m long, but it's a good one.

About 1 km below the slot, turn left and exit the canyon at one of 2-3 places heading north. Once out, route-find northwest between the slot and Cottonwood Peak back to your ropes and return the same way.

If coming up **Hackberry**, some may want to backpack to the mouth of Stone Donkey, then walk 400m up Stone Donkey to a spring. That's your waterhole. Camp there. Getting to that point should take about half a day. From the spring, continue upcanyon just over 1 km, route-find to the upper end of the slot and head downcanyon taking your ropes with you. Or, if you're a strong hiker, take a light pack and do it on a long day-hike. Or, if you skip the rappels, you can UC (don't forget **K&EP!**) the slot to the pool mentioned above & see 95% of it, making it an easier all-day hike from either trailhead.

Elevations Crossing of Rock Springs Creek, 1725m; Rock Springs Bench Trailhead, 2073m; high point of hike on Rock Springs Point, 2109m; exit in Stone Donkey, 1627m; junction of Hackberry & Stone Donkey Canyons, 1455m.

Time Needed To insure success, camp at the Rock Springs Bench Trailhead, then take from 9-13 hours for the round-trip day-hike. If coming up Hackberry, you'll likely need 1 1/2 days for the trip; but if you camp at the mouth of Hackberry, get an early start, strong hikers can do it in 11-14 hours round-trip. Read below.

Water Take plenty in your car and pack. There's normally some water in potholes and it's always at the spring in lower Stone Donkey. Take purification tabs for pothole water in the valley above the slot.

Maps USGS or BLM map Smoky Mountain (1:100,000) for driving; and Calico Peak & Slickrock Bench (1:24,000--7 1/2' quads) for hiking.

Flash Flood Danger High risk in the short 250m slot, but no danger elsewhere.

Best Time To Hike Any time the roads are dry, but best between late March or early April, through October. If going in the spring or fall, better take a wet/drysuit, because any little shower puts water into the slot from the surrounding slickrock. Late spring/early summer is best because of longer days.

Author's Experience On this first trip, he car-camped at Ott's Corral, then after an hour of wandering on top of Rock Springs Bench, he finally found the right road and got to the trailhead. His hike to the technical slot

Map 53, Stone Donkey Canyon, Paria River Country, Utah

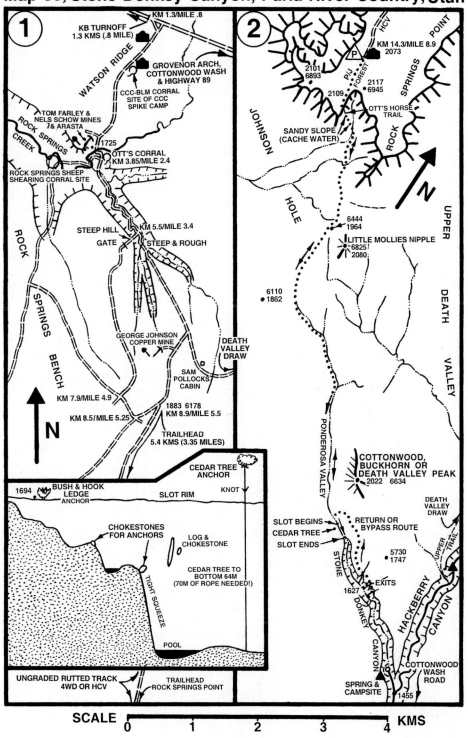

SCALE

0   1   2   3   4   KMS

247

(but not down it) and into the canyon below as far as the spring and back to his car took 8 hours.

On his second trip (9/23/2002), he carried a ton of ropes plus a wetsuit, then attempted to go down from the head of the slot. He attached a 60m rope to the bush, backed up by an Ibis hook, then HL'ed down to the last big drop above the pool. He wasn't sure if the rope reached the bottom, and it is a tight rappel; so being alone, he jumarred back up and went in from the south side as Scott Patterson & friends had done. After the slot, he retrieved his ropes (a 60 & 17m) and returned to his Chevy Tracker in 9 1/2 hours. He drank 4 liters of water on the hike which was an average day temperature-wise for September.

On a third trip (8/31/2003), he parked at the mouth of Hackberry, walked up canyon and UC'ed the slot to the pool. He returned via the Goulding Trail in 11 hours round-trip. See the author's other book, *Hiking and Exploring the Paria River* for better maps covering this area, including the Goulding Trail.

Right A pool in the lower part of the slot in **Stone Donkey Canyon**.

Below Left This is part of the **Lower Slot** just above Rappel 2 in **Deer Creek Canyon**.

Below Right Still in **Deer Creek Canyon**, Nat Smale has just made it down Rappel 2 from 1 B&H we installed in 10/2003. About 30m away is a nice DC into a pool or 2, then the canyon opens up into a wider and deep gorge.

Above Left Part of the nice slot in **Stone Donkey Canyon** just below the big rappel.

Above Right This picture was taken between the spring in lower **Stone Donkey, and Hackberry Canyon**. There is generally lots of good water in this section if you choose to backpack in to Stone Donkey's slot.

Left Nat Smale DC'ing in the canyon just below Rappel 1 in the **Upper Slot of Deer Creek Canyon**.

Deer Creek, Left Hand Fork & Oak C., Paria River Country, Utah

Location & Access Featured here is the **Main Fork of Deer Creek Canyon** and several tributaries including the **Left Hand Fork & Little Fork** (author's name), plus **Oak Canyon** which enters the middle part of the Upper Paria River Gorge just below Deer Creek and the CCC & Deer Trails Trails.

To get there, drive along Highway 89 about halfway between Page, Arizona & Kanab, Utah. Very near mile post 37, turn north onto the **Nipple Ranch Road**--this is the same road that runs toward Mollies Nipple and Kitchen Canyon. Drive north past Kitchen Point and the **CCC-built sheepman's supply depot** made of **rock** on the right (Km 9.1/Mile 5.65), then at **Five Pines Junction** (Km 16.4/Mile 10.2), and turn right or northeast and drive through a gate, *closing it behind you*. At Km 20.6/ Mile 12.8 is a pass between the Park Wash Drainage and Kitchen Canyon. From that point on you must have a 4WD with good clearance (if not, you'll have to park there and walk an extra 8 kms/5 miles). From the pass continue north downhill. At Km 24.5/Mile 15.2 is a turnoff to the right which takes you to Calvin C. Johnson's Nipple Ranch; but instead, continue north in deep sand across the head of upper Kitchen Canyon or Valley. In this area, you'll be in **deep sand** (drive fast!) for about 2 kms, then as you start to climb up toward **Oil Well Hill,** small rocks are mixed with sand making driving easier. At Km 28.6/Mile 17.8, you'll be on top of a ridge where the road turns east and runs out to an old oil well drill site. Stop at that curve where you see the very sandy ATV track heading northwest. Camp and/or park there.

Rating Main Fork of **Deer Creek: 3A (or B) III (or IV) PG SLOT; Left Hand Fork: 3A III; Little Fork: 3A II or III; Oak Canyon: 3A III.**

Equipment Main Fork of **Deer Creek**: RG, one 60m rope, K&EP, W&R/QL and a BK & 1 B&H just in case. If planning to rappel over all the dropoffs in the **Left Hand Fork** & **Little Fork of Deer Creek**, and **Oak Canyon**: RG, 60m rope & 60m pull cord, and W&R/QL. These last 3 canyons are not nearly as interesting as the Main Fork of Deer Creek, and none have slots.

Route Main Fork of Deer Creek From the 4WD parking place, walk northwest along a very sandy illegal ATV track. After 600m, and about 1 cm northeast of elevation **6139** (1879m) on the *Deer Range Point (DRP) 7 1/2' quad*, will be a **cow trail** heading due north downhill. This will be the end of the trail you'll use to get back to your car. But don't use it now; instead continue northwesterly on the sandy track toward DRP. Directly beneath this prominent peak, you'll cross a minor drainage, then veer northeast as the track fades. From there, continue north around the eastern base of DRP, then northwesterly on cow trails. When you're due north of the most-easterly peak on DRP, and at the head of the **West Fork of Left Hand Fork**, you'll come to many **toadstool or mushroom-like rocks**. These features all have a piece of weather-resistant iron-rich caprock on top which shelters softer Navajo Sandstone below leaving something that looks like a giant toadstool/mushroom or hoodoo. They're all over this region, but the best cluster seems to be north of DRP.

From the head of **Left Hand Fork,** continue west or northwest over a low divide near elevation mark **6273** (1912m) and route-find down into the head of the **Main Fork**. Walk north downcanyon. Soon you'll come to **Rappel 1**, a double drop of about 17m into a nice slot. Tie W&R/QL around a nearby chokestone and rappel; just below that will be an interesting **7m DC** in a vertical chute (see foto on page 249), then a short walk out of the **Upper Slot**. Below the upper slot are a couple of escape routes out to the east and several short shallow slots before coming to the **Lower Slot**. This is the good part. After about 400m of moderately cool dark slot, you'll come to **R2**, a drop of about **16m**. There you should find 1 B&H on the right. Rappel into a deep hole, then continue for another 30m to find a steep, **tight 8m DC** in the darkest part of the canyon. At the bottom is a pothole, possibly with water--no more than waist deep--and a 90° left turn, then a **2nd possible pool** (could be a swimmer?), and a 90° right turn. After that, walk out through a 100m-long, 2-3m-wide slot, after which it opens up into an impressive gorge.

Below R2, you'll walk along a sandy wash inside an open gorge. After 500m will be the **Northwest Fork** coming in on the left. Walk up this 200m to find another nice straight slot before coming to a 7m dryfall. There should be more slot & rappels above that, but getting to the upper end of that fork would be time consuming. One informant states there are lots of good toadstools just south of that slot.

In the 2 kms between the Northwest Fork and Rappel 3, you'll find one small escape exit to the south onto **Cad Bench,** and several exits to the north up to **Asay Bench** as shown on the map. Immediately above R3 will be a 40m-long slot which you can DC into. At the end, you could arrange a chokestone to rappel from, but it will likely be an awkward start for a 30m rappel.

Or, climb up on the bench to the right or south to find W&R/QL around 2 bushes & a small rock buttress. There you'll have an easy start for the **30m R3** right off the nose of a larger buttress. At the bottom will be springs, running water & trees. Walk downcanyon in or beside the small creek. In the lower end of the canyon are many campsites, **2 panels of pictographs** and a long panel of **petroglyphs,** as shown on the map. There's also a **deer & horse trail** out of lower Deer Creek up to Cad Bench which you could use to return to your car. Read more about it below under the **Left Hand Fork.**

Less than 1 km below where Deer Creek meets the Paria, pay close attention on the right side to find a cairn marking the bottom end of the constructed **Deer Trails Trail** zig zagging up the west side of the **Paria River Gorge** (this trail begins about 100m below the beginning of the **CCC Trail** on the east side

Main Fork Deer Creek Rappelling Section

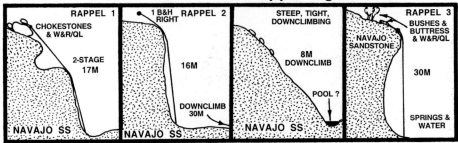

RAPPEL 1 — CHOKESTONES & W&R/QL — 2-STAGE 17M — NAVAJO SS

1 B&H RIGHT — RAPPEL 2 — 16M — DOWNCLIMB 30M — NAVAJO SS

STEEP, TIGHT, DOWNCLIMBING — 8M DOWNCLIMB — POOL ? — NAVAJO SS

RAPPEL 3 — NAVAJO SANDSTONE — BUSHES & BUTTRESS & W&R/QL — 30M — SPRINGS & WATER

Map 54, Deer Creek, Left Hand Fork & Oak Canyon, Paria River Country, Utah

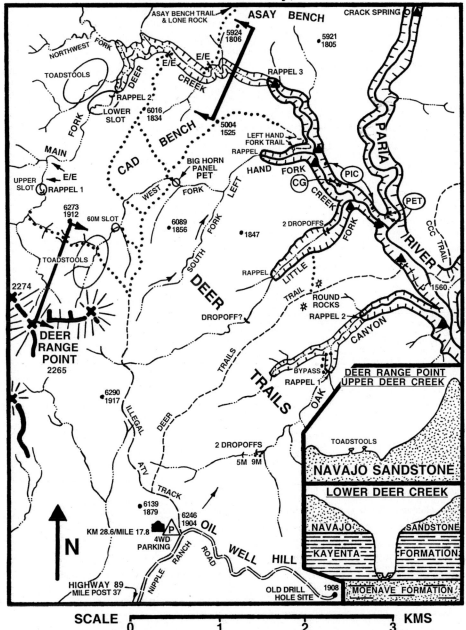

SCALE
0   1   2   3   KMS

of the canyon). After zig zagging up about 100m, this old cattle trail levels some, then heads southwest up a sandy hogsback ridge between **Oak Canyon** and **Little Fork**. Soon it passes between 2 prominent **round rocks** as shown on this & the 7 1/2' quad. This trail eventually meets the sandy ATV track near the main road as described above. From there it's 600m back to the main Nipple Ranch Road.

Left Hand Fork of Deer Creek This is the largest & longest of all the forks of Deer Creek except for the Main Fork. To get into it from the top, use the exact same route as if going to the Main Fork, but when you reach a point just east of DRP, follow that drainage (let's call it the **South Fork**) down to the north & northeast. Along the way are a number of dropoffs, all of which you can skirt around to the left or west.

You could rappel these, but there doesn't seem to be any slots of consequence, so rappelling seems pointless. When you reach the confluence of the **South & West Forks**, head up the West Fork skirting around 2 dropoffs. About 600m up the West Fork is one of the best petroglyph panels around. Wallace Ott of Tropic alerted the author to this. This **Big Horn Panel** is along the south side of a 2m-wide straight-sided slot. Most of the art is of big horn sheep which are 2-3m above the dry wash. Some markings are up as high as 5-6m. Those are much older than the lower ones.

You can also walk into the upper end of the **West Fork**, but after about 500m or so, you'll come to a 12m dropoff. The author still hasn't seen the top of this rappel, but has come up to it from below. Immediately below this rap is a nice tight 60m slot that's worth the effort. One full length rope and some W&R/QL to tie to a log should set up this rappel. About 1 km below that will be the Big Horn Panel.

Below the confluence of the West & South Forks, is one short slot a good climber can DC into, or skirt around it to the left and reenter just below. Finally, about 650m above Deer Creek, will be a big **55m** dropoff which you could rappel by using chokestones as an anchor (you should find W&R/QL there now). That would put you into the lower canyon with water & trees.

Or, get out of the drainage on the left or north, and route-find east toward the point between the Main Fork and Left Hand Fork of Deer Creek. As you approach the end of that point, be looking for signs of an old trail which was first used by deer. Let's call this the **Left Hand Fork Trail**. Look for some stone cairns. This trail zig zags southeast right off the point, then one bench above the floor of the main canyon, it turns west and follows the same bench along the Left Hand Fork for about 225m before heading down through oak brush to the creek below. Wallace Ott & Wallace Henderson modified this original deer trail so they could get their horses up or down--read all of Wallace Ott's stories in the author's *Paria River Guide*.

Little Fork This is the short canyon between the Left Hand Fork and Oak Creek. To do this canyon from the top end, start from the 4WD parking on Oil Well Hill, and walk down the **Deer Trails Trail** in the direction of the Paria. About 1 km above the 2 rounded rocks, turn left or north and work your way down into the upper end of the canyon. At the very beginning, you should come to at least 1 dropoff; you can either rig up a rappel--followed by some kind of slot or narrows--or skirt around this part on the right or left. About 750m below the first dropoff will be a big dryfall and rappel of up to 30-35m (?). The author hasn't done this, but has seen it from below. You may need two 50m ropes, or one 50 & pull cord.

Below this big dropoff is a dry & uninteresting V-shaped canyon. There are several possible exits, one of which the author used to escape south. One km below the big dropoff are **2 dropoffs** of 7-8m **each**, one after the other. You can DC the first, but it'll be tricky with crumbly-looking rocks on the right--so help each other. Or rap from boulders. Walk around the second drop on the left. From there it's 800m along a small stream to the mouth of the canyon and Deer Creek.

If you're in Deer Creek, you can walk up this short side-canyon and perhaps climb up the second dropoff on the left--it's tricky UC'ing, as part of that rock wall looks like it's ready to fall! Once in the middle part of the drainage, there should be several escape routes out to the north, and at least 1 exit to the south up steep slickrock as shown. That's the one the author used.

Oak Creek To do this canyon from top to bottom, walk north and a little east from the 4WD parking at **6246** (1904m). After about 1 km, you'll be in the main drainage. You'll soon come to a 5m drop; chimney down a crack on the left side. About 300m below that will be a 9m dropoff; skirt around this on the right and scramble to the bottom. After about 1500m of easy walking, you'll come to a big dryfall of about 20m. If you wish to rappel, you can install W&R/QL around some bushes or trees near the edge. If you don't want to rappel, then walk 200m along the left (northwest) rim and reenter the drainage just above a side-canyon. Below **R1** it's an easy walk of about 800m with at least 1 exit on the left or north, then **R2**, a drop of about **25m**. You can of course exit and head down the **Deer Trails Trail**, and walk up the lower end of Oak Canyon from the bottom. Or if rappelling, there are nearby trees or rocks you can attach W&R/QL to. At the bottom of this drop are springs, trees & running water all the way down to the Paria. Along the way are old signs that beaver have been there, but nothing new as of 2003. If you see fresh signs of beaver, drink water from the spring source only; or purify it. Oak Canyon doesn't have any slot or real narrow parts, so most people won't be interested in rappelling through this drainage. The best part of this green canyon is the lower end.

Elevations 4WD trailhead, 1904m; bottom of Deer Trails Trail along the Paria River, about 1560m.

Time Needed To rappel through the **Main Fork** and return via the **Deer Trails Trail**, will take from 9-12 hours. To go down the **Left Hand Fork**, view the Big Horn Panel, then rappel into the bottom--or walk down along the Left Hand Fork Trail, and return via the Deer Trails Trail, will take from 8-11 hours. The time it takes will depend on your fitness, route & side trips. To explore **Little Fork** or **Oak Canyon** will take from 6-9 hours, depending on whether you actually rappel or skirt the dryfalls, or take side-trips.

Water Running water is found in the lower parts of each fork of Deer Creek & Oak Canyon, but take it directly from the spring source to avoid contamination from cattle or beaver. Or treat it. Cattle may be here during the winter grazing season--November 1 to May 1.

Maps USGS or BLM map Kanab (1:100,000) for driving & orientation; and Deer Range Point & Bull Valley Gorge (1:24,000--7 1/2' quads) for hiking.

Flash Flood Danger High in the Lower Slot (above R2) of the Main Fork of Deer Creek, low elsewhere.

Best Time to Hike Spring or fall, but you can hike year-round.

Author's Experience On one scouting trip, the author went down the Main Fork as far as Rappel 2, then rim-walked other parts before going down Deer Trails Trail to the river, up Oak Canyon, up lower Deer Creek and back to his Chevy Tracker at the 4WD trailhead; round-trip took 9 1/2 hours. Another scout trip took him down Johnson Hole Canyon Trail (see the author's **Paria River Guide**) to Lone Rock, up Asay Canyon, onto Asay Bench, down into the middle part of Deer Creek, then returned, all in 10 1/4 hours. Another trip was with Nat Smale. We went down the Main Fork (setting up all rappelling stations), out the bottom end, up Deer Trails Trail and back to the Tracker & trailhead in 9 3/4 hours. Your trip will likely go fast now the canyon has been prepared, but check the W&R/QL closely, especially at the last rap because it's exposed to the sun. Later, the author explored the Left Hand Fork including the West Fork slot, the Big Horn Panel & Left Hand Fork Trail, plus explored Little Fork & parts of Oak Canyon, on 2 trips of about 8 hours each.

Above Left This is the pool at the bottom of the rappel if you come right down the very tight slot of **Stone Donkey Canyon**. Tom Martin was standing in this same place when he took the picture on the left.

Above Right This is the really nice Big Horn Rock Art Panel in the upper part of **Left Hand Fork of Deer Creek**. Look closely and you can see some of the petroglyphs, some as high as 5-6m off the present ground level.

Left Scott Patterson rappels into the upper chamber of **Stone Donkey Canyon** from the south side. (Tom Martin foto)

253

Above Left Typical rappelling scene in **Boundary Canyon**. This is one of the best canyons in Zion NP for technical canyoneering.

Above Right Keith Krumbach making the next-to-last rappel in **Heaps Canyon**. Notice the 4 B&H's in the wall; this is your anchor for the last **93m rappel**.

Right Byron Lemay swims a narrow pool below Rappel 2 in **Icebox Canyon**. Look closely and you might see the handline hanging down in the far background.

Map 55A, Area Map--Zion National Park Region, Utah

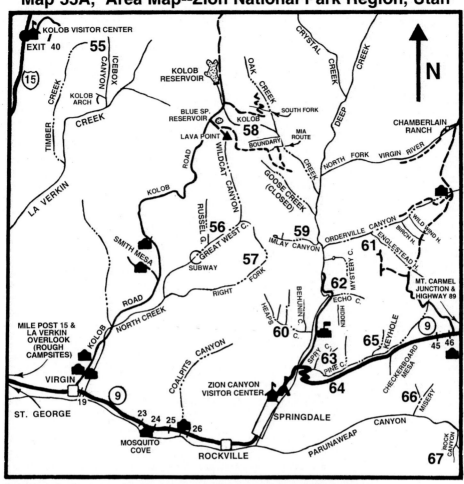

KOLOB VISITOR CENTER
EXIT 40
55
KOLOB CANYON
ICEBOX CANYON
CRYSTAL CREEK
CREEK
N
15
TIMBER CREEK
KOLOB ARCH
CREEK
KOLOB RESERVOIR
OAK CREEK
SOUTH FORK
DEEP CREEK
CHAMBERLAIN RANCH
LA VERKIN
BLUE SP. RESERVOIR
LAVA POINT
KOLOB **58**
MIA ROUTE
BOUNDARY
NORTH FORK VIRGIN RIVER
KOLOB ROAD
WILDCAT CANYON
GOOSE CREEK (CLOSED)
CREEK
WILD WIND H.
RUSSELL G.
GREAT WEST C.
56
59
IMLAY CANYON
ORDERVILLE CANYON
BIRCH H.
61
ENGLESTEAD H.
SMITH MESA
SUBWAY
57
RIGHT FORK
62
MYSTERY C.
MT. CARMEL JUNCTION & HIGHWAY 89
KOLOB ROAD
NORTH CREEK
HEAPS
BEHUNIN C.
ECHO C.
HIDDEN
60 C.
65 KEYHOLE
9
MILE POST 15 & LA VERKIN OVERLOOK (ROUGH CAMPSITES)
COALPITS CANYON
SPRY C.
PINE C.
63
CHECKERBOARD MESA
45 46
VIRGIN
9
ZION CANYON VISITOR CENTER
64
66
MISERY
ST. GEORGE
19
SPRINGDALE
CANYON
23 24 25 26
MOSQUITO COVE
ROCKVILLE
PARUNAWEAP
ROCK CANYON
67

Nat Smale pulling the rope and re-hanging it for the last segment of Rappel 1 on the **North Pass Route** in **Icebox Canyon**. At all of the rappelling stations for R1 in Icebox Canyon, be sure to tie yourself into the anchor with a meter-long runner or daisy chain.

Icebox Canyon, Zion National Park, Utah

Location & Access **Icebox**, or as it's sometimes called by hikers, **Waterfall Canyon**, is located in the northwest part of Zion National Park in what is known as the **Kolob Canyons Section**. Near the bottom end of the canyon is **Kolob Arch**, one of the biggest in the world.

To get there, drive along Interstate Highway 15 about halfway between Cedar City and St. George. At **mile post & Exit 40**, leave the freeway and drive east 300m to the **Kolob Visitor Center**. Stop there and pick up a permit, maps or other information about the hike.

From the visitor center, drive north, east, then south up the paved road to the parking place at **Lee Pass**, a distance of 6 kms (3.7 miles). If you'd like to look at part of the route before hiking, continue driving south toward the **Timber Creek Overlook** while observing the canyon to the east; the one between Beatty Point & Nagunt Mesa. Then return to Lee Pass and park.

Rating **South Pass Route** into Icebox Canyon rates a slightly more risky: **4B R III** because of 2 mid-wall rappelling stations with little room to stand. Group size there can only be 2 or 3, because everyone must use R/QD/DC to attach themselves to 2 rap stations while pulling ropes for the next rappel. The **North Pass Route** into the canyon rates a slightly easier & less-risky: **3B III**. The rap stations on this route have more standing room.

Equipment RG, two 60m ropes for the **South Entry Pass Route**; two 50s for the **North Entry Pass Route** (for either of these, you could also use a pull cord instead of a full length rope); also one 20m rope for R2 & R4, 2 R/QD/DC & 6-8 carabiners per/person, W&R/QL, drybags, and wet/drysuit for spring or fall. Ask backcountry office rangers when the last party went through and if you'll need a bolt kit--which is very unlikely.

Route From the parking lot at **Lee Pass**, walk north 75m to the beginning of the trail, then south along the path signposted for Timber & La Verkin Creeks, and Kolob Arch. After about 1 1/2 kms, you'll arrive at the bottom of the canyon marked 1700m on this map. From there, simply head north, then east up **Timber Creek**. Walking is easy, sometimes on a hiker's trail. Near the head of the canyon is a little drainage coming in on the left, but you turn right then climb up the left side of a smooth waterfall. Just above that, turn left or north and follow one of several minor trails north up through manzanita, sagebrush and oak brush on a **landslide slope**. Slowly you'll veer eastward, then as you near the top of this slope, you'll have a choice of 2 routes over a notch and down the other side into Icebox. To do the slightly more difficult one, turn south, get onto slickrock, then climb east up to the **South Pass**. To do the other, continue up the wooded slope to reach the other notch called the **North Pass**.

From the **South Pass**, continue east down the obvious slickrock drainage. After about 200m, and at the edge of the steeper part, you'll see **Rappel Station (RS) 1** down in front of you and a little left. You could use 1 B&H on the wall behind and HL down 18m, but if you have to use that, maybe you shouldn't be in the canyon! At RS1, which is 2 B&H's on a semi-flat place, rig your R/QD/DC to your rappelling harness and prepare for the next 2 rap stations which are on the canyon wall.

Rappel 1 is from 2 B&H & 1 piton down about **49m** to **RS2**, which is 3 B&H & 1 piton on the steep cliff face; not much room to stand there. All RS's have W&R/QL. That's where everyone tethers into the RS while pulling the ropes down to this 2nd set of anchors. Your R/QD/DC should be about 1/2m long. **R2** is about **53m** down a steeper wall to the RS3, which is 1 B&H & 2 pitons, and a more comfortable place to stand. Tether yourself in again, pull & re-rig your ropes, and make the **3rd Rap** down a near vertical wall about **45m** to the bottom.

Now for the **North Pass Route** which is best for those with 50m ropes, and for larger groups. From the pass, head east down the other side in a little broad slickrock valley. As you walk down, you'll see a total of 3 B&H scattered along the bottom of the drainage. We walked past all 3 to the 1st Rap Station, but if conditions are less than perfect--perhaps with ice (?), you may want to HL down, at least from the 3rd anchor.

The **1st Rap Station** is 2 B&H. Using two 50s, and instead of going straight down the low angle slope, veer southeast toward a small **3m-high ponderosa** pine tree maybe 35m away. You can almost HL down this part. Just below the pine tree is a ledge which will accommodate 5-6 people, or more. From this **2nd RS**, rappel straight down the slope about **34m**. There you'll find 3 B&H with another sizable ledge to stand on. For the **last big rap** of **about 45m**, be sure to tether your pack to the front of your harness because it's partly free hanging.

Once at the bottom of either rappelling route, head south downcanyon, walking in a green narrow canyon with occasional boulders to scramble around, and further down, some running water. After 700m (the small squares on the 7 1/2' quads are square kms), you'll come to 2 B&H on the right spaced about 4m apart, and the **10m R2** bypassing a deep pool. After another 300m, you may find a piece of rope/webbing from an old B&H on the right (**R3**). HL down **3m** into a very narrow swimming pool.

About 200m below that is a relatively new 15m-high rockfall which has dammed up the canyon. Climb over this, then after another km, the canyon opens up and you bench-walk on the right or west side to eventually find another old B&H and the **9m R4** down to trees (be aware, sooner or later, someone will install W&R/QL or B&H's so they can go right down the chute/stream course into several pools instead of bypassing it on the right). Walk down the creek a ways, then cross over to the left just above the **1st waterfall** and use a trail down a gully east of, and parallel to, the main drainage. Below that will be boulders and short jumps for a ways, then an easy walk in the creek bed down to where you'll see **Kolob Arch** high above on the **right**. From there, walk along the creek or trail 750m down to **La Verkin Creek**, then it's 10 kms west & north back to Lee Pass and your car.

Elevations Lee Pass Trailhead, 1850m; high point on the hike at either the North or South Pass, about 1963m; low point along La Verkin Creek near some cascades, 1536m.

Time Needed Between 8-12 hours, depending on group size, experience and fitness levels.

Water Take some to get started, then there's running water in Icebox and La Verkin Creek (treat).

Maps USGS or BLM map St. George (1:100,000) for driving; Kolob Arch (1:24,000--7 1/2' quad) for hiking; or the plastic Trails Illustrated/National Geographic map, Zion National Park (1:37,700) for driving.

Flash Flood Danger Low, but in a couple of narrow sections, you'll find moderate to high danger.

Best Time to Hike Because of the altitude, from late May or early June, through September. If you go in the early spring you may run into snow. This means in spring or late fall, you'll need a wet/drysuit.

Author's Experience The author and Byron Lemay car-camped near the La Verkin Overlook, between mile posts 14 & 15 on Highway 9, then got up early and drove to Lee Pass. It took 9 hours to do the entire hike using the South Pass Route. Later, with Nat Smale, we did the North Pass Route in 8 1/4 hours.

Map 55, Icebox Canyon, Zion National Park, Utah

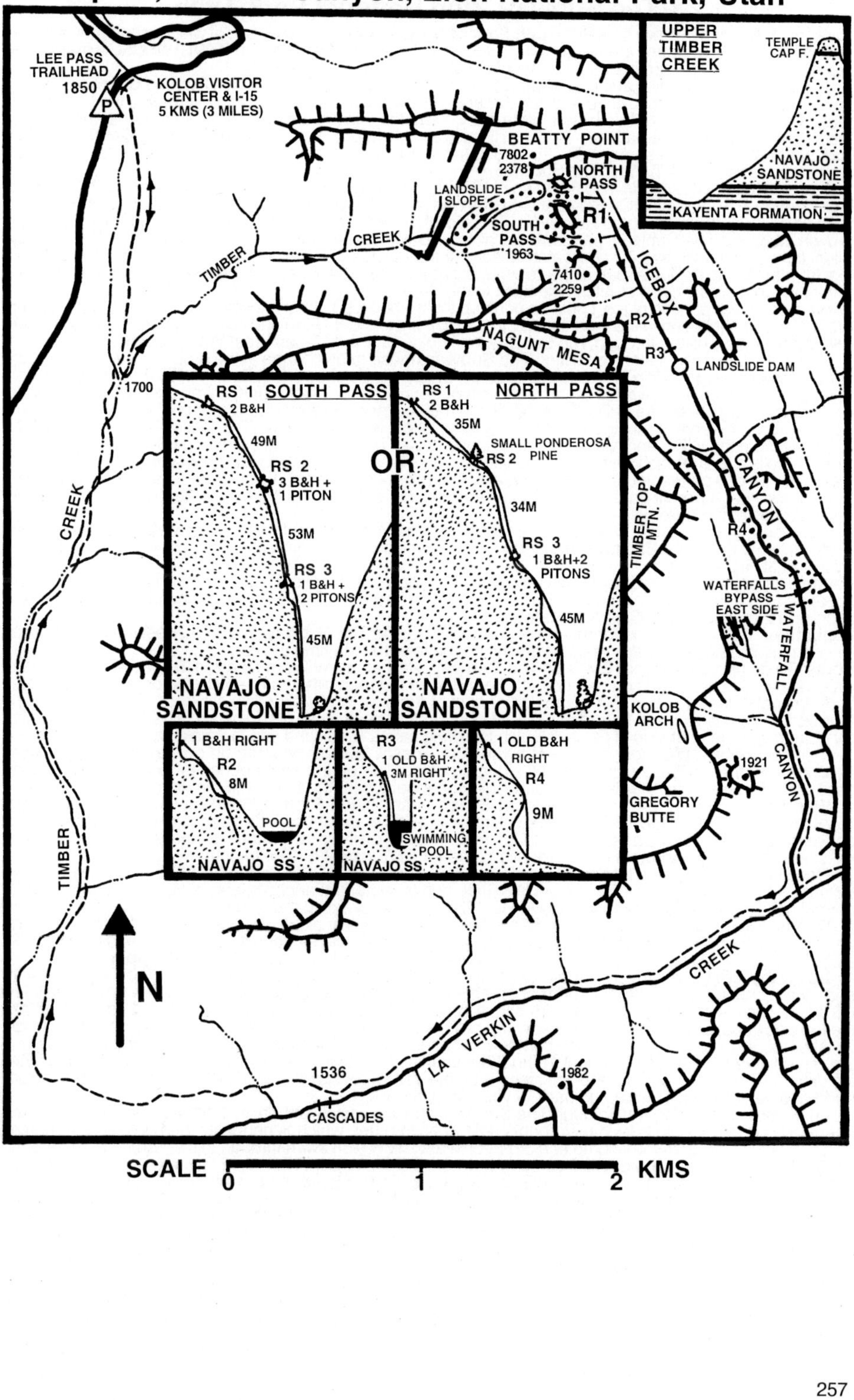

LEE PASS TRAILHEAD 1850

KOLOB VISITOR CENTER & I-15 5 KMS (3 MILES)

UPPER TIMBER CREEK

TEMPLE CAP F.

NAVAJO SANDSTONE

KAYENTA FORMATION

BEATTY POINT
7802
2378

NORTH PASS

LANDSLIDE SLOPE

R1

SOUTH PASS
1963

7410
2259

ICEBOX

R2

R3

LANDSLIDE DAM

TIMBER

CREEK

NAGUNT MESA

1700

CREEK

RS 1 SOUTH PASS
2 B&H

49M

RS 2
3 B&H + 1 PITON

53M

RS 3
1 B&H + 2 PITONS

45M

NAVAJO SANDSTONE

OR

RS 1 NORTH PASS
2 B&H

35M

SMALL PONDEROSA PINE
RS 2

34M

RS 3
1 B&H+2 PITONS

45M

NAVAJO SANDSTONE

TIMBER TOP MTN.

R4

WATERFALLS BYPASS EAST SIDE

WATERFALL

CANYON

1 B&H RIGHT
R2

8M

POOL

NAVAJO SS

R3
1 OLD B&H 3M RIGHT

SWIMMING POOL

NAVAJO SS

1 OLD B&H RIGHT
R4

9M

KOLOB ARCH

1921

GREGORY BUTTE

KOLOB CANYON

TIMBER

N

CREEK

LA VERKIN

1982

1536

CASCADES

SCALE

0   1   2   KMS

257

Upper Great West Canyon (Left Fork of North Creek)
& The Subway, Zion National Park, Utah

Location & Access Shown here is the **Left Fork of North Creek**, the lower part of which is commonly known as **The Subway**. This same drainage is also referred to as **Great West Canyon (GWC)** on USGS maps. This excellent canyon is located in about the middle of Zion National Park.

The Subway hike begins at the **Wildcat Canyon Trailhead (WCT)**, runs down **Russell Gulch** to the middle of Great West Canyon--Left Fork, down through the lower canyon and back to the road at the Lower Trailhead. **The Subway** part of this hike is quite easy, semi-technical, very popular in summer, and requires a special permit and/or reservations. It will be discussed only briefly here. The upper part of **Great West/Left Fork** starts & ends at the same WCT. To do this technical part of the canyon, you'll also have to have a permit, but it's easier to get than for The Subway.

To get to the trailheads, drive to the town of **Virgin** on Highway 9, which is a few kms west of the mouth of Zion Canyon & Springdale. On the east side of Virgin, and **between mile posts 18 & 19**, look for the turnoff heading north signposted *Kolob Reservoir*. This **Kolob Road** is paved nearly to the lake. If doing the entire canyon including The Subway, park one car at the **Lower Trailhead**. For that, drive 13.4 kms (8.3 miles) from Virgin & Highway 9 and watch closely for the trailhead on the right or east side of the road. Some hike from there into the lower part of the canyon, but most people park one car there which is normally the end of the hike (or use a bicycle or hitch hike back to the WCT).

From the lower parking place, continue north up the Kolob Road. After a total of 25.4 kms (15.8 miles) from Virgin, and at a tight bend in the road, turn right or south, drive about 100m and park at the **WCT**. This is where everybody starts if hiking through The Subway, or if doing only the Upper Great West Canyon.

Here's the latest regulations on hiking **The Subway** (these rules change at the drop of a hat, so check the website **www.nps.gov/zion** for the latest updates, or call the **Backcountry Desk** at 435-772-0170). Before doing this hike, stop at the visitor center just north of Springdale in lower Zion Canyon and get a **permit** from the **Backcountry Desk**. As of 2008, the NPS had a **80 person per/day quota** (some of which are walk-ins--some you can reserve in advance) on the number of hikers in The Subway. This includes those going all the way down from the WCT, and hikers beginning at the Lower Trailhead hiking up into The Subway from the bottom. There is a fee for each permit and you can charge it to Visa or Mastercard. Pick up the permit at the visitor center the day before your hike; or the day of your hike before 10am. If you're there during the week, you can normally get a permit on the spot without reservations, but perhaps not on weekends, and probably not on Saturdays or big holidays!

If you only want a permit to do **Upper Great West Canyon**, then get that the day before, or the morning of your hike, at the same **Backcountry Desk**. This permit costs the same as other permits but is easy to get if you exit at Russell Gulch.

Rating Upper Great West Canyon has up to 7 minor rappels and ranks **3B III** (same as The Subway but this upper canyon has more challenges). This entire canyon, including The Subway, is one of the most scenic & popular in Zion and on the Colorado Plateau.

Equipment RG, a 30m rope for the entire canyon, including The Subway, W&R/QL, hooking kit or tossbag for one possible KPH, drybags and wet/drysuit no matter what the time of year.

Route To do either the upper **GWC or The Subway**, start at the **Wildcat Canyon Trailhead**. From there, walk eastward about 2 kms on a good path (formerly an old road), then turn southeast on a trail signposted for **Northgate Peaks**. After 150m, veer left or east at the sign, and walk down slickrock along a developing cairned trail across **Russell Gulch** to the east side. At that point, if you're doing just **The Subway**, you simply head down the east side of Russell Gulch following cairns & trail and into the main canyon as shown.

If doing **Upper GWC**, veer southeast away from Russell Gulch and climb up onto an obvious slickrock bench. From there it's easy route-finding along a contour line on the same bench to the east using a compass and *The Guardian Angels 7 1/2' quad*. Walk at about where the trees & slickrock meet.

When you're **within 1 km** of the **Wildcat Seeps** near altitude 6118 (1865m) on the plastic *TI/NG Zion map*, walk about 100m down a steep slickrock ridge to another lower vegetated bench and turn northeast just above the slot. After that, and as you're heading northeast, you'll **cross 3 minor drainages**, then finally enter the slot immediately northwest of the elevation marked *6118 (or 6095/1857m on the 7 1/2' quad)*.

Once in the canyon, you'll encounter many potholes and minor DC's. In the **first 800m** or so, you'll have **5 rappels** or maybe HL or DC's, all from W&R/QL and from natural anchors, and a lot of short swims under normal conditions. A short distance above R5 will be a walkout exit to the left or south.

Rappel 5 is about **7m** into a possible **KPH**. The pack toss option didn't work for us (we didn't try very hard), so with low water & swimming, the author placed a hook in a crack, stepped into the etrier, and was out. On a 2nd trip, this KPH was full, and we swam right over it. This KPH marks the end of the **upper slot**.

Below the KPH, the canyon is more open and not so deep for the next 2 kms or so, then it gets deeper and more confined again. About **100m** or so above where **Russell Gulch** enters from the north, you'll find **R6 & R7**. The last rap **(R7)** used to be 7m from a log and into a long deep swimming pool, but as of 8/12/2006, the log was gone, and the anchor for a new rap of **11m** was from a tree stump about 20m to the right or north.

After the last rap, walk around the corner and head north into Russell Gulch. After another 100m, you'll zig zag up a very steep trail for about 100m (70m vertical rise). This is the same trail used by the people hiking down to The Subway. After the first steep part, the trail levels some and you'll follow cairns over slickrock north and out of Russell Gulch and back to your car at the WCT.

For those going all the way through **The Subway**, continue downcanyon, instead of retreating up Russell. Eventually you'll come to 2 swimming pools next to each other, which is generally the beginning of running water. A ways after that you'll come to a 3rd pool in a shallow slot; swim through this or climb up and around on the left or south side. Later you'll come to a waterfall & **5m rappel/HL** from **2 B&H** on the right down into waist-deep water. Just below that is The Subway part, and a **big log** that's been there for years. Great place for pictures, but take a tripod or lean against a wall for best results. Then a cascade, a couple of bends, and another waterfall. To get down just below that, you can sometimes slide down about 1 1/2m without a rope (if a flood hasn't scoured out the sand, otherwise it could be a 3m slide); or just below that on the left or south side, **rappel/HL** from **2 B&H's** using a 30m rope, or 2 shorter ones. Often times, you'll find a rope already there--but take your own anyway.

There are more good foto opt in the lower part of The Subway, as well as near the half dozen or so cascades below that. Then the long trudge downcanyon and up a trail to the right and back to the **Lower Trailhead**. Carry a map in hand and be alert for the sign & trail as it leaves North Fork Creek and begins climbing up to the north, then veers west.

Elevations WCT, 2134m; Wildcat Seeps, 1857m; end of Russell Gulch, 1695m; Lower Trailhead, 1540m.

Map 56, Upper Great West Canyon (Left Fork of North Creek) & The Subway, Zion National Park, Utah

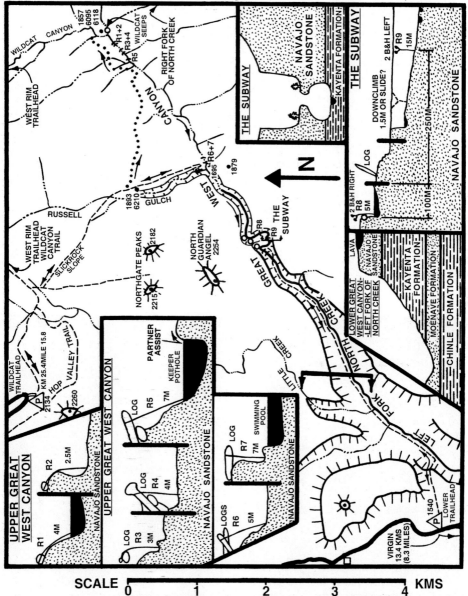

Time Needed If you begin at the WCT and walk through The Subway to the Lower Trailhead, plan on 5-8 hours. If doing just the upper GWC of Left Fork, allow for 7-10 hours round-trip from & back to WCT. If doing the entire GWC, from WCT to the Lower Trailhead, perhaps 9-13 hours. It's recommended you do this canyon in 2 trips, otherwise it may be a long day.

Water Upper GWC has many potholes & water, but it will need filtering or purification. Below Russell Gulch, is a year-round stream. Also good springs just above and below The Subway.

Maps USGS or BLM map St. George (1:100,000) for driving & orientation; The Guardian Angels (1:24,000- -7 1/2' quad) for hiking; or the plastic Trails Illustrated/National Geographic (TI/NG) map, Zion National Park (1:37,700) also for driving or hiking.

Best Time to Hike Because of all the wading and swimming, hike this in warm weather, from May through early September. A wet/drysuit is required any time of year in the upper GWC. In spring or fall, a thicker wet-suit or drysuit and maybe gloves are needed for all parts of the canyon.

Flash Flood Danger High in upper GWC, mostly moderate (high in places) in The Subway itself.
Author's Experience On one trip, the author left his car at the Lower Trailhead and mtn. biked up the paved road to the WCT. He then hiked downcanyon to his car. Bike-time was just over an hour; hike-time just over 5 hours. Total time, car to car, 6 1/4 hours. On a mid-October hike and wearing a shorty wetsuit (next to useless!), he did the same hike in 6 3/4 hours, but took more notes and fotos, then hitch hiked back to his car. Following a scouting trip, he and Byron Lemay, started at WCT and did upper GWC and returned, all in 7 1/3 hours. In 6/2004, with Keith Krumbach & Warren Egbert, did The Subway part only in a leisurely 5 3/4 hours. In 8/2006, with John Summerson, we did the upper GWC in 8 hours round-trip. On that hike, about a week after a flood, all potholes were full and we swam through 7 or 8 or more.

Commentary Concerning the latest regulations on the Left Fork, one ZNP ranger told this writer, over 800 people went through The Subway on one day in 1997. This is apparently the reason for the 80 hikers per/day quota. However, many of those were Boy Scouts, and it seems they were having some kind of "jamboree" in the area at the time. It also seems to this writer, that if the Boy Scouts are the problem and the reason for all the regulations, then why can't the NPS address the problem instead of making the rest of us jump through hoops just to do this hike? Or if the crowds seem too large, why not take the money from the fees collected, and improve facilities to handle more people? Why not enlarge both parking lots (they finally did build toilets at each trailhead in 2003), and create just one trail down through Russell Gulch, instead of several? Why not charge a higher price for hiking on weekends, especially on Saturdays? Why not allow permit holders to camp, or sleep in their vehicles, at each trailhead, instead of sending them down to Mosquito Cove on the Virgin River bottom on Highway 9 between mile posts 23 & 24? Sooner or later, that place will get too crowded as well, and "pooped" out because of lack of toilet facilities. Also, no one seems to be complaining about the paved trails in Zion Narrows, above Weeping Rock or to Angels Landing, so why not improve the situation along Russell Gulch & lower Left Fork to handle more people? There are thousands of other canyons on the Colorado Plateau where you can enjoy a *wilderness experience*. Going with up to 79 other people is not exactly a wilderness trip, so why use that argument as an excuse to create quotas and more bureaucracy!

Readers should understand that we are all shareholders in USA Inc., and we have the right to complain to those who set policy on public lands. Remember also, these are our national parks, not private reserves of the NPS, or a bunch of wilderness radicals. If you feel disenchanted about all the regulations created in the last few years in Zion and elsewhere, send a letter or email to the NPS (or BLM or USFS) and voice your opinion. That's how policy is set, with all of us voicing an opinion. In the opinion of this writer, who is a member of half a dozen conservation groups including SUWA, wilderness advocates have taken this hike and all the regulations one step too far.

Both fotos above are in the upper end of **Great West Canyon/Left Fork of North Creek**, and above **The Subway**. John Summerson is in both. In August or September, or after any flood, this part of the canyon will be full of water in potholes. On the right is Rappel 5 into a potential KPH. On this day, we swam right over the keeper, but the rope got stuck!

Above Left In the upper part of **The Subway** is the famous log, seen on many posters. **Above Right** Keith Krumbach HL's down what can be for some Rappel 9 in the lower end of **The Subway**. **Bottom** Just below R9 in that part of **The Subway** for which it got its name.

Right Fork of North Creek, Zion National Park, Utah

Location & Access The **Right Fork of North Creek** is the first major canyon south & southeast of **The Subway** (otherwise known as the **Left Fork of North Creek**). Both are in the western part of Zion National Park. To get there, drive along Highway 9 southwest of Zion NP. At the east end of the town of Virgin, and between **mile posts 18 & 19**, turn north onto the paved **Kolob Road** signposted for **Kolob Reservoir**. After **11.3 kms (7 miles)** you'll come to the **Lower Trailhead** for the Right Fork on the right or east side of the road. This is 800m inside the Zion NP boundary. You'll likely want to leave one car (or maybe a bicycle?) there. Then continue up the Kolob Road to one of 2 parking & starting places. First option, drive to Blue Springs Reservoir (32.9 kms/20.5 miles from Virgin), turn southeast onto a good graded gravel road, and drive 3.7 kms (2.3 miles) to the **West Rim Trailhead** which is just east, and downhill, from **Lava Point** (see **Area Map 55A--Zion National Park, page 255**). You must have the Zion NP map listed below which shows this driving route. You could also camp/park at Lava Point, but walking from there would add another 20-25 minutes to your hike.

The 2nd starting place, and the best one if it's early or late in the season and the road to the West Rim Trailhead is snowed in or muddy, is the **Wildcat Trailhead**. This is the same trailhead you use when going through **The Subway**, or doing the upper part of the Upper Great West Canyon/Left Fork of North Creek. The distance from Virgin to the Wildcat Trailhead is **25.4 kms (15.8 miles)**.

Rating The **Standard Route: 3B IV or V**. The Direct or **Upper Slot Route, 3B R IV or V.**

Equipment RG, one 50m rope, lots of W&R/QL, K&EP & drybags. Highly recommended are **wet/drysuits** if going in late spring or fall, or if you choose to do the **Upper Slot Route** in the main drainage. Before going, ask someone at the Backcountry Desk about the condition of the anchors and water in the potholes in the Upper Slot Route.

Route From the **West Rim Trailhead**, walk southwest on the **Wildcat Canyon Trail**. After 2 kms, the trail veers north for a ways and crosses the usually-dry **Blue Creek** stream channel. At that point head downcanyon along Blue Creek, otherwise known as **Wildcat Canyon**. The first km is a little brushy, then it's easy walking, sometimes in the dry creek bed or on hiker-made trails, usually on the right, or west side.

Where Wildcat Canyon makes a big right turn and heads southwest into Great West Canyon (and just above **Wildcat Seeps)**, route-find across the drainage if necessary and continue SSE

Now if you start at the **Wildcat Trailhead**, the recommended starting point, walk eastward along the very good trail signposted for the **West Rim Trailhead**. After about 2 kms, turn right or south onto the trail signposted for **Northgate Peaks**. After another 150m, veer left onto the trail heading down into **Russell Gulch**, the same path you use when going through The Subway. Soon after you reach the bottom of middle Russell Gulch, veer left or southeast and get up on a nearly flat bench. Aim for a point about 200m northeast of elevation **6210** (1893m) on *The Guardian Angels* quad. From there, contour eastward along the same **bench route** as if doing the entire hike down through Great West Canyon/Left Fork of North Creek. At one point you'll have to walk straight down a fairly steep slickrock ridge to a lower wooded-bench, then head northeast aiming for **Wildcat Seeps** & elevations *6118 or 6095*. Route-find in and out of the drainage and head SSE.

After the routes combine, and about 2 kms SSE of Wildcat Seeps, cross a divide, walk about another km in the dry creek bed, then veer right or west and get upon a minor ridge between the main Right Fork drainage and what this writer calls the **West Fork**, as shown on this map. This is called the **Standard Route**. Further along, drop off the low ridge into West Fork. At the bottom of West Fork, route-find down into the main drainage, always following the most-used trail. At the bottom will be a 3m rap/HL from a tree.

The other option would be to stay right in the main drainage, called here the **Upper Slot Route**. Less than 1 km from where you veer right to do the Standard Route, you'll come to lots of slickrock and potholes. You can skirt most of these in the beginning, then you'll have get down in the drainage and start jumping or sliding into potholes. This writer, who has one bad knee, had to be belayed into 2 of these. If they are full of water, this might be less threatening. Once fully into this upper slot, you'll have **2 short rappels/DC's? (4m & 5m)** from small logs (be ready to improvise, because the ones we found will be washed away). Shortly after that, it gets deep, then **R3**, a kind of **double rap** of up to **22m** from a log overhead. Soon, the slot opens momentarily, then comes the 2nd half of the Upper Slot and **R4** of 5m from a chokestone. After that it's easy walking nearly 2 kms to where the Standard Route comes in on the right or west.

Soon after the 2 routes meet, you'll DC under a huge boulder (maybe a swim?). It's here recent floods have created the need for the **7m R5** from a B&H on the left wall. Soon after that the canyon turns 90° right or west, then probably several long swims, including the **Black Pool** (on our last trip, we found this canyon had changed a lot. We didn't really see the Black Pool, although we did have some deep wading in one fairly long pool). Some have climbed to the left to avoid swimming; but most **rap 12m (R6)** from 2 B&H on the right. Later, veer left for **R7** from a tree & down **17m**.

Finally you'll come to the **Grand Alcove** with running water, and the best campsite around. Then **2 rappels (R8-14m & R9-6m)** at the lower end of Grand Alcove (or climb & slide right down the chute?) and **R10- -20m** over **Barrier Falls**. Below Barrier Falls there are no more impediments; except passing to the south side of **Double Falls**. Below that you'll find a trail along the creek all the way down to the Lower Trailhead.

Elevations West Rim Trailhead, 2271m; Lava Point, 2405m; Wildcat Trailhead, 2134m; Lower Trailhead and end of hike, 1372m.

Time Needed Depending on when you start, 2 average days, 3 shorter days, or 1 long day. If you can go in late May, June or July with lots of daylight, the 1 day trip from Wildcat TH with smaller pack is best.

Water Start with your own. Wildcat Seeps, but it may be stagnant or dry, and numerous potholes everywhere after rains, but they may be stagnant too. Good running water begins immediately above the Grand Alcove.

Maps USGS or BLM map St. George (1:100,000) for driving; The Guardian Angels & one tiny corner of Temple of Sinawava (1:24,000--7 1/2' quad) for hiking; and/or the plastic Trails Illustrated/National Geographic map, Zion National Park (1:37,700), also for driving.

Best Time to Hike Due to possible deep wading or swimming, do this one from late May through Mid-September. Or take wet/drysuits for hikes in April, May, late September and October, or anytime of year if going down the Upper Slot Route.

Flash Flood Danger Low, except for a couple of short narrow sections. High in the Upper Slot.

Author's Experience The author and Steve Tyler, both 55 and in 1998, left one car at the Lower Trailhead, then started hiking early at the West Rim Trailhead a day after some rains. We found pothole water everywhere and made it to the Grand Alcove in about 10 hours--but 1 hour was lost trying to locate the route down into the Right Fork along the standard route. We camped on a sandy bench in the Grand Alcove, then finished the hike the next morning in 4 1/2 hours. Total walk-time, 14 1/2 hours. On 6/22/2007, with Ryan Cornia, we did the Upper Slot Route from Wildcat Trailhead in one day, 10 3/4 hours.

Map 57, Right Fork of North Creek, Zion Nat. Park, Utah

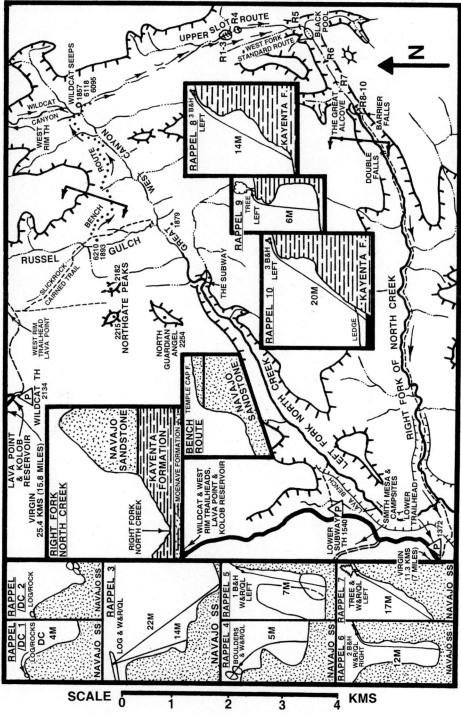

SCALE 0 1 2 3 4 KMS

Above Left The Upper Slot of **Right Fork of North Creek** has several 3-4-5m DC's or fast slides like this. Most of these have hard landings, but are not quite worth the effort of rappelling. Best to have a group of 3-4 hikers so you can help each other safely down.

Above Right Ryan Cornia prepares for Rappel 3 in the **Upper Slot Route** of **Right Fork**.

Right The Great Alcove near the end of the technical section of **Right Fork**.

Above Left The Great Alcove of **Right Fork**, this time as seen from just below Rappel 9 (in the lower right).

Above Right Starting the last rap, Rappel 10 in the **Right Fork.** (Ryan Cornia foto)

Left After **Double Falls**, it's a boring walk down to the trailhead.

Oak, South Fork of Oak, Kolob & Boundary Canyons, Zion National Park Region, Utah

Location & Access Oak Creek and it's **South Fork** are tributaries of upper **Kolob Creek Canyon**. These 2 short drainages are located just north of, and just outside, the northwest part of Zion National Park. They're also ENE of Kolob Reservoir. Oak Creek enters Kolob from the north right where Kolob begins to run southeast toward Zion Canyon. **Kolob & Boundary Canyons** are found just south of Oak and its South Fork, and east of Kolob Reservoir. **Kolob Canyon** begins below the reservoir and heads east, then southeast to enter Zion Canyon just below where the North Fork of the Virgin River and Deep Creek meet. **Boundary** enters the mix about 650m below where Kolob & Oak Creek meet.

For **Kolob**, you're officially supposed to get a hiking permit at the backcountry desk in the Zion Canyon Visitor Center (**NO ZNP permits** are needed for **Boundary, or any part of Oak Creek**), then drive south & west through Springdale and Rockville to Virgin on Highway 9. At the east end of Virgin and between **mile posts 18 & 19**, turn north onto the paved highway signposted for **Kolob Reservoir**. Drive the **Kolob Road 32.9 kms (20.5 miles)** and turn southeast just before **Blue Springs Reservoir** on the good graded gravel road signposted for *Lava Point Campground & West Rim Trailhead*. After driving just over 1 km, and at a 4-way junction, turn left, then right, and head east, and following the signs, downhill toward the **West Rim Trailhead**. This will be 3.7 kms (2.3 miles) from the paved Kolob Road. You'll park immediately west of a locked gate. This will be the **starting & finishing point for both Kolob & Boundary Canyons,** plus the end of **Oak Creek** and its **South Fork** hikes.

The **MAIN FORK OF OAK CREEK is presently not accessible due to private land issues, and a gate with a No Trespassing sign.** To get to the **South Fork of Oak Creek**, drive the **Kolob Road** toward Kolob Reservoir. About 2 kms northeast of Blue Springs Reservoir, and 35.1 kms (21.8 miles) from Virgin, you'll cross **Upper Kolob Creek** just before it makes a big drop. About 300m beyond that (**35.4 kms/22 miles from Virgin**), turn right or east onto a well-used graded county road running east downhill past many summer homes into Oak Valley. Follow this map and the *7 1/2' quads* very carefully and stay on the most-used road as it passes beside & through private property. You will be running east until you're 1 cm NNW of elevation **6963** (2122m) on the *Kolob Reservoir quad*. From there, and at the **1st 4-way**, turn left; at the **2nd 4-way** go straight; then you'll be heading northwest to a gate and a **NO TRESPASSING sign**. This gate is immediately after you cross the 1st of 3 west-side drainages of Oak Creek; in fact this is the **South Fork** (see map). In the recent past you could drive further, as shown on this map, but now they'll have to backup about 400m,. park, and bushwhack down to the bottom of South Fork. There are no No Trespassing signs on the south side of South Fork; so for the moment this seems our only legal way to start this hike. If or when this area gets posted (No Trespassing signs), then we'll have to assume all of Oak Creek is closed to hiking. If you get there, your exit from all canyons on this map will be up the **MIA Route** to the **West Rim Trailhead**.

Before hiking any part of Oak Creek, see Tom Jones' website *canyoneeringUSA.com/utah/zion* for possible updates regarding access to these canyons. He may have something new posted. The land in this area has been private for a long, long time, but until recently, no one cared about the few hikers who went there. But things are changing! Hopefully at some point in the future we'll have legal access again.

If you like, you could camp & park a car at the campsite on Lava Point, but you'd have to walk another km to get back to your car. This is a free campsite, but it has no water, and is on a first come, first serve basis. You can also camp around Kolob Reservoir, but at times, can be crowded & noisy with ATV's.

Rating Kolob: 4B (or C) R IV. This canyon has high risk potential, so at least one day before going, you must call the Washington County Water Conservancy District at 435-673-3617; open 8-5, M-F. They control the amount of water being released from Kolob Reservoir. You don't want to go if the flow is over about 3-5 cfs (cubic feet/second). In good water years, even if they aren't releasing water, you may have some flow in the canyon which is more or less normal. In dry years, like the summer of 2002, even when they release 3 cfs, the upper technical section may be without running water and half the big potholes dry! If possible, it might be best to go down to the first rappels and check out the water situation the afternoon before your hike; if the flow looks like a very small irrigation ditch, then you're OK. If it looks like a canal--don't go! Two Boy Scout leaders died here because they tried to go through with too much water. **Boundary Canyon: 3A (or B) III**, one of the best canyons in Zion. On 7/8/2006, this canyon was bolted-up after the 1st rap. But in the days after that, someone was threatening to go in and chop some of those bolts. Be aware of this and ask questions before going in as to the anchor situation. However, if there is no solid log nearby, bolts will not be pulled. **Main Fork of Oak Creek: 3B III or IV** and it has up to 12 rappels. **South Fork: 3C III or IV.** This one has running water and you'll rappel down through several waterfall in a deep narrow gorge.

Equipment Kolob: RG (for each canyon), two 50m ropes (or one 50m & pull cord)(and perhaps a 3rd 50 as backup), 1 or 2 shorter ropes, one extra figure 8 (a **Pirana**, read more below) or ATC per/group, thick wetsuit or drysuit (best), large drybags (and/or a keg) for everything, small plastic box for cameras, W&R//QL, one R/QD/DC/person, and 5-6 carabiners per/person. The bolts in this canyon should be good for a long time, but the first group down each spring or summer should replace or add to the slings--same as in all canyons. At the visitor center's backcountry desk, ask about the condition of the bolts and what the last group found. Consider taking one whistle, or more, so you can let people know when you're off rope (sometimes it's hard to hear over the sound of waterfalls). If there is running water, then a helmet and neoprene gloves & socks will keep you warmer. And of course a compass & map which are standard for all hikes. **Boundary**: (more often than not a dry canyon), basically the same gear as for Kolob, but two 60m ropes (or a 60 & heavy pull cord). In spring there can be running water, so wet/drysuits highly recommended, but there shouldn't be any deep pools.

Main Fork of Oak Creek: one 50m rope, and shorter 20-25m rope for short raps. **South Fork**: two 60m ropes (or a 60 & pull cord), wet/drysuit in both, W&R/QL and ascenders in all canyons just in case.

Route Kolob Creek From the **West Rim Trailhead**, walk north back up the road for 350-400m, and when the road turns west, you continue north into a meadow. After 60m, you'll come to a barricade and the beginning of a **former logging road** running down to the spring as shown on the *Kolob Reservoir 7 1/2' quad*. Walk down to near the spring, but circle it to the west, and continue up to a pass to the north. From there, this same old track double zig zags down (north) the obvious drainage (you can shortcut straight down without bushwhacking, but to save confusion, best to stay on the track as shown on the map). Follow the unused track, cluttered with deadfall, to where it crosses the drainage coming down from the southwest and the point marked **7822** (2384m). This same track continues northward, then doubles back to cross the southwest drainage a 2nd time, then parallels the dry creek bed north for 300m to Kolob Creek. Once you reach Kolob Creek, turn right or east and bushwhack a little until you can cross to the north side. There you'll come to an **emerging hiker's trail** which you can follow another 350m or so to a semi-level campsite; and just beyond, a pine tree

Map 58, Oak, South Fork of Oak, Kolob & Boundary Canyons, Zion National Park Region, Utah

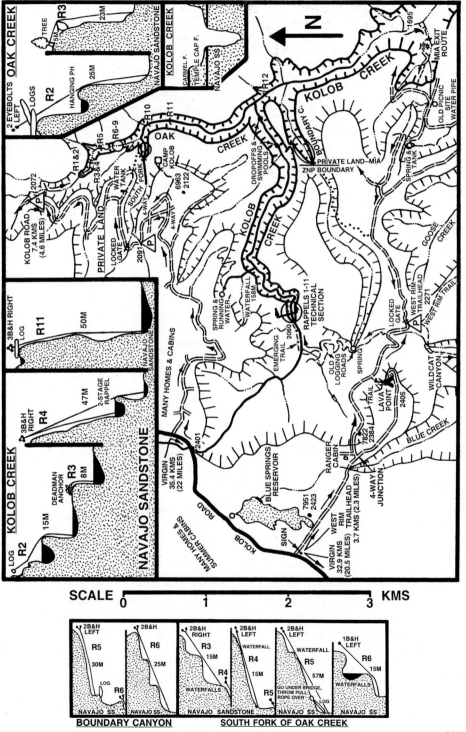

SCALE 0 1 2 3 KMS

BOUNDARY CANYON   SOUTH FORK OF OAK CREEK

with W&R/QL which will be **Rappel 1** (this anchor was in 3 different places on our trips).

The best and most exciting part of Kolob Canyon is the first 350m. This is where you'll make 11 or 12 rappels, one after another; and normally all but one will be over waterfalls into deep swimming pools with running water. This part of the hike is not very camera friendly, so be sure to carry yours in a small **waterproof plastic camera case on your belt** (Pelican 1010 case). A camera in the pack does not take pictures!

Another tip, if doing the entire length of Kolob in 2 days, travel as light as possible, because rappelling with a huge pack can be difficult. When doing these rappels, all of which will be with your feet on the wall, do so with your pack on--which is much easier and faster than trying to lower or throw it in first. A second option for those with large packs, and for the 3 longest rappels, is to tether your pack to the front of your harness with a **R/QD/DC** and let it hang down between your legs. Also, at the end of each rappel, simply lean back in the water against your floating pack and you'll have no trouble making a wet disconnect from your rappelling device & rope. This is especially easy for those using a drysuit because there's always air trapped inside keeping you afloat. If using a figure 8, buy & use a large wide-mouthed **steel biner** and loop the rope inside the biner as if using an ATC instead of wrapping it around the '8 as is the traditional way (buy and use a **Pirana figure 8** which is designed to be attach to biners for wet disconnects). When doing it this way in deep pools, you never take your figure 8 off the biner. This means no chance of dropping it.

Rappel 1 is a **25m** free rap from a pine tree on the left side about 60m downstream from the 1st waterfall (sometimes this anchor changes places, and some go right down the drainage rather than rapping in from the side), then walk 40m to a DC or jump in. Just below that will be another **DC**, or for some, a **4m rap**, perhaps from a fixed rope on a big tree root & log. The **2nd real rappel** is from W&R/QL around a log behind you; first down into a small pothole, then over its lip into a 2nd pool. If there's no running water, and the level is low in this small pool, you can still get out easy. This is the 1st of 3 potholes that you may have to work to get out of if the water level is low, **but none will be KPH's**. R3 will be from a deadman anchor (pile of rocks) and down into another pool. If this rock anchor is gone, be prepared to construct a new one with W&R/QL. Or, put 1 B&H on a side-wall--this would be better in the long run.

Rappel 4 is from 3 B&H and down about **47m** in **2 stages**. You go right down the waterfall and may be sprayed along the way; wearing a helmet and neoprene gloves will help. You'll land in a deep pool, but stay on-rope while swimming across to the **2nd stage drop**. At that point take up slack and wall-walk to your right (looking upcanyon) before descending straight down into the 2nd pool. Dragging a rope across a sharp edge under full weight will destroy it!

Rappels 5-9 are simple & short. The anchor for **R10** is from a ledge above on the right and over 2 drops of about **24m** total. About 35m below that is the **last rappel** of nearly **50m**. This marks the end of the riskiest part of the canyon. Below that, it's basically just a walk down a deep narrow gorge, with 2 pools where you can slide, jump in, or use a short rope. These 2 pools are close to where Oak Creek enters on the left. **Rappel 12**, not far below Oak Creek, is from an old B&H on the left; or from another B&H on the right.

About 350m below R12, **Boundary Canyon** comes in on the right or west. Let's forget Kolob for now and come down Boundary. From the **West Rim Trailhead**, walk east past the **locked gate** and down the road. After about 1200m, and on top of a high point, leave the road and head northeast down a minor hogsback-type ridge. Along the way, you can use old overgrown roads from the days they were cutting timber in the area. Near the bottom of the drainage, you'll have to route-find around some brush. Once in the bottom of upper Boundary, walk east down through meadows to the first dropoff.

Your first rap will be from a tree and right down the drainage; or from another tree on the north side (if you drop in from the north side, you'll need two 60m ropes--or a 60 & a 60m pull cord). **R1** is 20m down to a log with W&R/QL (or install a sling on a nearby tree), then down another **27m for R2**. Not far away are 2 short drops and **R3 & R4** from a B&H on the right. **R5** is from 2 B&H on the left and down a **30m** chute; immediately below that are 2 more B&H for **R6** and down **25m in 3 stages**. **R7** is **20m** from a single B&H, while **R8 is 6m** from a B&H. **Rappel 9** is from 2 B&H and down a **slippery slide 15m** or so into a shallow pool. From there to Kolob Canyon is a walk along a green canyon bottom. As stated above, some of these B&H may have been removed, so be prepared to **re-install W&R/QL** from a log or chockestone.

From the **confluence of Boundary & Kolob**, it's an easy walk through a deep sometimes narrow gorge. You can walk all the way down Kolob to Zion Canyon and out the bottom, but most people now do either canyon in **1 day** and exit up the **MIA Route**. This exit is about 2 1/2 kms below the confluence of Kolob & Boundary Canyons. Look for the tree-covered slopes to the right or west and a couple of cairns on top of large boulders on the right side. This can be easy to miss, so pay attention and have a map in hand! This climb will take around an hour, so have some water at the beginning on a hot summer day.

After climbing 50m or so up a steep hiker's trail, veer left or south and drop down to the main drainage, then it's straight up the dry creek bed (with one minor detour right) for 150m. At a dryfall, turn right and continue up a side-canyon. This will take you up to a pass and down the other side to the main drainage again. Once back in the main canyon, turn right at a major fork. You'll soon pass the remains of a very old car. Continue upcanyon for another 100m, but when the drainage turns right (north), head straight upslope to the west on an emerging trail. Parts of this trail are now marked with parts of ropes or plastic mesh and is getting better by the day. In places, you'll pull yourself up using tree roots or branches. Finally as the slope eases up, you'll come to an old logging road (a climb of 308 vertical meters). Turn left or south and walk 300m to find an old picnic site with water coming out of a plastic pipe. On all of his 7 trips there, the author found cold drinking water. From there, walk west upcanyon 600m to where the spring water is put into the pipe, then go up a dugway to the north & east, then west and eventually back to the **West Rim Trailhead** and your car.

South Fork of Oak Creek In the past, you used to drive and walk east along the north side of South Fork, but for now, you'll have to route-find down into the stream channel from the south and head east downstream. Eventually you'll run into water, and maybe some dropoffs (?); no one has produced any information. But after a ways you'll come to an old foot bridge at the very beginning of where the lower canyon begins to drop. At that point put on your wet/drysuit and perhaps look for a rope on a pine tree just below the first dropoff. After that you'll encounter at least **7 rappels** before reaching the **Main Fork of Oak Creek**. R1 will be from a B&H on the left and down **4m**; R2 is **5m** from a B&H on the right; R3 is from 2 B&H on the right and over 2 drops of about **15m** total; immediately below that is R4, 2 B&H on the left and down the middle of a waterfall of **15m**.

Rappel 5 is the **Big Rap**. It's from **2 B&H & chain** on the left. From there it's in or just to the side of a waterfall for **57m**. Near the bottom, slip down under a small **bridge**, but separate your pull rope/cord and throw it over the top of the bridge. Last person down make sure the ropes aren't tangled or overlapping at the top for an easier pull from the bottom. Below the big waterfall will be **R6**, another double drop of **15m** from a B&H on the left and down the middle of a waterfall--you should be getting cold by then. Our **last rap** was by looping a short rope over a log and down underwater for just **3m**. After that, you'll have some DC's but will soon be in the **Main Fork**. Let's stop here and go upcanyon.

Main Fork of Oak Creek From where you park (assuming you get there somehow?), walk north down the road maybe 200m to the creek bed which is dry most of the time, then head south downcanyon in the Main Fork. After about 30 minutes or 1 km, you'll come to slickrock and the first dropoffs. You'll likely have to install W&R/QL on a log for **Rappel 1** of **11m**. **R2** is just below that from 2 eyebolts on the left. You'll first rappel down halfway into a hanging PH, which could be a little difficult for one person to get out of (but it should never be a life-threatening KPH). Crawl out of the PH and do the 2nd part of this **24m rap**.

About 250m below R2 is **R3**, another long rap of **23m** from a nearby tree. In the next 600m or so, you'll have **6 more short raps** over chokestones; then coming in from the right or west, will be the **South Fork**. Water runs out of this canyon on a year-round basis and usually flows as far as Kolob Creek.

Shortly after South Fork will be **R10 & R11 (or R8 & R9)**, then a 1 1/2 km walk to Kolob Creek. Of the Main Fork of Oak Creek's 11 rappels, 6 will be from B&H's on the left. About 250m into Kolob Canyon will be the last **5m rappel** from a single B&H on the left or right. Below this, take off your harness. Read above under Kolob Creek about taking the **MIA Route** up and back to your car at the **West Rim Trailhead**.

Elevations Entry car-parks, about 2091m, 2072m & 2271m; bottom of MIA Route, about 1695m; West Rim Trailhead, 2271m; Lava Point Campground, 2405m.

Time Needed Kolob Canyon: 8-12 hours from the West Rim Trailhead; **Boundary Canyon**: 6-10 hours; **Main Fork of Oak Creek** to the West Rim Trailhead: 7-10 hours; **South Fork Oak Creek**: 6 1/2-9 hours.

Water Start with your own in all canyons, but filter or purify **Kolob, Oak & South Fork Creek** water. There should be good water next to the road & picnic site along the MIA Route (in hot weather, and if possible, check out this water source, and the route down the MIA, the day before your hike). Plan to carry some water up the MIA Route to the pipe.

Maps USGS or BLM maps St. George & Kanab (1:100,000); Kolob Reservoir & Cogswell Point (1:24,000--7 1/2' quad) for hiking and driving; or the plastic Trails Illustrated/National Geographic map, Zion National Park (1:37,700) for a look at the region, but not the back roads to Oak Creek.

Flash Flood Danger Overall about moderate, but high in a few place. **South Fork** has higher risks.

Best Time to Hike Late May to early July is the least-prone time for flash floods. Or with a good wet/drysuit, you can do the wet canyons anytime between April and October. Year-round for Boundary Canyon, if you can get close to the canyon in your vehicle.

Author's Experience On his first trip down **Kolob** in 2001, he went with Leroy Anderson & sons Matt & Craig, plus Newell White & Nat Smale. We got permits in the afternoon, drove one vehicle to where Kolob Creek crosses Kolob Road and parked. It took 1 1/3 hours to walk in from a northern route and camped next to Rappel 1 (forget that route--the southern approach is much better!). Next day it took about 6 1/2 hours to get through the technical section. Too long, take a smaller party and day-packs! Camped near the MIA Route, then down Kolob and Zion Canyons to Sinawava in 6 hours. About 18 hours total walk-time for the author and Nat.

Later, and just before the first rains of 2002, the author & Byron Lemay parked at the West Rim Trailhead, walked in from the south via the spring, and went through the upper section of Kolob in 3 hours. Later, they escaped via the MIA Route (1 hour 10 min. from the bottom to the road) then back to their cars in a total time of 10 1/5 hours. In 6/2003, with Keith Krumbach & Warren Egbert, and a late start, we hurried along and did the entire trip in about 8 3/4 hours. On 7/8/2006, along with Nat, we did **Boundary** in 6 1/2 hours from the West Rim Trailhead (47 minutes up the MIA Route).

After camping just outside the park on the **Smith Mesa Road**, Nat Smale and the author drove to the West

Rim Trailhead and organized everything. From there, one car was taken to the **Main Fork of Oak Creek Trailhead**. We found a fair amount of mud in low places down to the running water, then it was OK. Once in the canyon, we made it up the MIA in one hour, and back to the West Rim Trailhead in a total of 7 1/4 hours. On 7/9/2006, we again camped on Smith Mesa, drove to the car-park and did the **South Fork** (joined by Ryan Cornia in the Main Fork) and out the MIA in 7 hours total. On that trip, we did the 308m climb out the MIA Route in 42 minutes; that trail is getting easier to find & follow by the day.

Keith Krumbach & Warren Egbert at the top of Rappel 3 in **Kolob Canyon**. With every big flood, these cobblestones will wash away, so be prepared to reconstruct a new deadman anchor with W&R/QL--or install a couple of B&H's on the wall--this is the best idea.

Above Left Keith goes down one of several 2-stage rappels in **Kolob Creek's** Technical Section.

Above Right Starting down Rappel 11 in **Kolob Creek**. This rappel is from about 3 B&H's next to a big log that's been there for a long time.

Right This is **Boundary Canyon**, which is one big rappel after another. Boundary might be the most scenic technical canyon in Zion NP.

Above Left Another rappel in **Boundary Canyon**.

Above Right & Left This is Rappel 5 in the **South Fork of Oak Creek Canyon**. When doing this rap, slide right down the waterfall on a single rope, but throw your pull cord/pull rope over the natural bridge & log so it can be pulled more easily from the bottom. It's easier for you to rappel under the bridge.

Imlay Canyon, Zion National Park, Utah

Location & Access Imlay Canyon is located near the center of Zion National Park. Imlay drains into Zion Narrows from the northwest, and from there it's about a 5 minute walk downstream to the mouth of Orderville Canyon; about another hour's walk down to the end of the road and the Temple of Sinawava bus stop.

To get there, make your way to the **Zion Canyon Visitor Center** just north of Springdale, and just inside the park. After you get a permit at the backcountry desk and get everything organized, park nearby, then get into a **shuttle bus** and ride as far as **The Grotto**. That's where you'll begin hiking. At the end of the hike, get on a shuttle bus at the **Temple of Sinawava** and ride back to the visitor center and your car.

Another possible route requires a **car shuttle**. Leave one vehicle at the visitor center, then drive south and west out of the park on Highway 9 to the town of **Virgin**. Between **mile posts 18 & 19**, turn north on the paved road signposted for **Kolob Reservoir**. After 32.9 kms (20.5 miles), and just before Blue Springs Reservoir, turn right or southeast onto a graded gravel road running to the **West Rim Trailhead** (a distance of 3.7 kms/2.3 miles--see the Kolob Creek map). You'll need another map for this. The walking distance from The Grotto or the West Rim Trailhead, to the head of Imlay is roughly the same. But you drop 206m between the West Rim Trailhead and the head of Imlay; whereas you climb 755m from The Grotto to the pond at the head of the canyon. Fit hikers will choose The Grotto route and avoid the hassle of a car shuttle.

Rating For the **Shortcut Route**: 4B R IV or V; Or to do the **full length** of the canyon: **4B R V** or **VI**. This is one of the most difficult, challenging and potentially risky canyon hikes in this book. This adventure is for **physically strong & fit people only** because dragging full-sized packs in & out of potholes is not easy. Everyone should have experience with rappelling, chimneying, DC'ing and getting in & out of **KPH's**, those that are difficult to get out of if the water level is low or completely dry. Getting out of KPH's are one of the big challenges in Imlay. Between the head of the canyon just below the old stock pond & springs, and the Virgin River, are roughly 42 rappels.

Equipment RG, two 50m ropes, plus 2 or 3 shorter ropes for short rappels, and 1 PR per/person. Take enough short ropes & webbing to equal 50m in case you destroy one of your 50's. Also, K&EP, W&R/QL, 1 BK and several B&H per/group, a thick wetsuit, or drysuit (depending on the season), drybags, half a dozen carabiners and 2 R/QL/DC per/person, at least one pair of ascenders, hooking kit & G-pick per/group. See the Introduction to this book under *Hooking Kit* for a full explanation of this equipment and how to use it. Also one water filter or some purification tablets.

Route To do the entire length of Imlay, here's the description. From either trailhead mentioned above, make your way to the spring and **small pond** right at the head of the canyon. At that point you'll find an NPS campsite and several nearby springs with swampy water running downcanyon for 300m or so (use purification tablets). From the pond, head southeast on the left or northeast side of the canyon for about 60m. **Rappel 1** in this **Upper Section** is from a tree about **10m** down. For **R2**, use **two 50's** and you'll go over about **3 drops** & potholes. Before pulling your rope in any part of this canyon, the first person down should check downcanyon and see if several drops can be passed from one anchor.

At **R3** was a sad-looking piton on the right with no R/QL--use two 50's there. Shortly after that the canyon opens abruptly at a big wall & dropoff. To descend this wall, first climb up to the left, then down a little to a **pine tree** and the 1st of 3 parts to **Rappel 5**. The first part goes down **12m** to another pine tree, then after another **22m** is an rap station and **2 pitons & B&H** on the face with a small ledge for standing. This is where everyone needs a R/QD/DC to clip into the anchor. From there it's **50m** down to the bottom of the wall. Just below that is another rappel (**R6**) down to the shady canyon bottom. Getting down this Upper Section involves waist-deep wading, so you likely don't need a wet/drysuit for this part. Our group had wet and/or drysuits on and were hot as hell going down this southeast-facing **R5** exposed to the sun.

After the **Upper Section** and 6 rappels, you'll walk about 1 km with one short rap, before reaching the **Middle Section**. Here you'll pass **Rappels 8-14**. This section requires lots of hairy DC's which slowly destroys huge packs (you could lower packs, then DC, but it's slower). About 200m before the **Crossroads** is a pretty good campsite right where there's a 75m-long deep & dark slot. We camped there, then the next morning walked on the right side around the slot & next corner and rapped from a pine tree into the Crossroads area. From there walk 500m to **R16**--use one full rope to go over 3 potholes; or climb up to the left and HL down from a tree to avoid the potholes.

Shortcut Route Let's stop at the **Crossroads** area for a moment. Some hikers prefer a shorter route which can be done in **one very long day** by a small group (2 or 3) of very fit individuals who have been there before, and aren't interested in sightseeing. To do this in one day, get a hiking permit and a **special pass** to drive your car up to **The Grotto**. Start hiking the next morning before daylight using **headlamps**. Doing it in one long day means pushing every minute and maybe sacrificing the chance to take pictures. Also, do it in late **May, June or July** with long days. Or, it can be done in 2 days rather easily and at a less-hectic pace.

Now here's the route description. Get off the **bus** at **The Grotto** and hike up the **West Rim Trail**. At Scout Lookout, you'll go north up over a high point, then drop down to what might be called the **South Fork of Telephone Canyon (?)**. There you'll cross a **wooden bridge**. About 100m beyond this bridge, veer right and angle downhill going due north. You should see the beginnings of a hiker-made trail. Work your way down to where 2 little canyons meet, then head north along a minor ridge between 2 parallel drainages which is between the letters O & N in the word NATIONAL on the *Temple of Sinawava 7 1/2' quad*. You can go up the east or west fork, but our group went up into the east fork just north of those letters.

The **Shortcut Route** goes up to a pass and down the other side on a trail, then up to another pass with one steep short climb. Finally, you'll head down a little side-canyon into Imlay. About 200m before reaching the bottom of Imlay, veer right to avoid dropoffs and onto a rounded slickrock hogsback ridge with scattered ponderosa pines. From just above Imlay's bottom, you could veer left or west, cross over the bottom of the drainage you just came down, and DC to the bottom; or even better, **veer right** or northeast and head straight down toward a **big undercut alcove** immediately below **R16**, and carefully DC there. With big packs, you may have to help each other over one little steep place. Perhaps lower packs with a short rope. That alcove offers a great place to camp with pothole water nearby.

Now back to the main canyon below the Crossroads. After the alcove it's walking for nearly 1 km with a number of DC's, long swimming/wading pools, and **Raps 17-19**, before you reach the **Dark Slot** which contains **Rappels 20-29 in about 200m**. In this part you'll have some interesting challenges. Rap 22 is through a *window or bridge* on the left from another sad-looking piton (someone may have pulled this one out?); or you can send one person down each side of the window on opposite ends of one rope into the same pool countering each other's weight. Below **Rap 24** is the darkest part of the canyon. At that point the first person through may want to pull out a **headlamp**. Right after **Rap 28** is a **4m deep KPH** with holes already there to hook out with. Halfway down **Rap 29**, someone may have to **lasso a log** on the outlet side, then pull him/herself over to exit. On the author's 2nd trip, all the big potholes in this section were nearly dry, so we rappelled

Map 59, Imlay Canyon, Zion National Park, Utah

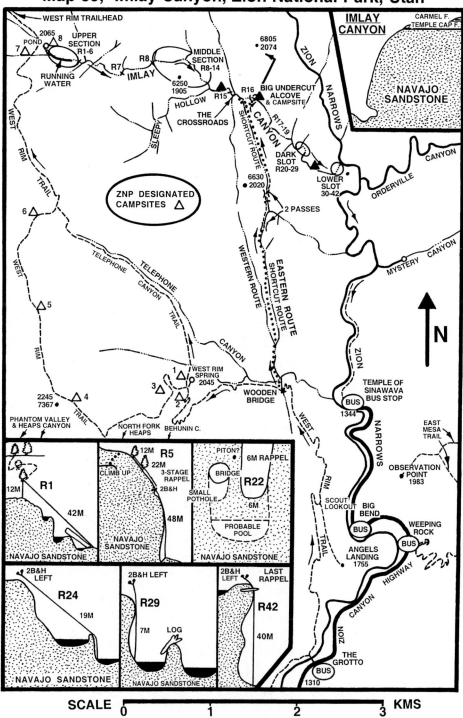

WEST RIM TRAILHEAD

IMLAY CANYON

CARMEL F.
TEMPLE CAP F.

2065 8
POND
7

UPPER SECTION R1-6

RUNNING WATER

R7

R8

MIDDLE SECTION R8-14

IMLAY

6250 1905

SLEEPY HOLLOW

THE CROSSROADS

R15

R16

BIG UNDERCUT ALCOVE & CAMPSITE

R17-19

ZION NARROWS

NAVAJO SANDSTONE

6805 2074

WEST RIM TRAIL

ZNP DESIGNATED CAMPSITES △

SHORTCUT ROUTE

CANYON

DARK SLOT R20-29

6630 2020

LOWER SLOT 30-42

ORDERVILLE

CANYON

6 △

2 PASSES

WEST RIM TRAIL

TELEPHONE

TELEPHONE CANYON TRAIL

WESTERN ROUTE

EASTERN ROUTE SHORTCUT ROUTE

MYSTERY CANYON

△ 5

CANYON

ZION

△ 1

WEST RIM SPRING 2045

3 △

2 △

WOODEN BRIDGE

TEMPLE OF SINAWAVA BUS STOP

2245 7367

△ 4

PHANTOM VALLEY & HEAPS CANYON

NORTH FORK HEAPS

BEHUNIN C.

WEST RIM TRAIL

NARROWS

BUS 1344

EAST MESA TRAIL

OBSERVATION POINT 1983

N

SCOUT LOOKOUT

BIG BEND

BUS

WEEPING ROCK

BUS

ANGELS LANDING 1755

CANYON HIGHWAY

ZION

RIM TRAIL

THE GROTTO

BUS 1310

Inset diagrams

R1 — 12M, 42M — NAVAJO SANDSTONE

R5 — 12M, 22M, 3-STAGE RAPPEL, 2B&H, CLIMB UP, 48M — NAVAJO SANDSTONE

R22 — PITON?, 6M RAPPEL, BRIDGE, SMALL POTHOLE, 6M, PROBABLE POOL — NAVAJO SANDSTONE

R24 — 2B&H LEFT, 19M — NAVAJO SANDSTONE

R29 — 2B&H LEFT, 7M, LOG — NAVAJO SANDSTONE

R42 — 2B&H LEFT, LAST RAPPEL, 40M

SCALE 0 1 2 3 KMS

into this pothole, then with a little push got out fine. The rest of the group was pulled over to the log with the rappelling rope. Just below that are several small, but deep potholes to jump over or crawl through, then it's an open walk for 500m to the **Lower Slot**.

The **Lower Slot** has **Raps 30-42** in about **250m**. In this section are **3 big KPH** about **5m deep**, so have your hooking kit ready. Thanks to your predecessors, there are now plenty of hook holes in place. Also, some people now use a **cheater stick** (a **long tent pole** will do) with an **Ibis hook taped** to the end along with an **etrier & biner** attached to the hook. The **3m-plus-pole** reaches up to the **W&R/QL** at the next rap station, then everyone climbs up. A pack toss or shoulder stand may not work here (?).

From where you get out of one possible KPH, there are several B&H's above to the east indicating some have gone up and around to avoid the KPH below. But this place was no problem for us; once when the KPH was nearly full of water, once when it was almost dry. Plenty of hook holes there too. Another option; with a Talon or Ibis hook in each hand, you might climb/hook out with the holes already there.

When you first begin to **hear the Virgin River**, there will be **6-7 more rappels** to make--some from B&H's, others from W&R/QL on logs. The last rappel is from 2 downward-slanting bolts on the left. The drop is just under **40m** and mostly a free rappel into water next to the Virgin River. Do this one with your pack tethered to the front of your harness. From there, it's about another hour down to the **Temple of Sinawava bus stop**, passing the mouth of Orderville Canyon and the shining water & mossy wall of lower **Mystery Canyon**.

Elevations The Grotto, 1310m; West Rim Trailhead, 2271m; pond at head of Imlay, 2065m; Temple of Sinawava bus stop, 1344m.

Time Needed To do the **full length** of the canyon, you'll need at least two 10 hour days, or maybe two 12's, or more. But for most, it may be more enjoyable, and with more foto opt, if taken in 2 1/2 or 3 days. The **Short-cut Route** can be done in as little as one 14 hour day, but only with a small & physically fit group who know the canyon, with little or no time for fotos. It's normally done in 1 1/2 days--camp at the alcove.

Water There's running water in the first 300m of the Upper Section, plus dozens of potholes throughout the canyon. Also, from the Virgin River and good water from lower Mystery Canyon.

Maps USGS or BLM map Kanab (1:100,000); Temple of Sinawava (1:24,000--7 1/2' quad) for hiking; and the plastic Trails Illustrated/National Geographic map, Zion National Park (1:37,700) for hiking & driving.

Flash Flood Danger High in the 4 slot sections, but in between are lots of places to get out of harms way.

Best Time to Hike It might be easier when potholes are full. Park rangers say May might be good, because running water from snow melt will fill the potholes. Or in the middle, or just after the monsoon season, and/or after there's been some good rains in the area. If the KPH's are full, you just swim across and you're out; if less than full, hooking out will slow you down. A cheater stick could be faster.

Author's Experience On the 1st trip, the author went with Bill Bees and Nat Smale. The evening of 9/21/2001, we left cars at the visitor center and took the bus to The Grotto. Hiked just over 2 hours, until just after dark using headlamps and bivouacked near the West Rim Spring. Next morning we hiked 1 1/2 hours to the pond & spring at the **head of the canyon**. From the first rappel to the campsite about 200m above the Crossroads took 6 1/4 hours--8 3/4 hours total for Day 2. Day 3, it took about 8 1/2 hours to get everyone to the bottom of the last rap, then about 1 1/4 hours to Temple of Sinawava and the bus--9 3/4 hours for the day. Total hike-time was 20 1/2 hours.

The 2nd trip was with Leroy, Craig and Byron Anderson. Got started hiking from The Grotto after 10am, then went in via the eastern branch of the **Shortcut Route**. The first day ended at the big undercut alcove below R16 discussed above; 6 3/4 hours walking. Day 2, it took our group 4 1/2 hours to get through the Dark Slot, then 5 hours through the Lower Slot. Day 2 took 11 1/2 hours; and a total hike-time of 18 1/4 hours for the trip. Our time may have been a little slower than average.

Nat Smale & Bill Bees tied into the big wall in the Upper Section of **Imlay Canyon**. If doing this rappel, be sure to have a couple of R/QD/DC's per/person to tie into the anchor while preparing for the 2nd half of the rap.

Above Left Bill Bees crawls out of a deep pothole in the Dark Slot of **Imlay Canyon**. It's here you can lasso the log on the other side, or rap to the bottom and climb up the log. Don't be surprised if that log is gone when you arrive; every big flood changes the canyon bottom.

Above Right Tim Barnhart using a 3-prong talon hook to get out of one of 3 KPH's in the Lower Slot of **Imlay Canyon**. Sometimes with a hook in each hand, and if you have strong arms, you can pull yourself out using exiting holes. (Jon Jasper foto)

Left Tim standing at the top of the last rappel in **Imlay Canyon**. About 40m below is the **Virgin River** and **Zion Narrows**. (Jon Jasper foto)

Heaps & Behunin Canyons, Zion National Park, Utah

Location & Access Heaps and Heaps Canyons are located in about the middle of Zion National Park, and just north and a little west of the Zion Lodge. Get there by driving into the park from the east or west on Highway 9. Immediately north of Springdale is the visitor center where you must get a permit, preferably the afternoon before your hike, at the **backcountry desk**. From the visitor center, where most people leave their cars, get on the shuttle bus heading upcanyon (you can also get on the shuttle at **Zion Museum** or **Canyon Junction**). Get off at **The Grotto**, which has a picnic site, toilets, drinking water and shade. This will be the beginning and end of both hikes. Or at the end of your hike, you could walk to Zion Lodge and catch a bus downcanyon from there.

Rating Behunin: 3A III or **IV**. This is quite an easy hike as technical canyons go, but with a big free rappel at the end it could up the rating to **R** (?). **Heaps Canyon: 4B R or X IV or V. Heaps along with Imlay, are perhaps the 2 most difficult & potentially dangerous canyons in this book. It's for experienced canyoneers only!**, people who are very fit and can do chin-ups out of dozens of potholes, plus people who can make a **93m** free rappel.

Equipment Behunin--RG with a variable speed control ATC, two 50m ropes (or a 50 & pull cord) that are really **50 meters long!**, leather gloves, R/QD/DC, W&R/QL, and a 15m rope for one short rap/HL. **Heaps**--RG with a variable speed control ATC, plus one backup rappelling device per group, one 100m rope (in a rope bag & drybag so it doesn't gain weight in water), a 50 & 60m ropes (perhaps a 3rd 50 as a backup), a 15m rope, thick wetsuit or drysuit, leather gloves, hooking kit and one set of ascenders/group, one sharp knife each, 2 R/QD/DC and 6 carabiners per/person, at least one helmet for the group leader (maybe one for each person?), lots of W&R/QL, water purification tablets or filter, duct tape, headlamps, drybags and an Otter/Pelican case for cameras. Also 1 BK & 2 B&H's per/group--ask a ranger about the condition of the anchors. If you have walkie talkies, take them; this will make communication easier at the last big rappel.

Route Behunin For both canyons, from **The Grotto**, follow the signs & trail west across the river bridge, then turn right or north onto the **West Rim Trail**. This takes you up **Walter's Wiggles**, Refrigerator Canyon, passes **Scout Lookout** near Angels Landing, then north along the West Rim of Zion Canyon. Further up, you'll turn westerly. Follow the map carefully, because at what this writer is calling **Behunin Pass**, you'll turn left or west off the main trail, and follow a recently-made hiker's track down into the head of Behunin Canyon.

Once in Behunin, you simply walk south down the normally-dry creek bed. In the upper part you may find a trickle of water and small pools, but a hiker's trail skirts these. About halfway through, you'll turn east and come to an open slickrock area. This is where the first 4 rappels are found, one after another.

At the start, as you look east down into a couple of big potholes, glance to the right or south, and you'll see a small oak tree (large bush) on a ledge. Just beyond that on the right are **2 B&H** and chain. See rappelling diagrams. Tie in there with one 50m rope and **rappel 10-12m** over the upper ledge. At the bottom, you could walk around the corner to the south to a pine tree with W&R/QL which is your next rappelling station; or use both 50's (or one 60m), make a longer rap, which makes an easier walk to the same pine tree. That tree is the anchor for **Rappel 2**. From there, it's down a low angle slope to another pine tree for **R3**. From there, you'll need all of two 50's to get down to **3 B&H** with chains and the beginning of **R4**, a drop of 30m.

After the 4th rap, it's an easy 10 minute walk south to the next dropoff and **R5**. Two B&H with a chain are on the bench to the right and you may need both 50's (or one 60 for the **26m drop**). If there's been rain just prior to your hike, you may land in shallow water. After R5, it's another easy walk through some narrows. After walking 7-8 minutes, you'll come to a log with W&R/QL and a drop of **12m** into a **swimming pool**. Or, if you don't want to swim, turn right at what could be **R6** and follow a trail up onto a bench and along the west side of the canyon. After 60m or so, and as you drop back down near the bottom, use a short rope around a small pine tree to return to the main drainage safely. Then more easy walking.

At the next drop, climb up to the left to a pine tree with W&R/QL. You could DC a steep slope, but it's equipped for the **7th Rappel**, so you might as well error on the safe side. After dropping down a dozen meters or so, you'll find another pine tree and more W&R/QL. Rap down **R8** carefully, jumping from side to side to avoid landing in a pool. At the bottom, crawl along a ledge on the right to avoid wading.

About 100m past R8 is the end of the canyon. DC on the right or west side of a large boulder to find a smaller chokestone & W&R/QL which is the anchor for **R9**. Using both ropes, it's an easy low angle rappel down to a large flattish ledge and the beginning of the next-to-last drop, **Rappel 10**. The anchor for this is to the left under another chokestone. This now consists of **4 B&H** and chain, plus **2 more B&H** & W&R/QL. From there, use both 50's and tether your pack to the front of your harness. Each 50 barely touches the rocks below. However, after rappelling about 8m down an inclined slope, you'll come to a small ledge and **3 more B&H** & W&R/QL. Either someone had short ropes; or they were worried about not being able to pull their ropes easily from the bottom (?). At that point on his first trip, the author couldn't see his ropes touching the rocks below, so he tethered himself to this lower anchor with a runner, retied his 50's and rappelled.

Here's something to remember; if you have a 50m rope cut at the factory, you'll have no problem here because they always add about 5m to compensate for shrinkage which will eventually come whenever a rope gets wet. The author cut his two 50's from a 200m roll, which made his ropes exactly 50m when new. After shrinkage, they were closer to 46-47m under full weight. Just one more thing to worry about.

At the bottom of the last rappel, separate the ropes, tie your 15m rope to the end of one 50, and walk east a ways for an easier pull. Another thing to think about before the last person starts this rap, make sure the ropes aren't twisted or crossing each other at the top--otherwise you'll have hell to pay trying to get the ropes down! Another way to avoid crossing the ropes is, separate the ropes before the last person comes down. Walk the pull side to one side, and have the last person come down a single rope using a beaner block & variable speed ATC (could be a fast ride down, so use leather gloves!). At the bottom of this last rap is the start of good running water, then it's route-finding & scrambling down to the **Emerald Pools Trail**, which takes you back to **The Grotto** (or the **Zion Lodge bus stop**) and a bus back to your car at the visitor center.

Heap Canyon There are **2 ways** to get through Heaps; down the **North Fork**, or the slightly longer **Phantom Valley** route. Either way, start at **The Grotto**, walk northwest from the bus stop across the foot bridge over the Virgin River, turn right and head up the **West Rim Trail** toward **Scout Lookout**. Once on the rim, continue north then west past the head of Behunin Canyon and tank up on water at **West Rim Spring**. From there, take the trail zig zagging up toward the higher plateau southwest from the spring. This main trail following the rim allows you to look down into **Behunin**, **North Fork** and finally into **Phantom Valley**.

Phantom Valley Route About 2 kms beyond the West Rim Spring, you'll come to the high point of the plateau in ponderosa pines which is very near bench mark **7367** (2245m) as seen on the plastic *TI/NG* map (see below). About 100m after you pass the trail to the right signposted to the **4th Campsite**, and about 75m before you come to a great **big stump & pine tree** that's been **cut down** on the left, turn left or southwest, and walk through ponderosas and a little brush to the rim. You should see a tent site on the south slope at

Map 60, Heaps & Behunin Canyons, Zion N. Park, Utah

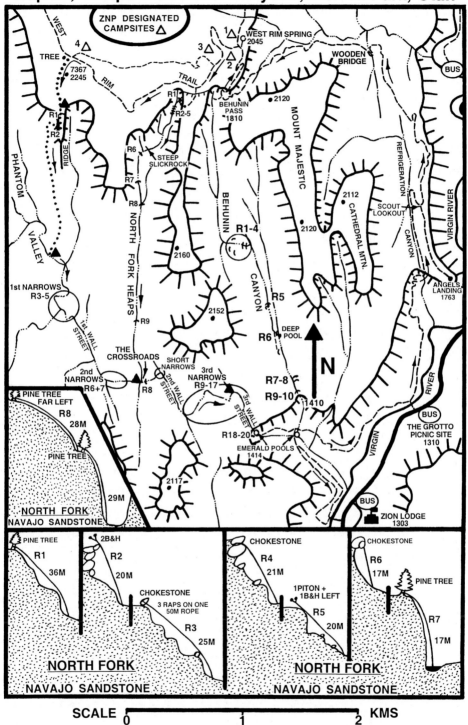

SCALE 0   1   2 KMS

the beginning of an obvious ridge, the first ridge east of upper Phantom Valley.

From there, head south down the ridge along an emerging hiker's trail 400m to **Rappel 1**, then after another 150m, do the **2-stage R2** down the west side of the ridge. At the bottom, walk south over a hump and down the slickrock ridge and into the bottom of Heaps dry creek bed. After that, the first km is easy, then it slots-up and you DC into pools. This is the **1st Narrows with rappels 3-5**. Rap 3 is down through 2 potholes, both of which could be **KPH's** if the water is low, so have your hooking kit handy. Or, with a short rope do a pack toss over the rim of the 1st pothole to the left and into a 3rd pool to have something to pull on. See cross-section of 1st Narrows. Do some preplanning & thinking about this possibility before you get there and before pulling your rope. This is the reason you may want to do this canyon just after spring runoff or during the monsoon when the potholes are full. The same situation could exist 2 potholes below the **first 2 KPH's**, the one just above the little bridge you swim under. See cross-section. This part seemed to be the worst section, but Heaps is full of **KPH's**, so go prepared with a **hooking kit** including **G-pick**, plus know how to do a **pack or bag toss**, and the time-honored **shoulder stand**, sometimes with water over your head.

Also, you'll be climbing up and over log jams and jumping or rappelling 2-3m into potholes. Before jumping, always belay one person down to test water depth, then others can slide or jump. There could be 50 or more potholes in this canyon!

After R5 comes the 1st of 3 straight walking sections--reminiscent of **wall street**. Raps 6 & 7 are in the **2nd narrows** with lots of logs to climb over & around, then a 200m walk to where the North Fork comes in on the left. This can be a camping area at what is commonly called **The Crossroads**.

North Fork Route Let's stop for a moment and introduce another way down to this point. If you decide to do the North Fork Route into Heaps, then follow the *Temple of Sinawava 71/2' quad* carefully. About **1 km** by trail southwest of the **West Rim Spring**, and immediately west of the word "west" on the *7 1/2' quad*, turn left or south and walk down the beginnings of a hiker's trail to flat ground 100m below. Once there, turn right or west and route-find through brush (should be a trail there now) to the northwest and the beginning of the North Fork. Locate a pine tree and W&R/QL and make the 1st of **9 rappels** in this canyon. About 100m after R1 will be **4 raps**, one after another, through a narrow slot. Then there's easy walking for 500m until you reach some steep sloping slickrock as the canyon turns west. Head up to the right to begin, then route-find down carefully doing some foot-sliding. Just after that the canyon turns south again with **3 rappels** in the next 500m or so. R8 is a long **double rap** down to the west from 2 pine trees. Using your 60 may get you down to a 2nd pine tree (if not, use 2 ropes); then the second part, again using the 60 as it may get you to the canyon floor again (?). After that it's easy walking with one more rap with some DC'ing and maybe a short swim or two to The Crossroads.

Now down the main canyon again. Just below The Crossroads is **Rappel 8** (counting from the Phantom Valley Route) from a tree on the right--or head straight down the drainage DC'ing through a couple of potholes. Not far below that is the **Short Narrows** shown on the map. It's here you'll find a couple of jump-ins, the 2nd of which will be a **KPH** if the water is low. You might hook out, or have one partner go under water and boost a second person up. This is what we did the 2nd time through.

After that is the **2nd wall street**, and finally the **3rd Narrows**; the longest. This one doesn't look too long on the map, but it seems endless. This includes **Rappels 9-17** and lots of pothole swimming. After R17, you'll walk straight again for 200m, then the bottom drops out on the left and you'll have to climb up a steep pitch to the right along a fracture line. Once on a high point, you'll find a flat place which will be your **staging area** for the last 3 rappels. Stop there to organize your group and gear; prepare ropes, attach 2 R/QD/DC per/person, etc, etc.

Once fully prepared, walk down a crack 15m or so to a pine tree with W&R/QL. R18 is about **14m** down to another pine tree. This is where you'll have to move slow to prevent dislodging small rocks, and where you'll want to use your R/QD/DC to secure yourself to the second pine tree. **Suggestion:** 2 walkie talkies around your neck would come in handy for communication between the last 2 raps and the Upper Emerald Pool below. This is also where you'll be looking straight down at the Pool about **140m below**. **Rappelling Station 19** is only big enough for 3 people at best, so plan & think this part out ahead of time. R19 is about **48m** from a pine tree straight down a fracture line with your feet on the wall.

The **last anchor & rappelling station** is 4 B&H on the right with a drop of **93 meters**. At this **bird's perch**, there's one good place for the group leader to sit or stand, & space for one other person in a crack--having 3 people plus packs would be a tight squeeze--4 is an absolute maximum. This is where everybody moves very methodically in near panic-mode and everyone must be tethered to the anchors with 2 R/QD/DC.

Once the leader is at that point, the 100 is secured and **lowered carefully--no knots!** (tossing a rope bag with the 100 inside might work the best), then others come down the two 50's, tether in momentarily, switch over to the 100, and move on down the single rope **with packs tethered to the front of their harness. Doing this keeps you sitting upright so you can use both hands as brakes if needed. Leather gloves are required for this one.**

After everybody is down R19, the last person/leader will pull the 50's and tie them on the other side of the 100. Tie the ropes together with a simple overhand knot, then pull up on the 100 & down on the two 50's (which will be your pull rope), until the knot is over the edge of the cliff by a couple of meters. Then have someone on the ground pull hard on the pull rope (the two 50's side) and either tie it to a tree, around a rock, or just belay. Then the last person comes down the single 100. With this game plan, everyone does a single rope rappel, and your knot is over the edge and can't get stuck. This makes it easy to pull from below.

To do what can be a fast rappel & a hot brake hand, use a **variable speed control ATC (Best)**; or with a regular ATC, tie a **munter hitch** to a locking carabiner on your brake-hand-side leg loop. Or from a regular ATC, wrap the rope around a locking biner twice on your leg loop--**practice this before doing Heaps and this 93m rappel!** This causes more friction to slow you down, but it twists the rope and you'll spin like a top going down! **Figure 8's are not recommended** here (or anywhere) because they twist the rope.

There are options to carrying a 100m rope through the entire canyon. This involves caching a 100 at the bottom of the last 93m rappel before the trip, then having one **EXPERT CLIMBER** rap down the two 50's tied together and passing the knot in the process. This is no easy task and **definitely not recommended!** Details are purposely withheld here because this situation is unlike any other most people will ever face.

Another option: carry a lighter 100m-long, 4-5mm cord through the canyon, then lower it to a friend on the ground (or throw it using a rope-bag) and the full-sized 100 attached to that. Then haul up the 100 and re-rigged. Or for a party of 3 or less, pull your two 50's at the birds perch, let one end fall to the ground, and have someone attach the 100. Use the two 50's as the pull rope--**but get the knot over the edge before the last person begins rappelling. The best way to handle this canyon & big rappel is to go with someone who has already been there/done that, and/or had lots of experience.**

Elevations The Grotto, 1310m; Behunin Pass, 1810m; bottom of last rappel of Behunin, 1410m. For Heaps, high point near Campsite 4, 2245m; bottom of last rappel/Emerald Pool, 1414m; Zion Lodge, 1303m.

Time Needed Behunin--Most groups make the hike in 7-10 hours. **Heaps**--For a strong & experienced group of say 3, one very long day of 14-16 hours (or more!) to do the **Phantom Valley Route.** Get a car pass to drive up & start in the dark using headlamps, and take 2 lunches--it'll be a long day. Another option would be to travel extra light and bivouac at the high point near Camp 4, then cache camping gear in the bush to be picked up later. Otherwise it might be best to travel light and spend 2 full days in the canyon.

Using the **North Fork Route** shortens the trip a little making it easier to do in one day with day-packs, maybe 12-14 hours. Or camp on top, as suggested above, and cache camping gear to be picked up later. Doing this in one day will mean a smaller pack, less time for crawling out of potholes and more time taking fotos.

Water The Grotto, but start with 3-4 liters in summer; you may find small trickles in Behunin. Or at the West Rim Spring (purify), in dozens of potholes and just above the Upper Emerald Pool.

Maps USGS or BLM map Kanab (1:100,000); Temple of Sinawava (1:24,000--7 1/2' quad) for hiking; or the plastic Trails Illustrated/National Geographic map, Zion National Park ((1:37,700) for hiking & driving.

Flash Flood Danger Behunin: low to moderate. **Heaps:** high in the 4 narrows, otherwise moderate (?).

Best Time to Hike Behunin: April through October (March or November might also work). **Heaps:** When the potholes are full, especially those in the 1st narrows, which might be just after spring runoff (late May or early June); or in the middle of the monsoon season and after rains. May, June & July would have more daylight hours than in late summer or fall. Consider this if doing it in one day.

Author's Experience Behunin: The author did this one alone on 10/12/1998, in 7 1/2 hours from The Grotto. On 7/14/2004, along with Jim Schnepel, we did the canyon from The Grotto with one 60m rope and a 5mm pull cord. All raps were down a single rope using a locking biner on the leg loop to slow things down (That does twist the rope! Using a variable speed control ATC normally works well by itself, but you'll need leather gloves on both hands. Or safer & slower still, set up a munter hitch with locking biner on your brake-hand side). This worked OK and we did the trip in 6 3/4 hours returning via the Zion Lodge bus stop.

Heaps: On 8/19/2001, with Bill Bees, who managed to get a special pass to drive to and park in the canyon, we started hiking from The Grotto at 6:10am, were at the high point on the trail at 8:40am and beginning to drop down the ridge into **Phantom Valley**. In the 1st narrows, we passed Bao ? and 4 friends who had camped below R2 and had full-sized packs. Soon after, Bill lost one 50m rope in a deep pothole--dry ropes do not float in water when they're old & coiled! Shortly after that, Bao's group destroyed one of their 50's--then it was one big group of 7 not-exactly-happy campers!. Going was very slow and a night was spent sleeping in wet/drysuits with lots of mosquitos about 400m from the last rappels.

Next morning it took 1 1/2 hours to reach the last 3 raps. The author was the first down the last rappel using Bao's **100m dynamic rope (Not good!)** Near the bottom, the outer sheath stretched & bunched up into knots making a bouncing yoyo-like descent which was scary as hell!), then he tied the 100m static rope we had hidden the evening before near the pool, to the other 100. The static 100 was hauled up and used for subsequent raps. It took 3 1/2-4 hours to get 7 people down the last 3 rappels. One ZNP ranger came to the pool with a radio and informed dispatch that everyone was accounted for and family & friends who had called were informed. We were back at The Grotto at 2:30pm on Day 2. Total hike-time, just under 21 hours.

A 2nd trip was with Blake Emett and Craig Anderson on 6/14 & 6/15/2002. We left in the afternoon and took 3 hours to reach a campsite at the head of North Fork. Next morning, one tent was left standing with gear inside and was picked up later by other Anderson family members. We went down the **North Fork** to The Crossroads in about 4 hours, then another 5 hours or so to reach the last raps. Getting everybody down took

about 3 hours, but we had one big rope tangle at the last pine tree, then a knot was discovered near the end of one 50 immediately above the **bird's perch**. What a mess! A light cord was lowered to fetch up the 100m rope from the bottom. Then after the author pulled the two 50's, one got tangled in the bush immediately below the last anchor at the **bird's perch**. Another mess and time lost! Even though it may be technically illegal, will someone please take a small hatchet or sharp pocket knife and chop & drop that small bush at the top of the last rappel so that doesn't happen again. This will also make better pictures! By the time we got to the bus stop near Zion Lodge, it was 13 3/4 hours for Day 2. Total 2-day walk-time was 16 3/4 hours.

On 8/9/2004, a 3rd trip was with Keith Krumbach & Warren Egbert. With a pass, we drove to The Grotto and started in the dark. It took nearly 3 hours to reach the 1st rap along the **Phantom Valley Route**. The pools were all full and things went well--except for leaky drysuits. This made the last narrows, the Endless Slot, pretty chilly! A partner on the ground (using radios) attached a 100m rope to our 4mm cord and the last rap went well. We were back at The Grotto after 14 1/3 hours.

If you're hiking down through **Phantom Valley** at the head of **Heaps Canyon**, this will be Rappel 3, a 2-stage rap into a 2nd pool just beyond the canyoneer shown. If water is low in either pool, they both could be KPH's! From where Warren Egbert is, you may have to do a pack/bag toss to the left and over the lip of the next possible KPH. Before going, ask rangers about water levels in these pools.

Heaps Canyon Rappels via Phantom Valley Route

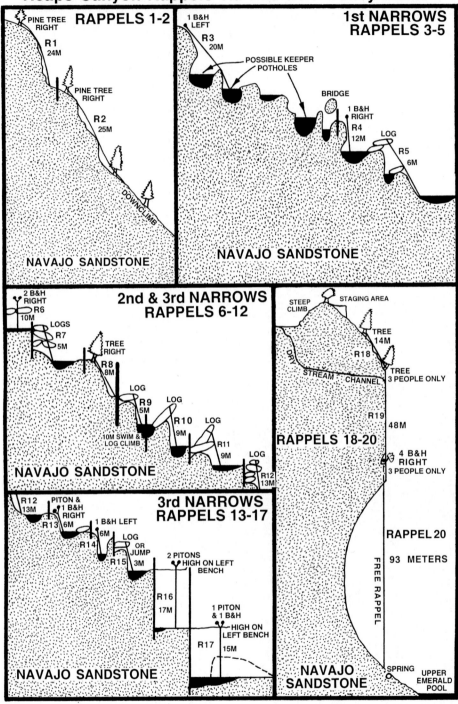

RAPPELS 1-2

PINE TREE RIGHT

R1
24M

PINE TREE RIGHT

R2
25M

DOWNCLIMB

NAVAJO SANDSTONE

1st NARROWS RAPPELS 3-5

1 B&H LEFT

R3
20M

POSSIBLE KEEPER POTHOLES

BRIDGE

1 B&H RIGHT

R4
12M

LOG

R5
6M

NAVAJO SANDSTONE

2nd & 3rd NARROWS RAPPELS 6-12

2 B&H RIGHT

R6
10M

LOGS

R7
5M

TREE RIGHT

R8
8M

LOG

R9
5M

LOG

R10
9M

LOG

10M SWIM & LOG CLIMB

R11
9M

LOG

R12
13M

NAVAJO SANDSTONE

3rd NARROWS RAPPELS 13-17

R12
13M

PITON & 1 B&H RIGHT

R13 6M

1 B&H LEFT

R14 6M

LOG OR JUMP

R15 3M

2 PITONS HIGH ON LEFT BENCH

R16
17M

1 PITON & 1 B&H

HIGH ON LEFT BENCH

R17 15M

NAVAJO SANDSTONE

RAPPELS 18-20

STEEP CLIMB

STAGING AREA

TREE 14M

R18

DRY STREAM CHANNEL

TREE 3 PEOPLE ONLY

R19

48M

4 B&H RIGHT
3 PEOPLE ONLY

RAPPEL 20

93 METERS

FREE RAPPEL

NAVAJO SANDSTONE

SPRING

UPPER EMERALD POOL

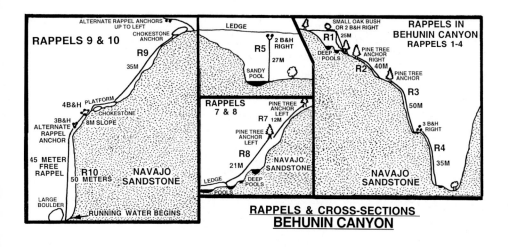

RAPPELS 9 & 10

ALTERNATE RAPPEL ANCHORS
UP TO LEFT
CHOKESTONE ANCHOR

R9

35M

4B&H PLATFORM

3B&H CHOKESTONE
ALTERNATE 8M SLOPE
RAPPEL
ANCHOR

45 METER
FREE
RAPPEL

R10

50 METERS

NAVAJO SANDSTONE

LARGE BOULDER

RUNNING WATER BEGINS

LEDGE

R5 2 B&H RIGHT

27M

SANDY POOL

RAPPELS 7 & 8

PINE TREE ANCHOR LEFT

R7 12M

PINE TREE ANCHOR LEFT

R8

21M

LEDGE DEEP POOLS

POOLS

NAVAJO SANDSTONE

SMALL OAK BUSH OR 2 B&H RIGHT

R1 25M

DEEP POOLS

PINE TREE ANCHOR RIGHT

R2 40M

PINE TREE ANCHOR

R3

50M

3 B&H RIGHT

R4

35M

NAVAJO SANDSTONE

RAPPELS IN BEHUNIN CANYON
RAPPELS 1-4

RAPPELS & CROSS-SECTIONS
BEHUNIN CANYON

This picture shows the first rappels near the head of **Behunin Canyon**. Rappel 1 begins in the upper right with Jim Schnepel getting ready. From there you can rap to the first bench down, then contour straight ahead over some steep slickrock, or go over a 2nd drop to the lower left, then contour straight to the big pine tree in the lower left-hand corner of this foto. That's your anchor for Rappel 2.

Above Left This is the 10m-long swim after **R9** in **Heaps Canyon**. In the upper left-hand corner is the log you'll have to climb, then make the next rappel from it. If that log is gone, washed away by a flash flood, you'll have to hook out because a pack toss won't be possible. Before you go, ask the rangers at the backcountry desk about all anchors & the water level of all KPH's.

Above Right Looking down from the top of the next-to-last rappel **(R19)** in **Heaps Canyon**. At the bottom of this 48m rap, is the bird's perch, 4 B&H's and the beginning of the last 93m rappel to Upper Emerald Pool.

Right This is the end of **Heaps Canyon** as seen through a telefoto lens from the Upper Emerald Pool. This picture was taken in early 3/2004 with water running down the stream course. Just to the left and above the top of the water-fall is the pine tree which is the anchor for **R19**. Hikers rappel down the crack below the pine tree and land on the bird's perch just above the undercut & shadows. From there it's **Rappel 20, 93m** straight down to the Upper Emerald Pool.

Above Left Starting Rappel 3 from a pine tree in the upper end of **Behunin Canyon**.

Above Right Starting the last rappel in **Behunin Canyon**. (Jim Schnepel foto)

Left Jim Schnepel setting up the rope for the last rappel in **Behunin Canyon**. There are 5-6 B&H's making up this anchor.

Englestead & Birch Hollows, and Orderville Canyon, Zion National Park & Vicinity, Utah

Location & Access Englestead & Birch Hollows are upper southeastern tributaries of **Orderville Canyon**, all of which are located along the eastern boundary of Zion National Park. To get to the trailheads or car-parks, first drive along **Highway 9** which connects Zion Canyon & Springdale on the west, with Mt. Carmel Junction & Highway 89 on the east. Between **mile posts 46 & 47**, turn north onto a paved road signposted for many places including *Zion Ponderosa Ranch & Resort and East Zion* (this is the same road you take if going to the **Chamberlain Ranch** and the beginning of the **Zion Narrows Hike**).

To reach the **Englestead Hollow Trailhead**, have the *Temple of Sinawava & Clear Creek Mountain 7 1/2' quads* in hand, then drive northwest from Highway 9 about 8.3 kms (5.2 miles) on the paved road, then turn left or west at the gate & sign reading, *Zion Ponderosa Ranch & Resort*. This private land development has a maze of dirt/graveled roads and summer homes & cabins. From the gate (Km & Mile 0) drive west & northwest staying on the most-used road about 1 km (.6 mile) to a junction marked **6495** (1980m). Continue northwest to Km 2.2/Mile 1.4 & T-junction marked **6590** (2009m). Turn right or north and drive to Km 3.5/Mile 2.2 and a 5-way junction marked **6530** (1990m) in an open meadow. Continue north, again on the most-used road to Km 4.5/Mile 2.8 and another junction. At that point you could turn left or west toward ZNP boundary and the **East Mesa Trail** running to **Observation Point & Weeping Rock**; but instead, drive straight ahead or north. From the last junction marked **6495** (1980m) at Km 4.8/Mile 3, continue north for 100m and park/camp under the shade of ponderosa pines.

If you park there at the normal beginning to Englestead, you'll have to leave a 2nd 4WD, car or mtn. bike at the head of Orderville Canyon--if you intend to go up Orderville at the end of the hike through Englestead. To do that, continue northward on the **Chamberlain Ranch Road** from the *Zion Ponderosa Ranch & Resort* turnoff for another 10 kms/6.2 miles (18.3 kms/11.4 miles from Highway 9) then turn left at the BLM sign reading *Orderville Canyon W.S.A.* Drive 260m to a corral and leave one car there. Read more below.

Or, if you intend to go down Orderville to Zion Narrows and leave the canyon via the shuttle bus back to the visitor center, then leave another vehicle at Zion Visitor Center or at Canyon Junction. Then drive back to retrieve your car at the Englestead Hollow Trailhead after your hike.

To reach the **Birch Hollow Car-park**, turn north from Highway 9 as described above, drive 13.7 kms (8.5 miles), and park where Birch crosses the road. If coming out via **Wild Wind Hollow (WWH)**, there is no need for a car shuttle. The unpaved part of the Chamberlain Ranch Road is very slick when wet, but is fine for cars when dry.

If you intend to go down **Orderville Canyon** only, then from Highway 9 drive a total distance of 18.3 kms (11.4 miles) and turn left or west at the same BLM sign mentioned above. Drive 260m to the corral; or if you have a HCV or 4WD, continue westward down a steep, rough & rocky track. After about 1300m you'll be at the bottom of the steep part & stream crossing heading west. Continue west for another 2 kms to the end of the road and the 4WD trailhead at 1737m.

To do the **Englestead/Orderville Canyon hike** with one car, park at the ZNP visitor center, take the shuttle bus to Weeping Rock & start hiking along the **East Mesa Trail** from there. You'll finish at Temple of Sinawava bus stop and take a bus back to the visitor center. You could also park at Englestead TH, do the hike, and finish by taking the bus down to Weeping Rock, then hike back on the East Mesa Trail.

Rating For **Englestead**, and continuing down through Orderville & Zion Narrows: **4B R III** (for some this may be closer to **3B III?**). This canyon has a rappel station 1/3 the way down a big 90m wall which is the entrance to the canyon. Below that are another 9 or so shorter rappels/HL's in one of the nicer gorges around. Once into Orderville Canyon, you can walk up to the corral or 4WD parking place; or downcanyon into one of the prettiest narrows anywhere ending in Zion Narrows and the Temple of Sinawava bus stop. **Birch Hollow: 3A II** or **III** if returning up **Wild Wind Hollow**.

To just go down **Orderville Canyon** into Zion Narrows and Temple of Sinawava rates an easy **2B or 3B III**; the same rating applies if going down to the Narrows and returning the same way to your vehicle (leaving 3-4 short ropes to HL back up).

Equipment Englestead Hollow--RG, variable speed ATC, 2 **full length** 60m ropes, W&R/QL, a 15m rope for short raps/HL's, 3 R/QD/DC and about 6 carabiners per/person, BK & several B&H just in case (shouldn't be needed), and drybags. If you're planning to head down Orderville Canyon & Zion Narrows to the Temple of Sinawava bus stop, you'll likely have 3-4 short swims, so take a wetsuit in cooler weather.

Birch Hollow--one 60m rope (be sure to mark the middle of the rope), W/QD/DC and compass (as always) if coming out WWH. If going up Orderville or WWH, you don't need a ZNP permit.

Orderville Canyon (they now give **50 permits a day** if doing just **Orderville**)--one 15m rope for HL'ing down over 3 dropoffs; or three 15m ropes to be left at each dropoff so you can UC back to return to your car. If doing this in spring or fall take a wetsuit for possible swims in the lower end of Orderville.

Route Englestead Hollow If you're starting from **Weeping Rock**, climb up the zig zag trail eastward. After about 2 kms, veer left and follow the **East Mesa Trail** toward **Observation Point**, then the east entrance to ZNP. From the boundary, follow the road (and this map) north to the trailhead parking.

From the trailhead/junction marked **6495** (1980m), walk east on a rough ATV track for 300m, then when that track turns right or south, veer left or northeast onto another ATV track which immediately passes the remains of an **old sawmill site**. Continue east toward the morning sun for another 250m or so, then leave the track and head northeast and walk through some brush. Soon you'll come to a shallow drainage; head northeast straight down the chute to the bottom of Englestead Hollow. Once there, turn left or NNW and walk 200m downhollow to the **big 90m dropoff**.

Once there, walk left about 30m to find W&R/QL around a pine tree which will be the anchor for the 1st rap. Rig up one of your 60m ropes and rappel down about 26m. As you rappel, and as you near the mid-wall rap station which consists of 2 B&H + W&R/QL, angle yourself to the right a couple of meters to actually reach the anchors. Once there, you'll have 2 nice foot stands in a corner, but you'll also have to use 2 R/QD/DC to attach yourself to the anchor. A 3rd R/QD/DC might be used to tether your pack to the front of your harness.

Once secured, your partner above will send down the 2nd 60. Perhaps the best way to do that is to tie an overhand knot & loop at one end of the 60, attach to it a locking carabiner and feed it down the double rope you just came down on. With the biner attached, you can't lose it! Once there, attach the biner to the R/QL and let the 2nd 60 drop. **Another option** would be for the 1st person down to just carry the 2nd 60 in a pack tethered to his/her harness for easy access. Attach a locking biner on one end before rapping.

With a rather small place to stand, as the 2nd person comes down the upper part of the rappel, the 1st person will then switch ropes and rap on down the single 60 using a **variable speed control ATC.** If this doesn't slow you down enough, you can also pull the rope around your back to your off-hand side to slow descent

Map 61, Englestead & Birch Hollows, and Orderville Canyon, Zion National Park & Vicinity, Utah

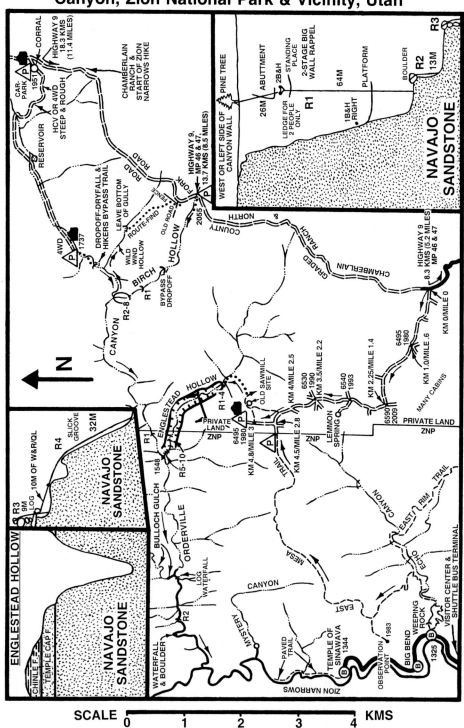

SCALE 0   1   2   3   4   KMS

(a pair of leather gloves help too). Installing a munter hitch on a biner on your brake-hand-side-leg-loop will also slow you down, but that's not really need here because your feet will be on the wall all the time. The last person down will pull the 1st 60, re-tie the 2 ropes together and do the 2nd part of this long rappel on a double rope. A word of **Warning**: we had 2 new 60's which we think measured 65m (?). As it turned out, both ropes were about 15 cms (6") short of the ground! So measure your ropes--the 2nd half of this big wall rap is about 62-63m! But of course, ropes stretch when weighted, and you could just rap off the end of the rope and drop a meter or two to the gravel surface.

   Options By 7/2004, someone had placed 2 B&H about 2/3's the way down this big wall. This means one person could likely do this rap with one 60 in an emergency, but this would include a tricky switchover! Or, you could likely do this with two 50's (?), because about 5/6's of the way down this wall, is a small flat place. On the wall to the south is 1 B&H; it seems someone's ropes weren't quite long enough!

   At the bottom of R1 is the **13m R2** from W&R/QL around a big chokestone. Just below that is **R3** from another chokestone; about 2/3's the way down this 9m rap is more W&R/QL attached to a big log. This is the anchor for **R4** which is about 10m away. **R4** is a 2-tiered rap down about **32m** in a smooth slick groove created by flood waters. This ends the **upper slot**. At times you could wade a little, but there should never be any swimming in any part of Englestead.

   Continue downcanyon. After about 1.8 kms, you'll come to the **lower slot**. In the last 500m, you'll do rappels/HL's 5 through 10, one or 2 of which will be from B&H's on the left; the others will be from logs, or W&R/QL from logs, plus a fair amount of DC'ing through log jams with short jumps; however **expect changes with every flood**. Immediately before arriving at Orderville Canyon, you'll climb up 2m to the left, then walk/DC 20m along a ramp to the bottom.

   Once in Orderville, you could walk upcanyon to the upper end and the 4WD parking, or the car-park at the corral--if you've parked a 2nd 4WD, car or mtn. bike at either of those places.

   Or if you choose to see the best part of Orderville on the same hike, continue downcanyon to the west. Along the way are **2 large chokestones/dropoffs** (below Englestead Hollow will be one) you can DC--or perhaps rappel or HL. In recent years, someone has placed 2 B&H & W&R/QL at each of those 2 impediments. As of 7/2004, you had 3 short swims in the lower section where there's running water and fotogenic waterfalls. Whether or not you swim, will depend on the previous flash flood, and whether it scoured out or filled in deep pools. **This and every canyon changes with every big flood!**

   Once in **Zion Narrows** along the Virgin River, walk downcanyon about 2 kms to the smooth & inclined mossy waterfall at the end of Mystery Canyon, another 500m to the paved tourist trail, then another 1 1/2 kms to the **Temple of Sinawava bus stop**. From there, jump on the next shuttle bus and ride down 2 stops and get off at Weeping Rock, then hike up the East Mesa Trail back to your car. **OR**, ride the bus down to Canyon Junction or the visitor center at the end (beginning) of shuttle bus service and your car.

   Birch Hollow Head straight down the dry creek bed, or on a hiker-made trail on the left side; there's little or no bushwhacking. Halfway through the canyon will be dryfall; bypass it easily to the left. About 2/3's the way down will be an **18m drop & Rappel 1**; there should be W&R/QL on a log or boulder. In the lower end of the canyon, it slots up in the Navajo Sandstone and you'll have 6-7 rappels in the last 500m or so. **R2** can be a DC if you have dry feet, or maybe a HL; **R3** is from 2 B&H's left. **Rappel 4** is the big one and the beginning of a nice slot. It's from a tree on the left and down about **31m**--know where the middle of your rope is so both ends are together at the bottom! Or use a **biner block**, add a short rope to one end of your 60, rap down the other side (not your pull-rope side) and pull the opposite end. In the last 100m or less, will be **R5, R6, R7 & R8** from 2 B&H on either side. This is where the canyon slots up good and is the best part of the hike. **R8** will put you within 30m of **Orderville Canyon**.

   From the bottom, you can go downcanyon to **Zion Narrows**; or upcanyon to the beginning of Orderville Canyon. Or walk up Orderville for 11 minutes with map in hand, and turn right into the **lower end of WWH**. The end of this canyon looks like one **big dry mud flow**. Walk up the bottom of this drainage until you come to a small dryfall, then turn right or southwest and get out of the bottom. Once out, look for a seldom-used hiker's trail running parallel to the canyon. Follow this trail or natural openings in the trees. About 2/3's of the way up, you'll have to veer right or south away from the main drainage. In this section you'll do a little bushwhacking, so long pants & sleeves would save scratches. After a steep section, the ground levels some and you should finally come to an old road; turn right. Soon you'll be on the main road about 400m from your car.

   If you're only going down **Orderville Canyon**, then walk (or 4WD) down the road from the corral & good campsite. From the 4WD trailhead, look for a hiker's trail going down the main drainage. After 500m, you'll come to a big dropoff and a rappel of 35-40m; but instead of doing that, turn left or south, and walk along a good hiker's trail down into this wide-open canyon. From there it's basically just walking all the way to the **Temple of Sinawava bus stop**, but with the **2 short raps/HL's** of about **4-5m** each, and up to 3 swims (?). There is no need for a harness as anyone reading this book should be able to DC/HL over the drops. If you're planning to return upcanyon to your car, then leave a rope at each dropoff. A ways below R2 is now another chokestone and waterfall. You can easily chimney or slide down into a swimming pool; or if you intend to return the same way, leave another rope tied to something to help you get back up. With friends, you can likely make it up this waterfall without a rope. The next big flood may change this. The lower part of Orderville is one of the nicest narrow canyons around.

   Elevations Englestead Hollow Trailhead, 1980m; junction of Englestead & Orderville Canyons, 1548m; Birch Hollow Car-park, 2055m; corral & 2WD parking in upper Orderville, 1951m; 4WD trailhead in Orderville, 1737m; Weeping Rock bus stop, 1325m; Temple of Sinawava bus stop, 1344m.

   Time Needed For **Englestead** and, depending on group size, about 7-9 hours from trailhead to Temple of Sinawava. From that point you'll get on the shuttle bus and either ride back to the visitor center, and drive another car back up to the trailhead; or take the shuttle bus downcanyon 2 stops to Weeping Rock, and walk the trail up Echo Canyon, turn left toward Observation Point, then turn right on the East Mesa Trail back to your car. Or better still, park at the visitor center, take the shuttle bus to Weeping Rock and hike from there. Doing this will take from 8-11 hours. For **Birch Hollow**, coming out WWH, should take 4-7 hours.

   For **Orderville** only, from the corral all the way down to the **Temple of Sinawava bus stop** can take 5-7 hours; about an hour less if starting at the 4WD trailhead. If returning upcanyon from the Virgin River to your car, may take 7-9 hours round-trip.

   Water Carry your own. The first reliable water will be in lower Orderville where Bullock Gulch enters.

   Maps USGS or BLM map Kanab (1:100,000) for driving; Temple of Sinawava & Clear Creek Mountain (1:24,000--7 1/2' quads) for hiking; and the plastic Trails Illustrated/National Geographic map, Zion National Park (1:37,700) for driving & hiking.

   Flash Flood Danger Moderate to high in Englestead, lower Birch Hollow & Orderville Canyon.

   Best Time To Hike In the heat of summer with dry roads, and you won't need a wet/drysuit. In spring or fall, you'd better have a wet/drysuit for lower Orderville, along with dry roads.

Author's Experience The author has been in & out of **Orderville** from both ends many times alone, the last time down & back was in May, 2002. He didn't have a 3rd rope, so he returned from the waterfall & pool. Round-trip time from the corral & 2WD trailhead was 7 3/4 hours. Later with Byron Lemay, camped on Mosquito Cove between mp 23 & 24 on Highway 9 between Virgin & Rockville, got up early, left one car at the visitor center in Zion Canyon, drove up to the **Englestead Hollow Trailhead** at 1980m, and entered the canyon. It took 1/2 hour to reach the 1st rap; another 3 1/4 hours to reach Orderville; 1 1/2 hours to reach Zion Narrows; and about another hour to reach the bus stop at Temple of Sinawava. About 6 1/2 hours total one-way hike-time. On 7/15/2004, with Jim Schnepel, we hiked from Weeping Rock to the trailhead in 2 3/4 hours; another 3 1/4 hours to Orderville; and another 3 hours to Temple of Sinawava. About 8 1/2 hours round-trip. With John Summerson, we did the loop-hike down **Birch** and up **WWH** in 4 3/4 hours on 8/10/2006.

BIRCH HOLLOW

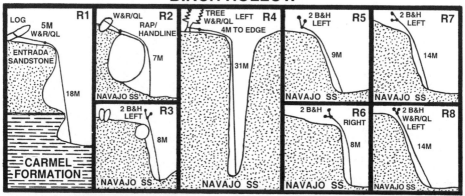

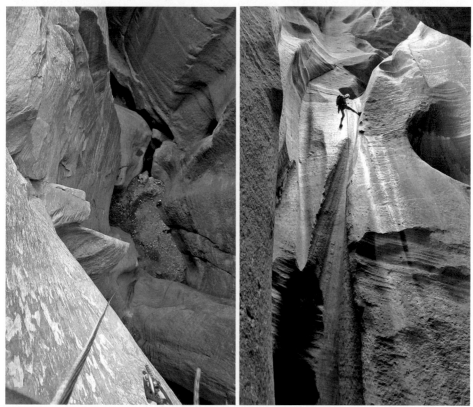

Left This picture was taken halfway down the 2nd half of Rappel 1 in **Engelstead Hollow**. The entire length of this rap is about 90m. **Right** This is Rappel 3, a slippery water-worn groove at the head of Engelstead Hollow.

Above Left Jim Schnepel standing at the rap-
pelling station 1/3 the
way down Rappel 1 at
the head of **Engelstead
Hollow**.

Above Right Jim mak-
ing a simple rappel or DC
in the lower end of **En-
gelstead Hollow**. There
were several log jams
like this in that area in
7/2004.

Right These B&H's are
located in the lower part
of **Orderville Canyon**.
(Jim Schnepel foto)

Above Left Starting Rappel 3 in **Birch Hollow.** Most anchors in this canyon are B&H's. (John Summerson foto)

Above Right Starting Rappel 6 in the lower end of **Birch Hollow**.

Left Rappel 8, the last rap in **Birch Hollow,** is a slippery slide, water-worn groove. And rather fotogenic.

Mystery, Echo & Hidden Canyons, Zion National Park, Utah

Location & Access These 3 canyons are located in the heart of Zion National Park. **Mystery** drains into Zion Narrows from the east roughly halfway between Orderville Canyon to the north, and the parking lot at the end of the road called **Temple of Sinawava**. The **Echo & Hidden Canyon** hikes begin & end at the **Weeping Rock bus stop**. Weeping Rock is the starting point for Mystery Canyon as well.

If you do either hike during the warmer half of the year when the shuttle bus is running--roughly the first week in April through October--then you must park at the visitor center near the park's south entrance (or at the Museum or Canyon Junction) and ride the shuttle bus up to Weeping Rock. When you're finished with **Mystery Canyon**, walk down Zion Narrows and pick up the shuttle bus at the Temple of Sinawava. A bus will take you back to your car.

Rating **Mystery & Hidden Canyons** are rather simple & easy as technical canyons go and are generally rated: **3B III & 3A III**. **Mystery** has a deep narrow section, 2 swimming pools, and a long slippery-slide rappel at the end. **Hidden** is a narrow canyon all the way, but with no real slot. **Echo Canyon** is normally: **3B II**. It's also possible to see the lower half of Echo's slot without ropes or rappelling. This writer ranks Echo with Pine Creek for scenery, but without the longer rappels.

Equipment RG, two 50m ropes for **Mystery** (or one 50 & a pull cord); one 15m rope for **Echo**; a 60m rope for **Hidden**, W&R/QL (especially if you're the first to go down in the spring or right after floods), K&EP, and dry-bags or kegs (Hidden should be dry or with shallow pools only). If you do Mystery in the spring or fall, take a wet/drysuit; but not during the heat of summer. If the potholes in Echo are full, then a wet/drysuit is recommended even in summer for a more enjoyable less-hurried hike (required in spring or fall).

Route **Mystery** Stop at the **Weeping Rock bus stop**, and walk up the **Echo Canyon-East Rim Trail** running toward Observation Point, Cable & Deertrap Mountains and/or East Entrance. The first part of the trail up Echo Canyon is paved with lots of foot traffic during the day. Higher up, it's a trail junction. After roughly 3 kms is a trail junction. Take the one veering left which is the **Observation Point Trail**. After another 1 1/2 kms or so, turn right onto the **East Mesa Trail**. After yet another 1 1/2 kms, and as you see the land falling away to your left, veer left or north onto a hiker-made trail. In just a few meters you'll be on the rim of upper Mystery Canyon. In getting to this point, **have a compass and follow your map closely**. One group got disoriented, went into the wrong canyon and had to be rescued!

There is now a hiker-made trail down into the upper forested Mystery Canyon, but it goes straight down with little or no zig zagging. You'll have to hang on to branches or roots in places to keep from sliding. The NPS really needs to do something about this situation as the next big storm to hit the upper canyon will send a wall of mud downstream.

As the slope becomes more gentle, there will be dropoffs/chokestones, but you can skirt around them to the right or east side on trails. Halfway through the canyon and about 200m before it makes an abrupt turn to the left or west, you'll come to the first of up to **13 rappels**. At that point you can rap off a B&H on the left and down 5m or so; or head up the right or east side on a trail to find W&R/QL around a large pine tree. Rap down the canyon wall about **22m**.

After another 125m will be **Rap 2** from 2 B&H on the left with a drop of **18m**. This is where you start down into a 250m-long fairly dark slot where many of the rappels are located. This is also the section where you'll want to go through quickly if storm clouds are gathering.

At what could be for some **R5**, is actually an easy **DC**. It's also possible to DC over **R6**, the one with an antique bolt & rebar, but it's best to use ropes to be safe. After **R7**, the canyon begins to open some with trees and brush. All anchors or bolts are well-placed and generally out of the way of flood waters. After **R10**, which is now a DC, you'll find a big landslide from a collapsed wall on the right or north. This **dam** must be 30m high, and after big floods, a small lake will appear above it. Below this dam there's little danger of flash flooding.

The most interesting rappel is the **12th**, a drop of about **40m**. You'll descend from 2 B&H's on the far left-side sloping shelf down onto a big chokestone. At that point, just above a year-round pool, you could pull your ropes (this could be done with one 50), then install W&R/QL to one of 2 small chokestones under the boulder and descend from there; **or better still**, don't pull your ropes, but continue down a crack to the south and into a cold swimming pool which is a walk-out. From there at the very bottom, it will be easier to pull your ropes, so it's best to use both 50's and do it in one long rappel (but don't cross your ropes at the top or you'll have to jumar up and retie!). If you're thirsty, this is the place to fill your bottle as there's running water from this pool down to the **North Fork of Virgin River**. Between Raps 12 & 13, will be some wading and another swimming pool immediately below a large boulder. You'll have to DC into the pool; or jump.

Finally you'll come to an overlook of Zion Narrows and **Rappel 13** from 2 B&H's on the left. You'll descend straight down over a waterfall which hugs a smooth wall of sandstone. From top to bottom, it's covered with slippery moss because of the year-round flow. Be careful not to slip and look stupid in front of many tourists down at the river who will be watching and taking pictures.

Once in the river, walk downstream about 500m to a **paved walking path** which starts (ends) at the end of the road and the **Temple of Sinawava bus stop**. This last part is about 1 1/2 kms and is safe from flash floods. The end of the road is where you'll catch the shuttle bus back to your car and/or visitor center.

Echo Canyon Head east up the trail from **Weeping Rock**. This is the same trail as the beginning part of the hike to Mystery Canyon. After about 1 1/2 kms, the trail runs along the bottom of Echo Canyon for a short distance. Continue eastward on the path until you come to the **trail junction** mentioned above, but this time walk down and to the right. This **Echo Canyon-East Rim Trail** will eventually take you to Cable Mtn., Stave Spring and/or Clear Creek, and Hidden Canyon. Just less than 1 km from the trail junction, and when the bottom of Echo Canyon drainage is easily visible on your right or south, walk down in. There's more than one place to enter.

About 400m into the drainage, you'll come to the 1st of **4 possible rappels/DC's/HL's** in the next 300m or so. If you're with a small group, you can likely help each other DC some or all of these. Some of these short drops may have W&R/QL from a log or chokestone. This part changed radically between 2002 & 2004.

About 500m below the upper dropoffs will be **Rappels 5-9** in a very short distance. **R5** will likely be from a small chokestone wedged into a crack on the right; **R6** is from 2 B&H on the left. Immediately below that the canyon makes an abrupt right turn and there will be a small, but deep pothole with **1 B&H & 1 piton** on the upper right. You'll have to crawl through one pothole before roping into the anchor. This puts you down **R7** and into a pothole that may be a little difficult for one person to get out of.

The author did this hike solo on 2 occasions--once when the canyon was nearly dry, the 2nd time a day after floods had filled all potholes. With 2 or more people, getting out of this pothole should never be a problem, but if you're not an experienced canyoneer, it may be best to take a partner.

What was **Rappel 8** in 2002 from a log with W&R/QL, was a simple DC in 2004--things change with

Map 62, Mystery, Echo & Hidden Canyons, Zion National Park, Utah

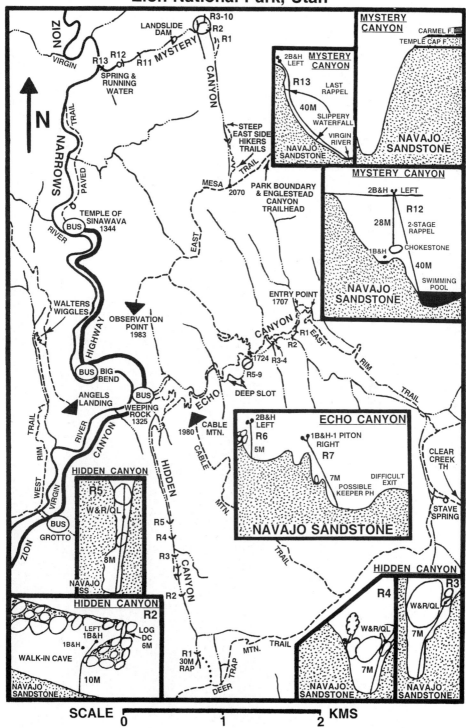

every flood! Just below that is **R9** from **1 B&H & 1 eyebolt** on the right. After R9, the canyon changes and there should be no more dropoffs. The last km of slot is deep & dark in places with several easy DC's. This part is similar to Pine Creek below the 18m rappel there. Walk and scramble down to the trail, then head back to Weeping Rock. You can walk & scramble up to R9 from where the trail crosses Echo Canyon Creek.

The **lower part of Echo Canyon** looks like a great technical slot, but you're forbidden by the NPS to make the last rappel over Weeping Rock. If you should decide to descend this part, you'll have to leave a rope at each dropoff and jumar back up. There are 4-5 dropoffs in that section.

Hidden Canyon Again, start at Weeping Rock and walk up the Echo Canyon Trail. After 10 minutes or so, you could turn right or southwest at the sign and walk into the lower end of Hidden. You can walk up about halfway before coming to the last rap coming down. But to do the entire canyon, continue up the Echo Canyon-East Rim Trail. Higher up and near **Stave Spring** turn right or southwest and head toward **Deer Trap Mtn**. Follow this map carefully so you get into the right canyon. At the head of Hidden, enter the upper end on the east side and work your way down the beginnings of a hiker-made trail to the first ledges & **R1**. This is about **30m**, then DC the last part. **R2** is in the best part of the canyon. This is where a bunch of chokestones have clogged the slot creating a cave about 35m long. You'll walk right over it, then DC/HL between a log & wall, then from **1 B&H** & standing on a middle tier of chokestones, rap **10m** back underneath. See cross-section. **Rappels 3, 4 & 5** are all about **7-8m** and down over huge chokestones. All together, there must be 30-35 chokestones in the canyon--all but 5 are easy to DC around. **After R5**, you'll begin to see footprints from people hiking/upclimbing from the bottom. Head downcanyon and back to Weeping Rock bus stop. After floods, you may get your feet wet, but never any deep wades or swims.

Elevations Weeping Rock bus stop,1325m; trail junction, 1724m; beginning of Echo Canyon narrows, about 1707m; head of Mystery Canyon, about 2070m; Temple of Sinawava bus stop, 1344m.

Time Needed Walking the trail to the head of **Mystery Canyon** will take most people 2-2 1/2 hours, then another 5 hours or so to go downcanyon to the Temple of Sinawava. Most people can do this hike in 7-10 hours. The hike to the entry point to **Echo Canyon** will take most people about 1 1/2 hours. Going down through the slot will take 3-4 hours, then another hour down to Weeping Rock. Perhaps 4 1/2-6 hours total round-trip. It will take about 3 hours to hike to the beginning of **Hidden Canyon**; perhaps 6-10 hours round-trip from Weeping Rock.

Water Always start each hike with a bottle or two, but there's drinking water at the visitor center, Weeping Rock (not in taps), lower end of Mystery Canyon and at Temple of Sinawava.

Maps USGS or BLM map Kanab (1:100,000); Temple of Sinawava (1:24,000--7 1/2' quad) for hiking; or the plastic Trails Illustrated/National Geographic map, Zion National Park (1:37,700) for driving & hiking.

Flash Flood Danger High in **Mystery** between R3 & R10, otherwise moderate; most of **Echo** is high as well; low to moderate in one place in Hidden.

Best Time to Hike May through September, otherwise in spring or fall take a wet/drysuit for Mystery (check the water level in the Virgin River before doing Mystery) & Echo; spring through fall in Hidden.

Author's Experience For **Mystery**, he got on the first shuttle bus going all the way upcanyon (leaves visitor center at 6:30am--check the times when you get there) and got off at Weeping Rock at 7am. Walking to the head of Mystery took 1 2/3 hours, then he was at the Temple of Sinawava after just over 8 hours total walk-time. He was alone which is normally faster, but he set up his tripod at a number of places to take pictures.

For his first hike in **Echo Canyon**, he got on a bus at the visitor center at about 7:30am, and started hiking at 8:20. It took 45 minutes to reach the trail junction, and another 20 minutes to get into upper Echo. It took about 3 1/2 hours to get through Echo's slot and back to the trail, but he had many camera stops with a tripod. Round-trip time took about 5 1/2 hours. His second trip was a day after floods and water was everywhere with half a dozen swims. Just before the last dropoffs, he was joined & passed by a local Boy Scout group, none of whom had rappelling harnesses; they had only 3 small day-packs between them. They simply HL'ed down or jumped into pools. They were cold and moving fast as none had wet/drysuits. That trip took the author 5 1/2 hours with a lot of foto stops.

He soloed **Hidden** in 7 hours round-trip on 8/14/2006. On that trip he found that using his 40 liter pack was easier to stuff his 60m rope into than a regular rope bag.

The last rappel in **Mystery Canyon**. The anchor is a couple of B&H's, and you literally slide down a slippery moss-covered slope to the Virgin River below.
(Jon Jasper foto)

Above Left This is one of the first rappels/handlines in **Echo Canyon.** The leader of this Boy Scout troop is DC'ing, but the rest HL'ed down this dropoff with a rope. This was a day after a flood and this pool was a swimmer.

Above Right The Deep Slot of **Echo Canyon** which is below the section with rappels. You can hike up to this point from where the trail crosses the stream channel in the lower end of Echo Canyon.

Left Rappel 2 in **Hidden Canyon**--see the cross-section on the map. The author has chokestones above, and is standing on more chokestones. He is about to rappel from 1 B&H down under this 2nd set of chokestones.

Spry Canyon, Zion National Park, Utah

Location & Access This drainage is located on the east side of Zion National Park and due north of lower Pine Creek and the Zion--Mt. Carmel Tunnel. **Spry Canyon** is the unnamed drainage (on most maps) running between Mount Spry, Twin Brothers and The East Temple.

After you get your permit at the visitor center, park one car at the **1st curve** where **Highway 9** crosses the bridge on **lower Pine Creek**, then drive up to the upper or **east end of the Zion--Mt. Carmel Tunnel** and park at one of several places in that area. Best to park about 500-600m east of the upper end of the tunnel and near where Pine Creek comes down from the north. As you start up the switchbacks from the lower canyon leading to the tunnel, be sure to look north and observe the lower end of the canyon and the route/trail from the last rappel down over 2 Kayenta cliff bands to the bottom. This will save time and frustration at the end of your hike.

If you have only one vehicle, plan ahead and make a small sign reading something like: *my car, east end of tunnel*, or whatever; then at the end of your trek, hitch hike from the Pine Creek bridge back up to your car. Getting short rides is easy in Zion, especially for hikers with a sign. Another option would be to contact someone in Springdale to drive you up to the beginning of the hike (leaving your car at the Pine Creek Bridge).

Rating Spry Canyon: **3A** (can be **3B**) **III**, and is pretty basic as technical slot canyons go.

Equipment RG, two 50m ropes (or a 50 & pull cord), W&R/QL, K&EP, and drybags just in case.

Route From the east end of the tunnel, walk eastward up the road about 700m (or park further east and walk less), turn left and follow the Pine Creek drainage north. Easy walking. While paying attention to the *7 1/2' quad*, and about 1 km from the highway, veer left or northwest and head up a slickrock slope with an occasional ponderosa pine. Some places are steep, but route-finding is easy. From the **pass** marked **1810m** between **Spry & Pine Creek**, continue northwest until you find a place to drop into the upper part of the entrenched Spry drainage. Once in it, head south downcanyon. After a ways, it opens up into a bowl.

From the bowl or basin, head west down the dry creek bed. Between Twin Brothers and The East Temple will be the **1st of 8 to 11 rappels** (some drops can be DC'ed). This is a very low angle rap and some people might be able to walk down to a pothole, veer left and finish by rappelling from a small tree into a pool. However, for safety you might as well use both 50's and rappel past both pools and keep your feet dry.

After another 500m, the canyon slots-up with a drop, but if you climb up on the right, you can avoid this 2nd rap. Either way, right of left, you can get back down into the slot with a number of entry/exit places. From what is normally the 2nd rap, down to the last, is about 500m. Once into the lower slot, just follow tracks or the trail of previous hikers. **R7** is over 2 possible pools, and if you use both 50's for **R8**, you can make it all the way down to a 2nd pool bypassing 1 B&H on the left. **R10** gets you out of the slot, and **R11**, the last rappel, is from a tree 4m from the edge of a cliff on the left. Don't be surprised if B&H have been removed, and W&R/QL left around a tree or chokestone. **Also, things change with every flood**, so go prepared; it's been a while since the author has been through this canyon.

From this **last rappel**, contour to the right or southwest along a hiker's trail. As time goes on, this trail is getting better. After about 300m, and as you approach an open treeless slope, follow tracks down and over the 1st of 2 cliff bands (or stay in the dry creek bed to cause less erosion). Below the 2nd cliff is Pine Creek. Follow it west until you reach the highway, then drive back to your 2nd car; or hitch hike back.

Elevations East end of tunnel, 1562m; pass & high point, 1810; Pine Creek Bridge--end of hike, 1245m.

Time Needed Roughly 7-9 hours from the upper highway to the Pine Creek Bridge.

Water Take plenty of you own. You'll likely find some pothole water in Spry, but it may need purifying.

Maps USGS or BLM map Kanab (1:100,000); Springdale East (1:24,000--7 1/2' quad) for hiking; or the plastic Trails Illustrated/National Geographic map, Zion National Park (1:37,700) for driving & orientation.

Flash Flood Danger No danger for 3/4's of this hike, but high in the lower slot between rappels 2 & 11.

Best Time to Hike Late spring through early fall. In cooler weather, take a wet/drysuit just in case.

Author's Experience He parked his car at the most-easterly of the parking places above the east end of the tunnel and did the hike alone in 7 3/4 hours-- but had to backtrack twice and set up the tripod several times for fotos. A couple of small or short pools on 8/17/2001 were near swimming depth. He waited 15 minutes for a ride back to his car.

The beginning of Rappel 4 in **Spry Canyon**. Below this rap is where the best slot begins.

Map 63, Spry Canyon, Zion National Park, Utah

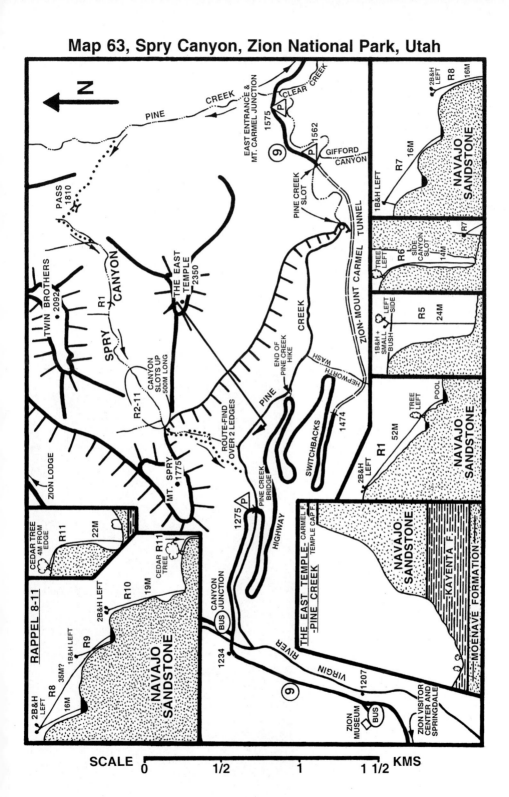

Pine Creek Slot, Zion National Park, Utah

Location & Access **Pine Creek Slot** is located in the eastern part of **Zion National Park** immediately below the **Zion--Mt. Carmel Highway Tunnel**. To get there, drive along **Highway 9** between Zion Canyon and the east entrance of the park. Stop at the small parking lot at the **east end** to the **tunnel**; there's a guard & toilet nearby. If you're with a group, and have a second car, park it at the 2nd switchback down from the west end of the tunnel, as shown on the map. That will be the end of your hike. Or, hitch hike back to your car.

Before hiking, stop at the backcountry desk in the visitor center in lower Zion Canyon just north of Springdale. They will give you a permit (there is a quota of **50 people per/day** in 2008) and the latest information on anchors and water in the many pools. For quota updates see the website **nps.gov/zion**.

Rating **Pine Creek**: **3B II**. This canyon is short & sweet with 5-6 rappels and a fotogenic slot.

Equipment RG, two 50m ropes (or a 50 & pull cord; or better still, one 60, but make sure it's a full 60m!), a 2nd short rope, R/QD/DC, W&R/QL, K&EP, wet/drysuit & drybags (ask a ranger about how much water is in the potholes, and prepare accordingly).

Route To enter Pine Creek, walk from the parking lot straight down into the drainage bottom immediately above the bridge. There's now a short trail. Walk under the bridge, then the first dropoff & rappel anchor is about 40m away. DC and wade through the first pool, then up a little on the right will be B&H with W&R/QL for **Rappel 1**. Use your longer rope and go over the first **2 drops** from the **same anchor**.

After about 50m will be 2 more rappels, **R2 & R3**. **Logs** will be used as an anchors for **R2**, then on the left will be **2 B&H** for the longer **18m R3** into a **cathedral-like cavern** with **2 natural arches/bridges**. This is where it gets darker and more interesting. Arrive at this point in late June at high noon for the best available light for taking pictures. At the bottom of this big cavern may be a short swimming pool.

From there, the canyon gets deeper and you'll have another **log rappel**, or two (?) then DC through a log jam into a long swimming pool. Be aware this section changes with every flood and you may find more/fewer rappels than the author found on his trips. Beyond **R4** are several more wading pools & DC's, then as the canyon begins to open a little, DC under large boulders. After that, the canyon opens to the sun and you can either walk right down the drainage to what is now the **5th or 6th rappel** (this depends on whether you like to DC or rappel some of the log jams higher in the slot), which is down a low-angle slope; or veer left and route-find up through brush & trees, as shown on the cross-section. Either way will get you to the **last rappel**.

The **last rappel** (**R6** for us in 2002) anchor is on the left with slings tied to **1 B&H & small arch**. This is a spectacular free rappel down through water-sculptured sandstone with a gushing spring & pool at the bottom. The spring is there because it's the contact point between the porous Navajo Sandstone above, and the impermeable Kayenta Formation below. If using 2 ropes for this rappel, the last person down should maneuver the knot over the edge so it can be pulled easily from below. If using one 60m, it'll be an easier pull. Also, since this is a free-hanging rap, it's best to tether your pack to the front of your harness. This way you'll be sitting upright in the saddle instead of leaning back.

After the last rap, simply head down Pine Creek which is full of huge boulders and pools. After about 1-1 1/2 hours, you'll see the highway curve on the left above. Climb up to the highway and your car.

Elevations Between 1562m & 1356m.

Time Needed Pine Creek Slot is about 650-700m in length, and the whole hike is only about 2 kms, but with all the rappelling, picture taking and coiling of ropes, it takes most people about 4 hours. To offset the cold swims and have the most light for fotos, start about 11am so you're there during the warmest part of the day.

Water Begin with a small amount, plus there's a good spring at the bottom of the last rappel.

Maps USGS or BLM map Kanab (1:100,000); Springdale East (1:24,000--7 1/2' quad) for hiking; and/or the plastic Trails Illustrated/National Geographic map, Zion National Park (1:37,700) for driving.

Flash Flood Danger High throughout the main slot, non existent after the last rappel.

Best Time to Hike Pine Creek Slot normally has bitter cold water so go in hot weather. Pre-monsoon--June to mid-July is best. You'll enjoy it more if you take a wetsuit, unless it's dry, which is rare.

Author's Experience The author once did this solo in 4 2/3 hours on an October 12 and about froze to death in a shorty wetsuit. Two other trips took 4 hours each--with a full length 3/2 wetsuit. His last trip was with Byron Lemay on 7/25/2002 when the canyon was bone dry! That entire hike took only 2 2/3 hours.

The bottom of Rappel 3 in **Pine Creek Slot**. The author calls this **Twin Arches Cavern**--but other people have different names. (Jon Jasper foto)

Map 64, Pine Creek Slot, Zion National Park, Utah

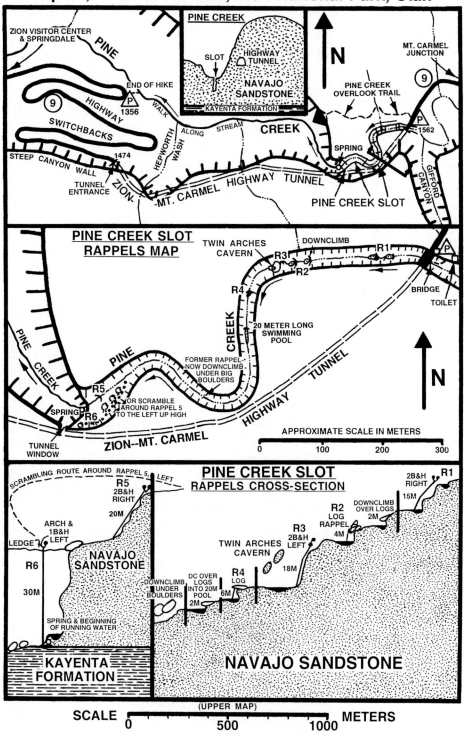

Above The last rappel in **Pine Creek Slot**. Water begins to flow at the bottom making this a good place to get a cold drink.

Right The lower part of **The Subway**, looking back upstream at approximately where you make the last rappel/handline to the bottom. (Jon Jasper fotos)

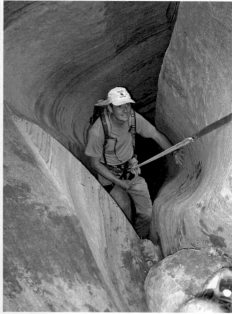

Above Left Ryan Cornia in the upper part of the **East Fork of Misery Canyon**. This section is very colorful.

Above Right Starting the 1st and only rappel in the **East Fork of Misery Canyon**.

Left The very first waterfall in upper **Kolob Canyon**. Some people rappel down beside this waterfall, others rap from a tree 30-40m further downstream.
(Jon Jasper stitched 3 separate fotos together using special software to create this picture)

Keyhole Canyon, Zion National Park, Utah

Location & Access **Keyhole Canyon** is located in the eastern part of **Zion National Park**. This short slot is an upper tributary of Clear Creek, which in turn is a major fork of Pine Creek. To get there, drive along Highway 9 which connects Mt. Carmel Junction & Highway 89 east of the park with Zion Canyon & Springdale to the west. The normal parking place is located about **3 kms (1.9 miles) west of the East Entrance fee booth**; and about 13 kms (8 miles) north & east of the visitor center & Zion Canyon's shuttle bus terminal. There's a second parking place located about 250m up the road to the southeast. Park at either place for this very short & popular loop-hike.

Rating Keyhole Canyon: 3B I. This a good one for beginners and is the shortest & easiest hike in this book, but you still need a hiking permit. In 2008, they were giving out **80 permits** a day for Keyhole Canyon.

Equipment RG, one 30m rope, drybags, and a wet/drysuit in cooler weather. In the heat of summer you can get by without a wet/drysuit.

Route From the main parking place, walk southeast up the highway about 300m and turn left or north into a short little open canyon. Follow a trail and/or cairns up this little valley to a **pass** or divide at 1720m, then down a steep pitch into the upper part of Keyhole. From there walk downcanyon in a shallow slot filled with small boulders. After about 75m, the slot opens for a short distance, then comes the technical section.

In 7/2001, there was a big log stuck in the beginning of the technical slot where there's a 3m drop. Some people slide down this log and drop to the bottom; others use a 30m rope attached to some W&R/QL around a pine tree about 10m east of the dryfall, then rappel. Once in, DC over 2 boulders into a pool. On the other side of the pool is a **4m rappel** from an **eyebolt** on the right. Soon after **Rappel 1**, which may be into shallow water, is **R2** with **1 B&H** on the right. This is a 2 stage rap which isn't very steep--some people can perhaps slide or DC this part. From the B&H, it's 8-9m down into a small pothole, but you can do it with a 15m rope. Climb out of the pool, then DC 2 steep places which ends the technical part.

Just below the technical section is a short walk in a tight slot to a 30m-long **swimming pool**. From the pool, it's another 100m or so to Highway 9 and the main parking place.

Elevations Trailheads, 1658m or 1665m; the pass or divide, 1720m.

Time Needed The loop-hike from either trailhead is roughly 1 km; about 1/4 of it along the highway. For most groups this will take only 1-2 hours.

Water In summer take a small bottle of water, but it may not be required because it's a short hike.

Maps USGS or BLM map Kanab (1:100,000) for orientation; Springdale East 1:24,000--7 1/2' quad) for hiking; or the plastic Trails Illustrated/National Geographic map, Zion National Park (1:37,000) for driving.

Flash Flood Danger High for about 200m in both the Upper and Lower Slot, otherwise no danger.

Best Time to Hike Summer, from June through about mid-September. Or any time with a wet/drysuit.

Author's Experience The author did this one alone in 2 hours, but he set up his tripod on several occasions.

Starting through **Keyhole Canyon.** This is Rappel 1 from a pine tree about 10m from the dropoff. This slot canyon is short & sweet, and many people start their canyoneering careers here. (Jon Jasper foto)

300

Map 65, Keyhole Canyon, Zion National Park, Utah

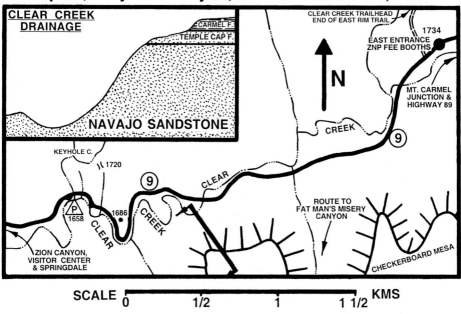

SCALE 0 1/2 1 1 1/2 KMS

Close-up of Clear Creek & Keyhole Canyon Slot

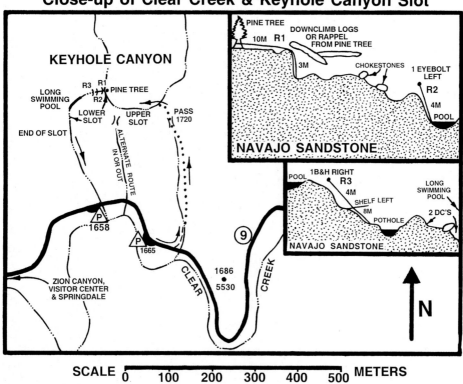

SCALE 0 100 200 300 400 500 METERS

Misery (Fat Man's Misery) Canyon, Zion Nat. Park Region, Utah

Location & Access Fat Man's Misery or **Misery Canyon**, is located immediately east of the southeast corner of Zion National Park. This drainage runs south into Parunuweap Canyon (East Fork of Virgin River) immediately east of the park boundary. Best way to get to this canyon is from just inside the park. Make your way to the East Entrance fee gate. From there drive southwesterly about 1.1 kms (.7 mile) to the first of 2 parking places or pullouts (or continue west another 300m to the 2nd parking place).

Rating Misery Canyon: 3B III. A good place for beginners; easy & fun for experienced canyoneers.

Equipment RG, one 30m rope, W&R/QL, K&EP, drybags, and a wet/drysuit for spring or fall seasons.

Route From the most-westerly of the 2 car parks, make your way into the canyon running north/south on the west side of **Checkerboard Mesa**. Follow the dry wash bottom, or hiker-made trail, south up to a pass and down the other side. Further along, veer left or east and pass around the south side of Checkerboard. As you head east on a pretty good trail, you'll cross 2 shallow drainages before coming to the north end of a prominent **white slickrock dome**. Once on that domed ridge, simply route-find east down into the next drainage, which is the **West Fork** of Misery. Walk south downcanyon.

After a km or so, you'll come to the first of 7 short slots (the first 5 slots can be bypassed on the left or east, but you'll miss half the fun). **Rappel 1** will be from a B&H to your upper left. Later, you'll find a single **eye bolt** (probably with a short rope attached to HL down) on the right around the corner, then immediately below that is a natural bridge. Throw your rope up through it and use that as an anchor for **R3**.

R4 (or a 2m slide/jump into sand) is from a small chokestone, then you can bypass several potholes on the east side just before the West Fork meets the East Fork (In 2004, on the author's 2nd trip down the West Fork, he made **only one rap--the others were DC's--experience does count**). At that point some people have rappelled over the last dropoff, but you can DC it; or, there's an easy walk-down route just to the east.

Now let's go back to the **slickrock dome** and the **East Fork**. From the north side of this dome, route-find due east. You'll have to look for a route across the upper end of West Fork, then enter the East Fork where 2 upper tributaries meet. From there it's down a walk-through canyon about half of which will be a nice wider slot. About 100m upcanyon from the confluence with West Fork, you'll slide down a short chute into a pool (or maybe sand). Immediately after that will be a **10m rappel** from W&R/QL on a chokestone.

Now the **canyon below** where the **East & West Forks** meet. About 150m downcanyon from the confluence, you'll come to the **last rappel** in the canyon. This rap is from a boulder on the left and down about **9m**. From there, it's DC'ing in several places in a deep canyon before the final slot which seems to be about 70m long. This is the darkest & best part. You'll have to slide into 2 usually-stinky pools, then walk under 2 natural bridges before coming to a **2m dropoff**. In 2004, there was a short rope attached to a small chokestone which is right next to a spring gushing from a hole in the wall. From there, it's 40m to the **East Fork of Virgin River**. Once there, if the stream isn't too large, you might as well explore the deep East Fork down to the **waterfalls** a little more than a km away, and return. The area below the falls it's closed to hiking.

To return, look for a trail running west up a steep slope covered with wild grapes beginning right at the bottom end of Misery's slot. You'll do some climbing on all-4's near the top, but it's an easy route/trail to the canyon rim. Or continue downcanyon about 800m and around to the south side of a gooseneck bend, to find another easier route up to the north as shown. From the top of either route, head north on an **emerging trail** marked with **cairns**. This route stays at or near the ridge top. From the north side of the slickrock dome, retrace the route/trail back to your car.

Elevations Trailheads, 1690m; Checkerboard Pass, about 1830m; end of Misery, about 1390m.

Time Needed From 6-9 hours roundtrip.

Water Carry your own; or from the spring in lower Misery--but it may be polluted?; treat or filter river water.

Maps USGS or BLM map Kanab (1:100,000) for driving; or Springdale East & The Barracks (1:24,000--7 1/2' quads) for hiking.

Flash Flood Danger High inside the short slots, but little or no risk elsewhere.

Best Time to Hike Warm summer weather; or take a wet/drysuit for the spring or fall seasons.

Author's Experience With Nat Smale, we made it to the river in 5 hours; roundtrip, 7 3/4 hours. On a second trip in 8/2004, the author rehiked the drainage but left Parunuweap Canyon via the second exit mentioned above. Round-trip time was 7 hours. In 7/2006, and with Ryan Cornia, we did the East Fork to the river as described above, and returned, in 6 1/2 hours.

Starting the last rappel in **Misery Canyon** below the confluence of the East & West Forks. If coming down the East Fork, this will be your 2nd rap. If coming down the West Fork, it could be your 5th, depending on whether you slide, DC, HL or rappel those dropoffs.

Map 66, Misery (Fat Man's Misery) Canyon, Zion National Park Region, Utah

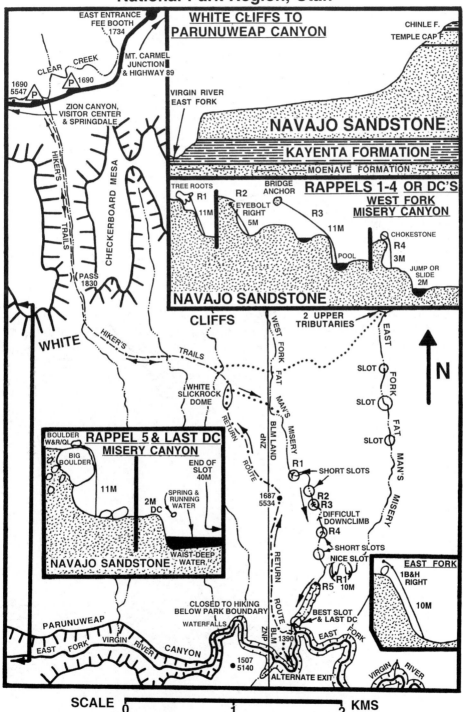

Rock Canyon, Zion National Park Region, Utah

Location & Access **Rock Canyon**, located southeast of **Zion NP & Misery Canyon**, flows north from Elephant Butte into the middle part of **Parunuweap Canyon**. You could get there from the south with a 4WD, but that country is one big sand trap! It's better to drive along **Highway 9** between **Zion NP & Mt. Carmel Junction**, and between **mile posts 47 & 48**, turn south onto a good, graded road. This runs over a hill (many good campsites), past at least one new house on the left, and down into **The Barracks** section of Parunuweap Canyon. Cars should stop at the **1829m parking**; otherwise stop at either of the 4WD trailheads shown. The track below that is very rough in places, then very sandy (if you have a car, consider taking a mtn. bike to shorten the hike time). Further down this old road is washed out. See map for distances.

Rating Rock Canyon: **3A III**.

Equipment RG, one 60m rope & 2nd at least 10m (or two 50's), W&R/QL, K&EP, and jumars.

Route From the 4WD trailhead, walk west & south along an old road. About halfway to the **East Fork of Virgin River**, you'll come to a **junction** (1615m) with 2 route possibilities. **Creek Route**--if you go straight south, you'll come to the East Fork; then walk & wade west downstream nearly 3 kms until you reach the mouth of Rock Canyon. **Petroglyph Route**--turn right or west at the **junction**. Shortly after that, and staying on an old track, you'll cross one major drainage. Near the top of the hill labeled **5522** (1683m) on *The Barrack 7 1/2' quad*, turn left or south and follow an **illegal ATV track** to a point very near the East Fork. From the end of that track, route-find down passing a pretty good east-facing **petroglyph panel**; then cross East Fork once before reaching the mouth of lower Rock Canyon. From there, cross **Rock Canyon Creek** and climb southwest up the west side of lower Rock Canyon, then walk south parallel to the canyon until near its head, and finally drop down in at 1539m.

If you enter the upper canyon right where a side-canyon enters from the west at 1539m, you'll come to a very short slot with **2 deep potholes** (see foto below). You can rap from W&R/QL on a log, or high-stem through this section, and rap the other side--about a **22m rappel**. Or, walk around R1 to the west, north and east and reenter the canyon at what for some might be **R2**. There's **1 B&H** on the right for a 30m very low-angle rap (this is really a DC for most except the first part which can be a little tricky). Not far below R2 is the **3m R3**. There's W&R/QL around a little arch to the right, but you could **DC** a little, then **slide** into sand or a pool. About 40m beyond that is the only real rappel in the canyon. This one is from 2 B&H on the right and down **33m**. To do this, you'll first need to mark the middle of your 60, then tie another piece of rope or webbing to one end--this will be your **pull rope**. Lower the rappelling side of the rope at least 5m, setup a **beaner block**, rap 5m to a little platform and make sure both ends are on the ground, then rappel the single line. Last person down separates the pull-rope making sure it doesn't cross the rap side for an easy pull from below. Try a test pull before the last person rappels.

From R4--the **Big Rap**, you'll be in a beautiful deep gorge for 1 km, then comes a **7m drop**. You could rap from a bush, but you can walk around this and DC on the right or east side. From there it's another km to East Fork along Rock Canyon Creek.

Elevations 4WD parking, 1780m & 1743m; Upper Rock Canyon, 1539m--lower end, 1455m.

Time Needed Depending on where you park (4WD's), 7-10 hours--more with a car & mtn. bike.

Water Start with your own, but there's a spring at the mouth of Rock Canyon, and in lower Rock Canyon Creek.

Maps USGS or BLM map Kanab (1:100,000) for driving & orientation; and The Barracks (1:24,000--7 1/2' quad) for hiking.

Flash Flood Danger Medium to high in the first 100m of the canyon; low after that.

Best Time to Hike Spring, summer or fall.

Author's Experience With John Summerson, we drove a little beyond the 1743m 4WD parking, then walked down using the creek route, did the entire route suggested above starting with Rappel 1, and came out via the petroglyph route. Total time was 7 1/2 hours.

This is the 1st optional rappel in **Rock Canyon**; or you can bypass it by walking around the bluff to the left and enter the canyon 40m below this place.

Map 67, Rock Canyon, Zion Nat. Park Region, Utah

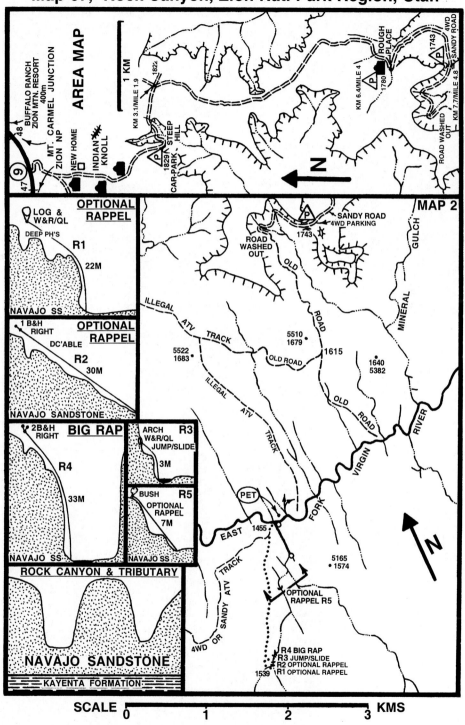

AREA MAP

BUFFALO RANCH
ZION MTN. RESORT
400m
MT. CARMEL JUNCTION
ZION NP
NEW HOME
INDIAN KNOLL
1 KM
1822
KM 3.1/MILE 1.9
STEEP HILL
1829
CAR-PARK
ROUGH PLACE
KM 6.4/MILE 4
1780
1743
4WD SANDY ROAD
KM 7.7/MILE 4.8
ROAD WASHED OUT
N

OPTIONAL RAPPEL
LOG & W&R/QL
DEEP PH'S
R1
22M
NAVAJO SS

OPTIONAL RAPPEL
1 B&H RIGHT
DC'ABLE
R2
30M
NAVAJO SANDSTONE

BIG RAP
2B&H RIGHT
R4
33M
NAVAJO SS

ARCH W&R/QL JUMP/SLIDE
R3
3M

BUSH
R5
OPTIONAL RAPPEL
7M
NAVAJO SS

ROCK CANYON & TRIBUTARY

NAVAJO SANDSTONE

KAYENTA FORMATION

MAP 2
P
SANDY ROAD
4WD PARKING
1743
ROAD WASHED OUT
OLD ROAD
MINERAL GULCH
ILLEGAL ATV TRACK
5522 1683
5510 1679
1615
1640 5382
OLD ROAD
ILLEGAL ATV TRACK
OLD ROAD
VIRGIN RIVER
FORK
PET
EAST
1455
5165 1574
OPTIONAL RAPPEL R5
SANDY ATV TRACK
4WD OR SANDY ATV TRACK
R4 BIG RAP
R3 JUMP/SLIDE
R2 OPTIONAL RAPPEL
1539 R1 OPTIONAL RAPPEL
N

SCALE 0   1   2   3 KMS

305

Above Left This is part of the petroglyph or rock art panel just around the corner to the northeast from the mouth of **Rock Canyon**.

Above Right John Summerson making the short 3m optional Rappel 3 in **Rock Canyon**. You can also slide or jump into soft sand, or with a party of 3 or more people, help each other down.

Right The end of the Big Rap, Rappel 4 in the upper end of **Rock Canyon**. See the person in the white shirt on the right? He is rappelling from the end of the upper narrows and into the beginning of the much larger & deeper lower **Rock Canyon Gorge**.

Above Left Ryan Cornia HL'ing down one of the 2 rappels in the **Lower Red Cave** in **Lower Sand Wash**.

Above Right Ryan coming down Rappel 2 in the middle part of the **Deer Creek Narrows**. This is the Deer Creek located in the middle part of the Grand Canyon. Once into these narrows, there's few places to get out of the water, so wear a wetsuit and have plenty of W&R/QL's. Also, for life insurance, take a BK & G-pick, because with every big flood, the boulders in this place are rearranged and anchors change. Some of the B&H's in this gorge are also getting pretty old.

Left Ryan stemming around one chokestone in the middle part of the **Upper Red Cave of Sand Canyon**.

Sand Wash (Red Caves), Zion National Park Region, Utah

Location & Access **Sand Wash & the Red Caves** are located just northeast of **Mt. Carmel Junction** and east of **Highway 89**. This mapped area is a short distance east of Zion National Park. To get there, and using a Utah State highway map, drive along Highway 89 between Panguitch & Kanab to Mt. Carmel Junction; or head east from Zion on Highway 9. Once there, your first stop should be the **Chevron gas station & Best Western Thunderbird Motel** next to the golf course. Ask for Tanya Milligan, co-owner in 2007. She can inform you about the access which crosses **private land**--this could be a problem in the future. To do the hike, start at the junction of Highway 9 and drive 800m (.5 mile) north on Highway 89. Where a little side-road turns east and goes through a gate, park between that fence and the highway (**Beware:** inside that gate is private land).

Rating Together: **3A or B II**; **Lower Red Cave** has 2 rappels: **Upper Red Cave** has UC/DC & water.

Equipment One 20m rope (body wrap/Dulfersitz the rappels or HL down), K&EP.

Route As of 2007, crossing some private land wasn't a problem, but it's best to walk, and leave your vehicle at the highway. Once through the gate mentioned above, walk southeast away from the highway and near the **East Fork of the Virgin River**. There are homes 300m south of the gate, so do your best to avoid them. Just behind or east of the closest home is a 4WD/ATV track crossing the creek heading southeast. Follow this to the top of the P/J-covered bench. Head south at the junction of Highway 9 and drive then east as shown on the map. After another 400m, turn left or north onto another well-used dirt track (near elevation 1677m). After a little more than 1 km, the road drops down into **Sand Wash**. Once there, turn right or southeast, and head up **Lower Sand Wash**. Walk 2 kms along an ATV track in the dry wash bottom. There, at a faultline, a red sandstone wall appears and you enter a slot known locally as **Lower Red Cave**. The best part is right there near the mouth. About 50m upcanyon is likely a **pool (?)** and **7m dryfall**. From there you could leave the slot and immediately climb out to the south, circle around the dryfall, reenter from above, and walk downcanyon. However, to do that you'll need one rope at least 20m long to do **2 rappels**; one is **9m** from a log/chokestone; the 2nd is down **7m** from 2 B&H's.

To reach the **Upper Red Cave slot**, you could walk about 1 km up **Lower Sand Wash**, route-find out to the north, and enter the next drainage which is the main **Sand Wash**. Or if cross-country bushwhacking isn't to your liking, you can walk along the dry wash bottom between the 2 Red Caves in 45 minutes. At **Upper Red Cave**, you can walk in about 100m passing several chokestones before coming to a long narrow year-round pool. Wade or high-stem this. With 2 or 3 people helping each other, you can upclimb the chokestones and get entirely through this slot (getting around the dropoffs is easier if coming downcanyon). Return to your car the same way. Or walk along the dry Sand Wash, which crosses more private land--back to the highway and your car.

Elevations Mt. Carmel Junction, 1584m; Trailhead, 1594m; high point on the bench, 1677m.

Time Needed Depending on how many fotos you shoot, it will take from 5-8 hours round-trip.

Water Carry your own, which you can get in Mt. Carmel Junction.

Maps USGS or BLM map Kanab (1:100,000) for driving & orientation; Mount Carmel (1:24,000--7 1/2' quad) for hiking.

Main Attractions Some pretty good fotogenic red Navajo Sandstone slots.

Best Time to Hike Spring, summer or fall; in cooler weather, take a wet/drysuit just in case.

Author's Experience On his first trip, he walked into Lower Red Cave, but it was too early for good fotos, so he walked in the dry wash to the Upper Red Cave and spent an hour. Later he returned to Lower Red Cave but was stopped by a pool & the 7m dryfall. He returned along the dry wash to his car in 6 2/3 hours. In 6/2007, he and Ryan Cornia drove to the confluence of Lower Sand Wash and parked his 4WD. We walked into Lower Red Cave from the top, rappelled, then walked up Sand Wash and completely through the Upper Red Cave slot and back. From the highway, the trip took 4 hours.

Coming down the 2nd of 2 rappels in the **Lower Red Cave** of **Lower Sand Wash**. This time the author is using the Dulfersitz method of rappelling--that is, not using a harness--just wrapping the rope around the body to create friction. (Ryan Cornia foto)

Map 68, Sand Wash (Red Caves), Zion N.P. Region, Utah

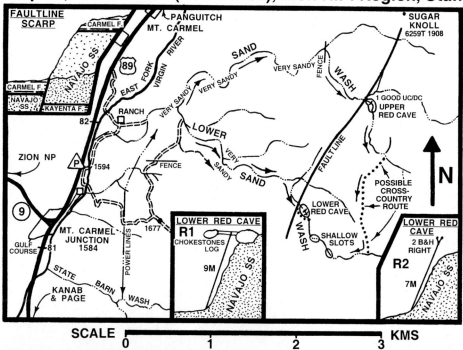

SCALE 0 ——— 1 ——— 2 ——— 3 KMS

Ryan Cornia in one of the swimming pools inside the narrows of **Deer Creek** (Grand Canyon).

Deer Creek Narrows, Grand Canyon, Arizona

Location & Access **Deer Creek** is located due south of Fredonia, Arizona, in the northwest portion of the central part of the **Grand Canyon**, and on the north side of the Colorado River. The easiest way to get there is to first drive to **Jacob Lake** on **Highway 89A** and stop at the Forest Service visitor center. Be sure to pick up the Kaibab National Forest North map or you'll never reach the trailhead! Then with map in hand, drive west from Jacob Lake on **Forest Service (FS) Roads 461 & 462** until you reach **FS Road 22** (this one begins just east of Fredonia, between mile posts 607 & 608 on Highway 89A, and heads southeast to meet Road 462). From that junction, continue south on FS Road 22 for about 19.3 kms (12 miles) to between **mile posts 32 & 33**, and turn right or southwest onto **FS Road 425**. Follow this for 18.9 kms (11.7 kms) to a **4-way junction** very near **Crazy Jug Point**. From that junction continue straight ahead to the west on **FS Road 292A**; after another 2.7 kms (1.7 miles) is the end of the road at **Monument Point** and the beginning of the **Bill Hall Trail**. Normally this route is good for cars.
Warning: Rumors are that Deer Creek will be closed to canyoneering in 2009? Write letters/emails!
Rating **Deer Creek** rates **4C V** (for some **IV**--a long one day hike from the trailhead).
Equipment RG, a 60m rope & 60m pull cord, a 2nd shorter rope--15-20m, **lots of W&R/QL**, 2 R/QD/DC per/person, helmet, K&EP, wetsuit and a BK & 2-3 B&H's just in case big floods have wiped out bolts in lower slot. Walkie-talkies would help with communications at the last thundering rappel.
Route From **Monument Point**, head southwest along the rim on the **Bill Hall Trail**. It soon drop off the top and zig zags down to meet the **Thunder River Trail** coming from Indian Hollow. From that trail junction, head south into Surprise Valley, then west to Deer Creek. Once there, head south along the year-round stream to the head of the Deer Creek Narrows. That's where the canyoneering begins.

At the beginning of the 300m-long narrows, cross to the east side, walk downcanyon to just beyond the first 2 waterfalls, and locate a large stone with W&R/QL. **Rappel 10m** into the stream. Wade downcanyon. After roughly 100m, you'll come to **R2**. We couldn't find an old bolt--must have been wiped out by a flood--so we rigged W&R/QL to boulders and rapped **6m**. Next comes the **2-stage R3**. Here you'll have a slippery climb to get to the **3 B&H** on the right to get started, then swim to reach the 2nd stage. This 2-stage rap is roughly **25m**. More swimming & wading, then **R4**. You'll have to climb to reach **2 B&H** on the right. Use the short rope to rap about **6m** to a very small landing next to the **last big 55m Rappel 5** to the bottom of Deer Creek Falls and the Colorado River. It's at these last 2 rappelling stations you can & should use your R/QD/DC to secure yourself while preparing for the final raps. Last guy down should use a **biner block** to secure the pull cord, and keep them separate for an easy pull from below. Halfway down the last rap, you'll be inside & under the waterfall until you hit the swimming pool at the bottom. For best fotos, be there after about 1:30 pm to be in full sun. Also, climb up to vantage points on the east side of the falls for a side-view of rappers and waterfall.

Elevations Monument Point, 2146m (high point, 2196m); Colorado River at Deer Creek Falls, 590m.

Time Needed One very long day of 12-14 hours. Or, start early on Day 1, drop big packs on Deer Creek, rap the narrows, camp, then return to your car in the cool of the morning on Day 2.

Water Cache extra water (and a light lunch/snack) along upper trail when going down, then tank up on water anywhere along on Deer Creek (the closer to Deer Spring the better).

Maps Kaibab National Forest North map (1:126,720); and Fishtail Mesa (1:24,000--7 1/2' quad).

Flash Flood Danger High in Deer Creek Narrows, non-existent elsewhere.

Best Time to Hike May & early June (best), and September to early October.

Author's Experience With Ryan Cornia we camped at Monument Point and left early on 6/24/2007. Left water at first trail junction, reached narrows in 4 hours. We did the narrows & had lunch in 3 hours; started back at 1:40pm. Reached cars after 5 hours; just under 12 hours for trip. Best to do in 1 1/2 days.

Deer Creek Falls. This is the end of **Deer Creek** as it falls into a deep pool about 60m from the Colorado River. Look closely and you can see Ryan Cornia in the middle of the picture and halfway down the 55m rappel.

Map 69, Deer Creek Narrows, Grand Canyon, Arizona

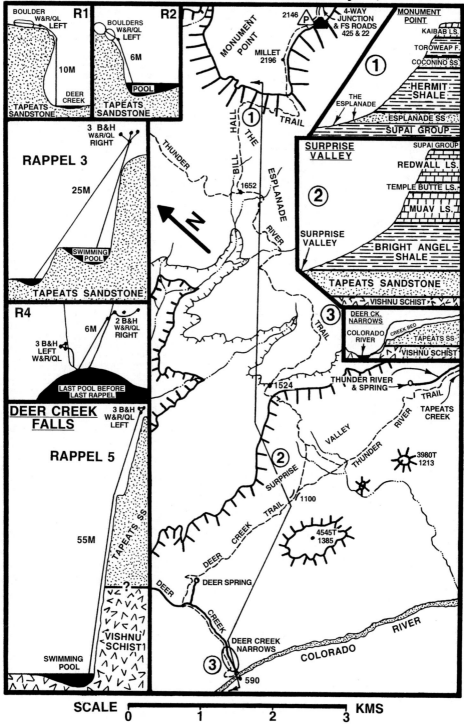

R1
BOULDER W&R/QL LEFT
10M
DEER CREEK
TAPEATS SANDSTONE

R2
BOULDERS W&R/QL LEFT
6M
POOL
TAPEATS SANDSTONE

RAPPEL 3
3 B&H W&R/QL RIGHT
25M
SWIMMING POOL
TAPEATS SANDSTONE

R4
2 B&H W&R/QL RIGHT
6M
3 B&H LEFT W&R/QL
LAST POOL BEFORE LAST RAPPEL

DEER CREEK FALLS
3 B&H W&R/QL LEFT

RAPPEL 5
55M
TAPEATS SS
?
VISHNU SCHIST
SWIMMING POOL

MONUMENT POINT
2146
P
MILLET 2196
4-WAY JUNCTION & FS ROADS 425 & 22

MONUMENT POINT
KAIBAB LS.
TOROWEAP F.
COCONINO SS.
① HERMIT SHALE
THE ESPLANADE
ESPLANADE SS
SUPAI GROUP

THUNDER
BILL HALL
THE ESPLANADE
TRAIL
①
1652
ESPLANADE RIVER

SURPRISE VALLEY
SUPAI GROUP
REDWALL LS.
② TEMPLE BUTTE LS.
MUAV LS.
SURPRISE VALLEY
BRIGHT ANGEL SHALE
TAPEATS SANDSTONE
VISHNU SCHIST

③ DEER CK. NARROWS
COLORADO RIVER
CREEK BED
TAPEATS SS.
VISHNU SCHIST

TRAIL
1524
THUNDER RIVER & SPRING
TRAIL
TAPEATS CREEK

②
SURPRISE
DEER CREEK TRAIL
1100
VALLEY
THUNDER RIVER
3980T 1213

4545T 1385

DEER SPRING
DEER CREEK
DEER CREEK NARROWS
③
590
COLORADO RIVER

SCALE
0   1   2   3 **KMS**

311

Lower Water Holes Canyon, Navajo Nation, Arizona

Location & Access **Water Holes Canyon** is located about 10 kms south of **Page, Arizona** and immediately west and south of the Navajo settlement of **Leche-e**. Water Holes runs northwest and ends in Glen Canyon and the Colorado River not far upstream from Lee's Ferry. To get to the lower end of the canyon, which is the technical part, drive along **Highway 89** south of Page and stop at **mile post 542**. This marker is next to the **bridge** spanning Water Holes Canyon. Park on the east side of the road just north of the bridge.

As opposed to some other hikes on the Navajo Nation, specifically along Kaibito Creek, Water Holes is officially open to hiking, but you're still supposed to have a permit. To get that, drive southeast & downhill from the middle of Page to a junction & signal light. This will be on Highway 98 which heads east to Antelope Canyon and Kayenta. But instead of turning east or west, continue straight ahead or south and onto **Coppermine Road**. After another 3 kms (2 miles) is **Leche-e** and Leche-e Chapter House (between mile posts 2 & 3 on the west side of the highway). Stop there to get a hiking permit at the **Navajo Parks & Recreation Office**, which is a small **white building** just behind or north of the chapter house.

Or, if no one is there, drive east out of Page on Highway 98 to between **mile posts 299 & 300**. There on the south side of the road will be the little white **entry booth** to **Antelope Canyon** belonging to the Navajo Parks & Recreation. Someone is there during the warmer 8-9 months of the year when tourists are going into Antelope. It may be best to call the Parks & Rec. Office first to see which place is open and/or the best place to get a permit. Tele. 928-698-2808 in Leche-e.

Rating **3A II** or **III PG SLOT**. This books concentrates on the lower end of the canyon west of Highway 89, not the upper walk-through & very fotogenic slots to the southeast. For more information on the upper canyon, see the author's other book, *Non-Technical Canyon Hiking Guide to the Colorado Plateau*.

Equipment RG, one 50m rope, several shorter ropes, ascenders, W&R/QL, K&EP, and if you intend to go all the way to the Colorado River, a BK, several B&H and two 60m ropes. Read more below.

Route From where you park at mile post 542, pass through the gate heading southeast over slickrock. About 100m from the gate, DC into the gorge, about 20m deep at that point. From there, walk west downcanyon under the bridge. After about 400m will be 2 dropoffs where Otis Manson installed 2 steel ladders when he was guiding tourists. If they happen to be washed away, you can still DC both places.

After passing through several short shallow slots, and about 1 1/2 kms from the bridge, you'll come to a place where the canyon begins to head due west. At that point is an **18m rap** from **2 B&H** on the left into a large pothole which could be a **KPH** at times. If rappelling down there doesn't look good, climb up to the right (north) a little to find **2 more B&H's** and W&R/QL. This avoids the big pothole.

Below the first rappel is a **DC** on the right (or an 8m rap from 2 new B&H's on the right), then a **15m rap** from 2 more new B&H's on the right. After that is the **5m Rappel 3** from a chokestone, and climbing over a boulder. Then it's walking for about 250m to a break in the wall on the right, or north, where you'll find a steep but easy scramble up an **escape route** to the rim. See the cross-section on the map. You can exit/enter there.

Below the escape route is a tight slot, then 3 steep places with 2 B&H's nearby. At one point you'll DC through a pretty little natural bridge (see foto page 314). Experienced hikers can get down & back up these pitches without rappelling or ropes, but for safety, consider leaving a short rope at each steep part, so you can climb or HL back up. Getting back up is usually easier and safer than DC'ing. Below the 3rd drop is a huge pool, always with water, then **R4** of about **13m**--but leave your rope there--don't pull it! Then comes the **Big Dropoff** which is said to be roughly 107m (?), with a rap station halfway down. It's at this Big Rappel where everyone turns back, and that's why you may need several short ropes and ascenders to get out of the canyon.

If going over the Big Rap, Otis Manson told the author there's a rap station halfway down on the cliff face, and more B&H's (?). **This is still unconfirmed!** This writer still hasn't done it, or talked to anyone who has. If this is so, then two 60's should get you down (?). However, even if you get down the Big Dropoff, there is at least one more short dropoff below that, then an easy walk to the river. But then what?

If you decide to go all the way, your choices of getting out of Glen Canyon are: (1) Have someone pick you up in a boat; (2) Flag down a fishermen heading back to Lee's Ferry; (3) Walk the southeast bank (with possibly some wading or swimming in one place?) all the way to a point opposite Lee's Ferry, then wave to someone in a boat to haul you across; (4) Carry a small, lightweight plastic/rubber raft and float down and/or across the river; (5) Or, make arrangements ahead of time to have one of the tour boats running between Glen Canyon Dam & Lee's Ferry pick you up and take you down to Lee's Ferry. **Absolutely do not attempt to swim the Colorado below Glen Canyon Dam!** Water coming from the bottom of Lake Powell just upcanyon is 7°-8°C and you'll likely die before you make it across! Like it or not, this stretch of river is now a premier trout stream. It seems best to stop at the Big Dropoff, and return.

Elevations Highway 89 bridge, 1395m; top of escape route, 1329m; Colorado River, 951m.

Time Needed Going down just to the Big Dropoff and return via the escape route, may take 4-6 hours.

Water If you're going down to the Big Dropoff and return on a hot summer day, take 3-4 liters of water.

Maps USGS or BLM map Glen Canyon Dam (1:100,000) for driving & orientation; and Lees Ferry (1:24,000--7 1/2' quad) for hiking.

Flash Flood Danger Overall, moderate risk, but higher in the tighter slots.

Best Time to Hike Spring or fall, but summers aren't so hot in the canyon bottom and it's a short hike.

Author's Experience The author spent half a day below the highway bridge on one trip. Then in 1998, he and Otis Manson & Tim Martin walked down the escape route and explored upcanyon first, then backtracked and went to just above the Big Dropoff, and returned. That took about 7 hours. Later, he and John Summerson entered at the bridge, got as far as the big pool, which is just above the Big Dropoff, and returned via the escape route, all in 4 1/3 hours.

Map 70, Lower Water Holes Canyon, Navajo Nation, Ariz.

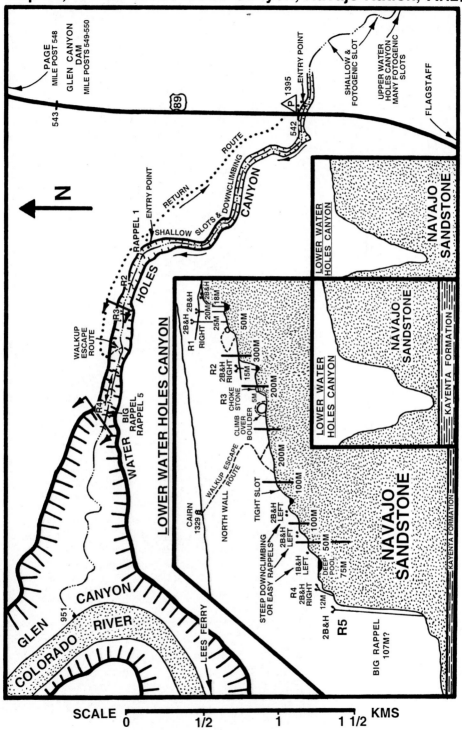

SCALE

0   1/2   1   1 1/2   KMS

Above Left John Summerson goes down the 15m Rappel 2 in **Lower Water Holes**.

Above Right Climbing back up in the lower part of **Water Holes Canyon**. There are old B&H's in place, but DC'ing & UC'ing are easy.

Right John DC'ing through a natural bridge in the lower part of **Water Holes Canyon**.

Above Left Looking down from the top of Rappel 4 in **Butterfly Canyon**.

Above Right Standing on the rim of the KPH in the Lower Slot of **Butter-fly Canyon**. If feet are dry, you can easily climb around this pool on the left side.

Left This is one of the nicer part of **Butterfly Canyon**.

Butterfly Canyon, Navajo Nation, Arizona

Location & Access Butterfly Canyon is located roughly 22 kms (14 miles) southeast of Page, Arizona on the Navajo Nation. Page is the town situated next to the Glen Canyon Dam & the southwestern end of Lake Powell.

ATTENTION: BUTTERFLY CANYON IS OFFICIALLY CLOSED TO HIKING!
There are many reasons for this closure, however, this canyon is so scenic & challenging, plus it is equipped with anchors along most of its entire length, that the author believes one day it will be re-opened. If or when that happens, here is the latest information. Anyone going to this canyon does so at their own risk!

To get there, start in **Page** and head eastward on Arizona **Highway 98** in the direction of Kaibito & Kayenta. After about 21 kms (13 miles), and 250m south of **mile post 311**, turn left or east (set speedometer at 0). Immediately after passing through a gate veer right and proceed eastward about 1 1/2 kms (1 mile) to the **big powerlines**; then turn left or northwest following the powerline road--this part is very sandy and you'll need a **4WD** for sure! When you reach **Km 4.5/Mile 2.8**, turn northeast and drive to the end of that road which ends on a point on the Carmel bench. This will be **10.6 kms (6.6 miles) from Highway 98**. The **gates between miles posts 309 & 310, and at 312, are now wired shut** making the only access from mile post 311. See the Peach Wash map for another look at the access route. The end of this road & parking place is the same trailhead as for Piñon Canyon.

Rating 3B III or **IV PG SLOT**. Butterfly has a total of 4 rappels (or at least 7 if you start at the very beginning of the canyon), and maybe 2 or 3 more if you continue right down to the very end of Butterfly and to the bottom of Navajo Canyon Inlet of Lake Powell. It has one good deep dark slot in the lower section.

Equipment RG, two 50m ropes (or a 50 & pull cord), plus a 15m, W&R/QL, K&EP, drybags, hooking kit & Ibis hook, headlamps, a BK & several B&H or baby angle pitons just in case. In spring or fall, take a wet/drysuit.

Route From the car-park, walk north then east down over the rim of the point which is the Carmel Formation & Page Sandstone (top part of Navajo SS). At the bottom, make your way due west as you route-find down over slickrock ledges. When you reach the sandy flats, continue west with compass & map. The first dropoff in Butterfly, and the point to aim for if doing the entire canyon, is about 1 km due west of the car-park. You'll have to rim-walk to locate the point where the bottom drops away. There you'll find a **50m rappel** from a cedar (juniper) tree about 2m from the rim. Install 2m of W&R/QL and rappel. Just around the corner at the bottom is **R2**; you must install **1 B&H** on the left and rappel down about 15m into a **KPH**. You'll need a **hooking kit** to get out of this, because a **shoulder stand** is not possible and a **pack toss unlikely**. Have a partner hold you against the wall for the 1st hole or two. Immediately beyond the KPH is another drop, then perhaps another soon after that (one group wasn't equipped to handle this part on their trip, so you may find this part a virgin! Be well prepared for this first 200-300m section, the part with missing information.

After the upper technical section, you'll find a more open canyon and boulders to negotiate. Soon you'll come to the 1st of at least **4 exits** on the right or southeast side. If you don't want to do the upper part, then use one of these walk-in routes to enter. Soon after that will be the 2nd escape on the right and on the down-canyon-side of a rincon-like circular feature. Soon after that will be a nice DC into a short slot. Then the canyon opens some and is a little boring for a ways.

R4 is about **34m** which puts you into a nice deep slot, then a 4m DC on the left side of a large boulder/chokestone. Some may want to be belayed. From there it's easy walking except for a DC or two, a section of wading and at least one small pothole which seems to have water no matter how long the drought. Soon after that are boulders choking the slot and the **10m R5** from a big boulder. This is the beginning of the **lower dark slot** which is the best part of the canyon.

Once in the slot, you'll have some tight places and likely a number of pools of cold water. In the spring or fall, better take wet/drysuits just in case, and one headlamp per person. At the lower end of the dark slot, is a big **KPH** with perpetual water and a small bridge immediately above it. Here you can climb around to the left on some downward-sloping ledges with some pretty good handholds. However, it's a bit tricky because you'll likely have mud on your shoes at that point. On the other side of the KPH is a **4m** vertical drop. One group punched out a hole for a grappling hook (Ibis hooks are better), HL'ed down **R6**, then flipped the hook out (the pioneer party through the canyon must have used the bridge as an anchor to get over this drop?). Immediately below the KPH will be a small pool of drinkable water.

After the KPH, you'll come to the same limestone lens which has created Piñon Falls a few kms up Chaol Canyon & Kaibito Creek. Where this lens ends will be **R7, a 14m low angle rap from a single piton** on the right. After that, it's scrambling except for one DC/slide into the 1st of 2 shallow slots, then you'll walk on the right side above a 2nd shallow slot until you come to an overlook of Lake Powell and/or Navajo Creek. At that point veer right and double back to the west and out onto flats to the south.

Or, if you stay in the last 100m shallow slot, you'll come to lots of big boulders and 2 more rappels (?). At the very bottom of that is a dropoff into Lake Powell; or onto sand flats if the lake is low, but locating a route back up from there might be more trouble than it's worth. It might be best to tie off a rope at each drop, rappel to the lake/creek/sand flats, and use ascenders to return the same way.

From the bench overlooking the lake/Navajo Creek, route-find generally southeast as shown on this map. After about 1 km, walk up a steep slickrock hogsback ridge, then veer southwest as shown. Have a map & compass in hand for this part. The car-park is about 6 kms from lower Butterfly.

Elevations Trailhead, 1555m; Lake Powell at the high water mark (HWM), 1128m.

Time Needed Depending on where you enter, and if there's mud or dry sand, plan on 6-8 hours in the canyon; then about 2 hours to hike back to the car-park. Plan on no less than 8-10 hours total hike-time; maybe longer for some. Still longer if you do the upper 300m slot.

Water Take your own. Or from potholes (there are at least 2 potholes with water almost guaranteed), or from Lake Powell/Navajo Creek (treat). Take 4-5 liters each if going in summer, as the walk back in the sun will be extremely hot!

Maps USGS or BLM map Glen Canyon Dam (1:100,000) for driving; and Leche-e Rock & Cedar Tree Bench (1:24,000--7 1/2' quads) for hiking.

Flash Flood Danger Moderate overall, but high in spots, with occasional wide places to run to throughout most of the canyon.

Best Time to Hike April or May, or late September or October; but take a wet/drysuit during cooler times.

Canyoneers Experience After several scouting trips over the years, Todd M., Stephanie Y & Mike R., tried the uppermost big drop, but didn't quite have the necessary bolts, etc., so they rim-walked to one of the 1st entries. Getting through the canyon took about 8 hours; about 10 1/2 hours round-trip.

Map 71, Butterfly Canyon, Navajo Nation, Arizona

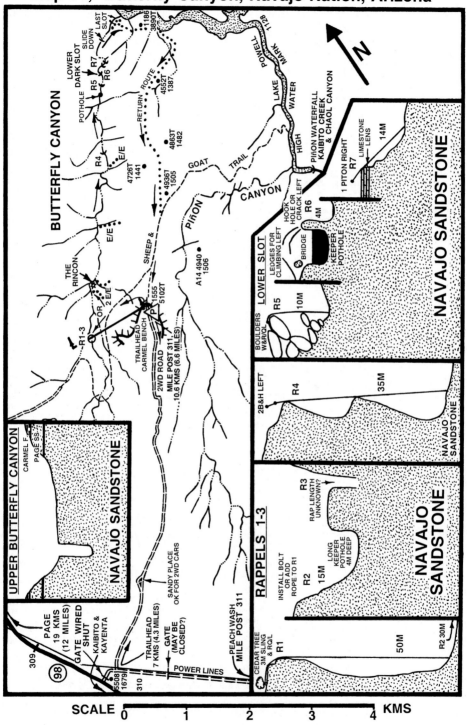

SCALE 0 1 2 3 4 KMS

Piñon Canyon, Navajo Nation, Arizona

Location & Access This drainage seems to have a Navajo name (that's unpronounceable), but it's been lost for the moment. However, since **Piñon Falls** (sometimes referred to as Kaibito or Chaol Falls) is located nearby, it's been decided to just call it Piñon Canyon, for lack of a better non-Indian name. Piñon Canyon is located roughly 22 kms (14 miles) southeast of Page, Arizona on the Navajo Nation.

ATTENTION: PIñON CANYON IS OFFICIALLY CLOSED TO HIKING!
There are many reasons for this closure, however, this canyon is so scenic & challenging, plus it is equipped with anchors along its entire length, that the author believes one day it will be reopened. If or when that happens, here is the latest information. Anyone going to this canyon does so at their own risk!

To get there, start in **Page** and head eastward on Arizona **Highway 98** in the direction of **Kaibito & Kayenta**. After about 21 kms (13 miles), and 250m south of **mile post 311**, turn left or east (set speedometer at 0). Immediately after passing through a gate veer right and proceed almost due east about 1 1/2 kms (1 mile) to the big powerlines. Turn left or northwest following the powerline road--this part is very sandy and you'll need a 4WD for sure! When you reach Km 4.5/Mile 2.8, turn right or northeast and drive to the end of that road which is a point on a Carmel bench. This will be 10.6 kms (6.6 miles) from Highway 98. The gates between miles posts 309 & 310, and at 312, are now wired shut making this the only access from mile post 311. See the **Peach Wash map** for another look at the access route. This is the same trailhead as for **Butterfly Canyon**.

Rating This canyon rates **3A or B IV +PG SLOT** with up to 13 rappels. Most of the first 250m section, along with 4 raps, is in total darkness, plus there are several easy but tight slots. In the lower half of the canyon, there are several big rappels and dark sections, plus the possibility of swimming potholes, one of which is a **KPH**. This canyon ends near Piñon Falls, a lovely fotogenic cascade.

Equipment RG, one 50m rope plus a 50m pull cord for the **Lower Canyon** (one 30m rope for the **Upper Dark Slot**), lots of W&R/QL, headlamps for the Upper Dark Slot, BK & several B&H per/group, ascenders, drybags, K&EP, wet/drysuit for spring or fall in the Lower Canyon. And of course a compass & map.

Route This canyon is divided into 3 parts: **Upper Dark Slot, Middle Section** and **Lower Canyon**. From the trailhead at 1555m, walk back up the road (southwest) 100m, then head southeast off the rim zig zagging to the bottom of the bench. Once there, turn right or south and walk a ways along a faint vehicle track; or just cross-country to the south. There are no good landmarks close by, so you must follow the 7 1/2' map with a compass until you reach the main drainage, which is the most-southerly of 3 upper forks. This shallow drainage has several twisting & convoluted slots in the upper sections, but none are more than 6-8m deep.

Once at the main drainage, look for a way in about 25m upcanyon from where an inner slot begins. There may be a small cairn on either side. This is where the real slot starts to cut deep into the Navajo Sandstone making it the **darkest slot** on the Colorado Plateau. This **Upper Dark Slot** is about 250m long.

Throughout this canyon, all bolt holes were drilled with a cordless drill and are high enough on the walls to remain out of the way of most future flood waters. Also, this part of the canyon system is surrounded by sand which quickly absorbs rain, so even though this is a tight slot, it sees a lot less water than further downcanyon. For water to run through this part, it will have to rain very hard in the 2-3 kms above this short slot.

Upper Dark Slot. Right where the slot begins is a chokestone underneath which you can chimney down 3m. After about 25m is 1 B&H on the left above for **Rappel 1** of **5m**; however, you can **DC** this easily into a narrow 1 1/2m-deep pothole. After another 5m is **R2**; place your rope over a **small buttress** on the left and **rap/HL 6m**. Put on your **headlamp** for the next hour and be alert for **snakes** in summertime. One group found one rattler and one that looked like a mountain king snake (?). **R3** is down **9m** from a B&H on the right. Next is a **steep 6m DC** (this may be difficult if the canyon is wet & muddy; in that case have an alternate--perhaps a bolt or rap from an Ibis hook). **R4** is from 1 **B&H** on the right. This goes down about 6m into a crack going left (this can be **DC'ed** easily), then a 90° right turn in a very tight section. Best to stay below some logs and foot-jam sideways dragging your pack behind.

R5 drops **12m** from 2 **B&H & chain** on the left; then after a tight walk of what could be 60m, you'll come to **R6**. This anchor is 2 **B&H** with W&R/QL to the left with a **14m drop**. After this last rappel, remove your rappelling harness to save wear as it's a tight squeeze & a walk of perhaps another 75-100m down to daylight, then some scrambling in and out of tight spots to where you can turn left and walk out of the canyon at **Exit 1**; or turn right and climb up a steep gully or couloir to **Exit 2** on the right. Both exits are steep walkups.

Middle Section Continue ahead downcanyon in a tight slot where in places you walk sideways and crawl under boulders. Where the canyon makes a sharp right turn at a dropoff, backtrack 40m, get out and on to a low bench to the left, or north, then walk & DC bypassing the dropoff.

After another 400m or so will be **R7**, a drop of about **10m** (maybe **DC** this?). Do this from a boulder sitting on top of the slot. After that, you could get out at **Exit 3** on the left or north (not totally explored, but it sure looks like an exit?); or walk down to **Exit 4**, which is a walkup. For some this would be a full-day's hike. If so, or if you get a late start, leave the canyon at that point. Walk east up slickrock to the sheep & goat trail shown, then head west and southwest back to your car.

If you have 3 strong hikers and with an early start, you can do the entire canyon in one long day. About 300m below Exit 4 will be the start of the **Lower Canyon** and R8 from 2 **B&H** on the left side. This is a free rap of **38m** into a dark & convoluted hole, the bottom of which can't be seen from above when the sun is shining.

A little further down, you'll walk through a dark tunnel-like feature similar to The Subway in Zion with a tight slot above. Be there at mid-day for great fotos. A little further along is a tight 30-40m section where you'll have to foot-jam sideways. A more rigid boot or shoe helps here.

On 3 occasions, hikers found a sand-filled pothole immediately above **R9**, but a 4th group found a a swimming pool! For this reason, always be prepared to swim and protect cameras and other goodies in a drybag or plastic box/case. R9 is **42m** at a low angle. There are 2 **B&H** on the left out on the face of the wall with a clear shot straight down, making it an easy rope pull from below.

From the bottom of R9, it's an easy 300m scramble among boulders down to **Exit 5** on the right which is a steep climb southwest up a big gully. Or turn left to find **R10**, which is a drop of about **20m**. One group rigged a long piece of W&R/QL around a boulder for the descent. Another group put 2 pitons & W&R/QL in the wall on the left.

After R10, you'll walk through a dark slot while the canyon gradually opens up, then comes **R11**. This is a semi-free rap into a sculptured hole about **26m** to a wide shelf. Next to it is a **KPH** which will always have some water. But don't pull your ropes yet!

To get over this **KPH** you have 4 choices: (1) If it's full of water, swim across and crawl out. (2) If water is low, the tallest person can stand on the bottom and push someone else up the other side. (3) An athletic per-

Map 72, Piñon Canyon, Navajo Nation, Arizona

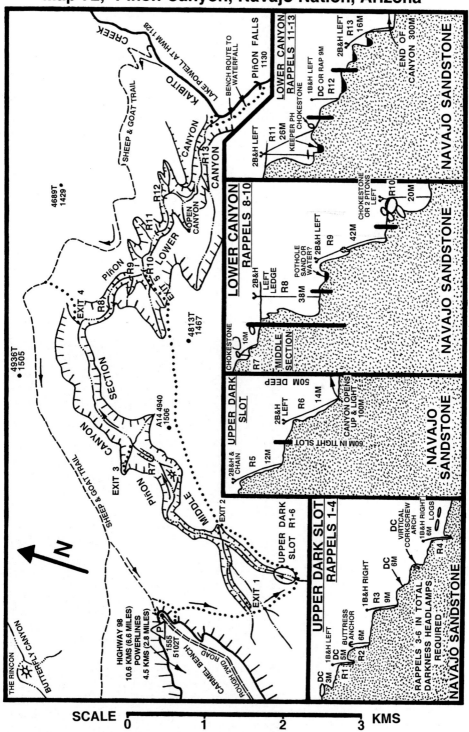

SCALE

0   1   2   3 KMS

son can jump over the pothole which is about 3m over while descending about 1m, but this could be risky as the landing isn't very big. This is what the first group did. (4) But the **best way** is to make a great big knot on one end of a 15m rope and throw it over the pool and have it jam in between the wall and a chokestone about 8m away (see foto on page 321). Then one person rappels about 1m into the KPH using a Prusik rope & knot, and pulls himself across the gap to the other side using the 15m rope. He then pulls everyone else across while hanging onto the 50m rappelling rope. This worked well on 2 occasions.

Not far below **R11** is a 2m-deep, 2m-wide round pothole. Most people take 2 quick steps around the left side on small ledges to get around this obstacle. **Because of these 2 potholes, never do this part of the canyon alone--always go with 2 people or more!** Just below the round pothole and just before the canyon opens up again, is a big pool; DC, then wade or swim. The first group through, which was about 4-5 days after a minor storm, found a swimmer.

About 30m below the big pool is a small pothole, then the canyon opens up again. At that point is **R12** down **9m** from a **B&H** on the left. Or, you could **DC** about **4m** and into a pool. Below that it's an open canyon for 350m, then you can climb up to the left bypassing 2 deep pools (to stay in the canyon bottom, you'll have to rig up an anchor for each drop). About 150m below that will be another pool or two, then the **13th & last rappel**. Use **2 B&H** on the left. This rappel will be down into one pothole where you'll have to kick your way over to the other side, and rap on down. This one drops about **16m**.

Below R13 will be a 250m walk on a limestone lens to the mouth of the canyon with Kaibito Creek below. To get down to the creek, veer left 40m and walk down a steep gully to the stream. Or walk along a trail on the right, which will take you south to **Piñon Falls**, about 500m away. Seeing, swimming below (in summer) and fotographing this cataract is required! The waterfall is there because of a thin limestone lens in the middle of the Navajo Sandstone.

From Piñon Falls, walk down the sandy Kaibito Creek for 1 km and when the creek veers to the right, climb left up on a low sandy bench with cottonwood trees to find a **sheep & goat trail** built during the 1930's by the CCC's. Follow this zig zag fashion upslope to the northwest, then west. In the middle sections of this trail, follow horse tracks or cairns closely. It's about 6-7 kms from Kaibito Creek back up to the car-park.

Elevations Car-park/trailhead, 1555m; Piñon Falls, 1130m.

Time Needed One day of perhaps 10-12 hours. After the canyon was prepared, 3 heroes went from the car-park to Kaibito Creek in 7 1/2 hours. After swimming and cooling off at Piñon Falls for an hour, and with all clothes wet, they walked back to their car in 1 2/3 hours in temps as high as 43°C. Round-trip took 10 hours. But some may want to do it in 2 days. For that, do the Upper Dark Slot and the Middle Section on Day 1; then on Day 2, reenter the canyon and start with Rappel 8. To do each half of the canyon on separate days may take 7-9 hours round-trip each.

Water Carry water from a culinary source. In summer take at least 4 liters each! There are no springs or running water until you reach Kaibito Creek, which may be drinkable (?). However, sheep & goats are occasionally taken into the canyon, so to be safe, treat of filter that water.

Maps USGS or BLM map Glen Canyon Dam (1:100,000) for driving; and Leche-e Rock & Cedar Tree Bench (1:24,000--7 1/2' quads) for hiking.

Flash Flood Danger Not many floods make it through the Upper Dark Slot, but rate the danger high; scattered parts of the Middle Section & Lower Canyon have high risk, but there are many open places too.

Best Time to Hike June and early July is the driest part of the year in these parts. If you have to swim at this time, you can do it without a wet/drysuit. If there's no water in the pools, then it can be done anytime of year with the proper clothing. In the cooler half of the year, better take a wet/drysuit through the Lower Canyon.

Canyoneers Experience Mike R., Julia R., Warren E., Keith K., assisted by Eberhard S., were the first to put bolts in the Upper Dark Slot (250m in about 2 1/2 hours--1 1/2 hours in total darkness!). They continued down to Rappel 8, but stopped there and returned via Exit 4 (7 1/3 hours round-trip). Next day, Warren E., Keith K. and Mike R. walked down the trail to Exit 4, reentered the canyon and made it to Rappel 10, but stopped. They left the canyon at Exit 5 (about 5 1/4 hours round-trip).

Later, Mike R., Ben B., Paul P. and Carter J. reentered the canyon at Exit 5 & R10 and completed the entire canyon for the first time (6 1/2 hours to Piñon Falls), then hiked out via the sheep & goat trail (just under 10 hours round-trip). Next day this same group went through the Upper Dark Slot (2 hours) to Rap 10 and returned to their car via Exit 5, all in about 7 1/2 hours.

It was the next summer when Mike R., Keith K. and Warren E. did the whole thing in 10 hours round-trip from their car. Next day, this same group plus Eberhard S. made it to Piñon Falls in 7 hours.

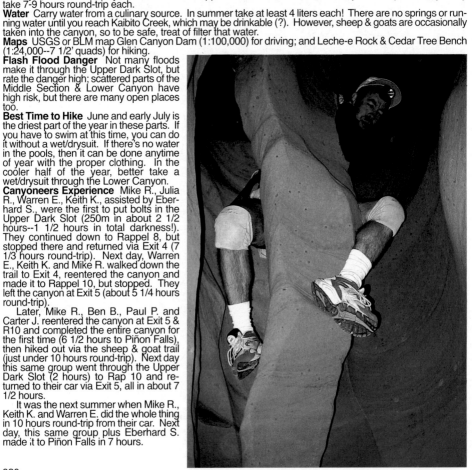

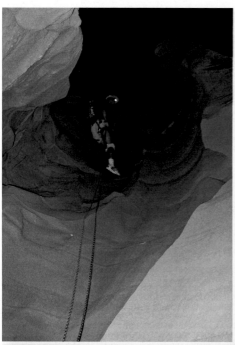

Opposite Page Crawling down a vertical bridge in the upper part of the **Dark Slot in Piñon Canyon**. It's totally dark at this point and headlamps are required.

Above Left Making the last rappel in the Dark Slot of upper **Piñon Canyon**.

Above Right One person has just thrown a rope, with a big knot on the end, across this big KPH and had it lodge in between the wall and a chokestone. Then he pulled himself across while still tied into his rappelling rope. This part is right at the bottom of Rappel 11 in the **Lower Section** of **Piñon Canyon**.

Left One of the more colorful & fotogenic parts of **Piñon Canyon** is in the **Middle Section**.

Peach Wash, Navajo Nation, Arizona

Location & Access Peach Wash is located roughly 26 kms (17 miles) southeast of Page, Arizona on the Navajo Nation. Peach is a tributary to the middle part of Kaibito Creek/Chaol Canyon.

ATTENTION: PEACH WASH IS OFFICIALLY CLOSED TO HIKING!

There are many reasons for this closure, however, this canyon is so scenic & challenging, plus it is equipped with anchors in critical places, that the author believes one day it will be reopened. If or when that happens, here is the latest information. Anyone going to this canyon does so at their own risk!

To get there, start in **Page**, which is near mile post 297, then head eastward on Arizona **Highway 98** in the direction of Kaibito & Kayenta. After about 21 kms (13 miles), and 250m south of mile post 311, turn left or east (set speedometer at 0). Immediately after passing through a gate veer right and proceed east about 1 1/2 kms (1 mile) to the **big powerlines**; then turn right or southeast and follow the powerline road for about 2 kms (1.2 miles) until you come to a **fence & gate**, the only one around. Park there if doing the **technical section**. You'll need a 4WD to get there. The gates between miles posts 309 & 310, and at 312, are now wired shut making the only access in this area from mile post 311.

If you want to do the **upper semi-technical slot**, proceed about another 3 kms (2 miles) to where the sandy road crosses the **upper part** of **Peach Wash**. Don't attempt this unless you have one of the maps below, and a 4WD! If you have a 2WD car, consider parking on the highway somewhere near mile post 312, then with a compass & map in hand, walk cross-country almost due east to the powerlines and look for the gate which is just south of **power pole #19**.

Rating 4B R III (depending on where you park). Peach Wash has 6 rappels and several **KPH's** or near-**KPH's** in the short **400m technical section**. Peach ranks fairly high in risk potential and is for experienced hiker-climbers only.

Equipment RG, two 60m ropes (or a 60 and a pull cord), a short rope or two, G-pick & Ibis hook, BK, several B&H's & W&R/QL, K&EP, and one or 2 etriers (aiders). Also, the person (group leader) going through the potholes first, or the person who will be in water the longest, should wear a thick wetsuit or drysuit. Everyone else should have a wetsuit of some kind.

Route For the **Upper Semi-Technical Slot**. From where you park in the broad dry creek bed marked 1582m, one option is to walk upcanyon. There you'll find a short twisting slot that's easy to get through. Return the same way. Below the road, the canyon slots up immediately and slowly gets deeper & tighter. After 300m, it gets even tighter for about 75m. This part includes a couple of short slides into potholes, then it opens for a ways. After another 100m or so, it closes in and gets dark. This is where you'll find a 5m DC into a pothole.

The slot opens up for good just before the **powerlines**. From there on down you can walk along a bench on the left all the way to the technical section; or you can DC or slide into several short slots with periodic escape routes, mostly on the left side. There is at least one short rap which could be done with an Ibis hook. When you come to the **Technical Section**, which will be a large pothole with water (and **R1** just below--see map), turn left and climb up the **short side-canyon** to the west, cross over the slot on a chokestone bridge, then head back down on the other side of the slot to just below the same pool. There you'll find a small cedar (juniper) tree on the west side which will be your anchor for **Rappel 1**.

Let's stop there and describe the route to just the **Technical Section**. From the **fence & gate at 1658m**, which is immediately beyond a shallow drainage, simply follow this same dry creek bed eastward down to Peach Wash. When you reach the rim of Peach where this **short side-canyon** makes a big drop, consider taking a look at the last rappel at the bottom end of the technical section before you jump in.

To do that, have a good map & compass in hand, then head north and a little west to enter the upper part of what will be your **escape route** once you've finished the technical slot. Head northeast down this faultline side-canyon. At the bottom, you'll see a big green colored pothole (this could be a little difficult to get out of depending on water levels, but is not quite a KPH) you'll rappel into at the very end of the line, **Rappel 6**. After you check this part out, climb back out this side-canyon escape route, veer right or northeast and walk along the **north-side rim** of the **same side-canyon escape route**. At one point you'll be directly across the canyon from the end of the technical section of Peach and will have a bird's-eye view of what you'll be up against. The challenges will be getting out of a KPH which is just out of sight inside the slot, and a **hanging near-KPH** on the upper part of the wall.

From the viewpoint, return to the dry creek bed you followed getting down to Peach Wash. At that point you have a choice of 2 entry places: (1) Head right or south about 200-300m and route-find down to a bench above the drainage bottom, then head north for 300m to the big pothole mentioned above. Climb up to the west and cross over the minor slot to get down to the juniper tree. (2) Or veer left or north from where the dry wash drops off and after 200m or so, veer northeast and route-find down to the same pothole & tree.

Once at the tree, install W&R/QL and **rappel** down about **7m** to a flat place between 2 potholes, then continue the rap into a chest-deep pothole which is easy to get out of. You'll make this 2-part rap from one anchor. From there, slide down 3m into a swimming pool and walk out of it.

A few minutes later you'll come to **R2** which is from a single piton placed right in the bottom of the water course (eventually this will have to be replaced, please put it higher up on a wall). After this **9m rap** into a shallow pool, there's a little more walking and DC'ing to **R3**.

You may find 1 piton at **R3** in the channel bottom, but there are now **2 well-placed B&H's** high on the right. This one has a drop of about **40m** into a **very deep, dark, ugly, ice-water KPH**. One group up to this canyon after a long, long dry spell and found the water level shoulder deep and about 2 1/2m below the top of a chokestone making this a KPH. The first person through this, and at first glance, about had a near-death experience! However, on the right side of the chokestone, between it and the wall, is a crack about 30-35cms wide, plus about 3 well-place, thank-somebody's-god-handholds! To get out, and while swimming, reach up to one handhold, then another doing chin-ups, and eventually slither out. At least one member of your group had better be a very fit, skinny, tough-ass climber who has good upper body strength and who can climb a wall using mostly arm strength. Your feet reach the opposite wall under the chokestone, but it's always wet & slippery.

There doesn't seem to be much debris or logs in this canyon, so that crack and the hand holds should be there a long time. If for some reason the crack is clogged, then have a G-pick handy to dig it out. Another possibility of escape is to tie an Ibis hook onto the end of a short rope and throw it over the chokestone above the crack and have it lodge in behind it. As for hooking out, forget it; everything slopes inward going up, so hooking is out of the question; except maybe you could reach up and place it somewhere (?). If the water level is high, then you can climb up the face of the chokestone. Another idea for helping less-fit hikers out, is to have one or 2 etriers (aiders) ready and tied to a short rope dangling into the pool.

After **R3 & the KPH**, is a little DC'ing to a short drop of **5m** with **1 B&H** on the left **(R4)**. Just below that is

Map 73, Peach Wash, Navajo Nation, Arizona

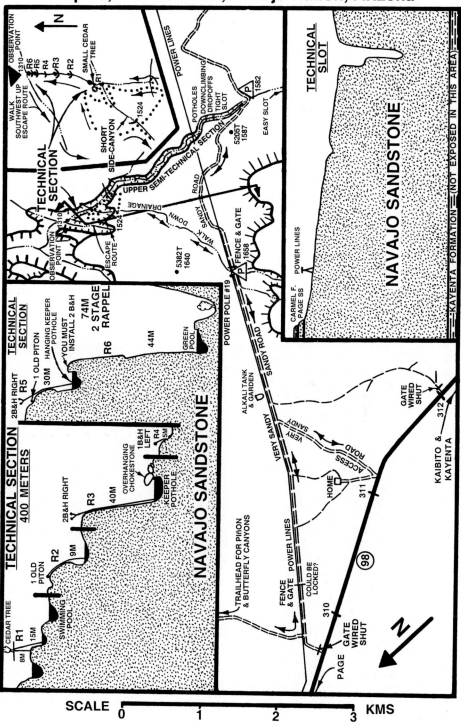

SCALE

0   1   2   3   KMS

the end of the slot and a big double rappel totalling about 70-74m. One group found a single piton and 3 stranded ropes going down about **30m to a hanging pothole** on the cliff face. Below that was a single rope with 3m of webbing which just reached the last green colored pothole. It appeared somebody didn't have ascenders or enough pitons, and had to make do barely escaping with their lives!

Now there are **2 B&H's** on the right at the top of this long **2-stage rappel (R5 & R6)**. One group used a 100m rope along with a 50m to reach the hanging pothole, then tied a 60 to the 50 at the lip of the pothole to get the rest of the way down. However, two 50's (or a 50 & 50m pull cord) will get you down the first stage easily, **then you must place 2 more B&H's on the lip of the hanging pothole, so you can make the 2nd half of this long rappel with just two 50's (make that two 60's just to make sure; or better still, a 60m rope & 60m pull cord). This will eliminate carrying a 100m rope.**

This hanging pothole has a flat shelf which is about 2 1/4m below the lip. Regardless of water depth, it should always be possible for 2 people to help each other out of this one. The lip is nearly 2m wide. At the bottom of **R6** is the large green pool for which you should have no problems getting out; but be ready for anything! One person had to do a beached-whale move to get out.

When everyone is out of the green pool, turn southwest and walk & climb up the steep side-canyon which must be along a faultline. Once out and on the rim, retrace the route back to your vehicle.

Elevations Gate near power pole #19, 1658m; creek bed parking for upper slot, 1582m; upper rim to technical section, about 1524m; bottom of last rappel & green pool, about 1310m.

Time Needed The technical part should take all day, 9-11 hours; because it's recommended you check out the pothole at the bottom of the last rappel, and/or view the end of the technical section from the viewpoint on the canyon rim across the way before taking the dive. If you start in the upper section, you might do the entire canyon in one long day--if you don't check out the technical part first. However, to do that you'd have to carry a rather heavy pack through a tight slot. It's recommended you do these 2 parts on different trips.

Water Take your own. Or, perhaps from potholes after being filtered or purified.

Maps USGS or BLM map Glen Canyon Dam (1:100,000) for driving; and Leche-e Rock & Cedar Tree Bench (1:24,000--7 1/2' quads) for hiking.

Flash Flood Danger Moderate to high in short sections with lots of slickrock in the immediate area, but there are many open places in the upper slot and a few in the technical section. Above the road and Peach's upper easy slot, it's short and vegetated with almost never any water coming down that part.

Best Time to Hike Warm weather, late spring through early fall. A wet/drysuit will extend the season.

Canyoneers Experience After several scouting trips, Julia R., Warren E., Keith K., Mike R. and assisted by Eberhard S., did the **technical section** in 10 hours round-trip from the gate. That included placing 5 B&H's, tearing apart & hauling out 3 long ropes, etc. You should do it quicker than this group; if you don't spend time looking at the canyon from the opposite canyon wall. Another time, Todd M., Stephanie Y. and Mike R. did the upper non-technical part, including looking into the technical section in 4 2/3 hours round-trip.

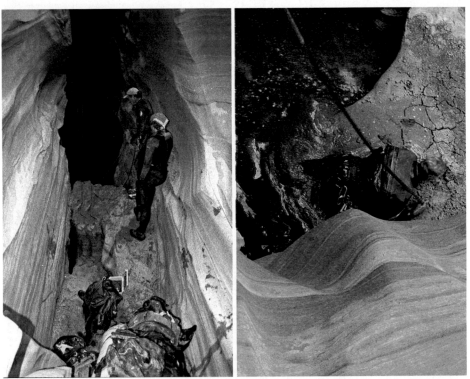

Left This is the exit from the **big ugly KPH** in **Peach Wash**. See the crack on the lower left? That's what one person will have to crawl out of. If all else fails, you could throw an Ibis hook over the top having it lodge somewhere. **Right** Crawling out of the Green Pool at the end of **Peach Wash**.

324

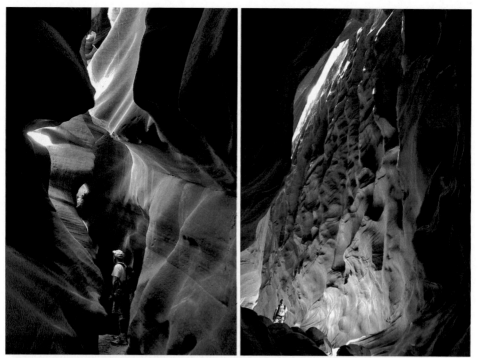

These are from the **Upper Kaibito Slot** above the 2 rappels. When the sun shines right down into the slot, it creates some amazing colors--but don't let a streak of sun in your image.

Starting the big 2-stage rappel at the end of **Peach Wash**. See the pothole & lip below? Someone must install 2 B&H's there to make a shorter 2nd half of this 74m rap.

Upper Kaibito Creek, Navajo Nation, Arizona

Location & Access This map features 3 narrow slot sections of **upper Kaibito Creek** located about 58 kms (36 miles) southeast of Page, Arizona. The parts of most interest are from 10 to 20 kms north of Kaibito & Kaibito Chapter House.

ATTENTION: KAIBITO CREEK IS OFFICIALLY CLOSED TO HIKING!

There are many reasons for this closure, however, this canyon is so scenic & challenging, plus it is equipped with anchors along its entire length, that the author believes one day it will be reopened. If or when this occurs, here is the latest information. Anyone going to this canyon does so at their own risk!

To get there, drive along **Highway 98** which runs between Page and Kayenta, Arizona. Proceed to Kaibito, a small Navajo settlement just south of **mile post 331**. From Kaibito, head east to **mile post 334** and turn north onto **Navajo Road #6330**. This road is on the east side of the canyon and is graded all the way to Tse Esgizii (Esgizii Rock) and beyond. Drive 8.4 kms (5.2 miles) to a junction (this is 1 km north of a spring & watering trough on the left). At the junction turn left and drive west and north 1.5 kms (.9 mile) to where you'll see a small rocky outcrop on the right, and 3 scattered juniper trees on the left, all marked **Rocks 1740m** on this map.

To do the Little East Fork (**LEF**) of Kaibito, continue north on the same road from the Rocks at 1740m, to Km 2.7/Mile 1.7 and park at the washed-out stock pond called **Dejolie Tank**.

You could also reach any west-side parking place by driving north from milepost 331, but many of those roads are very sandy and will require a 4WD.

Rating 4B R IV (or **V**) **+PG SLOT**. To do this entire canyon will take one very long day; or it can be done in 2, one-day stages. **The Upper Kaibito Slot** is dark and has some interesting DC'ing--this can be done in half a day; **Middle Kaibito** has a **75m free rappel**; and **Lower Kaibito** has 5 rappels over fotogenic waterfalls one after another (one pothole/pool is a swimmer), then an exit up an old sheep & goat trail (or out the East Fork of Starting Water Wash).

Equipment RG, 30m rope, K&EP and headlamps for **Upper Kaibito**; 2 ropes at least 80m long for **Middle & Lower Kaibito** (or an 80, and a 80m pull cord, plus a shorter 30m rope), leather gloves, a variable speed control ATC (for a single rope rappel) and perhaps a locking biner and/or a munter hitch on your brake-hand-side leg loop for this long free rap, W&R/QL, R/QD/DC, drybags, BK & several B&H's or baby angle pitons, K&EP, ascenders, and wet/drysuits for spring or fall seasons.

Route From the trailhead at **Rocks 1740m**, walk west to the canyon rim, then follow a trail-of-sorts and/or cairns, down over several benches to the shallow canyon bottom, a distance of about 400m. The last part is down some moki steps. If coming via the west side, the entry point is about 150m south (upcanyon) from the east-side entry route (or due west across the way from the moki steps).

The canyon at that point starts to entrench and for about 1 km, it's one of the best all-around slots on the Plateau. This is the **Upper Kaibito Slot**. As shown on the map & cross-section, you'll have 3 rappels before, and now just after, the slot opens up. The first **2 rappels** are a **single B&H** on the left going downcanyon (the B&H for R1 is high on a shelf); or use W&R/QL from a log with an awkward start. The 2nd rap is at the end of the dark slot. The 3rd rappel is now (as of 6/2005) about 200m below R2 and from W&R/QL around a chokestone--floods have really changed this canyon in recent years! Below this technical section, you can exit the canyon at the **Sandslide** or the **2nd Exit**, both on the west side. Once on top, rim-walk south to the entry points suggested above, cross the upper shallow canyon and return to your car.

To do this entire canyon, it's recommended you do the Upper Kaibito Slot in one day, then the **Middle** and **Lower Sections** on a 2nd day. You'll need a 4WD for the lower part (or park at Rocks 1740m, cross the upper canyon, walk north to the Sandslide, and enter). Best plan might be; make your way to the **Red Bluff Windmill** & stock tank and park somewhere nearby. Enter the canyon at the **Sandslide or 2nd Exit** and head north. Less than a km north of the **powerlines** it begins to slot-up. In the last 300m or so, you'll have to do **3 rappels**, and maybe swim or wade through one pothole, before arriving at the **Big Dropoff of 75m**. There are bolts & hangers in place, all on the left side, as shown on the cross-section. For this section take a shorter rope. Be aware, the author's information for this Middle & Lower Sections goes back to about 1998, and could change--go prepared.

When doing the Big Dropoff, hang your pack from the front of your harness with a R/QD/DC, as this is almost entirely a free rappel. As usual, and at this Big Dropoff, the last person down should take care not to tangle the long ropes so that pulling them from below will be easier.

From the bottom of the Big Dropoff, walk unobstructed along a small stream all the way to the **Lower Section**. There is no escape from the canyon between these 2 points. The Lower Section is only about 125m long, but there are 5 waterfalls. You'll bypass the first by rappelling from **2 B&H** on the right or east side just north of the waterfall & pool. Below the **2nd waterfall** you'll have to **swim**, then there are **2 more short 4m dropoffs**, before making an **11m rappel** over the last waterfall in the canyon. At each rappel station, there are 2 B&H placed high on one wall or the other. One report indicates you could bypass all 5 waterfalls by walking along the right or east-side ledge, then build a deadman anchor for one 30m (?) rappel straight down into the last pool.

From the last waterfall, walk downcanyon in the stream to the beginning of the **sheep & goat trail** which zig zags up to the west, then head south along a sandy track back to your vehicle. Another option for leaving this canyon would be to turn east, walk up **lower Starting Water Wash**, then exit via the **East Fork (Tse Esgizii Canyon)** as seen on the next map which is **Starting Water Wash**. From there you could walk eastward on **Navajo Road #6330**. Doing this would eliminate the need for a 4WD, but you'd have to park a car or mtn. bike at or near Road #6330 first.

The **LEF** has a fotogenic slot that isn't very long, then it drops into the gorge about 1 km upcanyon from the Sandslide Exit. The author went down part way years ago and was close to the big drop, saw a nice arch, but wasn't equipped with ropes or ascenders, and didn't see much. On one trip through Kaibito, a rope was seen hanging down the canyon wall but not touching the ground. No bodies or bones were observed. In recent years, it was found you can exit the slot to a little bench, where a couple of old B&H are found. It has to be close to 60m (?) to the bottom from there.

Elevations Trailheads, 1740m & 1639m; bottom of sheep & goat trail, 1300m.

Time Needed From the car-park marked Rocks 1740m, it's about 3 kms to the Sandslide Exit. To do just the **Upper Slot loop-hike**, returning via the west-side rim, should take 4-6 hours--perhaps more if you're less-experienced. To do the **Middle and Lower Sections** will take one long day.

Water Carry water in the Upper Section. There is running water from the bottom of the Big Dropoff all the way to the sheep & goat trail; it should be good to drink as is. There's also water in lower Starting

Map 74, Upper Kaibito Creek, Navajo Nation, Arizona

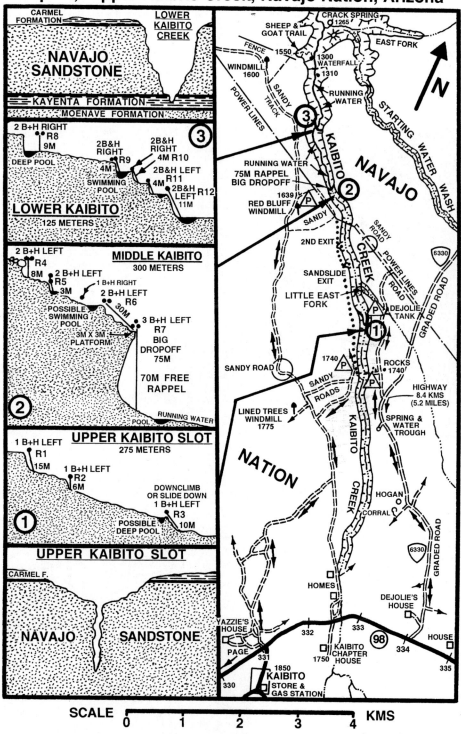

CARMEL FORMATION

NAVAJO SANDSTONE

LOWER KAIBITO CREEK

KAYENTA FORMATION

MOENAVE FORMATION

③

2 B+H RIGHT
R8
9M

DEEP POOL

2B&H RIGHT
R9
4M

2B&H RIGHT
4M R10

SWIMMING POOL

2B&H LEFT
R11
4M

2B&H LEFT R12
11M

LOWER KAIBITO
125 METERS

2 B+H LEFT
R4
8M

MIDDLE KAIBITO
300 METERS

2 B+H LEFT
R5
3M

1 B+H RIGHT

2 B+H LEFT
R6
30M

POSSIBLE SWIMMING POOL

3 B+H LEFT
R7
BIG DROPOFF
75M

3M X 3M PLATFORM

70M FREE RAPPEL

②

POOL

RUNNING WATER

1 B+H LEFT
R1
15M

UPPER KAIBITO SLOT
275 METERS

1 B+H LEFT
R2
6M

DOWNCLIMB OR SLIDE DOWN
1 B+H LEFT
R3
10M

①

POSSIBLE DEEP POOL

UPPER KAIBITO SLOT

CARMEL F.

NAVAJO   SANDSTONE

CRACK SPRING
1265

SHEEP & GOAT TRAIL

EAST FORK

FENCE
1550

WINDMILL
1600

1300 WATERFALL
1310

RUNNING WATER

③

STARTING WATER WASH

POWER LINES

SANDY TRACK

NAVAJO

KAIBITO

RUNNING WATER
75M RAPPEL
BIG DROPOFF

②

1639
RED BLUFF WINDMILL

P

SANDY

SANDY ROAD

2ND EXIT

6330

SANDSLIDE EXIT

LITTLE EAST FORK

POWER LINES ROAD

CREEK

P

DEJOLIE TANK

①

SANDY ROAD

1740
P

ROCKS
1740

P

HIGHWAY
8.4 KMS
(5.2 MILES)

GRADED ROAD

SANDY ROADS

LINED TREES WINDMILL
1775

KAIBITO

SPRING & WATER TROUGH

NATION

HOGAN

CORRAL

6330

KAIBITO CREEK

GRADED ROAD

HOMES

DEJOLIE'S HOUSE

YAZZIE'S HOUSE

332   333

PAGE

331

1850

334

HOUSE

335

98

KAIBITO CHAPTER HOUSE
1750

330

KAIBITO STORE & GAS STATION

SCALE

0   1   2   3   4   KMS

Water Wash.

Maps USGS or BLM map Glen Canyon Dam (1:100,000) for driving; and Cedar Tree Mesa, Horsethief Mesa & White Hill (1:24,000--7 1/2' quads) for hiking.

Flash Flood Danger High risk in the Upper Dark Slot, the slot just above the last Big Dropoff, and in the 125m-long Lower Kaibito waterfall section, but low to moderate in the rest of the canyon.

Best Time to Hike Upper Slot, spring, summer or fall; but because of swimming in the Lower Sections, between June and early September (or plan to use a wet/drysuit).

Canyoneers Experience Steve T. and Mike R. once parked at the beginning of the Upper Slot (east side) and did the loop-hike exiting via the Sandslide in 4 1/2 hours. The lower 2 sections were prepared in stages by various groups. Later, Eberhard S., Warren E., Keith K, and Mike R. once did the 2 lower parts in one long day (Warren & Keith actually did all 3 sections that day). In 2005, 2 others also did the Upper Slot in 4 hours.

Above In the 1990's, this party installed 3 B&H's for the 7th rappel, the Big Dropoff in **Upper Kaibito Creek Canyon**. This rappel is about 75m, most of which is in thin air, so suspend your pack from your harness.

Right Starting Rappel 1 in **Starting Water Wash**. Look closely and you might find a hole to squeeze down through behind the logs & rocks avoiding the use of a rope.

Above Left Starting Rappel 2 into the dark hole near the end of **Starting Water Wash**. A late report says that in 6/2005, and after some big floods re-arranged the canyon bottom, you could easily DC this slanted slot about 40m upcanyon from this rappel.

Above Right This is the last pothole near the end of the last dark section in **Starting Water**. In 2005, the water was chest deep, and hikers could just barely reach up and grab a handhold at the exit part of this pool. The small boulder in the middle wasn't there in the 1990's--or it was covered with sand (?).

Left This scene is just above the last dark slot in **Starting Water Wash**.

Starting Water Wash, Navajo Nation, Arizona

Location & Access **Starting Water Wash** is a major east-side tributary to the **upper part of Kaibito Creek** which is southeast of Page, Arizona. It's carved out of Navajo Sandstone, the same rock as Antelope and Kaibito Canyons, therefore is one of the best slot canyons around.

ATTENTION: STARTING WATER WASH IS OFFICIALLY CLOSED TO HIKING!

There are many reasons for this closure, however, this canyon is so scenic & challenging, plus it is equipped with anchors along its entire length, that the author believes one day it will be reopened. If or when this happens, here is the latest information. Anyone going to this canyon does so at their own risk!

To get there, drive **southeast** out of **Page on Highway 98** about 58 kms (36 miles) until you come to **mile post 331**, then turn south toward the village of **Kaibito**. Before leaving Kaibito, fill your fuel tank and water bottles at the store & gas station if needed, then consider a stop at the **chapter house** just downhill from the main part of town (see the previous map Kaibito Creek). Those people can update you on the canyon closure, road conditions, etc. Call 928-673-3408.

From Kaibito, continue east on Highway 98 to **mile post 334**, then turn north onto **Navajo Road #6330**. This is a graded washboard road heading north toward Tse Esgizii (Esgizii Rock), and is the same road you take if going to the narrows in Upper Kaibito Creek. Drive along this main road for **14 kms (8.7 miles)**. At that point you'll be in a wide valley which is the dry creek bed of Starting Water Wash (the powerlines you passed under were at Km 11.1/Mile 6.9). Park next to the bridge, or just to the north on a little sidetrack running northwest.

If you have a 2nd vehicle, you can save a fair amount of walking and shorten the trip up to 6.5 kms (4 miles). From where the road crosses Starting Water, continue north to **Km 16.6/Mile 10.3** (from Highway 98). There you'll see a Y junction. Veer left onto an ungraded track, then after another 100m or so, turn left or west again. Observe map. From there at Km/Mile 0, continue another **2.1 kms (1.3 miles)**, turn left onto a less-used 2WD track and drive a total of about **Km 3.9/Mile 2.4** from Road # 6330. Park one car in that area.

Rating **3B IV +PG/-R SLOT.** Starting Water Wash has 4 totally dark sections where you'll need a headlamp. It's normally dry, but you may have to wade or swim in the last dark part. There are some DC's for which you may need a partner assist to speed things along. As of 6/7/2005, this canyon had between 2 & 4 rappels.

Equipment RG, one 50m rope (or a 30m rope, plus pull cord); or a 30 plus webbing and PR's adding up to about 30m to be used as a pull cord), headlamp, K&EP, W&R/QL's, drybags & Ibis hook, just in case. In cooler weather, take a wetsuit.

Route From the bridge at 1600m, walk down the dry creek bed. Almost immediately you'll enter the upper slot which is shallow at first, but the further you go, the deeper it gets. It's narrow right from the beginning, often times barely shoulder width. Every so often will be a log jam or a chokestone you'll have to DC, and perhaps help each other by lowering packs.

Near the halfway point, you'll come to 3 places that are **totally dark.** For these sections, up to 75m in length, each hiker must have a **headlamp** (take extra batteries). Further along is an open area, another dark slot, then **Rappel 1** which is from 1 B&H on the right (you might be able to DC into a hole just before that?). Throughout the lower part of this canyon is lots of DC'ing, with skin loss, so K&EP are required. There are many places where, instead of rigging a rope for a HL, it's best to assist each other down with a helping hand. This can be a fairly strenuous hike.

About 100m before the slot ends, you'll enter a dark section and the **10m R2** from **2 B&H** on the left. However, 30-40m back upcanyon before that, you can as of 6/2005, DC to the right into a slanted slot bypassing the 10m rap; floods have really changed this part. Below that, it's total darkness & slippery, and there are 3 small potholes. The first 2 could be dry or with shallow water; the 3rd is deeper and at times may involve a deep wade or even swimming--in total darkness! In the middle of Pothole 3 is a mini butte or rounded boulder; on the exit side is one good handhold on the upper right-side ledge. Hopefully this should never be too difficult to get out of, but it can be slippery at times.

Just beyond Pothole 3, is the short **4m R3** (or HL) from **1 B&H** on the upper left. Or use an Ibis hook if the bolt is gone. From there, you'll enter a slanted walk-through slot, then sunlight as the canyon opens suddenly. Once out of the slot, DC on the left to a flat ledge where you'll find several small boulders, 1 B&H and lots of W&R/QL. From this rappel station **R4, rap 22m**. In the past, you could have walked to the east 30m and either made a difficult & risky DC, or rapped from a single B&H, but that bolt has been washed away by huge floods which have totally changed the last part of this canyon. Just rappel the 22m which is easy & safe, then DC a chute which used to be a rappel. Below that is a nearly-flat slickrock area.

Further down this open canyon, you'll come to a dropoff & pool, but stay to the right and bench-walk downcanyon while crossing one narrow ledge. Further along, you'll descend a steep slope into the lower end of the **East Fork** which is called **Tse Esgizii Canyon** on the *Cedar Tree Bench 7 1/2' quad*. Before leaving Starting Water and going into the East Fork, you could continue west another 300-400m and descend to the creek bed to find trees and usually running water. You could also continue down to Kaibito Creek.

As you near the upper end of Tse Esgizii Canyon, climb onto the slopes to the left to avoid dropoffs in the drainage bottom. Once on top, and with a compass & map in hand, walk due east. That bearing should take you toward the stock pond or the end of the 2WD sandy track marked 1690m. Road-walk back to your car on Starting Water (or leave a mtn. bike or 2nd car there before your hike and save some walking).

Elevations Bridge over Starting Water Wash, 1600m; stock pond & end of sandy track, 1690m.

Time Needed Going through this slot canyon is a little slow, with lots of pack shuttles and a helping hand with DC'ing, so count on taking 10-14 hours, round-trip. Two vehicles, one placed at each trailhead, could save about 1 1/2 hours walk.

Water Take water in your car and plenty in your pack (4 liters each in summer), but there is good spring water in lower Starting Water and Kaibito Creek.

Maps USGS or BLM map Glen Canyon Dam (1:100,000) for driving & orientation; and Cedar Tree Bench, Tse Esgizii, Horsethief Mesa & White Hill (1:24,000--7 1/2' quads) for hiking.

Flash Flood Danger Fairly high risk throughout, but with periodic wide places in which to run & hide.

Best Time to Hike June and early July, when conditions are normally dry and with 16 hours of daylight. But it has been hiked in winter. It's cool in the slot, even in summer, but the walk back across the flats would be more enjoyable in spring or fall (better have a wetsuit!).

Canyoneers Experience Some of the last people to go through were Steve T. and Mike R. Their loop-hike was completed in 10 1/2 hours. Later, 2 more hikers did it with a car shuttle in 11 1/3 hours and have updated previous information.

Map 75, Starting Water Wash, Navajo Nation, Arizona

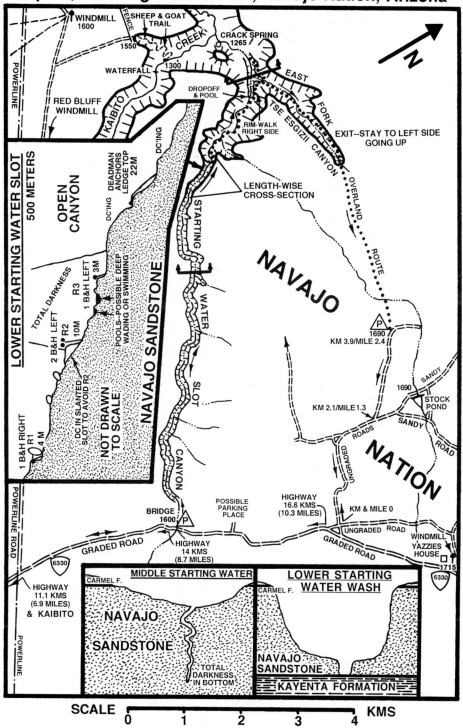

WINDMILL 1600

SHEEP & GOAT TRAIL

FENCE

CREEK

1550

CRACK SPRING 1265

WATERFALL

1300

POWERLINE

RED BLUFF WINDMILL

KAIBITO

DROPOFF & POOL

EAST FORK

DC'ING

RIM-WALK RIGHT SIDE

TSE ESGIZII CANYON

EXIT--STAY TO LEFT SIDE GOING UP

OVERLAND

LENGTH-WISE CROSS-SECTION

DEADMAN ANCHORS LEDGE TOP 22M

DC'ING

STARTING

NAVAJO

LOWER STARTING WATER SLOT

500 METERS

OPEN CANYON

POOLS--POSSIBLE DEEP WADING OR SWIMMING

ROUTE

TOTAL DARKNESS

R3

1 B&H LEFT 3M

R2

2 B&H LEFT 10M

DC IN SLANTED SLOT TO AVOID R2

NAVAJO SANDSTONE

WATER

NATION

P 1690

KM 3.9/MILE 2.4

1 B&H RIGHT

R1 4 M

NOT DRAWN TO SCALE

SLOT

1690

SANDY

STOCK POND

KM 2.1/MILE 1.3

SANDY

CANYON

UNGRADED ROADS

SANDY ROAD

POWERLINE ROAD

BRIDGE 1600 P

POSSIBLE PARKING PLACE

HIGHWAY 16.6 KMS (10.3 MILES)

KM & MILE 0

GRADED ROAD

HIGHWAY 14 KMS (8.7 MILES)

UNGRADED ROAD

GRADED ROAD

WINDMILL YAZZIES HOUSE

6330

HIGHWAY 11.1 KMS (6.9 MILES) & KAIBITO

POWERLINE

1715

6330

MIDDLE STARTING WATER

CARMEL F.

NAVAJO SANDSTONE

TOTAL DARKNESS IN BOTTOM

LOWER STARTING WATER WASH

CARMEL F.

NAVAJO SANDSTONE

KAYENTA FORMATION

SCALE

0   1   2   3   4   KMS

Geologic Formations of the Central Colorado Plateau, and Where They're Exposed

Below is a list of the geologic formations found in the central part of the Colorado Plateau and mostly in the Lake Powell & Escalante River area. The youngest rocks are the Straight Cliffs Formation, found at the top of Fiftymile Mountain/Kaiparowits Plateau. This is just west of the Hole-in-the-Rock Road and the Escalante River drainage. The oldest rocks, the Paradox Formation, are found in Canyonlands National Park and the bottom of Cataract Canyon.

Most of the technical slot canyons in this book are in the Navajo Sandstone, with several being in the Cedar Mesa Sandstone (some of which are in the San Rafael Swell and formerly designated the Coconino Sandstone), and a couple are in the Wingate Sandstone.

Straight Cliffs Formation Type Locality: The long straight cliffs on the northeast side of the Kaiparowits Plateau, sometimes known as Fiftymile Mountain, Kane County, Utah. This formation is a cliff-maker and is seen as the uppermost layer of rocks on top of the Kaiparowits. This is a light colored brown or yellow sandstone, sometimes mixed with shale and mudstone. This formation is the highest part of the flat plateau to the west as you drive along the Hole-in-the-Rock Road. In the southern parts where this formation is better exposed, it's divided into four members: Drip Tank, John Henry, Smoky Hollow and Tibbet Canyon Members. 333 meters thick.

Tropic Shale Type Locality: In Bryce Valley just east of Bryce Canyon National Park around the small town of Tropic, Garfield County, Utah. This is a dark gray marine shale, with thin layers of sandstone near the top, and thin fossiliferous limestone in the lower part. This formation makes a bench and is the first layer below the Straight Cliffs Formation on the Kaiparowits Plateau above the Fiftymile or Navajo Bench. However, it is buried by talus and landslide debris slopes, and almost never exposed at that location. 185 to 195 meters thick.

Dakota Sandstone Type Locality: Near Dakota City, Dakota County, Nebraska. This is generally a pale brown, course-grained sandstone with a few beds of mudstone and some coal seams. It's one of the upturned beds in the middle of the Cottonwood Wash Valley just west of The Cockscomb between Highway 89 and Kodachrome Basin State Park; and just east of the San Rafael Swell. It's a cliff-making formation, but thin. 10 to 30 meters thick.

Morrison Formation Type Locality: Near Morrison, Jefferson County, Colorado. Made up mostly of continental beds of sandstone, but with conglomeritic sandstone and mudstones. This formation is definitely a cliff-former, and is seen in the big cliffs of the lower end of Lake Powell, and just east of the San Rafael Swell near Tidwell Bottoms. This is the formation where most dinosaur bones in Utah are found. 0 to 155 meters thick.

Summerville Formation/Romana Sandstone Type Locality: Summerville Point, north end of the San Rafael Swell, Emery County, Utah: and Romana Mesa between Warm Creek & Padre Bays, Lake Powell, Kane County, Utah. The **Summerville** is found east of the Waterpocket Fold, whereas the **Romana**, which occupies the same niche or position, is found west of the Waterpocket Fold in the Lake Powell Country. Both are reddish to pale brown sandstone, mixed with shaley siltstone. The Romana is the narrow band found in the middle of the big cliffs formed by the Morrison above, and the Entrada below, and found in the lower end of Lake Powell. 5 to 25 meters thick.

Entrada Sandstone Type Locality: Entrada Point, north end of the San Rafael Swell, Emery County, Utah. It's mostly the reddish brown massive, or solid sandstone you see around Lake Powell; also some of the bluffs west of the Hole-in-the-Rock Road in the Escalante River drainage. It's also seen just north of the Paria River Visitor Center, and is one of the upturned beds seen immediately east of the San Rafael Reef, including the pinnacles and other features in Goblin Valley State Park. 185 to 230 meters thick.

Carmel Formation Type Locality: Mount Carmel, Kane County, Utah. These are thin beds of dusky red limy siltstone and reddish brown sandstone, with occasional pink limestones. It's the top layer of the benches along the Hole-in-the-Rock Road, in the Robbers Roost, North Wash, Poison Spring & Trachyte Creek areas and is always the formation immediately above the Page/Navajo Sandstone Formation. In Zion, it's just above the Temple Cap Formation, which in turn is just above the Navajo. It's also seen on the lower east side of the San Rafael Reef. 60 to 120 meters thick.

UPDATE: Carmel Formation, Page Sandstone and Navajo Sandstone In recent years more detailed studies have been made of the Carmel and Navajo Formations. It's been found in the Lake Powell Country around Page, Arizona, and the middle part of the Paria River in Utah, there are divisions in the lower Carmel and upper Navajo and geologists have come up with a new cross-section. On top is the **Carmel Formation**, which in some areas to the west has about 3 members. Next, and what used to be the top of the Navajo, is now called the **Page Sandstone** (type locality; Manson Mesa, Page, Arizona), which includes the **Thousand Pockets Tongue**. From all the author can gather, both of these 2 members are one in the same formation (?). Under the Page Sandstone (which is said to be reworked Navajo sands) in some sites is the **Judd Hollow Tongue of the Carmel Formation**. Then immediately under this thin layer is the main body of the **Navajo Sandstone**. Presently, only a few of the very latest geologic maps have any mention of this new classification which dates from the early to mid-1980's, and later. All other geology maps retain the older and still-used names of just the Carmel and Navajo Sandstone Formations. See the new book, and latest edition of *Geology of Utah's Parks and Monuments*, for the latest in the new geologic classifications.

Navajo Sandstone Type Locality: Navajo Canyon, Coconino County, Arizona. This thick layer of solid or massive sandstone is usually pale brown or buff colored. It's a cliff and/or dome & bluff maker, and the formation where the majority of the very narrow slot canyons on the Colorado Plateau are found. It's seen throughout the San Rafael Swell, Robbers Roost Country, Capitol Reef National Park, the Escalante & Paria River Country, all around Lake Powell including Rainbow Bridge and in Zion National Park. It has to be the most prominent geologic formation on the Colorado Plateau. 290 to 425 meters thick.

Kayenta Formation Type Locality: Just north of the town of Kayenta, Navajo County, Arizona. This is the reddish brown fluvial-made (created partly under water) sandstone, siltstone and shale layer which is always sandwiched in between the Navajo above, and the Wingate (or in Zion National Park the Moenave) below. This is a bench-former and often has several minor terraces wedding cake style. It can be seen everywhere the Navajo is found. 75 to 120 meters thick.

Moenave Formation Type Locality: Near Moenave and Tuba City, Coconino County, Arizona. Mostly a reddish brown sandstone with thin layers of siltstone and mudstone. This formation replaces the Wingate Sand-

Geology Cross-Section--Lake Powell & Kaiparowits Plateau to Cataract Canyon

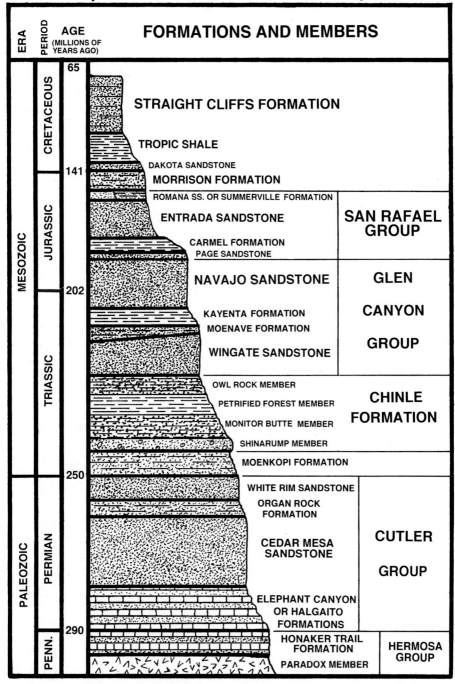

ERA	PERIOD	AGE (MILLIONS OF YEARS AGO)	FORMATIONS AND MEMBERS	
MESOZOIC	CRETACEOUS	65	STRAIGHT CLIFFS FORMATION	
			TROPIC SHALE	
		141	DAKOTA SANDSTONE	
			MORRISON FORMATION	
	JURASSIC		ROMANA SS. OR SUMMERVILLE FORMATION	SAN RAFAEL GROUP
			ENTRADA SANDSTONE	
			CARMEL FORMATION	
			PAGE SANDSTONE	
		202	NAVAJO SANDSTONE	GLEN
	TRIASSIC		KAYENTA FORMATION	CANYON
			MOENAVE FORMATION	
			WINGATE SANDSTONE	GROUP
			OWL ROCK MEMBER	CHINLE FORMATION
			PETRIFIED FOREST MEMBER	
			MONITOR BUTTE MEMBER	
			SHINARUMP MEMBER	
			MOENKOPI FORMATION	
		250	WHITE RIM SANDSTONE	CUTLER GROUP
PALEOZOIC	PERMIAN		ORGAN ROCK FORMATION	
			CEDAR MESA SANDSTONE	
			ELEPHANT CANYON OR HALGAITO FORMATIONS	
	PENN.	290	HONAKER TRAIL FORMATION	HERMOSA GROUP
			PARADOX MEMBER	

stone to the west in & around Zion National Park, and south into Arizona. The cliffs east of Lee's Ferry are mostly Moenave, with just a little Wingate. 0 to 125 meters thick.
Wingate Sandstone Type Locality: Cliffs north of Fort Wingate, McKinley County, New Mexico. Reddish brown to light brown, fine grained, and massive cross-bedded sandstone. This makes the big cliffs or walls

in the San Rafael Swell, Canyonlands and Capitol Reef National Parks, the eastern & southern Escalante River Country and in the lower Paria River Drainage. 70 to 105 meters thick.

Chinle Formation Type Locality: Chinle Valley, Apache County, Arizona. The Chinle is always below either the Wingate or Moenave Formations. There is great variation between the four main members of the Chinle, but mostly it's famous for its clay beds, petrified wood and where most uranium was mined during the boom days of the 1940's & '50's. In the central Colorado Plateau and around Lake Powell, the uppermost layer is the *Owl Rock Member* which is a reddish brown silty mudstone. Below that, the *Petrified Forest Member* has lots of petrified wood, as in the Petrified Forest National Park in Arizona. It was also in these clay beds that Charles H. Spencer tried digging gold at Lee's Ferry, and later at Pahreah, Utah. A reddish brown mudstone is the third layer or strata from the top and is called the *Monitor Butte Member.* The lowest member is the *Shinarump Conglomerate.* It's a light colored, course sandstone & conglomerate, which has lots of petrified wood. In some localities, it is known as the Black Ledge, because it's often covered with black desert varnish. This is also famous as one of the primary formations on the Colorado Plateau which contains uranium. In the northern parts of the Plateau, the *Moss Back Member,* which occupies nearly the same niche, was where lots of uranium was mined in and around the San Rafael Swell. 15 to 380 meters thick.

Moenkopi Formation Type Locality: Moenkopi Wash, Coconino County, Arizona. In some places there are some thin beds of limestone in this formation, but mostly it's a dark reddish chocolate brown, ripple-marked sandstone, with thin lenses of shale and mudstone. The Moenkopi is seen in the ledges across the channel from Hite, and throughout Lake Powell Country, in the San Rafael Swell, Capitol Reef, Canyonlands and Zion National Parks (east or west of Springdale) and in the Paria & Escalante River drainages. Most of the dark reddish-brown clay soil you see as you drive along White Canyon east of Hite is derived from the Moenkopi. 90 to 120 meters thick.

White Rim Sandstone Type Locality: The White Rim escarpment between the Green & Colorado Rivers in Canyonlands National Park, San Juan County, Utah. A white to yellow, thin, fine grained, cross-bedded, massive sandstone also found in the White and Dirty Devil River Canyon areas. 0 to 120 meters thick.

UPDATE: The Coconino Sandstone Type Locality: Coconino County, Arizona. This is a common formation in the top layers in the Grand Canyon. 0 to 120 meters thick.

Until recently, some geologists placed the **Coconino Sandstone** in the San Rafael Swell featuring it in such canyons as The Chute, the Upper & Lower Black Boxes, and in Upper Chute, Baptist Draw and Eardley Canyons. However, and with more research, a mistake has been found & corrected. In the San Rafael Swell, the **Organ Rock Shale Formation** is missing, thus the White Rim and Cedar Mesa Sandstone form one continuous layer of white sandstone. In the future, all references to the Coconino SS in the San Rafael Swell will be referred to as the **Cedar Mesa Sandstone**. Source: Helmut Doeling, Utah Geological Survey.

Organ Rock Shale Type Locality: Organ Rock, Monument Valley, San Juan County, Utah. This is a reddish brown siltstone with sandy shales. It forms a slope & ledge on the cliffs across the channel west from Hite Marina. Some of these rocks appear similar to the rocks seen in Goblin Valley next to Utah's San Rafael Swell. 90 to 150 meters thick.

Cedar Mesa Sandstone Type Locality: Cedar Mesa, west and northwest of Mexican Hat, San Juan County, Utah. This covers the top of the massive mesa south and southwest of the Blue or Abajo Mountains. It's a mostly white to pale light yellow, cross-bedded, massive sandstone. It's always a big cliff-maker which forms some of the finest slot canyons around in the White Canyon area, namely the Black Hole, Long, Gravel & Cheesebox Canyons. This is the same formation where all or most of the Anasazi ruins are found in the areas south of the Abajos, especially in Grand Gulch and nearby canyons. You can see this one as the massive sandstone walls throughout Cataract Canyon (top-most layer on those high walls). This formation and its name now replaces the Coconino Sandstone in the San Rafael Swell. 365 to 400 meters thick.

Elephant Canyon Formation Type Locality: Canyonlands National Park, San Juan Country, Utah. These beds are found in the big walls below the Cedar Mesa Sandstone in upper Cataract and Clearwater Canyons of Canyonlands National Park, and at the very bottom of lower Eardley Canyon in the San Rafael Swell. It's made of cherty limestones and some dolomites, interbedded with pale reddish sandstones. It forms a cliff, but with some slopes and terraces as well. In western Colorado these same beds are called the **Rico Formation**, Rico, Dolores County, Colorado and has similar stratigraphy. In the Four Corners region, this same formation changes to **Halgaito Shale,** Halgaito Springs, southwest of Mexican Hat, San Juan County, Utah. The Halgaito has more shale and is seen in the lower San Juan River Canyon at the end of Grand Gulch and around the Gooseneck Bend State Park near Mexican Hat, Utah. 125 to 450 meters thick.

Honaker Trail Formation Type Locality: Honaker Trail, Lower San Juan River Canyon above Lake Powell, San Juan Country, Utah. This formation is composed of dark colored thick limestone, interbedded with gray cherty limestone, and red & gray shales and sandstones. These rocks form ledges, terraces and slopes. You can see this one in Cataract Canyon (and in lower Clearwater Canyon) down as far as Mille Crag Bend on Lake Powell, and in the Gooseneck Bend area of the San Rafael River. In the past, some geologists have called this strata simply the Upper Member of the Hermosa Group. 300 to 500 meters thick.

Paradox Formation Type Locality: Paradox Valley, Montrose County, Colorado. You may not be able to see this one, as it's difficult to understand it with the member above, and much of it is covered with talus. This is a slope & cliff-maker and you may see it only in the upper end of Cataract Canyon near the mouths of Palmer, Gypsum and maybe Eastern Pasture Canyons (?). It's basically black shales and limestone, interbedded with salt anhydrite and gypsum. The author believes this is the blackish rock which has made Gypsum Falls in Gypsum Canyon (?). The Paradox Formation is a result of a large anticline in western Colorado and eastern Utah. In the middle parts of the anticline, the beds are composed of salts, a result of evaporation of a large shallow inland sea (perhaps similar to the Caspian Sea). But along the outer fringes, one sees mostly limestone beds. This would be the result of marine shellfish sedimentation during younger stages of the same salty sea. Since we see mostly limestone beds in Cataract Canyon, it's assumed this region was along the outermost edges of that ancient sea. 150 to 1500 meters thick.

In the **Grand Canyon** area (from top to bottom): Kaibab Limestone, Toroweap Formation, Coconino Sandstone, Hermit Shale, Esplanade Sandstone, Supai Formation, Redwall Limestone, Temple Butte Limestone, Muav Limestone, Bright Angel Shale, Tapeats Sandstone, then basement granites, schists & paragneises.

Deep in the **San Rafael Swell** below the Elephant Canyon Formation are found: Redwall Limestone, Ouray Limestone, Elbert Formation, Lynch Dolomite, Maxfield Limestone, Ophir Formation, Tintic Quartzite, then basement granites. None of these formation are exposed at the surface in that region.

Above Left Jon Jasper DC'ing back to ground level in the lower end of **Little Canyon** (next to and just east of the lower end of Brimstone Gulch in the Escalante River Drainage).

Above Right Tom Tolboys chimneys up and over a tight spot in the 125m-long slot in upper **Burr Canyon**. This part rates somewhere between +R and -X SLOT for difficulty in slot canyons.

Left This is the first rappel in the **Upper Dark Slot** in upper **Piñon Canyon**. It's from 1 B&H as shown, but it's really an easy DC of about 5m in a tight chute, ending in a 1 1/2m-deep pothole.

Other Guidebooks by the Author

Books listed in the order they were first published.
(Prices as of August, 2008. Prices may change without notice)

Climber's and Hiker's Guide to the World's Mountains (4th Edition), Kelsey, 1248 pages, 584 maps, 652 fotos, ISBN 0-944510-18-3. US$36.95 (Mail Orders US$40.00).

Utah Mountaineering Guide (3rd Edition), Kelsey, 208 pages, 143 fotos, 54 hikes, ISBN 0-944510-14-0. US$10.95 (Mail Orders US$13.00).

China on Your Own: and *Guide to China's Nine Sacred Mountains,* Kelsey, **Out of Print.**

Non-Technical Canyon Hiking Guide to the Colorado Plateau (5th Edition), Kelsey, 384 pages, 120+ hiking maps, **285 color fotos,** new ISBN 978-0944510-22-3. US $19.95 (Mail Orders US$22.00).

Hiking and Exploring Utah's San Rafael Swell (3rd Edition), Kelsey, 224 pages, 32 mapped hikes, plus History & Geology, 198 fotos, ISBN 0-944510-17-5. US$12.95 (Mail Orders US$15.00).

Hiking and Exploring Utah's Henry Mountains and Robbers Roost, *Including The Life and Legend of Butch Cassidy,* (Revised Edition), Kelsey, 224 pages, 38 hikes or climbs, 158 fotos, ISBN 0-944510-04-3. US$9.95 (Mail Orders US$12.00). **Momentarily Out of Print--Next Edition coming in 2009?**

Hiking and Exploring the Paria River, *Including: The Story of John D. Lee & the Mountain Meadows Massacre,* (4th Edition), Kelsey, 288 pages, 38 mapped hiking areas from Bryce Canyon to Lee's Ferry, 332 fotos, ISBN 0-944510-21-3. US$11.95 (Mail Orders US$14.00).

Hiking and Climbing in the Great Basin National Park--*A Guide to Nevada's Wheeler Peak, Mt. Moriah, and the Snake Range,* Kelsey, **Out of Print.**

Boater's Guide to Lake Powell (5th Edition), *Featuring: Hiking, Camping, Geology, History & Archaeology,* Kelsey, 288 pages, **263 color fotos,** new ISBN 978-0-944510-24-7. US$19.95 (Mail Orders US$22.00).

Climbing and Exploring Utah's Mt. Timpanogos, Kelsey, 208 pages, 170 fotos, ISBN 0-944510-00-0. US$9.95 (Mail Orders US$12.00). **Soon Out of Print for a few years.**

River Guide to Canyonlands National Park & Vicinity, Kelsey, 256 pages, 151 fotos, ISBN 0-944510-07-8. US$11.95 (Mail Orders US$14.00). **Out of Print**

Hiking, Biking and Exploring Canyonlands National Park & Vicinity, Kelsey, 320 pages, 227 fotos, ISBN 944510-08-6. US$14.95 (Mail Orders US$17.00). **Out of Print**

Life on the Black Rock Desert: A History of Clear Lake, Utah, Venetta B. Kelsey, 192 pages, 123 fotos, ISBN 0-944510-03-5. **Out of Print for a few years.**

The Story of Black Rock, Utah, Kelsey, 160 pages, 142 fotos, ISBN 0-944510-12-4. US$9.95 (Mail Orders US$12.00).

Hiking, Climbing & Exploring Western Utah's Jack Watson's Ibex Country, Kelsey, 272 pages, 224 fotos, ISBN 0-944510-13-2. US$9.95 (Mail Orders US$12.00).

Technical Slot Canyon Guide to the Colorado Plateau, 2nd Edition, Kelsey, ISBN 0-944510-23-X, or new ISBN 978-0-944510-23-0, 336 pages, **341 color fotos.** US$19.95 (Mail Orders US$22).

Distributors for Kelsey Publishing

Primary Distributor All of Michael R. Kelsey's books are sold by this distributor. A list of Kelsey's titles is in the back of this book.
Brigham Distribution, 110 South, 800 West, Brigham City, Utah, 84302, Tele. 435-723-6611, Fax 435-723-6644, Email brigdist@sisna.com.

Most of Kelsey's books are sold by these distributors.
Alpenbooks, 4206 Chennault Beach Road, Suite B1, Mukilteo, Washington, USA, 98275, Website alpenbooks.com, Email cserve@alpenbooks.com, Tele. 425-493-6380, or 800-290-9898.
Books West, 11111 East, 53rd Avenue, Suite A Colorado, USA, 80239-2133, Tele. 303-449-5995, or 800-378-4188, Fax 303-449-5951, Website bookswest.net.
Liberty Mountain, 4375 W. 1980 S., Suite 100, Salt Lake City, Utah, 84104, Tele. 800-578-2705 or 801-954-0741, Fax 801-954-0766, Website libertymountain.com, Email sales@libertymountain.com.
Treasure Chest Books, 451 N. Bonita Avenue, Tucson, Arizona, USA, 85745, Tele. 520-623-9558, or 800-969-9558, Website treasurechestbooks.com, Email info@rionuevo.com.

Some of Kelsey's books are sold by the following distributors.
Canyonlands Publications, 4860 North, Ken Morey Drive, Bellemont, Arizona, USA, 86015, Tele. 928-779-3888, or 800-283-1983, Fax 928-779-3778, Email info@clpbooks.net.
High Peak Books, Box 703, Wilson, Wyoming, USA, 83014, Tele. 307-739-0147.
Rincon Publishing, 1913 North Skyline Drive, Orem, Utah, 84097, Tele. 801-377-7657, Fax 801-356-2733, RinconPub@UtahTrails.com.
Recreational Equipment, Inc. (R.E.I.), 1700 45th Street East, Sumner, Washington, USA, 98390, Website rei.com, Mail Orders Tele. 800-426-4840 (or check at any of their local stores).
Online--Internet: amazon.com; adventuroustravelers.com; btol.com (Baker-Taylor); Ingrams.com; Bdaltons.com; borders.com (teamed with amazon.com).

For the **UK and Europe,** and the rest of the world contact: **Cordee,** 3a De Montfort Street, Leicester, England, UK, LE1 7HD, Website cordee.co.uk, Tele. Inter+44-116-254-3579, Fax Inter+44-116-247-1176.